My Life in the Golden Age of
Chemistry: More Fun Than Fun

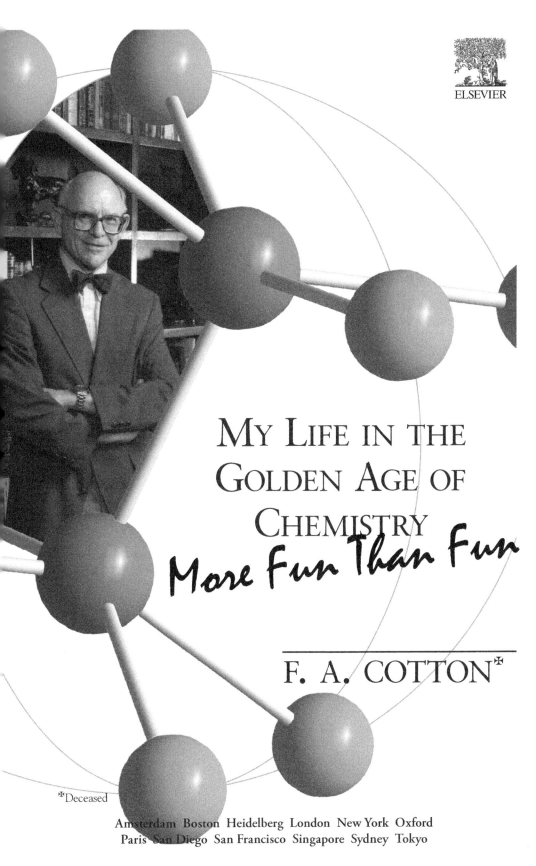

My Life in the
Golden Age of
Chemistry
More Fun Than Fun

F. A. Cotton✠

✠Deceased

Amsterdam Boston Heidelberg London New York Oxford
Paris San Diego San Francisco Singapore Sydney Tokyo

Elsevier
Radarweg 29, PO Box 211, 1000 AE Amsterdam, Netherlands
The Boulevard, Langford Lane, Kidlington, Oxford OX5 1GB, UK
225 Wyman Street, Waltham, MA 02451, USA

Notice

Knowledge and best practice in this field are constantly changing. As new research and experience broaden our understanding, changes in research methods, professional practices, or medical treatment may become necessary.

Practitioners and researchers must always rely on their own experience and knowledge in evaluating and using any information, methods, compounds, or experiments described herein. In using such information or methods they should be mindful of their own safety and the safety of others, including parties for whom they have a professional responsibility.

To the fullest extent of the law, neither the Publisher nor the authors, contributors, or editors, assume any liability for any injury and/or damage to persons or property as a matter of products liability, negligence or otherwise, or from any use or operation of any methods, products, instructions, or ideas contained in the material herein.

ISBN: 978-0-12-801216-1

Library of Congress Cataloging-in-Publication Data
A catalog record for this book is available from the Library of Congress

British Library Cataloguing in Publication Data
A catalogue record for this book is available from the British Library

For information on all Elsevier publications visit our website at http://store.elsevier.com/.

This book has been manufactured using Print On Demand technology. Each copy is produced to order and is limited to black ink. The online version of this book will show color figures where appropriate.

Working together
to grow libraries in
developing countries

www.elsevier.com • www.bookaid.org

Whoever has had the great fortune
To gain a true friend,
Whoever has won a devoted wife,
To the rejoicing let him add his voice.
—*Friedrich Schiller*

For Dee

F. Albert Cotton M.F.H.

Robert Douglas Hunter

1971

TABLE OF CONTENTS

FOREWORD

Immediately after the War, the approach to the study of Inorganic Chemistry underwent a revolution. This in part arose from the wide range of instrumentation that became available. The advent of relatively rapid methods for the determination of infrared, visible and ultraviolet, and Raman spectra coupled with the availability of new techniques such as n.m.r. and e.s.r. spectroscopy allowed for a wide range of applications to a variety of problems within Inorganic Chemistry. Many of these had not been possible or even conceived of before these developments.

This book not only provides an insight into the contributions of one of the major players in these developments but also encapsulates the atmosphere of the period when Government funding of science was generous and good research was assured of sponsorship. This situation has sadly declined worldwide and reflects the very much bigger scientific community that now exists and the increase in the financial requirements of modern-day research.

The author is well placed to discuss this position as he entered the field at the beginning of this period of rapid change. His own initial contribution was in the then novel field of Organometallic Chemistry, with work on the spectacular new range of "aromatic" inorganic compounds as typified by ferrocene. His subsequent work over the next 50 years provided leadership in many of the most significant developing areas of Inorganic Chemistry, and he has left a heritage that will last for many years.

He certainly excelled in his command of both the theory and practice of his subject, and his papers and written contributions reflected both his joy and efficiency in writing. Over a period of approximately 50–60 years, he published over 1600 papers and a variety of books that covered teaching at all levels—school, undergraduate, post-graduate, and research monographs. These are all excellent texts and as with, for instance, in books such as *Advanced Inorganic Chemistry* and his textbook on group theory, he provided a new approach and insight to the subjects under discussion. These books have become classics in their time. His contribution to the chemical literature was outstanding in both quality and quantity, and places him in a unique position within the chemical community.

This book not only relates to his contribution within research and teaching but also unfolds the wide appreciation and influence that he made to so many aspects of life in general. He enjoyed travel and visited many parts of the world, often to relate his chemistry but also to establish a wide range of contacts and friendships. This in part reflects the vast range of countries from which his research collaborators came. As is evident from the text, Al was a strong supporter of his students and built up a close relationship and concern for their future that often extended for life. He was very direct in his relationships, and this is evident in his coverage and his reflections on some of the people that are included in the book. I certainly enjoyed a friendship that stretched over more than 50 years. I find this book a constant reminder of so many aspects of Al and his attitudes across a large spectrum of interests. It will provide the public with an insight into how an outstanding scientist lived, worked, and thought.

It is something of a tragedy that shortly after finishing this book he should meet such an untimely death. He is missed by many but mostly by his family who were always paramount in his thinking and behavior. This book reflects the thoughts, attitudes, and reflections of a remarkable man who made major contributions to his chosen area of science and has certainly changed the way we view and study the subject.

—The Lord Jack Lewis
Cambridge, England

PROLOGUE

I began this book with considerable trepidation, indeed with the feeling that it might be wiser not to. Any attempt to recreate the past is bound to be an act of imagination as well as of recollection. Bias and subjectivity are inescapable. That particular effort to recreate the past known as autobiography must needs suffer from these distorting influences in the most extreme degree. In an attempt to mitigate the effects of an inaccurate memory, I have asked several friends to read the manuscript and point out errors of fact, and I have benefitted greatly from their perspicacity. Naturally, the responsibility for errors that remain is mine alone. I have discovered the truth of Ernest Hemingway's observation that for a writer it is necessary to know everything but to leave out most of it. As for my choice of events and my reconstruction and interpretation of certain events, they may well be colored by egotism. I have tried to minimize this, but I have no illusions about the impossibility of avoiding it completely. Caveat lector! If there are readers who recall certain things that are described here in a different light, I shall be neither surprised nor necessarily inclined to change my own view. I would, however, be glad to receive constructive criticism.

Whatever my misgivings about writing my own autobiography for this series, it was a call I could hardly have refused in view of my role in bringing it into being. Sometime in the latter part of 1995, after I had read a large number of the *Profiles, Pathways and Dreams* series of autobiographies of organic chemists, edited by Jeffrey I. Seeman and published by the American Chemical Society, I thought that the ACS might be interested in publishing a similar series of autobiographies of inorganic chemists. I also discussed the project with Stan Kirshner and made the suggestion that he would be a good editor, but he declined. I then turned to John Fackler and he carried on from there. He first found that the ACS was not interested (for financial reasons) and he then persuaded Plenum (now Springer) to be the publisher. Thus, when John proposed that I write one of the first few volumes in the series, I had to accept a task I had sort of brought on myself.

Choosing a title for this book was not easy. Had Jack Roberts not preempted *In the Right Place at the Right Time*, I might well have used it myself. I believe that in my life I have benefitted from a lot of good luck and not suffered from any unusual amount of bad luck. Some of the breaks I have gotten I think I made for myself, and there is, of course, the very true adage that "the harder I work the luckier I get." I must admit, however, that some things, like having both the mother and the wife I have had, were bounties that I did nothing to deserve. I was the beneficiary of unearned good luck as well at some other points in my life. Thus, to a significant degree, I have led what is often called a *charmed life*. At one point, I considered using that as a title. Other titles came to mind, but were also rejected. I was growing desperate, when I read in Hans Krebs' autobiography that Noel Coward had, supposedly, said, "Work is fun. There is no fun like work." Aha, I said, *No Fun Like Work* it is, but Krebs gave no source for the quotation and is no longer available to be asked, and I thought it would be nice to have a more direct attribution. After some searching, I discovered the Reverend James Simpson's book *Simpson's Contemporary Quotations*, in which Coward is said to have said "Work is more fun than fun." Regrettably, Rev. Simpson has no documentation other than a newspaper obituary. Probably Coward said either one or the other, and he may even have said both on different occasions. I decided that *More Fun than Fun* was to be my subtitle. As to the golden age mentioned in the title, I refer the reader to my Priestly Medal lecture, reprinted here as an appendix.

Finally, I should like to say to all, friends and foes, that in writing this book I am in no sense writing an epilogue to my life or even my career. I still get as much of a thrill as ever from learning something new that no one has ever known before, even if it's only a little thing. Seeing a beautiful new molecule for the first time still exhilarates me as much as it ever did, especially if the molecule has surprising, or better yet, puzzling features. Thus, while I cannot contest Caesar's observation that "death, a necessary end, will come when it will come," I do not feel in the least receptive to the idea, and hope to go right on doing chemistry for a long time yet. Moreover, I strongly believe in Andre Maurois' dictum that "Growing old is no more than a bad habit which a busy man has no time to form."

TO THE READER

On October 17, 2006, Al had gone out for his morning walk as he always did on Tuesday mornings. Al was a creature of habit: on Tuesdays and Thursdays, he would go on a walk before he went into the office. Monday, Wednesday, and Friday he went to aerobics with his wife, Dee. His route was predictable—with slight variations from time to time—bringing him home about 8:00 a.m. On this morning, though, things were different, Al didn't come back at his normal time.

Al was found at the end of the lane. He had been brutally assaulted.

We had been working on the design and layout of this book for many months and I had given him what I had hoped would be—though I was sure it would not— the final proof of the layout the week before. My husband, Carlos Murillo, was in the hospital on this fateful Tuesday morning, he had had an emergency appendectomy the Friday before. Al came to the hospital on Sunday to see Carlos and told me that he would not be back on Monday; he "had to get this book finished, I'm tired of looking at it." He told us that he would be back on Tuesday. He was back on Tuesday, but he was life-flighted into the emergency room in critical condition.

I was able to get the final proof back after the police combed his office for clues and evidence. In his usual manner, Al left me with just as many questions as answers. He would indicate that "we have a problem here" referring to an equation, figure, or line of text, but, since I am not a chemist, I had no idea what the problem was or how to fix it. That's just how Al was; he would mark something to remind him to go back and check on it and forget that he had marked it until I came back and asked him what the problem was. He put a lot of thought into what was included in his autobiography. The concept of including things just because they happened was not how he wanted to approach this. Always the teacher, he wanted what was included to be of use to the students that would be writing reports on chemists in the future. He was insistent that some things stay exactly the way that he wrote them. When I asked him if he wanted to include the English translation for a German quote that he had used he told me "Absolutely not! If someone wants to know what it says, I guess they need to learn to read German" and that was the end of that. Luckily, Carlos was able to answer my questions concerning the chemistry; Dee was able to answer my questions when the text referred to family and friends.

Al had an inner strength that would not let him give up. He beat the odds and regained consciousness. As everyone who knew Al would agree, he was a fighter. He fought to stay with us, to stay with the family that he was devoted to, to stay with the friends he thought of as family, but on February 20, 2007—four months after the assault—he succumbed to his injuries.

He didn't have a chance to look at the book one last time, he didn't have a chance to check the index, he didn't have the chance to see the cover. All of his chances were taken away when he was assaulted. What I can assure the reader of though is that every word in this book is Al's. Everything that is *in* this book and everything that is *not* in this book are due to the decisions that Al made. There were three chapters that Al had not indexed. That part was left to me. I have tried to follow what he did in the other chapters, but if your name is not in the index and you are in the book, please accept my apologies. Some names he indexed and some he did not. I never had the chance to ask him what his rhyme or reason was.

Al was a good friend to me. He opened my eyes to a world of things I would have never known. My life is all the richer for knowing Al and I miss him terribly. It is my wish that the reader will enjoy the last of Al's many books. To those that knew and loved him, it is my fervent wish that this will help to remind you of the wonderful, brilliant, caring man that he was instead of the tragedy of how he was taken from us.

As of the publishing of his autobiography, the person(s) responsible for his assault have not been brought to justice. I will never give up hope that someday I will be able to say that justice has been served.

—Debbie Murillo
College Station, Texas

ACKNOWLEDGMENTS

I would like to thank some very special people without whose help this autobiography would never have been published.

When Al succumbed to his injuries, Larry Falvello, Bruce Foxman, and Carlos Murillo were all determined, as was I, to see his final work through to publication.

Bruce had been Al's student at MIT; Carlos was a student of Al's at Texas A&M University, and he went on to run the Laboratory for Molecular Structure and Bonding, a responsibility that had been shared by Larry for 10 years prior to Carlos taking over. Al held all three of these men in the highest regard and treasured each one as a dear friend. They all published many papers with Al and knew exactly how Al would have wanted the final edits to be accomplished, what he would have agreed to revise, and what he would have fervently refused to do.

They all worked tirelessly through multiple reads and rereads — fact-checking, researching points that were brought up in the legal review, and putting polish on Al's last published work. I could not have finished this without them and I will be forever grateful.

I also want to thank Rick Adams, another dear friend of Al's, for putting me in touch with Katey Birtcher at Elsevier. She has been a pure joy to work with, and I am so thankful that she was willing to take on the task of getting this book published.

We hope you enjoy the story of Al's life, told the way he wanted it told.

—Debbie Murillo
College Station, Texas

In the lines above, Debbie Murillo thanks us for helping with the proofreading and fact-checking of this autobiography. This was an easy task, greatly facilitated by Al's characteristic accuracy and by his forthright manner and clear writing style. Although the three of us were direct colleagues of Al's for a cumulative total of thirty-three years, we understand that perfection is rarely if ever attained; nevertheless, we hope that any errors that may be present are few in number and minor in scope. We limited ourselves to the specific tasks of proofreading and checking facts. Similar to Debbie, we added nothing and deleted nothing before the text was sent to the publisher — the whole of this remarkable life story is told by Al, in his own way.

What Debbie does not say is that this book would have never seen the light of day without her involvement, perseverance, patience and professionalism. Words cannot express our heartfelt appreciation for her work on this project and for her tenacity in guiding it through to completion. While Al was preparing the text, Debbie did the work of an Editorial Project Manager, seeing to it that all elements of the process, other than the writing itself, were taken care of. A graphic artist by profession, she also designed the book, transforming the text and pictures into a finished work. After Al's death, Debbie saw the text through the fact-checking and proofing stages and found a publisher that would work with us. Those who knew Al Cotton and those who will come to know him through the pages of this book all owe Debbie a debt of heartfelt thanks.

—Larry Falvello   —Bruce Foxman   —Carlos Murillo
Zaragoza, Spain   *Waltham, Massachusetts*   *College Station, Texas*

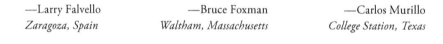

I would like to thank Debbie Murillo for her devotion to my father's autobiography and for seeing it through to publication. This book has had some challenges along the way but I know that my father would have been happy with the final outcome. My thanks also go to The Lord Jack Lewis for the prologue; to Larry Falvello, Bruce Foxman and Carlos Murillo for being associate editors; and to Rick Adams for leading us to Elsevier and Katey Birtcher. Most of all I want to thank my father, F. Albert Cotton, for all of his accomplishments that made this book possible.

—Jennifer Cotton
Bryan, Texas

CHAPTER 1

PHILADELPHIA

On the whole, I'd rather
be in Philadelphia.

—*W. C. Fields*
(ON HIS HEADSTONE)

I WAS BORN in the West Philadelphia Maternity Hospital (long gone, I think) at 10:04 on the morning of April 9, 1930. That is received information, of course, from my mother, who also enjoyed relating the fact that I emerged into this world mute until, after her anxious enquiry as to my initial silence, the doctor asked "Do you want to hear him cry?" and proceeded to give me a spank on the bottom. I cried lustily and haven't been notably quiet since.

The doctor who elicited my first oral communication was called Frank Abbot. He was a friend of my parents, and in his honor, they named me Frank Abbot Cotton, without, of course, finding out if I liked that name. It served very well until, at about the age of two and a half, I had an opportunity to do something few people do — to choose (from a limited list, however) my own name. My father, of whom more later, who died when I was not quite two years old, was Albert Cotton, and my mother asked me if I wouldn't rather be known as Albert instead of Abbott. I am told I agreed to this. Frank was retained and I became, then and forever, Frank Albert Cotton. This change has never been registered in a documentary fashion, but I believe that the change can be considered as sanctioned by usage. My father's nickname was Bert, but mine has been Al. Very early on, until about the age of nine, my playmates called me Sonny (ugh!) and briefly, in the Cub Scouts, F. Albert was contracted by some wag into Falbert, but after the humor of that grew stale, it too became history.

My family origins are only partly known. On my father's side, I know that the origins were in the environs of Manchester, England. My great grandparents were Simeon Cotton (b. 1821) and Elizabeth Magee (b. 1819). They were born in Portwood and Stockwood, respectively, both now incorporated into Manchester, I believe. They were married in 1842 and came to the United States soon after, settling in or near Chester, Pa. They raised a family of six children, one of whom, James Barton, was my grandfather. Simeon began manufacturing lace, and this must have gone well because, according to Ashmead's History of Delaware County (1884), he built a rather large spinning mill called the Centennial Mill in Chester. This mill had 3,000 spindles and spun raw cotton into cotton warp (whatever that was) at the rate of seven thousand

My parents shortly after their marriage.

pounds a week. Simeon was also reputed to have been a devout Methodist who often preached.

James Barton Cotton, my grandfather, was evidently not the businessman his father had been. Early in his life he had been very wealthy, married well, and had a family of seven children, the youngest of whom was my father, Albert (b. 1900). The older children enjoyed a privileged upbringing in a large house on a one-hundred acre farm where there were a tenant farmer, a coachman, a cook, and two maids. By the time of my father's birth, this wealth had been dissipated, and he grew up in much more modest circumstances, such that a college education was not affordable.

My father, however, was very keen to become an engineer, and after he married my mother in 1919, she worked to support him as he studied mechanical engineering, partly at night, at Drexel Institute of Technology (now Drexel University).

On my mother's side I know less of the family background. I can trace her family back to her grandfather, Lewis Taylor (a Quaker who died around 1900), and grandmother, Matilda (Hinkle) Taylor, who died in 1921, and then her father, Lewis Cook Taylor (1872-1947). The two men were watchmaker and metalworker, respectively. My maternal grandmother, Katherine Kippes, was born somewhere in Bavaria in 1872 and came to the United States in 1877. On leaving the ship, she fell off the gangplank and had to be fished out of the river (Delaware or Hudson, I don't know which). She and Lewis C. Taylor married in the 1890s and my mother was born in 1901, the younger of two sisters.

I did not choose the year of my birth very wisely — 1930, the start of a major depression — but I did choose my parents well. My father died when I was two, but he provided me with some good genes. In addition to being a mechanical engineer, he was an expert craftsman who made my set of nursery furniture himself out of mahogany (not an easy wood to deal with). He was also apparently an absolute nut about golf.

He once won the championship at the Tully-Secane Country Club, now long replaced by houses.

Because of my father's early death, my mother was the dominant influence as I grew up. She was a person of extreme practicality and determination. Though she had little formal education, she was an avid reader of "serious" books. She had left high school and married my father at the age of 16, to the great consternation of both families. The fact that my mother came from a "working class" family was also not to her advantage in the eyes of my very snobbish paternal grandmother.

At age nine months with my mother.

Since he was only 32 when he died, my father left only a modest amount of money, mostly insurance. My mother used virtually all of it to pay off the mortgage on our house, and proceeded to look for work in the midst of the depression. At first she did office work, but this was not well paid and she discovered that as a waitress she could earn a lot more. That was the work she did for nearly thirty years until she retired in 1962. She never had the least inclination to remarry.

There were probably some unhappy days during my nonage, but I simply don't remember them. I enjoyed my childhood. The death of my father when I was not yet two did not mar my childhood since I was entirely oblivious of it. My earliest memories go back only to the age of about four. I must have learned to read around that time (before going to school) and I remember being on a trolley car with my mother, on the way to or from downtown, when we passed a bakery, near the University of Pennsylvania, with a sign in the window that said *donuts*. I remember asking my mother what a dew-nut was and learning that the word was pronounced dough-nut. Then, of course, I realized what it was. I think I have always had an innate fascination with words, especially written words. I have always found it hard to understand how anyone could have difficulty with reading, spelling, or speaking, but of course, even some otherwise very intelligent people do. I've been lucky.

My mother and I lived in a working-class-cum-middle-class, mainly Irish-Catholic,

Ready for Halloween, 1935.

neighborhood in West Philadelphia and our street (the 6500 block of Regent St.) was full of kids my own age, give or take a few years. I spent lots of time playing with them, in the ordinary way that kids did then, games like tag, roller skating, stickball, kick-the-can, baseball, and touch football. We were adjacent to a very large undeveloped piece of land and also near Cobbs Creek Park, and that was about all we kids needed to be happy.

It used to snow a lot in Philadelphia in those days and we all enjoyed it. I remember at a very early age, and perhaps this is really my earliest memory of all, negotiating snow drifts at the end of our street that were deeper than I was high. I had a sled and this led to one singularly memorable occurrence. One day, when I was perhaps seven, I went sledding where my mother had strictly instructed me not to go. I went down a wooded hill and crashed into a tree, but in a reflex effort to protect myself, *i.e.*, to ward off the tree, I put my right arm out and broke it. I can still remember vividly how odd it looked, with a quarter-inch offset in the lower wrist. Still, I do not recall any great anguish except that I expected my mother would be furious — and she was.

The first school I entered, at the age of 5½ in January of 1936, was old and decrepit. However, because my mother and a neighbor had written a petition and obtained a large number of signatures, the Philadelphia Board of Education was persuaded to build a new elementary school just at the end of our street, on a corner of the undeveloped land, and after one or two years, the Joseph W. Catherine school opened and the old school was abandoned.

I remember practically nothing about elementary school. I do not remember being challenged by anything, except once when I returned after about a two week absence on account of chicken pox. The whole class had been taught to do long division and I had a short struggle to catch up, since the procedure seemed like an irrational ritual. On the whole, I was bored in school and did not behave very well. My marks were all good except for "deportment," which was frequently rated unsatisfactory. I also had measles and mumps, which provided additional welcome vacations, when I could stay home and build model airplanes.

My mother and several neighbors liked to play tennis and they somehow arranged to have two tennis courts built on the undeveloped land at the end of the street. How they accomplished this I don't know. Probably it was one of these neighbors in particular who saw to it, because he also made a croquet court next to the tennis courts. Thus, I was introduced to tennis at an early age.

At about age 12.

Growing up in a city, at least when I did, from 1930 to the end of WWII, has certain cultural advantages. There were many in the city of Philadelphia that I took particular pleasure in. Several, which I will now list in no special order, were close together, surrounding Logan Circle: the main branch of the Philadelphia Free Library (including its classical record collection and listening rooms), the Academy of Natural Sciences, and the Franklin Institute. The hours I spent in these three places must number in the hundreds. The library actually reached its zenith in my life a little later, when I was in my last years of high school and in college, but the two museums played a big part in my life from a very early age. As soon as I was old enough, perhaps ten, for my mother to permit me to "go in town" by myself on the trolley, these became my regular destinations.

In the Academy of Natural Sciences, it was the mineral and bird collections that especially fascinated me. A particularly favorite place was the small room where the fluorescent minerals were displayed, first in ordinary light and then under ultraviolet irradiation. I really think the beauty of this had a role in my eventually becoming an inorganic chemist.

The Franklin Institute was a never-ending source of fascination. I remember very vividly the splendid display of optics, where I was delighted by the demonstrations of prisms, lenses, and the phenomena of total internal reflection and interference patterns. When I was very young, the enormous steam locomotive was mesmerizing, and later, in my teens, I never missed a show at the Fels Planetarium. A planetarium show was not only a scientific experience but an esthetic one as well. As each one drew to an end, there was a graceful transition to a glorious sunrise during which the "Nocturne" (ironically) from Mendelssohn's incidental music for *A Midsummer Night's Dream* was played. My lifelong love of all of Mendelssohn's music unquestionably had its genesis in those planetarium shows.

The value and importance of these institutions in nurturing my interest in science cannot be exaggerated. I count the hours spent in them as among the happiest of my childhood memories. When my friend Zvi Dori created a science museum in Haifa, I was just delighted that he had conceived — and executed — such a wonderful idea. Science, like music, drama, and literature is an integral part of our culture.

Leary's bookstore on the east side of Ninth Street between Market and Chestnut streets is another of my cherished memories. It was in an old, narrow, six-story building crammed between two much larger ones, stuffed with second-hand books on all floors including a basement, books on every conceivable subject. The floors had ten-foot-high ceilings, the walls were lined with shelves, and there were parallel rows of shelves, back-to-back in between. These thousands of feet of shelves were insufficient and there were many piles of books on the floor as well. However, this superabundance was not chaotic. Each subject had its allotted place, and there were sales people who knew how to find things. I don't remember whether there was an elevator or not, but most up and down movement was *via* well-worn wooden staircases. The place had a phantasmagorical quality, reminiscent of scenes from Borges or *La Peau de chagrin* of Balzac. From the age of 9 or 10 until I left Philadelphia in 1951, I was a regular visitor and sometime customer. The building and one of its abutting buildings were razed in the mid-1950s to make way for more commercially attractive real estate, and the books, of which I would estimate there were no fewer than a million, dispersed to other dealers or destroyed.

While I was an eclectic peruser of Leary's vast trove, I was primarily interested in the science books. It was in Leary's that I bought Oliver Lodge's *Pioneers of Science* (1904), which covered physics and astronomy from Copernicus to the discovery of Neptune, and *Science and the Modern World* (1926), by Alfred North Whitehead (for $2.50), both of which, by some miracle, I still have. Other books I remember, but no longer have, were *The Universe Around Us* and *The Mysterious Universe* by Sir James Jeans and *The Nature of the Physical World* by A. S. Eddington. These books, which

I must have read between the ages of 12 and 15, introduced me to the concepts of relativity, quantum theory, and the probabilistic nature of the atomic and subatomic world. They were written for the intelligent layman and provided no mathematics, but that was just as well since I don't think I knew much math then. But even the qualitative presentation of concepts so much at odds with "common sense" filled me with wonder and excitement. I also still have a marvelous book, with wonderful illustrations, *Butterflies* (1924) by Clarence M. Weed, acquired for the attractive price of $0.75. Then there were *Microbe Hunters*, by Paul De Kruif; *This Chemical Age*, by Williams Haynes; and *Madame Curie*, by Eve Curie, all of which I read several times. I think that even today, these books would be good reading for an early teenager. In addition to these, I bought numerous books on elementary chemistry, especially those explaining how to carry out reactions.

Other preteen and early teen activities tended to be scientific and technical. I made innumerable model airplanes and became a devotee of the big red Radio Amateurs Handbook. But mostly I became fascinated by chemistry. A Gilbert chemistry set was soon expanded by my additions to provide a fairly diversified laboratory. When real heat was needed, I was allowed to use a gas stove that sat in an enclosed back porch. It was not very well ventilated, and the time I made nitrobenzene and then reduced it to aniline, it was made clear to me that such odoriferous experiments were off limits in the future.

Closely related to my activities in my home laboratory was an active interest in minerals. I made trips to quarries in Delaware County to collect specimens (such as garnets, which were very abundant in the mica schist) and also acquired others as gifts or by purchase. I think the beautiful shapes and colors of the crystals must have strongly, if subconsciously, directed me toward inorganic chemistry and crystallography. The marvelous display of fluorescent minerals at the Philadelphia Academy of Natural Sciences, which I have already mentioned, doubtless enhanced this interest.

I should also not fail to mention that I joined a Cub Scout den at age nine and went on to the Scout troop at age twelve, where I remained for about three more years. I don't remember very much about the Cub Scout activities, though clearly I must have enjoyed them, since I remained a member. The Boy Scout activities I remember quite well, and I did, indeed, enjoy them thoroughly. I met a much more diverse group of boys than those living in my immediate neighborhood, and the troop I belonged to

A tenderfoot boy scout, ca. 1942.

was very active, with lots of hikes and camping trips on weekends. The highlight of each year was a period of two weeks at Treasure Island (TI). This was a permanent summer camp on an island in the Delaware River between Frenchtown and Millersville, N.J. and was used by many Scout troops in and around Philadelphia.

Each troop was assigned a campsite and occupied its own tents — tents with four cots, high enough to stand up in. We had a swimming area in the river and spent much of our time on nature hikes, boating and canoeing. The major canoe activity was a canoe hike, in which we first went north about ten miles using an abandoned canal that paralleled the river on the Pennsylvania side. The canal had several locks that required portaging. It was the upstream trip that was arduous and limited the length of the hike. We returned in the river where the going — with the current — was easy and fun. The canoe hike was the major event each year.

I was fairly diligent about passing tests to rise through the ranks from tenderfoot to Scout second class, Scout first class, and then, by earning merit badges, to Star Scout and Life Scout. I never became an Eagle Scout because I dropped out at the age of about fifteen in favor of the musical activities I shall describe later. I don't know what being a Boy Scout is like today, but I am a 100% supporter of the scouting movement as I experienced it. It was wholesome, very educational, and tremendous fun. I have the fondest memories of it and feel privileged to have had the opportunity to be a Scout. Were more opportunities like that available today, we would have fewer gangs and high school shootings.

High School (Jr. and Sr.) Years

In Philadelphia in the 1940s, the second six years of public schooling were split into junior high school, grades 7–9, and senior high school, grades 10–12. At the age of ca. 11

years, I entered the William T. Tilden Junior High School, which was located about a mile, perhaps a bit more, from home, with no connecting public transportation.

My Tilden years were marked by few significant events, but there were some worth mentioning. I made a new friend, Alan Cook, and we remained friends until our early college years, when our careers diverged. Alan was the first contemporary, so far as I can recall, with whom I enjoyed a sense of intellectual equality. I don't recall that either of us had much to do with the other students, whom we both considered as intellectually challenged (to use an expression not then in use). Alan intended to become a mechanical engineer, which he subsequently did.

Both of us had birthdays in the spring and thus we had both entered school for the spring term, beginning in February, and were therefore destined to graduate in January. We both recognized the awkwardness of this with regard to entering college expeditiously. Since, in addition, we both found the pace and level of study boring, we put our heads together toward the end of our time at Tilden and decided to ask permission to move at an accelerated pace by taking an extra course each semester, with the object of making up a half year. We were permitted to do this and thus we both graduated from high school in June of 1947 instead of January of 1948. I have thought a number of times since, it is a shame students were not permitted to proceed at a rate commensurate with their abilities instead of being governed by what I call "democracy by common denominator." Since I was later, on several occasions, in the right place at the right time, owing directly to having shortened my stay in high school to five semesters, perhaps Shakespeare had it right that "there's a divinity that shapes our ends, rough-hew them how we will." As with most of what happens to us, there's no control experiment, so who knows?

I also remember receiving my first formal instruction in science (aside from math) in a course taught by a very well qualified female teacher in the ninth grade. The highlight of this course for me was extracurricular. This excellent teacher realized that I already knew more chemistry than she was going to cover, and so she lent me a college textbook she had used. I devoured it. It was in this book that I first learned about the beauty of the gas laws. I can still remember how impressed I was by the logic that if two volumes of gas A react with one volume of gas B to give two volumes of gas C, it is certain that if both A and B are diatomic, C must be triatomic. The real case which illustrated this was the reaction $2NO + O_2 \rightarrow 2NO_2$. While I was already fascinated by descriptive chemistry, this was my first realization that chemistry could also entail the use of simple but rigorous argument to learn about the nature of individual molecules. It presaged my lifelong interest in the structure and bonding of molecules and with reactions rather than with the bulk properties of matter.

Without any doubt, my most satisfying and memorable experience in high school was being a member of the debating team. There was a very active inter-school debating league at the time and the competition was keen. The question of who won a debate was determined by a judge, and since the judgment was necessarily subjective, it was important to have well-qualified judges. Fortunately, we did. The debating team at John Bartram High School, which I attended, was coached by one of the history teachers, Mrs. Elisabeth Gentieu. Mrs. G., then in her mid-twenties, was brilliant, vivacious, and stimulating. There is no doubt that without her our team would not have been as successful as it was.

In my senior year, our team won the league championship. The final match that clinched the championship was against the team from Central High. Central was then a magnet school for superior students from all over Philadelphia. A significant number of the faculty members had Ph.D. degrees! Needless to say, their football team didn't amount to much, but their debating team was perennially the one to beat and did not get beaten very often.

A debate consists of three sections: Presentations of the pro and con positions concerning the resolution (for example, Resolved, that the country should have a comprehensive, publicly funded health care system), followed by cross-examinations. Then, after a recess during which each team went into conference, there were rebuttal presentations, in which each team tried to recoup any losses inflicted on them in the cross-examination, follow up on any damage they felt they had done to the other side, and finally restate their case as convincingly as possible.

My role on this occasion was to do the presentation and rebuttal, while someone else did the cross-examination. To beat the formidable Central team, all embryonic shyster lawyers in my opinion, was a thrill, especially as it was decisive in giving us the championship.

Another activity to which I dedicated a lot of time in high school was playing in a dance band and several "combos." I had learned to play the guitar, in a classical way, in my early teens, but I was a somewhat indifferent pupil. Only at the age of about

With my mother in 1945.

sixteen did I get really enthusiastic about it when I was asked to join a small combo. We had fun, but so far as I recall we did not get any jobs. I was then asked to join a full-fledged orchestra, consisting of about four saxes, two or three trumpets, trombone, clarinet, drums, bass, and guitar. Every Saturday night we played in a dance hall called Chez Vous. It had a large ballroom, a bar, and drew a big crowd. We played standard repertoire using sheet music for the tunes popularized by the Dorsey bands, Glenn Miller, and others. It was also the era of fast and athletic dancing, called jitterbugging, by young men in zoot suits (baggy pants with peg bottoms, long coats, and really loud neckties), and it was fun to watch. The band also played at high school dances, including my own Prom. Because I spent my dance evenings in the band, I didn't learn to dance very well, and still don't today.

The band I played in consisted of second-generation Italian and Polish Americans. This led to very interesting jobs for a small combo derived from the band. In the Italian community, it was the custom for the bride to be serenaded on the eve of the wedding. We would play outside for a while and then be invited in to share in the refreshments before playing some more. The food was delicious and more often than not wine was the homemade type known colloquially as dago red. Though I was hardly a connoisseur at the time, I thought it quite good.

In the Polish community, we got jobs playing at the party following a Polish wedding. This was a more exciting experience, because large numbers of people came, had lots to drink, and were in no mood to go home just because the bride and groom had left for their honeymoon. The leader of our group, one Joey Mack (whose real Polish name, which I can no longer remember, was long and unpronounceable owing to a surfeit of c's, y's, and z's) was a shrewd businessman. At midnight, he would very histrionically order us to pack up. This immediately evoked passionate entreaties from the bride's father, whose party it was, to continue. Joey would then remind him that he had only paid us until midnight, after which he would extract more money from him and we would unpack and go on for another hour, whereupon the whole charade might be repeated. At Polish weddings, they don't like to quit early.

College Days

During the spring of 1947, Alan Cook and I both decided to apply for admission to Drexel Institute in Philadelphia, he for mechanical engineering and I for chemical engineering. I would probably have chosen chemistry but Drexel, at that time, did not offer degrees in chemistry (or physics) although it does today. Moreover, I didn't think there would be a big difference. This, of course, was a gross misapprehension that soon required corrective action. We were both accepted, and enrolled in September. The first year of engineering was the same for everyone, so Alan and I had the same classes, which were chemistry, physics, calculus, English, and ROTC. Drexel required two years of ROTC for all students.

The first year, during which I had a tuition scholarship, went well. The chemistry and physics courses were fun and I thoroughly enjoyed the English course, which entailed both literature and essay writing. The ROTC was not arduous, and although I griped a lot, I rather enjoyed it. In addition to drill, which included learning to disassemble, clean, and reassemble an M1 rifle, even in the dark, and a little shooting, there were also lectures on military strategy and tactics. The only thing that was not pure fun that year was calculus, which was taught in a fast, problem-solving kind of style. In retrospect, I can't fault Drexel for teaching math to engineers, or other non-mathematicians, in this way, because for such people mathematics is not an end in itself but a useful tool.

It was in my second year at Drexel that clouds appeared, in several forms. Integral calculus the Drexel way gave me just as much trouble as had differential calculus, although I enjoyed analytical chemistry, both qualitative and quantitative, and of course, English, and ROTC went on as usual. The new wrinkles were a course in civil engineering, required of all engineers, and the advent of work periods. Drexel, like my wife's alma mater, Antioch, was one of the pioneers in what is called cooperative education. What this meant was that beginning in the second year, one worked in an industrial concern every other quarter. I'll say more about the work program presently, but my main problem was with the civil engineering course and its implications.

As I noted earlier, when I started Drexel, I naively thought that chemical engineering and chemistry were practically the same thing. I began to realize, as that year went on, that chemical engineering was much more engineering than it was chemical and that I was interested in chemistry rather than engineering. The required course in civil engineering really brought things to a head. All we did in that course, as I remember it, was to learn methods for calculating stresses and strains on beams in more and more complex cross-braced structures. We also learned how to make engineering drawings. I realized that while I could do these things, I didn't want to do them. I wanted to be in a chemistry laboratory. Thus, early in my second year, I recognized that I had to transfer to a college or university that offered a degree in chemistry.

In this period of time, that was a major problem. All institutions of higher learning in the country were jammed full of students who were veterans on the GI bill, in addition to the normal intake of recent high school graduates. I wrote to a number of colleges and universities who made no offer whatever, but Temple University offered to admit me, in September of 1949, and to accept my Drexel credits. Thus, it appeared likely that I could complete my degree in two more years, except possibly for one thing. Temple required three years of a foreign language, and I had had, as yet, none. I resolved to deal with that problem by studying over the summer for an advanced placement examination in German.

During my second year at Drexel, one of my co-op jobs had been in the analytical laboratory at the Philadelphia Rust-Proof Co. (PRC), one of the larger electroplating and coating shops in the Philadelphia area. The "man-in-charge" was Dr. Samuel Heiman, who had earned his Ph.D. at Penn while holding down a full-time job supervising the lab at PRC. Sam was a wonderful man, of whom I shall say more later. However, during the summer of 1949, for reasons I cannot remember, I was employed at the Gulf oil refinery, not in the laboratory, but in a work crew charged with maintaining the catalytic crackers and other units. The work included removing spent catalyst, cleaning the reactors, and other such laborious and unpleasant jobs. In addition, the men with whom I worked were not exactly Nature's gentlemen. Come to think of it, I may have taken this job assignment simply because it paid better than any other. It did give me a lasting sense of how hard some people have to work for their living and why, therefore, retirement is so eagerly awaited by many people.

This job allowed a lot of time to study German, which I did on my long commutes to and from my home in the extreme northeast of Philadelphia to the refinery in the extreme southwest, about the longest possible intracity journey. Happily my efforts paid off. I was able to get advanced placement in German so that I could start with the third-semester course. In fact I took another three semesters and had a quite decent knowledge of literary German when I graduated. Two excellent courses I took were a study of *Urfaust*, Goethe's preliminary sketch of his full-length masterpiece, and *Die Deutsche Romantik*, a survey course in German romantic literature, both taught by a wonderful professor named Schuster, who communicated his own love for the material.

My two years at Temple University were wonderful. Though Temple was already a rather large university, it had the feel of a small college. The chemistry faculty consisted of only about eight professors, but they were all stimulating, dedicated teachers, and several were particularly memorable. The head of the department, William Rogers, was a physical chemist with a British accent, a scholarly demeanor, and a wonderful way with words. I can still remember one of his observations concerning the fact that to accommodate the heavy influx of students, Temple had built a number of "centers" in, or close to, various suburbs. He remarked that Temple was in the odd position of having many centers that were all on the periphery.

One who was to have a major influence on my life was William T. Caldwell, who was a professor of organic chemistry, but also Dean of the College of Arts and Sciences. Dean Caldwell was a commanding presence, tall, bald, and elegant. He was no longer doing research, but he continued to teach the introductory course in organic chemistry, for which he used a textbook he had written himself. I found it to be an excellent book in a purely technical sense, but its real distinction lay in its elegant literary style. In what other textbook would you find (as the opening of the chapter on acetoacetic ester and malonic ester) a sentence like this: "How the discoveries attendant upon geographical exploration have often differed from those anticipated and planned is commemorated in the name of the American Indians or of the rapids in the St. Lawrence near Montreal, whose La Chine recalls Champlain's ironic humor or disappointment when he realized at last that he was sailing up no river to Cathay."

Other professors were Edgar Howard, Floyd T. Tyson, and Francis Case (organic); M. Wesley Rigg (physical); and Hazel M. Tomlinson (analytical). It was in Professor Rigg's course that I realized what a beautiful thing physical chemistry is. We used as a text whatever edition of Daniels and Alberty was then current. I also did some small project in Rigg's laboratory, and it was there that I first met my present-day colleague Abe Clearfield, who was doing a master's degree with Rigg.

There was one small experience that I had in a one-semester course in qualitative organic analysis that I'll never forget. One of the unknowns I was given had as loathsome a smell as I've ever encountered, before or since. I determined what it was in due course: thiobutyric acid. With the addition of a sulfhydryl group to the parentage in butyric acid, the outcome in terms of odor was understandable. In connection with the qualitative organic analysis course, we all had access to the Huntress and Mulliken encyclopedia of organic compounds and their properties, in which odors, among other properties, were mentioned. It occurred to me to look up the odors of thiols, butyric acid, and thiobutyric acid, to see if there was consistency in how their odors were described. Indeed, there was. Thiols were generally described as having *disgusting* odors. Butyric acid was said to have a *foul* odor. Finally, for thiobutyric acid it said, *odor insufferable*. I was very much impressed by the care that had been taken to scale up the adjectives.

I also found that among my fellow students in chemistry and physics there were some excellent people with whom I formed very satisfying friendships. Even though I was a newcomer to the chemistry club, they elected me president in my senior year. Actually, I was one of only two in my senior class of about a dozen chemistry majors who planned to remain in chemistry. All the rest were premeds, except for two women who did not go on to graduate school. One of the premeds told me candidly that he really would have liked to go on in chemistry, but found the higher income of a physician irresistible. Two of my best friends among the premeds really did want to be physicians and they have had distinguished careers, one as a surgeon and the other as a professor of endocrinology at the University of Michigan. There were also two or three physics majors whose company I particularly enjoyed.

I had no need to work terribly hard in the courses I took at Temple (the math courses were very agreeable) and I took lots of humanities courses. Although no Phi Beta Kappa chapter existed during my time, one was established later and I was elected an *ex post facto* member. Besides the German I have already mentioned, I particularly remember a course in esthetics that was a lot of fun and also a course in the history of Western art taught by a world-famous scholar named Herman Gundersheimer. I also squeezed in a course in biochemistry, taught at night by Sidney Weinhouse (who later became an NAS member). At one point I got annoyed by the amount of rote learning required in this course and had the temerity to write a boorish letter to Professor Weinhouse that I have regretted ever since.

I was also required, for distribution, to take a course in either biology or psychology. The latter held absolutely no attraction for me, so by default I registered for a biology course. The main feature of this course was the laboratory, where we dissected, in the following order, an earthworm, a crawfish, and a frog. While this was sort of amusing, the course as a whole was quite boring, focusing as it did on facts and more facts, with little in the way of breadth or general principles. I think I scraped through with a weak B.

One final word on courses has to do with a required course in history. I had to take this during the summer because I could find no room on my schedule during the regular terms. During my time at Temple, I continued to work part time for Sam Heiman at PRC during the year and full time during the summer of 1950. Sam was kind enough to let me go over to Temple each morning for a month to take this course, and I stayed later in the evening so as to make up the time. The course, which I took simply because it was the only one available early in the morning during a summer term, was on African history, precolonial and colonial. The whole thing was a crashing bore, much of which concerned all the bloody tribal battles in precolonial times. It seemed that Africans of that time were nothing but barbarians whose main activity was killing each other. Some years later I read a book about early English history and discovered that the early inhabitants of the British Isles, from whom I descend, were quite similar, essentially barbarians whose main activity was killing each other.

Apart from enjoyment of my courses and my fellow students at Temple, I continued to derive great pleasure from the cultural resources of Philadelphia. In this connection, I particularly remember the Art Museum, including a little offshoot, the Rodin Museum, the Philadelphia Public Library, and the Philadelphia Orchestra. The Art Museum was and is a great museum, with an impressive variety of collections, including wonderful works from the late nineteenth and early twentieth centuries. The Rodin Museum was a gem, with most of Rodin's most celebrated works. These museums unquestionably formed my lifelong taste in the visual arts, which does not incline to the post-WWII trends.

In addition to books, the main branch of the Philadelphia Public Library had an extensive collection of phonograph records and comfortable listening rooms. This facility was an adjunct to my Saturday nights at the concerts of the Philadelphia Orchestra. The Philadelphia Academy of Music — which isn't an academy but a concert hall in the La Scala style — has at the top a gallery with hard wooden benches, and for each Saturday evening concert, seats in this location were available on a first-come, first-served basis for seventy-five cents. I had formed a friendship with the son of one of my mother's friends, a young man who had a scholarship at the Curtis Institute of Music to study with Marcel Tabuteau, then first oboe in the orchestra. This chap, Charles Edmunds, and I regularly got in line on Saturday nights during the orchestra seasons of 1949 and 1950 in order to climb the many flights of stairs to the thin air and hard seats at the top. Charles did much to increase my enjoyment of these concerts because of his musical knowledge and sensitivity. Apart from his irrational aversion to certain pops-type pieces, his taste was excellent and I learned a lot from him. Another friend of my mother's was Sol Ruden who was first chair in the second violin section for many years. He was a wise and patient counselor who taught me what to look for in symphonic music. These cultural enrichments made the city of Philadelphia a marvelous place to grow up.

There was never any doubt in my mind that I wanted to have a career in academia, so in the fall of 1950, I began to look seriously at graduate schools. I had a feeling I might like to do radiochemistry so I decided to apply to Washington University in St. Louis because they had a leading program in that area. I believe that I also chose another one or two places (although I no longer remember which ones they were) but the important thing was that Dean Caldwell told me to apply to Harvard — and of course I did. I heard from Washington University first, with an acceptance and the offer of a teaching fellowship, and immediately, in a state of elation, reported this to Dean Caldwell. He congratulated me politely, but said, "You will, however, be going to Harvard." It seems he knew someone there personally and had already been informally assured that I was to be accepted. He had sent a student there several years earlier who had acquitted himself well with Woodward on the quinine synthesis (one of Woodward's early triumphs), and so his recommendations had credibility. Shortly thereafter I received the official word from Harvard, and I don't even remember if there were any other acceptances, because I no longer cared.

One final observation about Temple University, or more exactly about its chemistry department. The undergraduate curriculum it provided may have been a bit classical, but it was thorough. The professors who gave the lectures also spent a considerable amount of time in the laboratories, although they did, of course, have graduate teaching assistants to supervise continuously. While a graduate of their program may have lacked a few of the most up-to-date topics, the grounding provided in fundamentals was admirable. The best evidence of this, in my personal experience, is that in my first week at Harvard I had to sit four examinations given to check on the adequacy of the undergraduate backgrounds of the incoming students. In a group of about twenty, in 1951, I was one of only two who passed them all. I should have been gracious enough to write back and thank my teachers at Temple for what they had done to make this possible, but with the thoughtlessness of youth, I never did.

I did, however, return for visits in the years following and maintained contact with several of the faculty, in particular Edgar Howard, Francis Case, and Hazel Tomlinson. No doubt it was due to such people that I received my first honorary degree about ten years later (1962) from Temple.

I would like to end this chapter by noting that my first scientific publication was published before I left the fair city of Philadelphia. While I was working at the Philadelphia Rust-Proof Company, Sam Heiman, the Chief Chemist there, discussed with me the importance of knowing accurately and reproducibly the ammonia content of bronze electroplating baths and the uncertainty as to how best to do this. Numerous methods had been given in the literature and a critical evaluation was needed to find out which was the best, considering convenience, accuracy, and reproducibility. We planned a study that I would carry out to provide the desired information. I did most, if not perhaps all, of the lab work on my own time, staying at nights. We obtained good results and they were published in September of 1950 [1] in the journal called *Plating*, published by the American Electroplaters' Society. I am still proud to have this as number one on my publication list. The paper is very well written (by Sam mostly) and the content is thorough and accurate. I can honestly say that my career in research started with a practical, useful investigation, and I am grateful to Sam Heiman for offering me the opportunity to do the work.

CHAPTER 2

HARVARD YEARS

YOU CAN ALWAYS TELL
A HARVARD MAN,
BUT YOU CAN'T
TELL HIM MUCH

— *Anonymous*

I ARRIVED AT Harvard to begin the fall semester in 1951, settled down in Richards 201, and was quickly caught up in the fascinating business of making new friends and finding out what I was supposed to do. I was able, in spite of (or probably because of) my very classical preparation at Drexel and Temple, to perform sufficiently well on the four qualifying examinations that P. D. Bartlett was able to tell me that I was not required to take any specific courses, but could choose a few that I liked. However, he suggested that since I had not studied thermodynamics beyond the level of introductory physical chemistry, I should take a half-course (Harvard terminology for a one-semester course) in that subject. I also agreed that I should take a course in quantum mechanics. I then made what proved to be a very fateful move by saying that since I had not had a course in inorganic chemistry, perhaps I should take one. P. D. (as all we students called him, among ourselves, that is) said he thought that was a very reasonable idea and signed me up for the fall semester of a full course, which was to be taught by a brand new Assistant Professor named Geoffrey Wilkinson (GW).

I had been accepted at Harvard with the promise of a Teaching Assistantship and so I next went to the Chairman's office to find out about that. The first words I got back from the secretary were: "Oh what a lovely Pennsylvania twang you have" — this from someone with a classic "Bahston" accent. After this blow to my self-esteem, she looked up my assignment and informed me that I was to be a teaching assistant in physical chemistry. I was sent to see another new young faculty member, William Chupka, who would be in charge of the P-chem lab course. I was quite pleased that I would be a TA in P-chem and, moreover, I found Bill Chupka to be a great guy. We soon got to be — and still are — good friends. We once went to a dance together at a girl's school called the Academie Moderne, and Bill removed his eyeglasses in the belief this would improve his appeal to the opposite sex. Since he was very nearsighted, he then said, "Point me in the direction of the cutest girl you can see." I did not yield to the temptation to point him toward the homeliest.

The first two sessions of P-chem lab were devoted to teaching the students to do some simple glass blowing — butt joints, T joints, and finally a ring seal. This was

Geoff Wilkinson as an Assistant Professor doing work in a glove box, ca. 1953.

an interesting challenge to me, since I had never learned to do *any* of these things myself. I immediately recognized that under no circumstances could I let myself be inveigled into demonstrating. Right after the first lab session, I spent quite a few hours practicing T and butt joints, but I still refused all suggestions that I *show* someone how to do it. Incidentally, there was another purely physical danger to avoid. The students had the charming habit of taking their work right out of the flame and bringing it over for my inspection, handing me the hot part while holding onto the cold part. After these glass-blowing preliminaries were over, I enjoyed the job. In the following three years, I no longer had to be a teaching fellow because I had a succession of industrially supported fellowships.

Getting back to course work, the thermodynamics course was taught by Paul Doty in his well-known discursive style, but I got through it. The main problem was that the graduate students were competing with a group of very sharp seniors (*e.g.*, Steve Berry), still possessing undergraduate fervor for getting good grades, who made it very hard to get an A. I don't think I did. The inorganic chemistry course was the one that really appealed to me, and by late in the fall semester, I had, in fact, decided to do my thesis research with Wilkinson.

I began my Ph.D. research with a project that was not destined to go far. Wilkinson suggested that I find a convenient method to make anhydrous metal chlorides, beginning with those of the rare earths. I was then to try making organolanthanides by reacting them with Grignard reagents. It was an interesting, if modest, challenge.

Rare earth chlorides are normally obtained from aqueous solution as hydrates:

$$M_2O_3 + 6HCl(aq) \rightarrow 2MCl_3 \cdot nH_2O$$

These hydrates cannot be thermally dehydrated because on heating the following reaction occurs instead:

$$MCl_3 \cdot nH_2O \rightarrow MOCl + 2HCl(g) + (n-1)H_2O$$

It was possible to get the desired result by heating the hydrate mixed with NH_4Cl:

$$MCl_3 \cdot nH_2O + NH_4Cl \rightarrow MCl_3 + nH_2O + NH_4Cl$$

but this is rather messy and inconvenient.

What I thought of was to reflux the hydrate with acetic anhydride, with the intention of obtaining the following reaction:

$$MCl_3 \cdot nH_2O + n(MeCO)_2O \rightarrow MCl_3 + 2nMeCO_2H$$

I tried this first with $ScCl_3 \cdot 6H_2O$ and it appeared to work beautifully. A fluffy white solid was obtained that seemed likely to be not only pure and water free but also reactive. However, I soon discovered that it contained no chloride. What had actually happened was:

$$ScCl_3 \cdot 6H_2O + 6(MeCO)_2O \rightarrow Sc(MeCO_2)_3 + 3HCl(g) + 9MeCO_2H$$

I had found a very simple way to make anhydrous rare earth acetates. I tried this with one or two other $MCl_3 \cdot nH_2O$ compounds and it seemed universally applicable. When I left for the Christmas holidays, we were very happy with this unexpected development and had some plans for things we might try to do with these new materials.

However, soon after I returned in January, a momentous development occurred. GW as well as R. B. Woodward (RBW) saw, sometime in January 1952, the paper by T. J. Kealy and P. L. Pauson, *Nature* **1951**, *168*, 1039, reporting an extraordinarily stable "dialkyl" of iron, $(C_5H_5)_2Fe$. The authors had made this compound by the reaction:

$$FeCl_2 + 2C_5H_5MgBr \rightarrow (C_5H_5)_2Fe + MgBr_2 + MgCl_2$$

and they proposed to explain its remarkable thermal and chemical stability by an ionic-covalent resonance:

This had instantly struck GW as being an entirely insufficient explanation and he immediately repeated the preparation and began to investigate the compound in more detail. RBW had independently also concluded that Pauson's ionic-covalent resonance proposal could not be correct. They discussed their shared incredulity and came up with the "sandwich" structure. After a few experiments to show that the compound was even more stable (up to at least 470 °C) than Kealy and Pauson had realized, that it had a very simple infrared spectrum and that it reversibly formed a stable blue cation which had one unpaired electron while $(C_5H_5)_2Fe$ was diamagnetic, they proposed the now famous pentagonal antiprism structure for it, in a Communication to the Editor of the *Journal of the American Chemical Society*. This communication (*J. Am. Chem. Soc.* **1952**, *74*, 2125) was received March 24 and appeared in the April 20 issue. I believe it is fair to say that this paper is the most seminal one in the entire history of organometallic chemistry.

In the face of something this exciting, interest in anhydrous metal acetates vanished, at least at that time. As I shall relate later, GW subsequently returned to the subject (1960) with significant consequences.

The extraordinary sandwich structure and the extreme thermal stability of $(C_5H_5)_2Fe$, which Mark Whiting, one of Woodward's postdocs, christened *ferrocene,* got me to thinking about its heat of formation. I found out that George B. Kistiakowsky had a complete apparatus for making heat of combustion measurements and with GW's approval I borrowed this, set it up in a cubicle next to GW's lab-office, and with a little advice from GBK, went about measuring the heat of combustion. From the ΔH_f^0 value, 33.8 ± 2.0 kcal mol^{-1} and some auxiliary data, I estimated that the combination of resonance within the rings and the iron to carbon bonding amounted to about 113 kcal mol^{-1}, which we felt "accords with the unusual stability of ferrocene." By the end of the Spring semester (June 6 of 1952), I had submitted my first publication from Harvard to the *Journal of the American Chemical Society*. It was published as a Note in the Nov. 20 issue [2]. Geoff and I did not have Woodwardian clout! I shall say no more about the early part of the ferrocene story but there is a lot more to be said [1488].

This heat of formation work led to my first personal contact with R. B. Woodward. He was a godlike figure to all of us graduate students (including his own) and was seen by us only at a distance. When I had my work done and had final figures, Geoff said, "Go over and show this to Woodward." So, with some trepidation, I went to his office to make an appointment. When his secretary learned that I had research results to show him, she said, "Oh, in that case, go right in." We sat down to look over my results and he asked a lot of questions about how I got from the raw data to the final result. I was amused to find that like me and other mortals, he had to labor a bit over the question of which ΔH was positive and which was negative. It was a thrill for me to be able to help the great man to understand something. I also drew two lessons from this experience. One, that discussing research results should always

As a graduate student, August 1952.

take priority over other things and, two, that the professor should always go into things in detail and not simply accept the final results from his younger coworkers. I have always tried to adhere to both of these practices myself, and I have seen several cases where an eminent but busy professor got into embarrassing difficulty because he failed to follow the second practice. It must be very sobering and painful to have to retract published work because not enough questions were asked as to how the results were obtained.

Meanwhile GW and other graduate students of his, along with Mark Whiting and a Woodward graduate student (Mike Rosenblum), were continuing to work on the chemistry of ferrocene. For me, however, there was a summer-long hiatus because I had accepted a position at Los Alamos National Laboratory as a summer research student. GW was also out for much of that summer to be married in Denmark.

A Summer at Los Alamos, 1952

My summer at Los Alamos was a great adventure and very educational. To someone who had never before been farther west than Hershey, Pa., the trip alone was exciting, and the New Mexico landscape was exhilarating. I was picked up by a laboratory car at the Santa Fe railway station and conveyed to the entirely closed and guarded Los Alamos community. Security was rather strict, so my first few hours were spent being questioned, photographed, finger-printed, and equipped with a photo ID. I had, of course, been investigated by the FBI months earlier before I was formally offered the job. I was then taken to my accommodations in a long-term visitors' dormitory where I unpacked, and finally to the laboratory (CMR-5, for chemical and metallurgical research) where I was to work under the direction of a young staff member, R. N. R. (Bob) Mulford.

The work I was assigned to do dealt mainly with studying a small corner of the P *vs* T diagram for the interaction of plutonium with hydrogen. This was very instructive because I learned a lot about the use of high vacuum technique, with Töpler pumps, vacuum gauges, mercury vapor pumps, and so on, as well as the use of inert atmosphere boxes. Generally speaking, my work routine consisted of cutting off a 5–10 mm segment from a spool of plutonium wire, and transferring it from the box to my vacuum line, where it was then reacted with H_2 at certain temperatures and pressures.

All of this went well except for the infamous day when I dropped a piece of plutonium on the lab floor. I was just bending down to pick it up (with tweezers, of course) when an alarm went off and I was hustled out of the lab. Very soon a crew of men arrived all decked out in white smocks, head covers, face masks, gloves, and booties over their shoes. They removed the offending piece of plutonium and replaced some tiles on the floor. After about an hour they left, a much chastened (though not chastised) young summer research student went back to work, resolved to be a whole lot more careful in the future.

Actually, I was not expected to work very hard, but rather to read, talk with other people, and in general, learn as much new chemistry as possible. Perhaps an ulterior motive in all this was to encourage young people to consider making a career at Los Alamos later. It was a great life. Bob Mulford taught me a lot, and was also very kind and friendly. The eminent crystallographer W. H. Zachariasen, from the University of Chicago, was also there for the summer, and I began to learn some structural chemistry.

When I arrived I had been scheduled to have Tom Lehrer as a roommate, but his arrival was delayed and other arrangements were made. We had had a casual acquaintance at Harvard, where he was still a graduate student in mathematics. He and some friends had put on a show, just before Christmas, called the Physical Revue, in which they performed songs written by Tom. I had gotten to know this group and occasionally joined the Lehrer table in the Commons for breakfast or lunch. Tom would read *The New York Times* at breakfast and sometimes pipe up and say "listen to this." He would then read a story aloud, except that he creatively and humorously embellished it in the process so that it came out as a burlesque of the actual story. By the end of 1952, he had made his first record, a ten-inch LP that I and others helped sell on the Harvard campus. One of the songs on this record is descriptive of his summer at Los Alamos and includes the lines "Yes, I'll soon make my appearance (soon as I can get my clearance)" which explains his late arrival.

Weekends during that summer were devoted to camping and exploring. In this way, I met a Nobel prize winner for the first time. One of my camping buddies was another summer research student, Rod Smythe, son of the Princeton physicist, Henry Smythe, who wrote the official explanation of the nature of the atomic bomb that was published by the U.S. government sometime after the end of the war.

Rod and I and others had planned a two-day camping and climbing expedition for a weekend in July, and put up notices that persons other than ourselves were welcome to join us by just showing up at 6 am on Saturday morning (suitably dressed and equipped) at the post office parking lot. After the car pooling had been sorted out and all the other cars had left, Rod and I were about to leave in his big convertible when we spotted a small, nondescript man who was still waiting quietly. Rod and I did not recognize him, and I remember that we speculated as to whether he might be a janitor. We invited him to go with us and he took the middle place on the front seat (bucket seats were not then the norm) and off we went.

Once on the open road, he spoke up in an accented voice and asked us about ourselves. I identified myself and explained that I was there as a summer research student in chemistry and then he turned to Rod who said "I'm here for the summer in physics, my name is Rod Smythe." Then the small man asked him whether his father was Henry Smythe. Rod said "yes," and he and I had just a second to glance at each other sharing the new realization that whoever our guest was, he was obviously no janitor, when he said, very quietly, "and I am Fermi." After we recovered from almost running off the road as Rod turned to stare at him, we settled down to a wonderful three or so hours with the world's greatest all-around physicist to ourselves. I think we were so overawed that we were inhibited, but in spite of that, it was an experience I'll never forget.

Fermi did not make it to the top of the mountain, but no one thought much about this, considering his age. He was, of course, only 51, but to us 22-year-olds, that seemed old. Only later did I learn that even then the cancer that eventually took his life (in 1954) was already developing. Later, in the Fall of 1952, he came to visit at Harvard, where one day he came to lunch at the Harkness Commons. I astounded the group I was in by going over to the table where he was and having a brief chat with him. It was clear that he remembered me and he shook hands warmly. My friends showed me new respect — for a while.

The Pace of Research Quickens

On my return to Harvard in September of 1952, I found that GW was already back, working like one possessed, and had shown that $(C_5H_5)_2Ru$, $(C_5H_5)_2Ru^+$, and $(C_5H_5)_2Co^+$ could be made. Also, during that summer, E. O. Fischer had been heard from, with the same proposal (quite independently) of the sandwich structure, as well as $(C_5H_5)_2Co^+$ and there was every reason to believe that he, too, would be proceeding to make other $(C_5H_5)_2M$ compounds. The race was on to see who could get there fastest with the mostest.

Another remarkable development was that Peter L. Pauson, who had started the whole thing with his paper in *Nature* in December of 1951, arrived at Harvard that fall, as a result of plans made before the ferrocene story broke, to spend a postdoctoral year sponsored by DuPont.

I immediately got involved in preparing and physically characterizing more compounds, and by the end of 1952, a Letter to the Editor by Wilkinson, Pauson, Birmingham, and Cotton had been prepared for *J. Am. Chem. Soc.* [3], in which we reported $(C_5H_5)_2Ni$ (and the then surprising fact that it had two unpaired electrons) as well as $(C_5H_5)_2TiBr_2$, the $(C_5H_5)_2Ti^+$ ion, $(C_5H_5)_2ZrBr_2$, and $(C_5H_5)_2VCl_2$. The nickel work was mine; the Ti, V, and Zr compounds had been made by John Birmingham. The long range significance of the Ti, Zr, and V compounds as Ziegler catalysts was, of course, not foreseeable at that time. Nevertheless, one fine day, some forty years later in a court of law, where I was testifying as an expert witness in a patent case concerning so-called single site catalysts, I stretched the truth. A rather bumptious attorney for the other side was trying to convince the judge that I should be thrown out. He asked me what I had ever done with such catalysts, and I coolly replied "Not much other than to have discovered them." The judge decided "I'd like to hear what Professor Cotton has to say."

The pace of work during 1953 was very swift, and I coauthored a paper reporting the $(C_5H_5)_2Rh^+$ and $(C_5H_5)_2Ir^+$ ions [4], as well as one giving more physical data on and new preparations of $(C_5H_5)_2Ni$ and $(C_5H_5)_2Co$ [5]. The latter was submitted in September of 1953. In the meantime, GW, who was now married to Lise Skou, a Danish plant biologist, and I were preparing to spend the Spring semester in Copenhagen.

My First Trip to Europe

One of the best things about an academic career in science is that it affords the opportunity to travel, frequently, far, and in the best possible manner. Wherever you go you are received by friends or soon-to-be friends who provide guided tours that beat anything money can buy. My first experience with travel as a scientist came when I was still in graduate school.

It was one of Wilkinson's privileges, as an Assistant Professor at Harvard, to take one semester of sabbatical leave, and he elected to take it in the Spring of 1954. Since his wife was Danish (they had met in Berkeley when Lise was in Calvin's laboratory and Geoff was with Seaborg), he chose to spend it in Copenhagen, and Jannik Bjerrum, the head of inorganic chemistry in the Technical University there, agreed to accept him as a guest in his laboratory.

I had a strong yen to get to Europe and as good fortune would have it, the departmental fellowship I held at that time (the Coffin Fellowship) allowed travel. Had I received the Coffin and Swope Fellowships in the reverse order, there might have been a problem. There's nothing like good luck! Geoff was quite willing to have me along so that we could both work in the laboratory on cyclopentadienyl metal chemistry. Thus it was that on a cold, gray day, January 29 of 1954, I departed New York on the S.S. Stavangerfjord for Copenhagen, calling in at Bergen, Norway.

The trip was in itself memorable. Eight days in the middle of the north Atlantic ocean in the dead of winter is not something I can recommend. It was bitterly cold and the little Stavangerfjord (13,000 tons) was an old ship with none of the stabilizers that had by then appeared on newer ships, and it pitched and rolled badly. The first night out I joined my tablemates, a Norwegian couple and their son, and enjoyed a hearty dinner. They were from Seattle, where he was a commercial fisherman. That first night

I slept fairly well but still I felt a trifle queasy at breakfast and a little worse at lunch. By dinnertime I had little desire to eat, and learned why each table had a little metal railing about an inch high around the perimeter. Eating soup was a tricky business.

The second night I did not sleep well and by the next morning I was too nauseous to want a meal. In fact, I had a few negative meals during the day and did not stir from my cabin. During the third full day out, I managed to consume a saltine for breakfast and two for lunch. I was not keen on having dinner but felt I should try to eat something and made my way along the outer passage of the heavily tossing and rolling ship to the dining room. There I was greeted by my table companions with a hearty "well, young man, where have you been the last few days?" It appeared that these scions of Scandinavian seafaring stock had never experienced *mal de mer* and were genuinely unable to believe that there was such a thing. I made the following entry in a diary I kept of the trip: "The merit of a firm determination not to be seasick is now proven: it doesn't do a damn bit of good."

Things gradually improved, and I enjoyed the rest of the trip. The boat seemed to be full of Scandinavians who had emigrated to the States and were returning for visits to the countries from which they came. Predominantly older men, they spent evenings drinking akavit in the classic toss-it-down way, talking, and playing cards.

On the sixth day, we passed the Hebrides Islands and on February 5th, at nearly midnight, made first landfall at Bergen in Norway. A Danish girl who had spent the last two years in the States and I got off and took a walk in the quiet streets of Bergen, where the heavy mist put halos around the lights. This girl, who was returning to Copenhagen to take final examinations for a degree in English language and literature, told me she was worried about her ability to pass the examination in spoken English. I was astonished to hear this since she spoke fluently, in a natural and melodious way, with only a slight and charming accent. Then she explained that she was required to speak with what was then called a BBC accent, that neutral, synthetic blend of upper bourgeois and Oxbridge pronunciation heard in those days on British radio. The problem was that after two years in the States she now sounded almost like an American. I never learned whether she passed or not, but was annoyed by the insularity (pun intended) of the requirement.

Late the next afternoon we arrived in Copenhagen where I was collected by Geoff and Lise Wilkinson and taken to Pension Gotha, across from the south end of the

Botanical Garden, on the north end of which was the Technical University of Denmark, where we would work. After having dinner I went for a walk in the neighborhood and was captivated by how "European" everything was. I was immediately fascinated and charmed by Copenhagen, and still am today.

My sojourn in Pension Gotha was not to be a long one. The next evening we were served large, ugly slabs of beef liver for dinner. I struggled to eat a modicum of what I considered — and still consider — to be inedible. Two nights later, this ghastly experience was repeated. When it happened for the third time only a few days later, I betook myself to a restaurant just a few steps away and pitched into a lovely beefsteak with a slice of parsleyed butter on top — what the Danes call franskboef. The next morning I interrogated Fru Winter, a large dour woman who presided over Pension Gotha, concerning the frequency with which she was wont to favor her guests with revolting slabs of beef liver. When I learned that this outrage occurred two or three times a week, my course was clear. For the next week or so, during which I sought and found a room elsewhere, with no *pension*, I popped into the restaurant for dinner whenever the inedible was on offer at the Gotha.

In my new digs, it was a pleasure to explore all the bakeries in the neighborhood to obtain breakfast and thus I became an aficionado of Danish pastry (which is infinitely superior to what is called "Danish" in this country). Even better, I became a regular at several superb restaurants in downtown Copenhagen and sampled the wares of many more.

Life in Copenhagen was rich, even for one knowing but a few words of Danish. The city itself is a work of art, but it also offered a full range of musical and other activities. The Royal Danish Ballet was one of the world's best. The opera too was very fine, but offered a remarkable experience: hearing one's favorite arias sung in Danish took some getting used to. As is still common in Europe, operas were sung in the language of the country in which they were being performed. I shall never forget the performance of Pagliacci, since hearing that warhorse *Vesti la giubba* in Danish wakes up the ears.

Just before I left Denmark on June 19, I had the opportunity to see an English company perform *Hamlet* on site, so to speak, namely, at Elsinore Castle, which is not too far north of Copenhagen. The cast was "a little above average," having in the leading roles Richard Burton, Claire Bloom, Fay Compton, Michael Horndern, and Robert Hardy. I've had some great nights at the theater since, but this was beyond duplication.

I must now make a conscientious effort to stop writing about all the delights of Copenhagen — and they are many — and get back to my experiences in chemistry. Geoff and I were placed in a corner laboratory which was spacious and sunny, albeit not well endowed for the preparative work we wanted to do. There was but one fume cupboard of no great efficiency and very little in the way of glassware of the sort we required. Fortunately, a supplier of chemical glassware was located about a half mile away and Geoff's Guggenheim fellowship provided the purchasing power, so that we made frequent visits to this emporium. Ordering it through the Danish bureaucracy would have been a hopeless enterprise. In addition, Geoff was a very capable glass blower, and so we were able to get some serious work done.

It was during this period that we fully realized that E. O. Fischer was going to be a serious competitor for supremacy in the new field of cyclopentadienyl metal chemistry. This, of course, was far more troubling to Geoff than to me, and he spent his "sabbatical leave" working extremely hard.

Life in Kemisk A, the Department presided over by Jannik Bjerrum, was, for me, delightful in every way. To begin with, the lab in which I worked was full of historical significance. I worked on the very bench where Sophus Mads Jørgensen had done a great deal of his work, which, along with that of Alfred Werner in Zurich, laid the foundations of coordination chemistry. Right outside our laboratory door was a huge cabinet containing hundreds of Jørgensen's own compounds, each one with a label, written elegantly by Jørgensen in India ink.

There were many very kind and interesting people working in Bjerrum's group who were always helpful and provided very pleasant company. It was the custom for all of us to eat lunch together around the conference table in the library. All of these people spoke English very well, even those who had never been out of Denmark for any length of time. The Danish educational system was very efficient. Since Geoff and I were the first American visitors they had had since the beginning of the war, we always became involved in discussions about all sorts of cultural and political questions, and the range of our discussions was extremely broad. I learned a great deal about Danish attitudes and actions during the German occupation. The Germans had subdued the Danes easily, without significant bloodshed, and settled down to run a benign dictatorship over people whom they were pleased to regard as their little Teutonic brothers. This picture, idyllic from the German point of view, was spoiled by the fact that the Danes stubbornly rejected the role of "little brother." The Danish and German national personalities, be it said, are not similar. The Danes immediately set about making things as frustrating as possible for their "big brothers" and even in 1954 they well remembered and happily recounted their successes. Of course, one of their major successes was getting practically all Danish Jews over to neutral Sweden, where they were safe. Danes are the kind of people who will never have much of an army, but, when necessary an excellent underground.

The *pater familias* to the happy band in Kemisk A was Jannik Bjerrum, then about 50 years of age. Jannik was the son of the great physical chemist Niels Bjerrum. The Bjerrums are an old Danish family of scholars and intellectuals; several of Jannik's sons are respected chemists today. Jannik, I would describe, with no intent to be disrespectful, as benevolent but scatty. About his own chemistry he was very focused, inventive, and practical, but he was a very unworldly man. Although on paper he had the same sort of autocratic power as a German Herr Professor, he was not puffed up and entered freely into the give and take around the lunch table. He was, of course, accorded a much greater degree of attention and respect than anyone else and thus when he announced one day that the solution to the race problem in the United States was very simple, the group fell silent to listen. He then gave it as his recommendation to "send them (Negroes) all back to Africa." No one else felt free to respond to this but I, who had little to lose, felt impelled to say, as politely as I could, that it would be absolutely impossible to do such a thing, for both moral and practical reasons. He insisted that it could be done, and that this would be the best thing for all concerned, but discussion soon turned to another subject. Jannik was not in the least malevolent, only unrealistic.

The two most interesting people I met in Kemisk A were Christian Klixbüll Jørgensen (no relation to Sophus Mads) and Carl Johann Ballhausen. Both were then working toward a Danish D.Phil. degree, which was taken by and only necessary for those who wanted someday to fill a professorial chair. These two young men, a little older than myself, had nothing in common except a shared interest in the application of a theoretical scheme called Crystal Field Theory to the interpretation of the spectra and magnetic properties of transition metal complexes. They were among the first (others being Leslie Orgel in England and H. Hartmann in Frankfurt) to recognize that this arcane theory, created by Hans Bethe in 1929 and further developed by

J. H. van Vleck in the late 1930s, could be expected to provide tools needed to deal with the electronic structure of transition metal complexes that would be more powerful than Pauling's valence bond method. Ballhausen and Jørgensen were not personally attracted to each other and not very long after this their paths diverged, but during that year, they were working together under Jannik's auspices to measure and interpret the ultraviolet and visible spectra of complexes.

Crystal field theory is more sophisticated than the valence bond theory *à la* Pauling, but far more suited to the task. It requires a much greater understanding of quantum mechanics and entails the use of group theory; for these reasons, it had not quickly found favor with chemists. However, I was so impressed by what it could do that I immediately began to learn what I needed to know in order to understand it. This effort bore abundant fruit when I went to MIT, as I shall explain later. Equally important, it began a lifelong friendship with Carl Ballhausen.

Carl became a theoretician in an unconventional way. While he was pursuing a conventional university degree in chemistry, which in Denmark involved long hours in the laboratory, he had the misfortune to be hit by a car while on his bicycle. This caused great damage to one knee and he was confined to bed for two and a half years, unable to do laboratory work. To pass the time, he devoted himself to three things: listening to opera, mainly Wagner; reading Agatha Christie; and learning quantum mechanics. He subsequently built up an enormous collection of classic recordings as well as an encyclopedic knowledge about Wagner operas and all of the great singers. He also became the only person I have met who knew at least as much about the Christie stories as I do (or once did). But most important he became dedicated to theoretical chemistry. In this he was greatly aided by his lifelong friend Aage Winter, a theoretical physicist whose own career was spent at the Bohr Institute.

Though helped by Winter, Carl was essentially an autodidact. To fulfill his potential, he needed to work with an outstanding theoretical chemist, and thanks to Geoff Wilkinson, he was able to do that. Geoff financed, out of his own research money, a one-year sojourn for Carl at Harvard with William (Bill) Moffitt, then the anointed wunderkind of theoretical chemistry.

Thus, one fine day in late July or early August of 1954 Carl arrived in Cambridge, Massachusetts. Throughout the remainder of that summer, he was a frequent visitor to the only ranch house in Cambridge, on Humboldt Street, which Barry Shapiro, Bob

Naylor, Al Moscowitz, and I had rented for the summer. It was a marvelous summer. We enjoyed getting Carl "Americanized" and we really learned about the joy he took collecting old records, *really* old records. I remember the time Carl made a special trip to New York to purchase a small disk with grooves on only one side, made in the earliest days of disks. When he returned we were all very eager to hear it, so Carl got out a very special phonograph he had brought with him from Denmark, which could be adjusted to any speed by use of a stroboscopic device. Evidently his new treasure was played at some unusual speed. The needle was placed on the disk and we heard a periodically repeated swosh, swosh, swosh, …, with, now and then, a glimmering of a male voice. When it was over, in a matter of a few minutes, Carl beamed with pride of possession. He also told us that he had got it for "only $50," or maybe "only $100," and that it was one of only five known to still exist. We were all very puzzled, but politely said no more.

This was also the summer that Al Moscowitz decided that *his* natural diurnal cycle and the earth's didn't jibe. Instead of being awake 16–18 hours and sleeping for 6–8 hours, he decided to alternate 24 hours of activity with 8-hour sleeping periods. He persevered with this curious routine, whereby he rarely slept when the rest of the world slept, but was usually trying to sleep while the rest of us were up and making noise, for some weeks. He eventually gave up and tried for a while to sleep only every other night. This too was soon abandoned and he reluctantly rejoined the conventional world.

To get back to Carl (who was later responsible for getting me as well as Al Moscowitz, Steve Berry, and Stuart Rice elected to foreign membership in the Royal Danish Academy), his year at Harvard set him firmly on the road to a distinguished career. He and Bill Moffitt wrote an elegant article for *Annual Reviews of Physical Chemistry* (1956), on crystal field theory, which was highly praised by van Vleck and remains very useful to this day. Carl later wrote *Introduction to Ligand Field Theory*, McGraw-Hill, 1962, which is still the best rigorous introduction to the subject that exists. During the year, 1955–56, Carl and I did a little theorizing on the fine structure of X-ray absorption edges [15]. If we had only gone a bit further, we might have recognized the phenomenon now called XAFS, but we didn't.

Carl returned to Denmark, but two years later he came back to the United States. to take a position at Bell Laboratories, where he and A. D. Liehr carried out the first quantitative calculations of crystal field splittings, preceding those of Tanabe and

Sugano that have since become standard textbook material. In the Summer of 1958, Carl became an Assistant Professor at the University of Chicago. He was on the verge of receiving tenure there when he received an irresistible call to the chair in physical chemistry in the University of Copenhagen, which he took in September 1959. He enjoyed a distinguished career there, and in 1992, it was my pleasure, as a member of the International Affairs Committee of the NAS Council, to promote strongly and successfully his election as a Foreign Member of our National Academy of Sciences.

To return now to the rest of my stay in Europe, Geoff and I worked hard until mid-June of 1954. I had been getting tutoring in everyday, spoken German (not at all like the German I had learned in my college studies of high romantic German literature), and was prepared to spend two months traveling in Germany, staying largely in the *Jugendherbergen* (JH, youth hostels). I was still, by nearly a year, under the cutoff age of 25 for using the JH network. My journey commenced with an automobile trip from Copenhagen *via* Holland to Aachen with Bill Gwinn and his wife. Bill was a Berkeley physical chemistry professor who had been on sabbatical in Denmark.

Arriving in Aachen on June 23, I immediately purchased *Lederhosen* and then registered at the JH. This, like every other German JH I stayed in subsequently, was clean and efficiently run. The efficiency was typified by the blanket each guest received. The blankets were clean, but were not necessarily washed every time they changed hands. Therefore, the top and bottom ends had, woven in, the designations *Fussende* and *Kopfende*. This attention to detail made me proud of my German heritage.

My first day in Aachen was warm and sunny and I remember lying on the grass in a park enjoying la dolce far niente, when suddenly I became aware that a man in a blue uniform (a *Schuhpolizei*) was striding very purposefully toward me, with a fierce scowl on his face. He said "können Sie nicht lesen" and pointed at the sign "Den Rasen nicht betreten." I had simply not noticed it, and I began to explain to him in my halting German that I was very sorry and would certainly not trespass again when his expression began to change and he asked "Sind Sie Amerikaner?" I admitted I was (my Anglo accent always gave me away) and then he broke into a broad, warm smile and said in excellent American English, "Welcome to Germany, I'm sorry I was so rude." He explained that he had spent several very happy years in Arizona and that he loved America. I asked him how come, and he replied, still very cheerfully, that he had been a prisoner of war.

Old Lederhosen never die! They just get a little tighter. It's the same pair, 1954 and 2004, honest.

What had misled the *Schuhpolizei* was my nordic appearance and my *lederhosen*. My ability to pass for a young German (until I spoke) was often a great advantage, especially when I traveled by *autostop*, *i.e.,* hitchhiking, which I did a good bit of, to save money. I was fairly successful at this perhaps because it was only after I was settled in the car that my benefactors realized that I was not a German youth on the traditional summer *Wanderung*.

I have very vivid memories of my six weeks or so in Germany, especially of the warmth and friendliness of the people. I especially enjoyed the company of young people, mostly much younger than I, that I slept and ate with, all over West Germany. One day I hitched a ride with a young man in a three-wheel truck, who, upon blowing my cover, began talking fluently in a rich southern accent. It seems that shortly after the war he had been employed at a U.S. Army base run by troops from Georgia, where he had learned English by exclusively auditory means. On another occasion I made a trip to see the *Eiserne Vorhang*, just north of Nuremberg, where I won over the initially inhospitable border guards with flattery and (more to the point) a pack of American cigarettes. They took me up in their tower and gave me a lengthy explanation of the whole setup. Finally, I must note that I was able to see two performances (of Siegfried and Die Walküre) at the Festspielhaus in Bayreuth. A fellow graduate student and I were not properly dressed, nor could we afford, to partake of the expensive refreshments and meals served between the acts, but we managed to do something more memorable. While the stage settings then used by Wolfgang and Wieland Wagner were mostly stark and minimalist, two exceptions were a huge, sparkling Fafner who roared and gaped menacingly and a live ring of fire to surround Brünnhilde. My friend and I simply strolled into the backstage area during one of the intervals and inspected the enormous *wurm* which was covered with multicolored bits of glass. We also saw the great metal gas ring which provided the live ring of fire around Brünnhilde in both operas. We were having a wonderful time until some crew member spotted us. With shouts of "Hei, Buben, was tun sie her, gehe heraus" following us we heraused hastily into the July sunshine outside.

I must, however, refrain from relating any more about this magical summer in Germany, with the exception of my visit to Munich. I went there, by prior arrangement, to meet what Geoff referred to, more or less seriously, as "the enemy." E. O. Fischer turned out to be a wonderful host. After all, I was only a graduate student, but he treated me as a respected colleague, and I spent two nights at his house, south of Munich. His English (later to become fluent) was exiguous at the time, but we managed in German and then with the help of Erwin Weiss, whose English was very good, as a translator. I don't remember whether Erwin (who, like Fischer, had been lucky enough to survive long service on the Russian front during the war) was then still a student or a dozent. He was, and is, an extremely able chemist and a fine man, who recently retired from a chair at the University of Hamburg.

Although neither one (certainly not GW) would recognize it, GW and EOF were very similar in their approach to chemistry. Both were intuitive geniuses. Neither had much knowledge of nor much interest in theory, and neither made any pretense about that. Both began their academic careers as inorganic chemists relatively late, because of the war, and were in very much of a hurry (in a controlled way) to get moving up the academic ladder. They were, of course, rivals, but it did not seem to me that EOF had as hard an edge toward GW as GW had toward him. In any event, I was totally charmed by EOF from the time I met him, and have always regarded him as, according to the standard cliché, a gentleman and a scholar.

Following my tour of Germany, I spent about a week in Switzerland and a week in France. In neither country were the youth hostels, les auberges de la jeunesse, useful to me. In Switzerland they were booked in advance and in France they were overcrowded and not clean. I therefore stayed in hotels every night and by the time I arrived on a boat train at Victoria Station in London, I was flat broke. I therefore checked into the fairly posh hotel that formed a part of Victoria Station, had a hot bath, a good night's sleep, and a hearty English breakfast, charged to the room, of course. I then enquired how to get to the American embassy.

At the American embassy I explained my situation. I had no idea what they would do, and was astonished to hear that they would make me a loan of $100. I then sent a telegram to my mother, and she sent me some more money. Thus, I completed the remainder of my trip in comfort. I did, immediately, move from the expensive hotel at Victoria Station to a more modest one.

Back to Harvard

I returned to the United States, leaving England on August 19 on the Greek Lines ship Olympia. This was a trip utterly unlike my trip the previous January. The Olympia was a relatively small ship, but totally modern and the Summer weather was perfect. I met some very nice people, played some bridge, and generally relaxed.

In Copenhagen, GW and I had finished work on the $(C_5H_5)_2Mg$ and $(C_5H_5)_2Mn$ compounds and submitted a preliminary communication [6] to *Chemistry and Industry* in early February. By late February we had also submitted a paper [7] to *Zeitschrift für Naturforschung* in which we described a new and better preparation of $(C_5H_5)_2Cr$ (which Fischer had already made in very small yield) as well as the new $(C_5H_5)_2MoCl^+$, $(C_5H_5)_2MoCl_2^+$, and $(C_5H_5)_2WCl_2^+$ ions, thus bringing for the second time an entire vertical group of elements into the realm of bis-cyclopentadienyl chemistry.

GW got back to Harvard in August and the whole group was soon working full tilt again. One project in which I participated concerned compounds containing the "sandwich-bonded" C_5H_5 ring and carbonyl groups. Several of these had already been reported by GW and by EOF and Hafner. This work played a very important part in my development, because it gave me my first opportunity to work with metal carbonyls and to think about questions concerning their electronic structures, bond energies, and particularly, their infrared spectra. Most of the preparative work was done by another student, T. S. Piper, who had transferred into the group from the

Rochow group in the Summer of 1954. I became involved in a way that aptly illustrates the adage, "In the land of the blind, the one-eyed man is king."

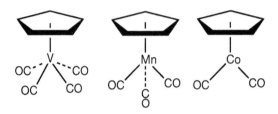

GW and Piper had recorded the infrared spectra of $(C_5H_5)V(CO)_4$, $(C_5H_5)Mn(CO)_3$, and $(C_5H_5)Co(CO)_2$, and they were puzzled by their observations that these molecules displayed 3, 2, and 2 CO stretching bands, respectively. They believed that these molecules should have structures of the types shown above, in each of which the CO groups appear to be equivalent, but to them, the appearance of multiple bands in the infrared spectra seemed to be inconsistent with this equivalence. They had expected to see only one CO stretch in each molecule, since "all equivalent CO groups should vibrate at the same frequency." While it is tempting to be supercilious about such naiveté, it must be remembered that inorganic chemists, and organic chemists as well, in those days did not, except in very rare cases such as my own, have the slightest idea about how to do a vibrational analysis of a polyatomic molecule. There was nothing unusual about the ignorance shown by GW and Piper in this regard. In fact, they even pointed out in an empirical defense of their position that $Cr(CO)_6$, in which there are *six* equivalent CO groups, has only *one* CO band in the infrared.

Because I had learned a few basic things about group theory and its application to vibrational spectra, I was able to supply an explanation and assure them that all was well, namely, that the structures they had conjectured were fully consistent with the observed spectra [9,10]. However, for me there was more to it than that. I began to think more about the nature of metal—CO bonding and how CO vibrations coupled with each other to give the observed differences between the frequencies of different modes. It was not until later at MIT, however, that my concern with these questions began to bear fruit.

Apart from one more paper [14] written in early 1955 on $(C_5H_5)_2Mn$ and other $(C_5H_5)_2M$ compounds, I did no more work on sandwich compounds and wrote no further papers with GW while I was at Harvard. I had already begun to work independently on other things. I don't know whether GW was displeased or not, but he never expressed any displeasure, and we remained friends and, from time to time, research collaborators for many years. Moreover, in 1956, after his return to the UK and my translation to MIT, we started working on the first edition of *Advanced Inorganic Chemistry*, about which I shall say more later.

In 2001 Rick Adams organized and edited an issue of the *Journal of Organometallic Chemistry* (Volumes 637–639) in which I [1488] and a number of others still alive, such as Mike Rosenblum, Peter Pauson, E.O. Fischer and Reinhardt Jira, summarized the activities in various laboratories concerning the early explorations of metallocene chemistry. These articles are a unique source of historical facts about this pathbreaking period in transition metal organometallic chemistry.

During my last year, 1954–55, at Harvard, I spent most of my time on several other projects. One was a comprehensive review of the bizarre history of transition metal alkyls and aryls [12], which was published in *Chemical Reviews* in early 1955. It was my first "solo flight" in the chemical literature. To my considerable surprise, it drew a huge number of requests for reprints and the supply of 100 that I had ordered (and my industrial fellowship paid for) was soon exhausted. I arranged to have another 100 printed, but for these I charged about one dollar to cover the cost of buying them and postage. These, too, went like hot cakes. Interest in the organometallic chemistry of the transition elements was already much more widespread than I had imagined.

Other interests that I pursued on my own concerned soft (Kα) X-ray absorption edges and the thermodynamics of chelation. The former had started earlier, when I thought that the Kα edge of ferrocene might give useful information about the bonding in ferrocene and I persuaded GW to foot the bill for me to go to the University of Texas and make the measurements in the laboratory of Harold P. Hanson, then head of the Physics Department, whom I had contacted and discussed the project with. He was interested in collaborating and when I came he showed me wonderful hospitality. Harold was, to employ the Yiddish, a mensch.

In fact the structure seen on the Kα absorption edge of ferrocene did not give any specific or useful information regarding the bonding in ferrocene and nothing was ever published on that. However, for comparison, I measured the edges of other complexes and found them more informative. This led to an extended stay in Austin, during which I measured some 30–40 other coordination compounds, and eventually Hanson and I published three experimental papers in *J. Chem. Phys* [16, 23, 26]. As noted earlier, I also did a little theoretical work with Carl [15] during his stay at Harvard in the Fall of 1955, by which time I was already at MIT.

Work on the thermodynamics of chelate formation was also done entirely independent of GW. It happened, in part, because I broke a leg skiing in the winter of 1954–55 and this project was more easily handled by one on crutches than preparative chemistry. Besides I had become quite interested in the question of how best to understand the chelate effect. I made a number of measurements of formation constants of chelate complexes which seemed to support the idea already proposed by others, that the chelate effect was mainly entropic [11]. Much more interesting, however, was a theoretical study of ring-closure probabilities that Frank E. Harris and I carried out on one of the earliest computers ever built for scientific computation. This beast, Howard Aiken's Mark IV, contained thousands of vacuum tubes, was fed by punched paper tape, and occupied hundreds of cubic feet in a building adjacent to the chemistry department. It rarely ran more than an hour before it went down with one or more blown tubes, or perhaps a torn paper tape. What we were doing was a Monte Carlo calculation of the distribution of end-to-end separations for aliphatic chains with 4, 5, 6, and 7 links. Fortunately, such a calculation, if interrupted, could be restarted from the point of interruption, but nonetheless, Frank spent many hours baby-sitting the Mark IV in order to get the results. In the end we got the results from which we calculated relative entropies of ring closure in decent agreement with experimental data [18]. I think this may have been the first machine Monte Carlo calculation carried out on a chemical problem. In any case it was my first experience with an electronic computer, and I found it quite exciting.

During the academic year 1954–55, GW and I discussed my future, about which I had only one certainty: it was to be academic. Geoff mentioned that he was sure Glenn Seaborg would take me on, but I said I was not inclined to radiochemistry or nuclear chemistry, so we dropped that. GW then talked to Charles Coryell and John W. (Jack) Irvine at MIT and it was arranged, with Arthur C. Cope's approval, that I be appointed an Instructor in Chemistry (the normal entry level rank in those days) at MIT. Half my salary was to be paid out of the large AEC contract that Coryell and Irvine had, and I was to have a lab in, and charge my research expenses to, their project. This sounded fine to me and I accepted immediately. The salary was a princely $3,600 per annum. These were simpler times when all the bureaucratic arglebargle involved today in hiring a new young faculty member was undreamt of. One professor called another professor and a deal was quickly made.

Before concluding this chapter of my story, I must comment on the exceptional importance of these four years at Harvard in shaping the rest of my life. The intellectual and cultural life at Harvard had an enormously nutritive effect on my young mind. To put it simply, I had discovered my proper environment, where I was stimulated, nourished, and satisfied. In short, I learned where I belonged, and I felt fully at home. Later, I was to have some more nuanced thoughts about some of it, but at the time, I had none, and that's the important thing. I'm basically a happy person anyway, but those years were carefree bliss.

A major factor in the ambiance at Harvard was the extraordinary quality of my fellow graduate students in chemistry. Perhaps this is naïve, but I wonder if one could have encountered a more brilliant collection of graduate students during a period of only four years at any time, anywhere. I offer in evidence the following partial list of those who became academics and who overlapped me in the period 1951–55, with the year in which they obtained the Ph.D. in parentheses.

Guy Ourisson (52)	Alfred Holzer (54)	Martin Saunders (56)
Jerrold Meinwald (52)	John Kice (54)	Fred Anson (57)
Earl Muetterties (52)	Myron Rosenblum (54)	Gideon Fraenkel (57)
Albert Bergstahler (53)	James Hendrickson (55)	Andrew Kende (57)
Fred Greene (53)	Stuart Rice (55)	Albert Moscowitz (57)
Dan Kivelson (53)	Dietmar Seyferth (55)	William Agosta (58)
Norbert Muller (53)	R. Stephen Berry (56)	Dudley Hershbach (58)
Alex Nickon (53)	Ronald Breslow (56)	Herbert Kaesz (59)
Wayland Nolan (53)	William Graham (56)	Thomas Katz (59)
Richard Hill (54)	J. C. Martin (56)	

To be less parochial, I also benefitted from the company of extremely talented fellow graduate students in other fields. I was privileged to get some help and encouragement in learning to play the recorder from Jeremy Bernstein, who went on to become a very distinguished science writer. I shall also never forget a Polish student of mathematics who did much to support the notion that while you don't have to be eccentric to be a good mathematician, it helps. Alex, whose last name I have forgotten, turned up one day with a black eye and other facial bruises. I should also say that Alex was very stocky and muscular. Naturally, I wanted to know what had happened to him. He explained to me, very matter-of-factly, that from time to time he felt like having a fight and since he was a good fighter he took mercy on his academic friends and betook himself to South Boston, where there was a ready supply of young Irish toughs. He would walk up to the first one he met who was about his own size, give him a punch, and enjoy the benefits of the ensuing fight, which he normally won. Once in a while, however, he lost, and I was now seeing the outcome of such an occasion.

On a more benign note, there was the annual celebration of Arbor Day, a Massachusetts tradition not observed in many other states, when small trees or nuts are planted. The dormitory complex in which I lived, which had been designed by Walter Gropius, was built in a very pleasant but sparse contemporary style and had an inner court. The only adornment in this inner court was a highly abstract "tree" made of shiny stainless steel rods and called the Ygdrasil tree (the world tree of Norse mythology). Each year on Arbor Day, a few students would conduct the ceremony of planting some equally shiny ball bearings close to the Ygdrasil tree. I doubt if this would get a laugh on Saturday Night Live, but it's my kind of humor, and I felt very much at home in a community where people did whimsical things like this.

CHAPTER 3

MIT
1955–60

ALL'S WELL THAT
BEGINS WELL.

— *Anonymous*

I HAVE ALREADY explained how it was arranged between Geoff Wilkinson and his former hosts at MIT, Charles Coryell and Jack Irvine, that I should go to work in their laboratory as an Instructor on the MIT faculty. Charles and Jack were marvelous supervisors: they suggested some problems of interest to them that I might look at, but also said, or at least implied, that I should feel free to pursue ideas of my own. Since I had lots of the latter, I am afraid I did very little to advance their program, but they never complained. Indeed, they were very helpful and supportive of what I did, even though it had nothing directly to do with their objectives. By 1957 I had been promoted to Assistant Professor (as a direct result of an offer from Henry Taube to come to the University of Chicago at that rank) and I was soon able to secure my own funding from the AEC. In those days right after Sputnik, federal support was abundant, at least relative to the demand.

When I began at MIT in September of 1955, the inorganic group consisted of five full professors: Charles Coryell and John W. (Jack) Irvine, who were almost entirely engaged in radiochemistry and nuclear chemistry; E. Lee Gamble, whose research in main-group chemistry had ceased many years before; Ralph Chillingworth Young, then in his early sixties; and Walter C. Shumb, also in his early sixties. Let me remind the reader that retirement age then was sixty-five, so the last three were, in effect, lame ducks.

Lee Gamble was a true southern gentleman. He was not ambitious and no longer research active, but kind and helpful. I think he did all his teaching in Freshman chemistry. Ralph Young was a true northern gentleman. He, too, was very kind, modest, and generous. Walter Schumb was a Dickensian curmudgeon, whose understanding of chemistry had been arrested before quantum mechanics. He was a rather small, prissy man, with a dogmatic attitude about everything and I think his general unenlightened pedantry is captured in an anecdote that Walter Stockmayer told me.

Stocky (as everyone called Stockmayer) was an MIT undergraduate in the early 1930s and was required to take Schumb's course in inorganic chemistry. Schumb gave this course (and still did in 1955) by means of a box full of 3 × 5 cards on which the salient features of compounds (mp, bp, color, preparation, etc.) were written. He would take about a centimeter-thick helping of these cards with him and give a lecture of immensely soporific character based on what he read from the cards. One day, however, he arrived in a state bordering on (for him) exaltation, ignored his cards, and began: "Gentlemen, we all know the importance of the electron in chemistry, and how important it is to know as much about it as possible. So far we know only its mass and its charge. I think any chemist would also be interested in knowing its color, and now we know that! It is blue." He was referring, of course, to the work of Charles A. Kraus of Brown University, who had been studying the blue solutions of sodium in liquid ammonia and had recently announced evidence that they consisted of solvated Na^+ ions and electrons. Since $Na(NH_3)_n^+$ ions are colorless, it seemed evident to Schumb that this proved that electrons are blue.

Although most of my research has been concerned with the chemistry of the transition metals, I have always found main group chemistry interesting. In fact, the first Ph.D. thesis I supervised dealt with main group chemistry. When I arrived at MIT, there was a young Princetonian named John W. George who was in his third year of Ph.D. research with Professor Walter C. Schumb. The problem Schumb had set him was to prepare and isolate a mixed halide of boron, and he was at his wits' end because nothing he had tried worked. At first I couldn't believe it. Once I realized that this situation was real, I had a serious talk with Jack George and explained to him that his task was impossible, and why. Mixed halides of boron can never be isolated because equilibria of the type

$$BX_3 + BY_3 = BX_2Y + BY_2X$$

are rapidly established. Jack was both relieved and alarmed to hear this: relieved because failure was not due to his own incompetence, but alarmed because, apparently, Schumb was insistent that it could be and had to be done before he could get his Ph.D.

I retain no recollection at all of how I went about it (the 26-year-old upstart dealing with the 63-year-old senior professor), but sometime in the Spring of 1956 I convinced Schumb to make me co-supervisor of Mr. George's research. Since Schumb paid little further attention, I soon had Mr. George doing some interesting work on SF_4. At the time, this was a new molecule and its structure had been shown by IR and Raman spectra to have C_{2v} and not T_d symmetry (Fig. 3-1). It was tricky

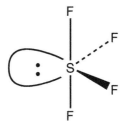

Figure 3-1. The structure of SF_4 showing the position of the pair of non-bonding electrons.

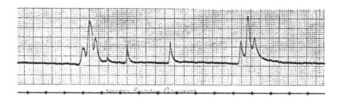

Figure 3-2. The ^{19}F NMR spectrum of SF_4 as measured at MIT in 1958 (F. A. Cotton, J. W. George, and J. S. Waugh, *J. Chem. Phys.* **1958**, *28*, 994). The three peaks between the triplet are due to impurities.

to prepare, but George prepared it, and with the collaboration of John Waugh, we obtained a beautiful and definitive corroboration of its structure [29]. The pairs of ^{19}F nuclei in the axial and equatorial positions split each other's resonances into triplets, as shown in Fig. 3-2. I reproduce this spectrum here for its nostalgic value. In those days the recorder worked on the basis of a hot wire melting a thin coat of light wax off of the moving chart paper.

This now seems about as quaint in comparison to a present-day digitally stored record as an Edison tubular phonograph record would to a laser disk. The structure itself was more significant than one might realize today because SF_4 provided the first evidence that the VSEPR theory predicted correctly the structure of a molecule of the type AB_4 with an additional unshared electron pair. George also studied some of the chemistry of SF_4, especially that of its "adducts" with Lewis acids, such as "SF_4BF_3." The latter had been postulated by Neil Bartlett, in his Ph.D. thesis with P. L. Robinson, to be a conventional donor-acceptor adduct. George's work led us to propose that it is, in fact, an ionic compound, $SF_3^+BF_4^-$, as indeed it is [30]. Thus, in May of 1958, John W. George became my first Ph.D. recipient. He went on to a teaching career at the University of Massachusetts in Amherst from which he retired some years ago. I have not seen him for many years.

As for teaching at MIT, I was immediately assigned a rather formidable responsibility: lecturing to the freshmen. There are several reasons why I call this formidable. One is that the class was quite large — about 200 — and was held in a very large and imposing lecture hall. There were also lecture demonstrations to be done. There was a very able person who prepared these demonstrations, but still, one had to make them work or suffer the derision of the students, many of whom were unhappy campers taking chemistry only because they were forced to. Another factor was that MIT freshmen were, and no doubt still are, very bright, very full of themselves, and quite intolerant of anyone who didn't provide what they considered a good show. The first time around (it was a full year course) was stressful, but in the next two years, in which I continued to do it, it was not a bad assignment. I did not have to make up exams or grade them.

After the first year, the lecture demonstrations were usually enjoyable. There were some very ordinary ones such as placing a rubber tube in liquid nitrogen and then shattering it on the bench top. There were also several that were imaginative, two of which I remember vividly. One involved the highly exothermic radical reaction of H_2 and Cl_2 to produce HCl. This was set up so that there was a clear glass, wide-mouth bottle filled with an equimolar mixture of the gases and held in a metal jig. The bottle was stoppered by a black rubber squash ball and aimed, like a mortar, at a broad, two-story wall about 15 feet from the bench. On the same jig was an electric bulb and a switch. I would throw the switch and after the bulb had been on for about five seconds, the squash ball would be fired against the wall with a sound like the report of a pistol. One year I succeeded in catching the ball as it came off the wall (like a left fielder at Fenway Park) and received an enthusiastic ovation.

The other demonstration that I remember very well from the first year I did it was the thermite reaction. This was set up with a small clay flower pot at the top containing a mixture of aluminum powder and powdered ferric oxide, with a magnesium wick. Below this were several metal pie pans and, finally, on the bench top was a large battery jar full of water with about two inches of sand in the bottom. I lit the wick which then set the reaction going in the flower pot, and soon molten iron was pouring out of the hole in the bottom, burning its way through the pie plates and going into the water generating great clouds of steam and lots of noise. The students loved it. However, the really memorable part, which happened the first time I did it, occurred before it was actually done. In an effort to be chatty and down to earth, I remarked that what I was about to show them used to be seen often in public because this reaction for generating molten iron was often used to weld segments of trolley tracks together. To my total surprise, this offhand remark sent the students into a paroxysm of laughter.

After the lecture I went to one of my senior colleagues to see if I could find out what caused the unexpected laughter. He explained to me that one of the most renowned examples of mischievous behavior by MIT students had occurred not many years earlier, when trolleys still ran on Massachusetts Avenue, with a terminus at the back edge of the MIT campus. One night, when the motorman had left his trolley to go into a little building for a break, MIT students who had witnessed this lecture demonstration rushed out to the trolley with their own thermite generator all ready to go and welded one wheel of the trolley to the track. When the motorman returned and threw the trolley into gear it simply sat there while the electric motor, or some linkage to the motor, burned up. My freshmen had evidently heard this glorious story and were extremely amused when I inadvertently reminded them of it.

Once Walter Schumb retired, I took over the junior or senior level course in inorganic chemistry, taken by chemistry and chemical engineering majors. I also invented a course for first-year graduate students in which I tried to give them a more sophisticated understanding of chemical bonding and an introduction to ligand field theory. Since group theory was needed for the latter, I included a smattering of that as well.

Group theory proved to be addictive and I gave it more emphasis each year until, by about 1960, I had changed the course structure so that it began with group theory and then took up its various applications in chemistry, particularly organic pi electron systems (which my witty colleague Walter Thorson called "micropiology"), ligand field theory, and vibrational spectroscopy. This evolved into the book *Chemical Applications of Group Theory*, published in 1963.

I have already mentioned my first Ph.D. student, Jack George, whom I rescued from Walter C. Schumb. During my first year, 1955–56, I also picked up three incoming students, Richard H. Holm, Roch R. Monchamp, and Douglas Meyers, and the following year three more, John P. Fackler, Jr., William D. Horrocks, and Joseph R. Leto. I also had two of Geoff Wilkinson's students at Harvard,

Albert K. Fischer and L. Todd Reynolds, to supervise. Thus, along with teaching and the research I was still doing with my own hands, I had a busy schedule right from the beginning. In the two years 1956 and 1957, I published six papers giving MIT as my address as well as seven more giving Harvard.

The work I did in the period 1955–57 was transitional, much of it derivative from interests begun while I was still a graduate student. After Geoff went off to England, he never set eyes on Albert K. Fischer again, and I was Fischer's real though unofficial research mentor. The work done by Fischer at Harvard followed on my interest in the bond strengths in arene compounds and metal carbonyls. By combustion calorimetry, Fischer and I obtained the heats of formation of $Cr(CO)_6$, $Mo(CO)_6$, and $W(CO)_6$ [17]; then $Ni(CO)_4$ [22]; and finally, $Fe(CO)_5$ [39]. The mean thermochemical bond energies were estimated to be about 27 kcal for the Cr and Fe compounds, about 35 kcal for the Mo and Ni compounds, and 42 kcal for $W(CO)_6$. We also obtained the heat of formation of $(C_6H_6)_2Cr$ and showed that it was consistent with the considerably lower stability of $(C_6H_6)_2Cr$ compared to $(C_5H_5)_2Fe$ [38]. Obtaining all these results was quite laborious, and looking back, I'm not sure it was the best investment of a graduate student's time, but in that era, it seemed exciting.

In the spring of 1956, I obtained a Guggenheim fellowship that enabled me to spend the summer in England and Copenhagen, where Geoff and I wrote some papers. I also wrote a paper that I still look back on with pleasure. It was already recognized that, in principle, ligand field stabilization effects (*i.e.*, thermodynamic effects) should be quantitatively related to spectroscopic effects (*i.e.*, to the *d*-orbital splittings responsible for the visible spectra of transition metal ions). Apart from an observation by Orgel in 1952 that the irregular trend in hydration energies of M^{2+} ions could be explained in this way, I do not think anyone in 1956 had quantitatively demonstrated such a correlation. A way to do so occurred to me that summer while I was in Copenhagen.

I culled thermodynamic data from the literature and calculated values for the following processes in the gas phase

$$M(NH_3)_6{}^{2+} = M^{2+} + 6NH_3$$

for M = Ca, Mn, Fe, Co, Ni, Cu, and Zn. The results were in good agreement with theoretical expectation, as shown in Fig. 3-3, taken from the paper [19].

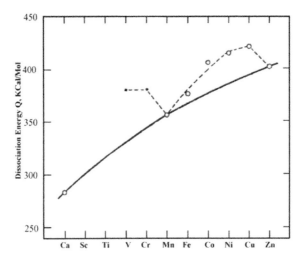

Figure 3-3. Energy in kcal/mole of the gas phase dissociation of $M(NH_3)_6^{++}$ ions. Calculated points o; – – – values predicted by adding ligand field stabilization energies to points on solid curve [19].

A few years later, Philip George and Don McClure made a comprehensive survey of such thermodynamic/spectral correlations in an article I asked them to write for Volume 1 of *Progress in Inorganic Chemistry*.

As I said above, I was very busy indeed during my first years at MIT, and yet I took on still another activity, namely, creating and editing the annual review series *Progress in Inorganic Chemistry*. The initiative for this came mainly from the co-owners of Interscience Publishers, Maurits Dekker and Eric S. Proskauer, who had already talked with me about the possibility of writing a textbook for Interscience (which, of course, was later done). We had a conference one morning at an ACS meeting in 1957 (perhaps early 1958) and laid out the general plan of action. Our intention was to call the series *Advances in Inorganic Chemistry*. Only a few months later there appeared the announcement of a forthcoming series from Academic Press to be called *Advances in Inorganic Chemistry and Radiochemistry* and edited by Harry Eméleus and Alan Sharpe of Cambridge University. Their advertisement preempted the title I had intended to use and thus my series was rechristened *Progress in Inorganic Chemistry*. After editing it for ten years, I decided it was time for a change and arranged for my former student Steve Lippard to take over, which he did for Volumes 11–40. After three decades, he, too, felt the need for change and in 1993 turned it over to his former student Ken Karlin. Thus, it continues in my academic family.

In May of 1956 I was tremendously flattered to be invited to speak in a Symposium on Valency and Chemical Bonding organized by the Division of Physical and Inorganic Chemistry of the ACS and held at the University of Wisconsin. I was asked to speak on Structure, Bonding and Properties of the Metal Sandwich Compounds, which was still a very hot topic at that time. The reason I was so flattered to be invited was that the entire speakers list was a Who's Who of the most eminent physical and theoretical chemists, most of whom I had never met before. A nice photograph was taken of this collection of celebrities, plus me. All but perhaps six of us have since died, so this picture may be of historical interest. The topics were all hot at the time; for example, Hydration of Cations (Taube), Structure of Boron Hydrides (Lipscomb), Electron Deficient Compounds (Rundle), How to Understand Chemical Bonding (Mulliken), and Applications of Self-consistent Molecular Orbital Methods to pi-Electrons (Pople).

I have already mentioned that my promotion from Instructor to Assistant Professor at MIT in 1957 was triggered by an offer of an Assistant Professorship at the University of Chicago. Henry Taube was then chairman of the Department at Chicago and was

Speakers at the Symposium on Valency and Chemical Bonding, University of Wisconsin, 1956. **First Row:** *Farrington Daniels, Henry Eyring, J. L. Franklin, Virginia Griffing, F. A. Cotton, Simon Bauer, Henry Taube.* **Second Row:** *W. T. Simpson, John Pople, R. S. Mulliken, Don Hornig, John Wertz, A. Frost.* **Third Row:** *G. Glockler, John Platt, R. A. Buckingham, Bob Rundle, Bill Lipscomb.* **Last Row:** *John Margrave, F. H. Field, W. W. Beeman, Sam Weissman, Per-Olov Lowdin, Joe Hirschfelder.* **Speakers not in this photo:** *Paul Cross, Bright Wilson, and Michael Szwarc.*

personally interested in adding an up and coming young inorganic chemist with a physical orientation to his department. He made the offer during the academic year 1956-57. Art Cope responded immediately and I became an Assistant Professor in September of 1957.

Art Cope was a very efficient chairman who had come from Columbia to MIT in 1945 with a mandate to modernize and upgrade the chemistry department. He had gone about this vigorously and with considerable success, especially in the organic area, which was his own. He was not, however, a great fan of inorganic chemistry and I got the impression (and I believe I was meant to get the impression) that upon the retirement of the older inorganic chemists, which was due in just a few years, I was to be the only inorganic chemist in the department.

I believe that Art Cope took a liking to me and felt I was the right person for this future responsibility and thus he responded very promptly to the offer from Chicago. Another thing that probably worked in my favor was the frustration, still fresh in Cope's mind, of losing Jack Roberts in 1953 to Cal Tech. Art did not like to make plans and then have them disturbed. He was an orderly and systematic person.

Cope's satisfaction with me helped lead him, I think, to change his mind about having only one inorganic chemist. He was a regular consultant to the Experimental Station at DuPont and there he encountered Earl Muetterties, who made a very good impression on him. One day, in 1957 I believe, he called me to his office and asked what I thought of Earl and whether I would like to have him as a colleague. I assured him this was a great idea, and he said he would pursue it. For whatever reason, this did not work out, and Cope again talked to me about getting another inorganic chemist. I gave him a suggestion that did work out.

My friend Dietmar Seyferth, who was my exact contemporary in graduate school, had finished his Ph.D. with Eugene Rochow and gone to Munich for postdoctoral work. His efforts from Germany to obtain an academic job back home were unsuccessful and he came back to an industrial job. However, he knew that this was not what he wanted, and he left it to take a postdoctoral position with Gene Rochow while looking hard for an academic position. I recommended him to Cope, who interviewed him, liked him, and offered him a job. Once again, the procedure was simple, direct, and entirely unimaginable in today's bureaucratic, politically correct, and legalistic world.

But, could MIT have found two better prospects than Dietmar and me? I doubt it. We have both been elected to the NAS.

In 1959 I was offered a promotion to the tenured rank of Associate Professor. As far as I can recall, this was done without any specific trigger in the form of an outside offer. This promotion took effect in September of 1960, but no sooner had that happened than I was offered a comparable position at Harvard University.

It had been my ambition to become a Harvard faculty member ever since I was a graduate student, Harvard having at that time the generally recognized number one chemistry department in the country. Therefore, when the offer was first made, I felt I had all but reached the promised land and it was just a matter of settling some details. Then I rediscovered the old Shakespearian adage that "all that glisters is not gold."

Detailed discussions with Frank Westheimer, who was then chairman of the Harvard chemistry department, soon led to some serious misgivings on my part. At MIT I already had a very nice office, whereas Harvard appeared unwilling to provide anything comparable. It was suggested that "when you become more senior" that can be taken care of. Things got a bit worse when I was told I should not be so demanding because after all I hadn't "discovered anything." I asked for an example of what this meant and the reply was along the line of "well, like Peter Pauson discovered ferrocene." I suppressed the temptation to suggest that perhaps the job should be offered to Peter Pauson and said merely that I thought a rather bizarre criterion was being invoked, or something to that effect, but I was stung and angry. There were a number of other matters that also irritated me.

With my previous naively idealistic image of Harvard tempered, it was not surprising that when Art Cope offered me a Full Professorship (to which Harvard did respond by raising their offer to the same level) I felt comfortable in deciding to remain at MIT. Thus, in September of 1961, six years after my two-mile move down the Charles River, I became the youngest full professor of chemistry ever at MIT (a record later eclipsed by Mark Wrighton, again recently by Kit Cummins, and perhaps, in the future by others).

I do want to say that in subsequent years I have come to know and admire Frank Westheimer very much. He is a very smart and very decent man. His negotiations with me in 1960 were simply not his finest hour. He snatched defeat from the ashes of victory.

If I had gone to Harvard in 1960 there are three things I would probably have had a long time ago: an unshakable belief in my own divinity, a Nobel prize, and a private toilet. The latter is something no Harvard professor has, or so I was told by a Harvard professor (immediately after using his). On politely inquiring whether I had just, out of ignorance, deposited feces in a place where they were not meant to go, and only fantasized that I had used a roll of toilet paper, I was assured I had not. I had used what is known in the Harvard bureaucratese as (*not* a toilet) a "special room." I believe they are not uncommon among the Harvard pantheon. As long ago as the early 1950s, R. B. Woodward had a special room.

As a newly minted full professor in 1961, in a three-piece suit made on Saville row and photographed by Bachrach.

The years 1955-60 at MIT were exciting in many ways. I marvel, looking back, at the energy I apparently had in those days of golden youth. They say that "youth is wasted on the young." I am immodest enough to say that I believe that in my case, at least, it was not wasted.

By 1960 I had already seen six students through to their Ph.D.: John W. George, who transferred to me from Prof. Walter C. Schumb; Richard H. Holm, M. Douglas Meyers, and Roch R. Monchamp, who all entered MIT in 1955; Joseph R. Leto (dec. 1974), John P. Fackler, Jr., and William DeW. Horrocks, Jr., who entered in 1956. What all these young men had in common (despite some differences in ability) was great enthusiasm and the kind of appetite for work that is relatively rare today. Together we produced thirty-six papers. Dick Holm's work led to nine of these. It is worth noting that from 1958 to 1960 I also had six postdocs, four from the UK (Eric

Dick Holm and his wife Florence.

Bannister, R. V. Parish, David Goodgame, and Margaret Goodgame) and the first two of a run of excellent Italians (Franco Zingales and Fausto Calderazzo). These people, especially the Goodgames and Calderazzo, were also very enthusiastic and productive.

The work I chose to do in my early years at MIT was based on a simple principle: Try everything that seems new and interesting. I do not recommend this approach to young faculty members today. The reasons it worked (more or less — after all getting a tenure offer from Harvard after only five years is a sign that one was doing something right) were that, first, I never had to think about money, since Coryell and Irvine, and then the AEC directly, were generous. Second, I quickly enlisted a number of exceptionally good coworkers. Neither of these advantages seems to go with an entry level academic appointment today. The fact that I worked 60-70 hours a week was also helpful, of course.

During the 1955–60 period, I was able to publish a total of 78 papers. In addition to the 36 mentioned above that were based on the work of my graduate students, I wrote several based solely on my own work; others were coauthored with postdoctorals and with a couple of graduate students that Wilkinson left at Harvard for me to supervise, as I have already mentioned.

These publications ranged over quite a spectrum of topics, of which the principal ones were metal carbonyls, ligand field theory, organometallic compounds, complexes of phosphine oxides and sulfides, and applications of infrared and NMR spectroscopies to inorganic problems. All of these were actively emerging fields at that time. Let me describe just a few of our projects.

In the NMR area, one study [35] in particular stands out. In 1958, the structure of the $Fe(CO)_5$ molecule was still controversial, with trigonal bipyramidal and square pyramidal models (Fig. 3-4) being the leading possibilities. It occurred to me that the ^{13}C NMR spectrum, at the natural abundance level, could perhaps be observed because $Fe(CO)_5$ is a liquid, and that this spectrum should decide unambiguously between the two structures. In both, there are two kinds of CO groups, but in different ratios, 3:2 and 4:1.

At that time, ^{13}C spectra were by no means routine even for organic compounds, and no one had yet examined a metal carbonyl. My colleague John Waugh was able to find conditions that gave us the spectrum, but to our great surprise, it turned out to be a single line. After considering several ways to explain this result, we proposed that

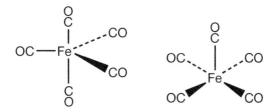

Figure 3-4. Two possible structures for $Fe(CO)_5$.

it was due to a chemical shift difference that was too small to observe at the relatively poor resolution (ca. 40 cps) then available.

As all the world knows today, this was not the right answer. As soon as he saw our note, Steve Berry wrote me to propose that a rapid intramolecular scrambling process — now called the Berry pseudo-rotation — could explain the appearance of only one line. Steve very generously suggested that John and I join him in publishing a note on this, but we, quite properly, declined because the idea was entirely Steve's. I'm sorry that John Waugh and I missed the boat in this case, but I don't feel particularly culpable. The Berry proposal was not obvious unless one had been thinking about intramolecular deformations. Steve mentioned in his letter that he had been, and thus we have a good example of chance favoring a prepared mind. A more detailed account of this incident will be found in an article I wrote for the 100th volume of the *Journal of Organometallic Chemistry* [441], in which Steve Berry's letter is reproduced.

In spite of the frustration in trying to determine the $Fe(CO)_5$ structure by ^{13}C NMR, I continued to think about the problem and a few years later, working with R. V. (Dick) Parish, a postdoc from England, it was shown that the infrared spectra of a number of $LFe(CO)_4$ and $L_2Fe(CO)_3$ compounds were consistent only with their having the substituted trigonal bipyramidal structures, with the substituents at one or both of the apical positions [58]. This (as we said) did not prove that $Fe(CO)_5$ itself has a trigonal bipyramidal structure. Later work has shown that it does, but that the barrier to Berry pseudo-rotation is so low (≤ 2 kcal mol^{-1}) that we must conclude that the square pyramidal structure is barely less stable than the trigonal bipyramidal one.

One of my most sustained efforts in research, which began in the period 1957 to 1960 and continued for several more years after that, was the study of the spectra and magnetism of tetrahedral complexes of the Mn^{2+}, Co^{2+}, and Ni^{2+} ions. Rather little was known about these species and ligand field theory offered, for the first time, the opportunity to gain a full understanding of their electronic structures. Another motivation for the work was to establish spectroscopic and magnetic criteria for inferring whether the Co^{2+} and Ni^{2+} ions were in tetrahedral (*vs* planar or tetragonal six-coordinate) coordination environments. These were days when simply performing an X-ray crystal structure, as we would do today, was not a practical option, so that indirect methods (what I once referred to as "sporting methods") of getting at structures were much sought after. In another ten years, this motivation no longer existed, but the electronic relationships remained interesting.

The whole project was begun sometime in 1957 or 1958 by Dick Holm, who concentrated on cobalt complexes. He began with the simplest cases, the CoX_4^{2-} ions (X = Cl, Br, I), of which the first two were already known from X-ray work to be tetrahedral. This study [47] provided clear experimental evidence for several theoretically predicted relationships and thus made an early contribution to validating ligand field theory as the best way (instead of Pauling's ideas) to understand electronic structures of tetrahedral — and all other — transition metal complexes.

Ligand field theory predicted that in a tetrahedral complex of the Co^{2+} cation the energy levels should be as shown in Fig. 3-5a, and this has many implications. Our purpose was to explore all of the implications and see how well the whole picture held

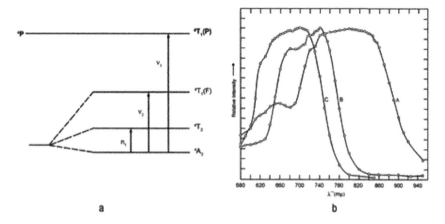

Figure 3-5. (a) Simplified energy level diagram for Co(II) in a tetrahedral ligand field, ignoring the $4T_1$–$4T_1$ interactions and spin-orbit coupling. (b) Reflectance spectra of $(C_9H_8N)_3CoX_4$. A: X = I⁻; B: X = Br⁻; C: X = Cl⁻.

together. We began by establishing with certainty the correct assignment of the big band in the visible spectrum that accounts for the intense blue or green colors of most tetrahedral cobalt(II) complexes. Some of these bands are shown in Fig. 3-5b. Note all the little circles along the band contours. The young researcher of today may well be puzzled by these and surprised to learn that these are the actual data points, each one measured manually on a wonderful spectrometer called a Beckman DU, which lacked any form of automation or recording device.

These spectra show, *inter alia*, that the order of the halide ions in the spectrochemical series is Cl⁻ > Br⁻ > I⁻. They also, a bit less obviously, provide a basis for the phenomenon that Klixbüll-Jørgensen called the nephelauxetic (cloud-expanding) effect. These bands all appear at appreciably lower energies (about 70%) than would be expected from the energy difference between the ^4F and ^4P states of the free Co^{2+} ion. The separation of these two states is caused by interelectronic repulsion, which is lessened in the CoX_4^{2-} complex. The reason it is lessened is that the d electrons are partly delocalized onto the ligands and are thus farther away from each other. It is as if the cloud of d electrons were simply expanded. If all of this sounds a little vague, I can assure the reader that it can actually be given satisfactory mathematical expression in terms of the covalence of the Co–X bonds. It was a point of real interest and novelty at the time, though we take it for granted today.

In this same paper (and in others that followed), the magnetic moments (μ) of the complexes were measured and shown to be in accord with the equations:

$$\mu = g\,[S(S+1)]^{1/2}$$

$$g = 2.00 - \frac{8\lambda'}{\left(\frac{4}{9}\right)\Delta}$$

in which λ' represents the effective value of the spin-orbit coupling constant and Δ represents the d orbital splitting parameter. Since λ' is a negative number for the more-than-half filled shell of the Co^{2+} ion, all g values will exceed 2.00 and all μ values will exceed the spin-only value of 3.88 Bohr magnetons (BM). Moreover, the greater the ligand field splitting, Δ, the lower will be g and hence μ. Thus, for the series $CoCl_4^{2-}$, $CoBr_4^{2-}$, CoI_4^{2-} the μ values should increase, and we showed that they did (~4.75, ~4.86, ~5.01). Moreover, in order to reconcile the measured values quantitatively with those calculated, it was necessary to give λ' a value of only about 70% of the free-ion value, λ. This is another consequence of the nephelauxetic effect or, more realistically, of the covalent nature of the Co–X bonds.

It may be noted in Fig. 3-5a that the ground state of the tetrahedrally coordinated Co^{II} is without orbital angular momentum and the rather large orbital contributions to the observed moments come in entirely from mixing with other states through a perturbation mechanism. This idea was, of course, not new, but it was an unfamiliar one to chemists at the time.

In summary, in this first excursion into this sort of work, we were able to establish in a beautiful way the quantitative internal consistency of a body of theory that had previously been *terra incognita* for the majority of coordination chemists. In papers that soon followed [55, 64], the coverage was broadened to a much larger array of tetrahedral cobalt(II) complexes. Many of these were not truly tetrahedral, but consisted of two neutral and two anionic ligands arranged in an approximately tetrahedral fashion, CoL_2X_2. It was demonstrated there that even though the spectrochemical series had been established on the basis of Cr^{III} and Co^{III} octahedral complexes, it was valid for the tetrahedral Co^{II} complexes as well. All this early work on cobalt was done by Holm as part of his Ph.D. research.

The work on tetrahedral Co^{II} complexes set the stage for a series of studies of tetrahedral Ni^{II} complexes by David Goodgame, who began postdoctoral work with me in the Fall of 1959, along with his wife Margaret. I will say more about both of them later.

Compared to those of cobalt(II), tetrahedral nickel(II) complexes were still curiosities, although the NiCl$_4^{2-}$ ion had been recently established structurally by Peter Pauling in his Ph.D. work with Nyholm at University College, London. I felt that it should be possible to make a lot more of them and to prove that they were tetrahedral by means of spectroscopic and magnetic data, correlated by means of ligand field theory. This is exactly what David Goodgame did over a period of two years. As a result of his work [69, 77, 78, 81, 97], tetrahedral nickel(II) complexes became numerous and well documented, as they never had been before. Again, this may seem unremarkable now, but it was all new and exciting (at least to us) in 1960.

In the meantime, Margaret Goodgame continued the study of tetrahedral cobalt(II) after Holm finished in 1959 and took an Assistant Professorship at Harvard. Her work, in some of which David also shared, vastly broadened the realm of tetrahedral Co^{II} chemistry and further emphasized the importance of ligand field theory in understanding and correlating the facts [87, 88, 96, 98, 103, 104, 108].

One more interesting experimental confirmation of theory resulted from the study of an unorthodox tetrahedral cobalt(II) compound, Co(DPM)$_2$ (Fig. 3-6). This compound has a reddish color and absorbs rather weakly, and yet, in one of the earliest X-ray studies carried out by my students, Roger Soderberg was able to show that the CoO$_4$ core had a distorted tetrahedral arrangement. The low intensity of the absorption band and its unusually high energy (which accounts for the red color)

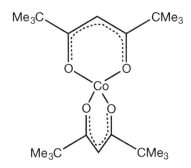

Figure 3-6. The Co(DPM)$_2$ molecule.

At Lee Gamble's, 1960. Top to bottom: Margaret Goodgame, David Goodgame, FAC, Dee, Emmanuel Yellin, Nancy Dusenberry.

got me to thinking about still another relationship that was implied by ligand field theory but had not previously been pointed out or demonstrated.

The high energy of the absorption band was shown to be due to the fact that in this compound there was an exceptionally small nephelauxetic effect (*i.e.*, not much delocalization of cobalt *d* electrons into the ligands). Now Ballhausen and Liehr had previously shown that for tetrahedral complexes, a major cause of the higher intensities of *d–d* transitions, compared to the intensities in octahedral complexes, was the mixing of *d* orbitals with ligand orbitals. Thus, it appeared that the high energy and the low intensity had an understandable relationship, but the question was: is this a general relationship? With the abundant data we had accumulated on tetrahedral cobalt(II) complexes, we were able to show conclusively that it is. A plot of band intensity *vs* the decrease in inter-electronic repulsion) gave a monotonic relationship, except for the most extreme values, where scatter was evident [102]. Again, we had strong support for the value of ligand field theory as the best way to understand the bonding in transition metal complexes.

One final phase of this work on tetrahedral complexes concerned those of Mn^{II} [94, 99]. These were even less well known than those of Ni^{II}. Indeed, they were almost unknown, although Mn^{2+} doped into Zn_2SiO_4 was a technologically recognized green phosphor. Work done with David and Margaret Goodgame provided the first systematic understanding of tetrahedral Mn^{II} complexes, as well as greatly enlarging the number of well-characterized ones. In octahedral complexes, Mn^{II} was already known to give very weak (because doubly forbidden) spectra with some very sharp bands. Thus the pink color of Mn^{II} in octahedral ligand fields is so pale that they commonly appear to be white. The non-centric character of the tetrahedral field removes one source of forbiddenness so that the tetrahedral complexes are clearly colored, but now yellow-green. Another peculiarity is that they are frequently triboluminescent. David Goodgame discovered this when he was using a spatula to poke one of them out of a glass tube after a magnetic susceptibility measurement. He immediately called me over to the dimly lit room where he had been doing this work and we were both so excited that we rounded up everyone else in the lab to come and see the show. We then checked all of our other tetrahedral manganese(II) compounds and found that several more were triboluminescent, while a few were not. As far as I could find out at the time triboluminescence was neither very common nor well understood. The color of the emitted light seemed to be the same as that in the normal fluorescence. We did not then take the time to look into it any further, but I returned to the subject in the year 2000.

Over the 3–4 decades since the triboluminescence (TL) of some (but, importantly, not all) MnX_4^{2-} compounds was discovered by David Goodgame, quite a bit had been done to characterize the TL phenomenon. Some of the leading researchers, such as Jeff Zink at UCLA, had tentatively adopted the view that TL resulted because extremely high local electric fields were created across cracks as crystals were crushed. However, for this to occur, it would appear to be necessary that the crystal structures be non-centric. Thus, the experimentally testable hypothesis that only non-centric crystals could display TL came in for serious consideration. It occurred to me that by study of an extensive series of tetrahedral manganese(II) complexes, this hypothesis could be tested. Therefore, in 2000, eleven compounds, which necessitated doing six more crystal structures, were examined [1480], and they all passed the test.

Thus, it seemed that the truth of the hypothesis was plain and simple. However, as some anonymous philosopher has remarked, the truth is rarely plain and never simple. My graduate student, Penglin Huang, and I looked at many other compounds [1519] and were disillusioned to find that there appear to be other physical bases for TL and no universal rule concerning its appearance. The noncentricity rule probably works most of the time, but there must be more facets to the phenomenon than one can deal with by that simple rule.

David and Margaret Goodgame were the first, but not the last, husband and wife team of postdocs I have had. They arrived very well trained and very well recommended by Luigi Venanzi, both having been Oxford students all the way. I was greatly amused by David's asking me, along with many other questions about preparing for life at MIT, whether he and Margaret should bring their gowns. I was able to tell him with certainty that this would be unnecessary.

It was a further illustration of the small but amusing cultural differences that existed between England and New England, that when David and Margaret invited us to dinner at their flat during their first winter at MIT, we arrived to find the place stifling hot. They quickly noticed that we were uncomfortable and expressed a concern that we might not be warm enough. Obviously they had formed the impression that Americans (and it was certainly true of many) liked their dwellings overheated in winter, whereas the English (especially during and after the war) had high fuel prices and generally kept their homes quite cool. We soon sorted out the problem and to the great relief of all of us turned down the heat, opened the windows, and got the temperature down to about 68 °F.

Another amusing incident occurred when David's parents came to visit. They were very alert and lively people who were on their first visit to America and interested in everything. I therefore took them to a baseball game between the Boston Red Sox and the New York Yankees. As usual between these intense rivals it was a good contest and therefore exciting. I enjoyed explaining the goings on, which surely seemed arcane and even bizarre to any first-time spectator. All of the Goodgames, young and old, seemed to be getting the hang of things pretty well when Mrs. Goodgame suddenly burst out in sheer amazement: "Surely, that old gentleman isn't going to play, is he?" What caused her bewilderment was that the Yankees had decided to yank their pitcher, who was in trouble, and the manager, Casey Stengel, who was in his late sixties and looked every year of it, had creaked out to the mound to make the change. He was, naturally, dressed exactly like the players, and it must have seemed to Mrs. Goodgame that the "old gentleman" was about to come into the game.

David and Margaret both did marvelous work and set a standard that not all of my subsequent postdocs have been able to meet. After their two years with me, both obtained positions as lecturers at Imperial College, where today David is an emeritus Professor and Margaret a retired Senior Lecturer. They were pillars of Geoff Wilkinson's Department for decades and at the same time raised a family. They represent an early and eminently successful example of solving the now common problem referred to as the "two body problem."

I might mention a few more of the research subjects that engaged my attention prior to 1960. The vast potential for dimethylsulfoxide (DMSO) and other sulfoxides to serve as ligands was entirely unrecognized in the late 1950s. It seemed obvious to me that just as phosphine oxides could be ligands (**1**) sulfoxides could be, too (**2**). However, for a sulfoxide there was also the possibility of what was called ambidentate behavior, meaning that there were two possible donor sites, one being the oxygen atom and the other the sulfur atom (**6**).

$$R_3PO \longrightarrow M \qquad R_2SO \longrightarrow M \qquad \overset{O}{R_2S} \longrightarrow M$$

$$(\mathbf{1}) \qquad\qquad (\mathbf{2}) \qquad\qquad (\mathbf{3})$$

Working mainly with Ronald Francis, but also on the more physical aspects of the work with Bill Horrocks (both Ph.D. candidates), I looked into this. It was quickly shown that DMSO is a superb ligand, capable of coordinating with all metal ions [65]. It was also soon shown that both O- and S-donation could indeed occur and that there is a simple infrared spectroscopic criterion to distinguish them [76].

The criterion was not only empirically sound but it had a logical rationale because the S–O bond has considerable π character resulting from O to S dative π bonding. Thus, when the sulfoxide is O-bonded to an acceptor, this π-bonding is reduced, the S–O bond is weakened, and the S–O stretching frequency is lowered relative to that in the free sulfoxide. When the sulfoxide is S-bonded, electron withdrawal from the sulfur atom enhances O to S π-donation and the S–O stretching frequency goes up. This criterion is one of the most reliable infrared criteria of structure in the field of coordination chemistry and has been used for hundreds of sulfoxide complexes that have been made and studied since our first papers were published. Needless to say, we [195, 228, 232] and many others later published many crystal structures that fully supported the conclusions drawn from the infrared criterion.

The two students who did this work were of quite different character. Ron Francis was a rather introverted, secretive young man from rural Maine who took a terribly long time (1956-63) to actually write his Ph.D. thesis. He then bounced around in various jobs with more or less applied character, but eventually went to the Department of Photographic Science at Rochester Institute of Technology. Here he apparently found his metier, for he was soon head of the department and Weidman Professor of Imaging Science. When I last saw him, he was a highly regarded and very happy authority on the chemistry of photography. Sadly, in 1987 he died suddenly of a massive heart attack. Bill Horrocks, on the other hand, was an outgoing, self-confident person, who finished his Ph.D. quickly (1956-60) and became a junior faculty member at Princeton. During this time, Princeton seemed to award few if any new appointments to tenure. This caught Bill at the critical time and he moved to Pennsylvania State University, where he has had a fine career doing very sophisticated work in several areas of physical inorganic chemistry.

Another interest of mine that had begun while I was a graduate student and continued into the early 1960s was the nature of metal carbonyl hydrides. Walter Hieber, who discovered the first such compounds, had suggested that the hydrogen atom(s) was (were) attached directly to the metal atom, but there was no proof of this. Just before Geoff Wilkinson and I left Harvard in 1955, I had shown that for $H_2Fe(CO)_4$ there is an ^1H resonance at about 16 ppm upfield from that for H_2O, which is similar to what had just previously been reported for $HCo(CO)_4$ [20]. We also presented arguments in favor of the IR band at 703 cm^{-1} in $HCo(CO)_4$ being due to a Co–H stretching vibration. It had been argued by others that in $HCo(CO)_4$, the H atom was not bound directly to the cobalt atom but was bound to the C and O atoms of three CO groups. It occurred to me that a good argument against this arose from the fact that in $H_2Fe(CO)_4$ the first and second acid dissociation constants differed by about 10^9, very much more than in H_2SO_4 or H_3PO_4 where the differences are only about 10^4–10^5. However, in another well-known case where both H atoms are attached to the same atom, namely, H_2S, there is also a very large difference, namely, 10^8.

In 1958, I published a theoretical discussion of the $HCo(CO)_4$ problem [32], which led me to conclude that "Co–H bonding is quite important and that the hydrogen atom lies within 1.2 Å of the Co atom." I went on to note that "the sum of the Co and H radii is 1.2–1.4 Å." Some ten years later, Davison and Faller [210] deduced by solid state NMR that the distances in both $HMn(CO)_5$ and $HCo(CO)_4$ are around 1.25 Å. Ultimately (1969), it was found by neutron diffraction that in $HMn(CO)_5$ the Mn–H distance is 1.60(2) Å, but the fact that the H atom is directly bonded to the metal atom and not in any way to the CO groups was conclusively established.

Before concluding this chapter, I would like to say something specific about several more of the outstanding students I had in the period I am discussing (1956–60), and one in particular, John P. Fackler, Jr. John was principally responsible for initiating studies of β-diketonate complexes, which were, of course, known then but not well understood. In addition to undertaking studies of their reactivity and electronic structures, he also began to uncover the fact that several of them, Ni(acac)$_2$ and Co(acac)$_2$, were oligomeric in solution. The study of the electronic structures was later expanded [182, 184, 215] by John J. (Jack) Wise (Ph.D. 1965), and the oligomeric structures were further investigated [165, 177] by Roger Soderberg (Ph.D. 1962), Richard C. Elder (Ph.D. 1964), and Roger Eiss (Ph.D. 1967).

One of the interesting things about many M(acac)$_2$ compounds, which had not previously been recognized, is that they are not simple mononuclear molecules. Instead, because of the tendency of the M^{2+} ions to have a coordination number higher than 4, they tend to form oligomers in which some of the acac oxygen atoms are bridging. We first realized that something like this must be going on because while "Ni(acac)$_2$" is green and paramagnetic the sterically hindered compound shown in Fig. 3-7a is red and

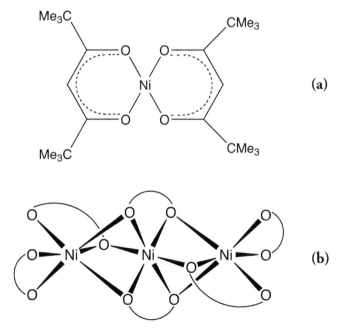

Figure 3-7. (a) The sterically hindered monomer. (b) The trinuclear structure of Ni(acac)$_2$.

John Fackler in grad school.

diamagnetic. At the same time, at University College, London, a study of the crystal structure of "Ni(acac)$_2$" was being carried out by Ron Mason and Peter Pauling, and it revealed that the true molecular formula is [Ni(acac)$_2$]$_3$, with the structure shown in Fig. 3-7b. Evidently, the CMe$_3$ substituents in the compound in Fig. 3-7a are too large to allow oligomerization. However, when a group of intermediate size, CHMe$_2$, is used, beautiful temperature- and concentration-dependent equilibria between the monomer and the trimer can be characterized both magnetically and spectroscopically as John Fackler showed [93]. Later on, when we were set up to do crystal structures, others in my laboratory showed that oligomerization occurs for Co(acac)$_2$ [165, 177] and Zn(acac)$_2$ [204].

John Fackler went to Berkeley after finishing his Ph.D., but a few years later, he became one of a group of bright young people who accepted the challenge of creating a vigorous, high-quality chemistry department at Case Institute of Technology in Cleveland. There he did some brilliant work dealing with complexes with sulfur ligands and with the Jahn-Teller effect. He was department head for several years and showed signs of being both able and interested in moving to a higher level of administration. In 1982, Texas A&M started a search for a Dean of Science, and I persuaded John to apply for the position. He joined us in that capacity in 1983.

John caused considerable surprise here when he told the Provost that he didn't want simply the title of Professor of Chemistry but actual laboratory space for research. He was advised that this would be unnecessary because as Dean he would be kept far too busy to do research. John insisted and took over several labs in my area, where he proceeded to do some of the best work of his entire career, concerned with cluster compounds of the coinage metals and the remarkable structural and photophysical properties of binuclear gold complexes.

As a Dean, for eleven years, John was terrific. The level of the entire College of Science rose by leaps and bounds under his guidance. John had the knack of combining realism with idealism. It was during his deanship that the superconducting supercollider (SSC) issue came up. I served on the State Commission charged with producing a competitive proposal to the federal government, and in fact, the Texas proposal was the winning one. The SSC was to be built in Waxahatchie, a small town some thirty miles south of the Dallas-Ft. Worth area. Several Texas universities, most particularly Texas A&M, made commitments to strengthen their physics departments in directions related to high-energy physics, particle theory, and the like. Since UT-Austin already had Steven Weinberg, John decided to try for Sheldon Glashow. This didn't work out, but Shelly was a great help in finding others, and, in short, John was able to give our Physics Department a tremendous boost. It was a very exciting period, but unfortunately, the physics community's political clout eventually ran out and Congress killed the SSC. John returned full-time to the Chemistry Department in 1994, where he has continued to teach, do research, and to play a major role in a wide spectrum of activities vital to the welfare of the department and the university.

As to several other people mentioned above, Roger Soderberg spent a successful career at Dartmouth University, Richard Elder at the University of Cincinnati, and Roger Eiss in government administration and business. Jack Wise rose to Vice President for Research and Development at Mobil Corporation and was elected to the NAE. There is no question that those who worked with me in different phases of my research on β-diketonate complexes all make me proud to have had a role in getting them started in their careers.

During the period covered by this chapter, I continued, as I had done in graduate school, to regard the social aspects of my life as no more than occasional diversions from the important business of getting on with my career as a scientist. I enjoyed parties and the company of young women but didn't take any of them seriously until some time in 1957. I was living a bachelor, workaholic life, occupying a small flat on Gloucester Street in Back

Dee about the time of our marriage. Not an easy woman to ignore.

Bay (Boston). My friend Peter Verdier, who was also responsible for introducing me to Peter Pauling, had another friend from his Cal Tech undergraduate days, Jess Weil, a physicist. Jess had a delightfully unconventional girl friend, Nancy Dusenbery, who in turn shared an apartment, near mine in Back Bay, with her former college roommate at Antioch. Thus, I met Diane (Dee) Dornacher, who took up occupational therapy following her B.A. in biology. Dee was then working in Massachusetts General Hospital, at the other end of Back Bay from where we lived.

A wedding day display of Edwardian elegance.

This little social circle of Dee, Nancy, Jess, and Peter Verdier and his wife Marilyn gradually took up more of my time. Our main activities were folk music, Mahjong, and generally fine-tuning our intellectual snobbery. Over a few months, I found myself

The first minute of married life.

A honeymooner's supine approach to Alpine mountaineering.

becoming more and more attracted to Dee, and very happy in her company. Finally, in early 1959, I proposed, was accepted, and we were married, June 13, 1959.

Arranging the wedding entailed a bit of improvisation, since we wanted to have a formal ceremony, but neither of us had any religious affiliation. We decided to be married in the non-denominational MIT chapel, which provided a pleasing, dignified, but not obtrusively religious setting, but we also needed a "man of the cloth" to perform the ceremony. Dee solved this problem by persuading the pastor of the Sudbury Unitarian Church with whom she had become acquainted through some youth group at the Arlington Street Church in Boston. We asked him to give the least religious service that he would find acceptable. Dee chose her twin sister, Gretchen, and I chose my friend Frank Harris (who by then was on the faculty at Stanford) to be our "supporting personnel." We also decided that formal dress was to be worn, and thus Dee wore a bridal gown, while Frank and I wore daytime formal attire, as the wedding took place at 5:00 pm.

That evening we had a superb dinner, in a private room at the Locke-Ober restaurant, with guests including Frank, his mother, the Verdiers, Andy Liehr, and our own family members. Following this we went directly to Martha's Vineyard for a short first honeymoon. In August we went to London and Switzerland for a longer second

Dee and I in our apartment at MIT in the Fall of 1961, with our siamese cats, Yum Yum, Ko Ko San (usually called Coco Sam), and Dimetri.

one. Our policy on honeymoons followed the adage attributed to Edwin Land that anything worth doing is worth doing to excess.

In September of 1959, we moved into an apartment on the MIT campus in the dormitory complex known as The Parallels, where we served as faculty residents. We had the use of this very nice apartment rent-free and our only duties were to be nice to the boys (as all the denizens of the parallels then were) and this included having them in for sherry on Friday afternoons. Thus, we had a very nice way to save money for the later purchase of a house, while I was only a few minutes walk from my labs and Dee was only five minutes walk from the MTA which took her over the Charles river, *via* the "Salt and Pepper bridge" to the Massachusetts General Hospital.

In those days, students were much less obnoxious than they became only a few years later and, generally speaking, politeness and tranquility prevailed. Occasionally, there would be a so-called hi fi war in which students in each of the parallel buildings would aim speakers fed by high wattage amplifiers at the opposite building and max out the volume. We simply fled until a truce was declared. One other favorite trick of the students was helping themselves to free telephone service. The telephone booths, one per floor, all had several dozen emerging wires, each going to one of the dorm rooms. At the end of each wire was a cheap second-hand telephone. Exactly how this "tap-into-Ma-Bell-for-free" arrangement worked, I don't know but it seemed to be common knowledge among the students.

On December 11, 1961, our daughter Jennifer was born and we decided that it was time to go out and face the rigors of home ownership. By then we were horse owners and were spending a lot of time in the town of Dover where we kept our horses. We decided that we would either live in Cambridge and commute to the horses or live near the horses and commute to Cambridge. Living halfway in between and commuting everywhere was not appealing. It soon became evident that in Cambridge there were no houses we liked that we could afford (not to mention those we *didn't* like *and* couldn't afford), so we started checking out Dover, and its sister town, on the opposite bank of the Charles river, Sherborn. We soon found a suitable house on the western edge of Sherborn with six acres of land suitable for pasture and a three-horse barn. In June of 1962, we formally became hayseeds and have been ever since.

CHAPTER 4

MIT
1961–71

IT WAS THE BEST OF TIMES;
IT WAS THE WORST OF TIMES.

—*Charles Dickens*

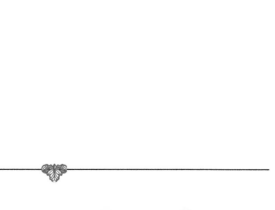

I THINK THIS must have been the busiest and most productive decade of my career. However, living in Sherborn, twenty miles from MIT, I learned what it is like to be a commuter, sometimes by car, sometimes by train, and I didn't like it. Commuting by train meant driving to Wellesley and leaving the car or having Dee drive me to Framingham and then pick me up again in the evening. Over the ten-year period we lived in Sherborn, the roads got much more crowded and the train service went from poor to abysmal. This was certainly the least attractive aspect of our life style. The rest of life in the country suited us very well, however, and it has never occurred to us again to live in a city or town.

My professional life in this decade became rapidly more rewarding. As I shall explain later, the size and scope of my research grew vigorously. It was also a period during which my level of recognition rose a great deal. In 1962, the new journal *Inorganic Chemistry* was inaugurated with Robert W. Parry as Editor. I was one of those whom Bob asked to be on the first editorial board, along with Earl Muetterties, Henry Taube, John Bailar, and six others. I was very pleased by this genuine form of recognition, and also by the fact that inorganic chemistry had at last emerged from obscurity in the United States. I always hoped that someday Bob Parry would write up the full story of how *Inorganic Chemistry* came into being. I know that he and John Bailar played key roles and that they had to fight hard to overcome the opposition of W. Albert Noyes, Jr., then the editor of JACS, who took the position that inorganic chemistry was simply a minor part of physical chemistry and had no need of a journal of its own. I also know that during the first year, at least, Bob Parry read every word of every paper published to make sure that

Bob Parry around the time he founded the journal Inorganic Chemistry.

Noyes or others who had opposed the new journal would get no opportunity to snipe at its quality. When in 1965 Bob Parry received the first ACS Award for Distinguished Service in the Advancement of Inorganic Chemistry, I felt, and still feel, that that was perhaps the most well-deserved ACS award ever given.

The year 1962 was full of other very satisfying forms of recognition. I spent three weeks in February as the Gooch Lecturer at Yale University, with two long weekends back home, since it was an easy trip by car. The Yale faculty were all extremely hospitable, especially Bill Doering and my two friends from graduate school, Martin Saunders and Steve Berry, who were Assistant Professors.

The Yale chemistry department was mainly housed in a mediaeval castle, the worst excuse for a chemistry building I have ever seen, although charming in its eccentricity. One day, Lars Onsager, a theoretician of legendary status, showed up for my lecture, which happened that day to be on the ligand field splittings in square complexes of nickel(II) and copper(II), and the spectroscopic and magnetic properties resulting therefrom. He paid very close attention and, after the rest of the audience had left, started asking very detailed and probing questions about ligand field theory in general as well as about what I had said. It appeared he had not thought about the subject before but wanted to get it all sorted out in his mind then and there. I might well say that he was the most brilliant pupil I have ever had. He would not, however, accept anything until he had gone into every possible aspect of it. This was all well and good, but Bill Doering had made a dinner reservation at Morrie's (the place in the whiffenpoof song) for dinner and was getting visibly nervous about whether we would lose the reservation. However, Onsager, like many other great forces of Nature (tidal waves, volcanic eruptions, etc.), went on without regard to anything around him. Finally, he said, "Yes, I think that is all OK," and we were able to leave.

On one other occasion, there was a party at someone's house and Bill Doering, Marty Saunders, and I got into a discussion of whether astronauts would have trouble swallowing in a weightless environment. To settle this, Marty Saunders was held upside down and he showed that he could drink a glass of water (through a straw, of course) in that position.

There was another memorable consequence of my Gooch Lectureship, due to the fact that, by the standards of the time, it paid handsomely. I decided that it was time I had a sports car and here was the money to get one. It was Dee who had the excellent

Jennifer and the Morgan.

taste to propose that we buy a Morgan, the most classic of classic English sports cars. In the Fall, I was scheduled to visit England to give a lecture at a meeting of The Chemical Society in Leeds and I therefore preordered my car. I picked it up at a dealership in London and then drove out to the Malvern hills where the factory was, to see how these handmade cars were actually put together. My visit to the factory is still vivid in my mind. I began by having tea with the president of the company (a Mr. Morgan, of course) and then visited each department. While the chassis was, as expected, made of steel, most of the superstructure was made of wood, one car at a time, no assembly line. The final stage was the ragtop, each of which was sewn together by a little old lady with an electric sewing machine. Before I departed for home, I arranged to have the Morgan shipped to Boston where I subsequently went to the dock and picked it up.

Owning a Morgan was not an unalloyed blessing. It was relatively simple to repair, but it needed repairs quite often. Parts usually had to be ordered from the factory. Still it was a lot of fun to drive. The phrase "feeling the road" doesn't fully capture the sensation. When we left for Texas in 1972, I sold it. I still have mixed feelings about doing that, and if I had the decision to make again, I'd keep it.

In 1962, I was elected to membership in the American Academy of Arts and Sciences. While this was an honor not to be disparaged, it was not quite so significant as it would be today because the Academy then was a rather regional organization and most people at New England universities tended to get in, and to get in when fairly young.

More important in 1962 was my selection as the first recipient of the new ACS Award in Inorganic Chemistry. The inauguration of this award in 1962 was another concrete sign that the American chemical establishment was realizing that inorganic chemistry is a valid and viable branch of the science. I remember, in 1961, Art Cope telling me about the creation of the award and my being bold enough to ask him if he could get me on the selection committee. He smiled and said something like, "Al, the objective is not to *pick* the winner, it's to *be* the winner." The certificate I received stated, in part, that I had been chosen for "bringing physico-chemical methods to bear on important inorganic problems . . . ." This did accurately reflect the thrust of my work. I had recognized as a graduate student that there were very good inorganic chemists who lacked knowledge of physical chemistry and that physical chemists in general knew so little about inorganic chemistry that they were unaware of the opportunities that existed for bringing their skills to bear. It had been my conscious strategy to take advantage of this disconnect.

The year 1963 saw more recognition, partly in the form of single-lectures named lectureships, namely, the S. C. Lind Lectureship of the ACS East Tennessee Section, the Glidden Lectureship at Iowa State, and the McGregory Lectureship at Colgate. To one only 33 years old, such invitations were impressive. I was also the Reilly Lecturer at Notre Dame, which lasted for about a week and was a very enjoyable experience. However, the biggest events of that year were the award of an honorary doctorate by my alma mater, Temple University, and the Baekeland Medal of the New Jersey Section of the American Chemical Society. The latter was, and still is, a major award and did a lot to enhance my status at MIT. I remember the presentation vividly because that morning I awoke in Newark, New Jersey, with a severe cold, perhaps the flu, and kept feeling worse as the day went on. I begged off of the visits and interviews that had been on the agenda during the day and stayed in bed, but at about 5 p.m. I dragged myself out, got into my tux, and toughed it out as best I could. I had to meet a lot of people at the reception and dinner and give a talk after dinner and somehow managed to do this before I could get away and collapse into bed. Amazingly, I felt much better the next morning. This experience taught me the valuable lesson that it is possible to work through colds and that giving into them is just a waste of time. They don't go away any sooner just because you lie in bed and eat chicken soup.

In about 1963, I was invited to consider joining the new University of California at San Diego, located in the northern suburb of La Jolla. Dee and I found the idea attractive enough to go and have a look. There were not as yet very many people there in chemistry. Among them were Jim Arnold, Ted Traylor, and Joe Mayer. The Scripps Oceanographic Institute was then newly opened in La Jolla.

We were very warmly entertained and they nearly succeeded in convincing me to come. I have a vivid memory of having dinner one evening with Joe and Maria (Goeppert) Mayer, because of our preprandial drinks. The Mayers and we shared a fondness for dry martinis, but Maria suggested that we might like to try a new and innovative drink called a margarita. She explained that it was a way of making the Mexican peasant's custom of swigging tequila, licking salt, and sucking on a lime (in what order was never clear to me) into a socially acceptable potation. This was done by chilling a mixture of lime juice, tequila, and orange liqueur, and placing it in a cold glass with salt around the rim. Well, we came, we tasted, we were conquered.

When we returned to Massachusetts, we found that this wonderful concoction was, as yet, unknown to bartenders there and therefore we could enjoy them only at home. However, it was just a matter of months before margarita madness swept the country. On the other hand, Dee and I remained unconvinced that California was our kind of place and I did not accept the offer.

The La Jolla caper did have a lasting effect (besides a lifelong delight in margaritas) on my life style. Before I had made up my mind not to go west, MIT saw reason to agree that I might deserve a bigger and better office. A laboratory on the corner of the fourth floor of building six was gutted and converted into an office with oiled walnut paneling and a built-in wash basin, which is as close as I have ever come to having a Harvardian "special room." This may have been the most elegant office there was at the time for a mere professor (*i.e.*, one with no administrative title). It made me a very happy camper for some years to come. When I left it, was inherited by Alan Davison.

In the years immediately following my 1961 promotion to Full Professor, the pace of research continued to accelerate. Before going into any specifics of the chemistry, I want to describe one of the most important developments, namely, X-ray crystallography, because this was enabling for nearly everything I undertook thereafter.

Starting around 1960 I had begun to think seriously about X-ray crystallography. It had fascinated me since my first year in graduate school when I audited a course given in the geology department by Clifford Frondel, a well-known mineralogist. It now became more and more evident that the kind of research problems I liked nearly always involved structural questions. In those days, there was no such thing as a staff crystallographer. The only way to get a structure done was to ask a faculty colleague to do it, and it was a big job. Most crystallographers had their own interests and were unlikely to do a structure for someone else very often. Learning to do it myself was going to be the only way to meet my needs. But how?

Good luck came my way. My colleague Alex Rich in the biochemistry department rang up one day, in the Spring of 1961, and asked if I could supply him with a crystalline sample of a metal complex of an amino acid, any amino acid. I said I could probably do that quite easily, but I asked why he wanted it. He said he had a new graduate student, R. H. (Bob) Kretzinger, who was going to work on a biological structure problem and needed hands-on training in small-molecule crystallography first. He also had a postdoc, R. F. (also Bob) Bryan, who would teach him the ropes. I realized instantly that here was my chance. I told Alex that in return for supplying the crystal, I would like to become Bryan's pupil as well.

This arrangement was ideal. There was a lot of sheer drudgery in doing a crystal structure then. What would now take a day took months of boring and tiring work then. Hundreds of reflections had to be collected *via* numerous Weissenberg photographs and then the intensities of the spots on the film had to be estimated by eye on a lightbox by comparing each spot with a set of graduated reference spots, for weeks and weeks. After that, the position of the heavy atom had to be determined *via* the Patterson function — which was calculated and plotted by hand. Then began an extremely lengthy process of calculating electron density maps and looking for peaks that might be real, *i.e.*, that represented other atoms in the molecule. Such peaks were used to improve the next map and then, with luck, one or a few more real peaks could be introduced. Of course, especially in the early stages, some of the first peaks found might be spurious and their inclusion in the structure factor calculation would make the next map worse rather than better. It could take months to finally get the full structure.

My good fortune was that I could participate in each of the time-consuming phases of the work only until I understood how to do it, and then the graduate student, Kretzinger, would do the remaining 90% of the dogwork. Only when it was time to proceed with the hunt for the entire structure, after we had found the heavy atoms, did I begin to share more fully in the labor. Even here I could leave all the repetitive work to Kretzinger and focus on the interesting part.

Once we had the data set and the Patterson map, Bob Bryan returned to his own work and pretty much left Kretzinger and me to our own devices. We set up a large crude model which took up most of my office and began working on the solution. Neither of us realized then that we had taken on an unusually hard job, because I had made crystals of the zinc complex employing naturally occurring *l*-histidine. This meant that the space group was, of necessity, non-centric, and that, in turn, meant that the process of solution was very, very much more difficult than it is for a centrosymmetric case. Had I used racemic histidine, we would have had a far easier task. We struggled, day after day, on good days taking two steps forward and one step backward, on bad days doing just the opposite. The trouble was that until you got enough atoms right to be able to recognize a portion of the structure, you had no way of knowing if you had *anything* right. I remember several times calling on John Slater's wife, Rose Mooney Slater, who had been doing crystallography from the earliest days. She came and gave us both technical counsel and moral support; she told us we weren't doing anything stupid, which was very encouraging.

Ever so slowly we progressed, until one night about 11:00 pm we had it all. I called Dee at the flat across the way and said I hope there's some beer in the fridge because two happy warriors are coming to celebrate a great victory. The next weeks were spent learning from Bob Bryan how to refine the structure, and finally, in February 1962, I submitted my first crystallographic publication [105], a preliminary report in *Proceedings of the Chemical Society* (the precursor of *Chemical Communications*), followed a year later by a complete report in *Acta Crystallographica* [117]. I mentioned above that our task was unusually hard because we had a non-centrosymmetric crystal to deal with. Ironically, the centrosymmetric structure of the racemic compound appeared, from another laboratory, at about the same time as our full paper. I have always wondered if we would have published first or been scooped had we chosen the same compound.

The solving of this structure, with an asymmetric cell and 12 light atoms phased initially by only one zinc atom was not of the same order of difficulty as the solution by Dorothy Hodgkin of the vitamin B_{12} structure with 102 light atoms phased by only one cobalt atom, but it was the same type of nasty problem, and gave me a very personal sense of how great was Dorothy's achievement.

Finishing the zinc histidine structure was a tremendous shot in the arm and a more important turning point in my career than I could possibly have imagined at the time. I quickly acquired the very modest equipment then required (a Weissenberg camera, a lightbox, and a darkroom) and tackled two crystallographic problems of my own. One was the structure of dibenzene chromium, on which we began work in the Fall of 1962. The interest here was to try to settle the question of whether the benzene rings had equal C–C distances or whether there was significant alternation. E. O. Fischer and coworkers had postulated the latter, but I was in favor of essentially hexagonal rings.

Perhaps I should say a little more about the motivation for this work. E. O. Fischer, who had discovered dibenzene chromium, proposed to explain the metal to ring bonding in the same way he had earlier proposed to explain the bonding in ferrocene. He referred to both molecules as "durchdringungskomplexe," which translates roughly into English as "penetration complexes." This, in turn, was a way of designating what Pauling called covalent complexes and ligand field theory called low-spin complexes. Fischer's concept was that $C_5H_5^-$ and, in a more perfect way C_6H_6, provided a triangle of three electron pairs and that the sandwich compounds could be viewed as analogous to octahedral complexes, as shown below:

From this it followed that in $(C_6H_6)_2Cr$ the rings might display bond length alternation. This concept, incidently, also seems to imply that the D_{5d} and D_{6h} structures are preferred electronically over the D_{5h} and D_{6d} ones.

On the other hand, the molecular orbital approach, with d_{xz} and d_{yz} metal orbitals matching the e_g orbitals of the rings, does not imply that there should (nor even could) be any bond alteration, nor does it imply any electronic preference for D_{5d} or D_{6h} structures over D_{5h} or D_{6d} structures (although a steric preference is not precluded).

I was concerned, at that time, to "prove" not only that the MO approach is *preferable* to Fischer's hybridized orbital approach, but that the latter leads to an incorrect structural prediction. That is why I undertook the study of the structure of $(C_6H_6)_2Cr$ and why I found the result we obtained so satisfying at the time. In fact, of course, in spite of what we found, an advocate of Fischer's concept could easily salvage it by invoking resonance, in exactly the same way as one does with benzene itself, so I don't think today that our work really disproved the Fischer concept at all. I think this is an example of what is known as becoming older and wiser.

Spectroscopic studies by H. P. Fritz had proved inconclusive and a crystal structure published in 1960, of very low quality, had led to the conclusion that there were alternating bond lengths, 1.439 and 1.353 Å, with standard deviations claimed to be as low as 0.014 Å for each. The crux of the problem was that the crystallographic symmetry was required to be D_{3d} which allowed but did not require bond alternation. It was all a question of accuracy, *i.e.*, the quality of the structure. This was a good learning experience in how to collect the best possible data and carry out the best possible refinement. My coworkers were Wayne Dollase (a student of Martin Buerger's) and John S. Wood, a postdoc from England. Our result was obtained by full-matrix least squares refinement with inclusion of hydrogen atoms, neither of which were routine procedures at that time, although today they are. We found all C–C distances equal at 1.387 Å with standard deviations of 0.017 Å [115]. This result has never been challenged.

The spectacular crystallographic event of 1962-63, however, was the determination of the structure of what had been known in the literature as $CsReCl_4$, and postulated by some to contain the only tetrahedral complex ion in a low-spin ground state. No other such species had ever been found (nor has there been one since[1]), and the initial objective of the work was to see if this was really the case. If ever there was to be one, this would be it, since it had everything going for it: it contained a metal ion of the third transition series, where ligand field splitting would be maximal, and it had exactly the optimum number of electrons to give a closed e^4 ground state.

I gave the problem to J. Aaron Bertrand, who had joined my group as a postdoc in the fall of 1961. Wayne Dollase, mentioned earlier, also spent some time on this work but it was Aaron who really sweated it out. All of the intensity data were collected photographically by employing a Buerger precession camera. It rapidly became clear on indirect grounds that no $ReCl_4^-$ ion was present, but solving the structure proved to be extremely difficult in that virtually pre-computer era. Today's direct methods were non-existent, and I think any young crystallographer of today might be astonished to read our papers [114, 122] and see what we had to go through (the so-called Patterson superposition method, with Patterson sections hand-drawn on transparent sheets) to solve a structure that today would be done in a day — or less. Those were the days of wooden ships and iron men, as the saying goes.

The reason this structure was so singularly important in my career was that it set me on the road to the most prominent theme of my research for the rest of my life. What we found was that instead of a simple $ReCl_4^-$ ion, or any other conventional or traditional type of structure, such as a polynuclear one held together solely by shared chlorine atoms, there was a triangle of metal atoms in which the metal atoms were

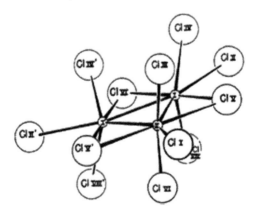

Figure 4-1. The $Re_3Cl_{12}^{3-}$ structure, exactly as originally published [122].

[1]Compounds such as $Ru(c\text{-}C_6H_{11})_4$ and $OsPh_4$ are tetrahedral and diamagnetic, but can hardly be called complexes.

astonishingly close together, ca. 2.50 Å, as shown in Fig. 4-1. This may be compared to the Re–Re distance in rhenium metal (2.76 Å). This was unprecedented chemistry. Moreover, from general electronic considerations, I saw no alternative but to propose that the Re–Re bonds were double bonds, which was also unprecedented.

Not only did the $Re_3Cl_{12}{}^{3-}$ structure have a seminal effect on metal atom cluster chemistry, but it also shone a bright light on a whole range of rhenium(III) chemistry that had previously been a conundrum. Rhenium(III) chloride itself had strange properties, such as being very insoluble when freshly prepared, but much more soluble after exposure to air. It then reacted with a great variety of neutral and anionic ligands to produce compounds with uninterpretable compositions.

With the critical participation of Joel T. Mague, all the confusion was soon cleared up [139, 141, 146]. We found that "$ReCl_3$" had a structure in which Re_3Cl_9 clusters of exactly the same kind as those found at the core of the $Re_3Cl_{12}{}^{3-}$ ion were linked into infinite sheets by the sharing of Cl atoms, as shown in Fig. 4-2a. We also quickly obtained the structure of $Re_3Cl_9(PEt_2Ph)_3$, shown in Fig. 4-2b, which made it clear

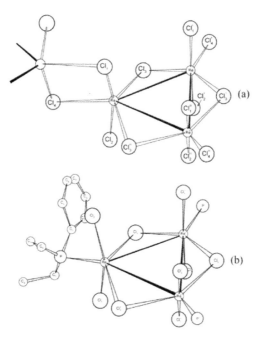

Figure 4-2. (a) A portion of the structure of "$ReCl_3$" [141]. (b) The structure of the tris-phosphine complex $Re_3Cl_9(PEt_2Ph)_3$ [139].

that the key to understanding all the chemistry derived from "ReCl$_3$" was to recognize that the Re$_3$Cl$_9$ units therein could be pried apart and by addition of ligands to the three outer positions all manner of complexes could be formed. In most cases, all three outer positions would become occupied by identical ligands, as for Re$_3$Cl$_{12}^{3-}$ or Re$_3$Cl$_9$(PR$_3$)$_3$, but these sites could be incompletely occupied, as in Re$_3$Cl$_{11}^{2-}$ or occupied by a set of nonidentical ligands. The initial low solubility, and generally inert behavior, of "ReCl$_3$" was clearly due to the dense, tightly knitted structure of (Re$_3$Cl$_9$)$_\infty$. Exposure to the atmosphere allowed water molecules to break the Cl bridge bonds and replace μ–Cl by H$_2$O in some of the outer positions. Thereafter further solvation or attack by other ligands could take place relatively easily.

Steve Lippard and I then went on to show that "ReBr$_3$" was structurally similar [147, 152] and thus to explain its chemistry. The problem of "ReI$_3$" proved recalcitrant. Bruce Foxman tried diligently to get crystals, but the stuff just seemed bent on dying a non-crystalline death. Finally, I gave Bruce the OK to give up. He was so disgusted he didn't even have the heart to clean up the glassware, so he pushed it to the back of his bench and ignored it. Several months later he appeared in my office, smiling like the cat that had just swallowed the canary, and asked me to come down to his laboratory to see something interesting. There was a small beaker containing the grungy residue of a solution but also lots of big, beautiful, black crystals. Benign neglect had won the day. The structure of "ReI$_3$" also had trinuclear clusters, but linked in a different way so that one Re atom in each remained coordinately unsaturated.

Bruce Foxman took a keener interest in crystallography *per se* than most of my students and eventually made solid state chemistry his main research interest at Brandeis University where he has been since 1972. I have always loved Bruce for his wonderful dry, self-deprecating humor. With regard to the position of Brandeis faced with such giant neighbors as Harvard and MIT, Bruce used to say that he was at "Brand X" university.

An interesting byproduct of our work on the chemistry of Re^{III} was the highly unorthodox discovery of the compound ReCl$_4$. The preparation of "ReCl$_3$" was tedious, and when a small company (Shattuck Chemical Co.) offered it for sale at an appealing price, we bought some. It came as a dark, highly crystalline material, but we soon found that it showed none of the expected reactions for "ReCl$_3$" [206]. Because of its crystallinity we were able to solve the problem by determining the structure [218,

347]. It proved to be $ReCl_4$ and to have a structure then (and still) unknown for any other MX_4 compound.

Out of deference to an earlier claim to an amorphous form of $ReCl_4$, we called ours β-$ReCl_4$, but I now believe that the previous claim was erroneous. Subsequent claims of other forms are without structural support. I don't know of any other inorganic compound that was discovered in this peculiar way. Contacts with the Shattuck Co. did not directly reveal how their synthesis of $ReCl_4$ had occurred, but eventually it was shown that a comproportionation of $ReCl_3$ and $ReCl_5$ had occurred and there are now several other methods known to make β-$ReCl_4$.

It was the discovery of the Re_3Cl_9 structure that triggered my active involvement in the study of compounds with direct metal-metal bonds. It caused me to look more thoroughly into the existing literature to see if there might be other examples, and of course, I found a few, namely, those with octahedral groups of metal atoms. Among these were the so-called dihalides of molybdenum, which had been shown to consist of the units depicted in Fig. 4-3a linked together by other chloride ions. In this case, the Mo–Mo bonds could be straightforwardly described as single bonds; just as in the case of Re_3Cl_9, the assignment of double bonds is also straightforward based on the number of metal d electrons available and the number of nearest neighbors. In the molybdenum case, the number of electrons is 4 and the number of adjacent Mo atoms is 4, so a bond order of 1 results. In the rhenium case, the number of electrons is again 4 but the number of adjacent Re atoms is only 2, so bond orders of 2 result.

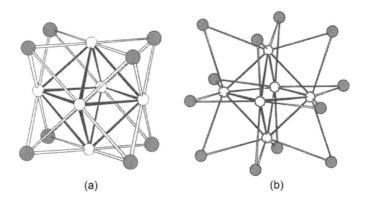

(a)   (b)

Figure 4-3. Central units in (a) the Mo_6Cl_8 structure and (b) the $[Nb_6Cl_{12}]^{n+}$ structure.

However, there are also $Nb_6Cl_{12}^{2+}$ compounds (Fig. 4-3b) where things do not work out so neatly. Here the six Nb atoms are left with only 16 electrons, *i.e.*, only two-thirds of the 24 that would be needed to make twelve Nb–Nb single bonds. What this said to me was that in order to understand the entire class of what I proposed [142] to call *metal atom cluster* compounds, it would be necessary to develop a molecular orbital method of treating their electronic structures.

While today's young inorganic chemists may be surprised to hear it, in 1964 it was not feasible to do better than the crudest sort of molecular orbital calculation on even the simplest metal atom cluster species. And yet anyone doing experimental work on these compounds craved some understanding of their electronic structures. Working with a graduate student, Terry E. Haas, I found a method of doing this that depended heavily on symmetry properties and used as its only quantitative tool the rough relationship between orbital overlap integrals and bond energies. This method was developed and applied, first to the M_6X_8, M_6X_{12}, and M_3X_{12} systems [126] and soon thereafter to the M_3X_{13} systems [142]. Despite its simplicity, this method captured the essence of the problem and its results remained useful for many years until more sophisticated methods became feasible [636]. In particular, we were able to explain why the $Nb_6Cl_{12}^{2+}$ cluster is stable and diamagnetic.

As will be explained in the next chapter, where I shall return to the development of metal-metal bond chemistry, the work on the Re_3Cl_9 cluster soon led to work on the $Re_2Cl_8^{2-}$ ion with its quadruple bond. However, I did little further work on metal atom clusters until about a decade later at Texas A&M University. At this point, I shall continue and complete the story of my involvement with X-ray crystallography.

With the discovery of the astonishing $[Re_3Cl_{12}]^{3-}$ structure, I was totally sold on the value of having a major crystallographic capability as part of my research program. While several more structures were done employing film methods, and I acquired a second Weissenberg camera and another Buerger precession camera, I looked to the future regarding technical capability and acquired a GE quarter-circle diffractometer (a so-called single crystal orienter), with which the intensities of reflections could be directly measured in a radiation counter. The first structure [129] done on this machine (by R. C. Elder) was that of trioxo(diethylenetriamine)molybdenum(VI), shown in Fig. 4-4. Because of the higher accuracy of the intensity measurements, this structure was refined to a final R value based on F of 0.060, which in 1964 was remarkably low

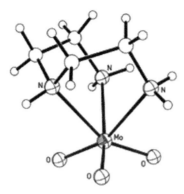

Figure 4-4. A perspective view of one molecule of MoO$_3$(dien) [129].

for such a structure. However, although the accuracy was excellent, the labor was still enormous. A human operator had to set the crystal and counter to the right positions for each reflection and then measure the intensity as well as the background counts on both sides of the reflection, and punch these data onto an IBM card. This procedure took at least five minutes for each of, in this case, about 600 reflections. This amounted to 15-20 hours of monotony. It was not until several years later, ca. 1970, that a fully automated, four-circle diffractometer came on the market, and I acquired one as soon as I could raise the money.

During all this time, machine computations steadily improved, but were still extremely slow and limited in scope. For any structure with more than about ten atoms, it was impossible to do full-matrix least squares refinement, and we had to separate the full matrix into small blocks along the diagonal (so-called block diagonal refinement). The only practical method for solving the structures was to use Patterson functions to first locate the metal atom and then, in a series of difference electron density maps (two-dimensional ones, mostly), gradually develop the rest of the structure. Finally, there was the problem of making drawings, by hand in India ink, of the structure for publication. It took a good student at least two months to complete one project involving a relatively small (*i.e.*, about 20 atoms) structure. The only thing that kept us going was that the rewards, in terms of marvelous new structures not knowable in any other way, were tremendous.

Before leaving this topic to return to the mainstream of the story, I should mention a few of the coworkers who bore the brunt of this early (pre-1965) crystallographic work. The graduate students, Roger H. Soderberg, Richard C. Elder, Joel T. Mague, John G. Bergman, Stephen J. Lippard, and the postdoctorals, J. Aaron Bertrand and John S. Wood, were among the very first. Another participant in these early days was Wayne Dollase, a student of Martin Buerger, who worked on a part-time basis. Professor David P. Shoemaker and his wife Clara were also very generous with advice and wise counsel. Incidently, the Patterson function that we depended on so heavily was actually devised in an office just a few yards beyond the area of my own labs by A. L. Patterson when he spent some time with my physics colleague Bert Warren. I remember how impressed I was when Bert showed me the very desk where he said the Patterson function was devised.

An important event that occurred in 1964 was the addition of Alan Davison to our MIT faculty. With all the older people retired and gone and with Coryell, Irvine, and a young Seaborg protégé, Glenn Gordon, being totally concerned with nuclear chemistry and radiochemistry, Dietmar and I persuaded Art Cope that inorganic chemistry deserved another faculty slot. Alan Davison was a Welshman who had finished his Ph.D. with Wilkinson at Imperial College in 1962 and gone directly to Harvard as an Instructor. We had gotten to know him and Dick Holm, who was there then as an Assistant Professor, gave him a very strong recommendation. So once again by the magic of direct, word-of-mouth contact, MIT was able to make what turned

Davison, Seyferth, and I in the early 1960s. Thin neckties were "in."

out to be another excellent appointment. Alan was such a wonderful colleague that about ten years later Arthur Martell and I tried very hard to attract him to A&M. This was before he did his now well-known work on heart imaging agents based on technetium.

A very important change in the chemistry department took place in 1964. When Jerry Wiesner returned from Washington (where he had been Science Advisor to President Kennedy) in 1964 he was made Dean of Science. I believe this was intended to be only a temporary assignment until he could take a higher administrative position (which he did, becoming Provost a few years later). Nonetheless, Jerry took an activist role and began rattling some cages. He was not pleased with the chemistry department and decided that Art Cope, who had been head for 19 years, should be replaced. After consulting with several senior members of the department, who generally approved his plan, he went to Cope's office and proposed that Cope should announce that he was going to give up the headship as soon as a replacement could be found on the grounds that (to use the standard cliché) he wanted to devote more time to his research. He would then be made an Institute Professor, the highest rank, and rarely bestowed, that MIT had to offer.

Cope's answer to this, as Jerry reported to several of us later, was "If you want me out of this job you'll have to fire me. And I don't want your (bleeping) Institute Professorship." So, that's what happened. Although it was, indeed, time for a change, this scenario was a tragedy and everyone was saddened by it. Only two years later, Art was taken ill while at an ACS dinner in Washington and died, at age 56, en route to a hospital. Art was in so many ways an admirable man; it was a shame he made the one important mistake of not knowing when it was time to go.

The Sporting Life: Horses and Hounds

As a small child I joined the neighborhood boys in playing baseball and touch football, but never took these sports seriously. I do not have any better than average hand-to-eye coordination and thus games in which it is necessary to hit a moving ball have never given me the satisfaction that comes with proficiency. In Argentina I even had a brief go at polo but very quickly gave it up. The relative speed of ball and mallet is probably the greatest for any sport and thus my worst bête noire. The horse I could handle, but not the ball.

I did become a swimmer, under my mother's tutelage, at the age of five or six. I continued swimming, as well as a little gymnastics and running at the Philadelphia YMCA for a number of years and I wasn't bad. I remember working very hard to get a Red Cross senior life guard certification when I was about fifteen. Serious competitive swimming never interested me, however, because it would have taken time I preferred to spend otherwise. In high school, for example, I found the debating team far more interesting.

My father, so I am told, was a golf fiend. In winter, if there was a light coating of snow, he and some of those similarly afflicted, would paint their balls with red nail polish and play. I have often wondered what they did about putting. Perhaps they just charged themselves two putts per hole, which might deny them an occasional birdie but doubtless saved them numerous three-putt bogies. My father belonged to the Tully-Secane Country Club, where he once won a silver cup. I remember one day after the war I took a notion to go look at the place but found it had become covered with Levittown style tract housing. *Sic transit gloria mundi.*

However, his passion was passed on to me to a degree. Beginning with the set of clubs I inherited from him, clubs which would look unbelievably quaint today, I began hacking around at public courses, even as a ten- or twelve-year-old. I also took up caddying and found that to be a good source of supplemental income until I finished college. I played quite a bit through graduate school days and stopped only when I got married and transferred my sporting enthusiasm to horses and foxhunting, of which more shortly. My last efforts to play golf were during my first few years at Texas A&M, when I joined the local country club and played once a week with Barry Shapiro. Barry and I were beneficiaries of the local, nearly universal, devotion to churchgoing by playing on Sunday mornings when we virtually had the course to ourselves from 8 to 11 am. There are times in this life when being an outsider is not all bad. After a few years, however, I lost interest, resigned from the club, and I haven't played since.

My only serious sporting interests have been horseback riding and foxhunting. As a very young child I had loved horses and had taken a few lessons, but it was only when I met Dee that things became serious. She and her former college roommate Nancy Dusenberry, who later became my first very own (*i.e.*, not shared) secretary at MIT induced me to drive them out to Dover, Ma., on Saturdays where they rode, and soon I was riding too. I soon decided that, apart from the delightful company, this was a

Perfect example of more guts than skill over a fence. No, I didn't fall off. I rarely did.

very enjoyable activity and I began to take it seriously. Soon I was taking jumping lessons and I was hooked. I gave Dee a horse for a wedding present and we made up our minds to move, as soon as we could afford it, out to the Dover-Sherborn area, southwest of Boston, where keeping and riding horses was convenient. We eventually chose Sherborn, right across the Charles River from Dover, for our first house, a place with room for horses and adjacent to miles of riding trails.

When I reached a certain level of proficiency at jumping, and had bought a reliable jumper of my own (Clarence), the couple who kept the barn where I took lessons suggested I might care to have a go with the local hunt, the Norfolk Hunt, founded in 1897. I shall never forget that first day. Apart from being bewildered and dazzled by the company, especially the older members, many riding gorgeous thoroughbreds and wearing full hunt regalia, not to mention the lively pack of hounds, there was the excitement. The excitement was felt keenly by my horse who had hunted before. As soon as the hounds found the scent and streaked off in full cry, so did my "good old Clarence." He very nearly unseated me at first, and throughout the day he, not I, was in charge. But I was thrilled and resolved to master the challenge. I had ridden in a party of non-jumpers, consisting of other novices and some of the oldest members, with a person who knew the country well as a leader. I was glad of it on that occasion, but I resolved to move into the regular field (the "field" being the group of riders who followed the hounds and the staff under the leadership of a "Field Master") as soon as I could.

Riding in the regular field was another challenge because the Norfolk country had lots of jumps. Few were over three feet, but they came thick and fast at times, because of the many stone walls that farmers had been erecting since the seventeenth century. I shall never forget the time at the end of a long Thanksgiving Day hunt the next year when I was near the end of the field with only one post and rail fence to clear to join the rest who had finished. Clarence, for some reason, refused this fence. I turned him round, tried again, and again could not get him over. I made one more unsuccessful attempt, and then heard the Master's commanding voice boom out, "Thank you, Mr. Cotton. Please use the gate." It was a great day otherwise, but I knew I had a long way to go.

All of the above takes me up to about the end of 1963. I continued hunting, got to be a much better rider and also began to know something about hounds. In 1965, as will be described, the family and I spent the months of January through August in Buenos Aires. This provided me an opportunity to make a major improvement in my horsemanship. Through the very kind offices of Baron Alexis Wrangel, of whom I shall say more later, we found an excellent place to keep the horses we bought and also obtained some fine instruction from a retired Argentine army cavalry sergeant. He also improved our Spanish because he spoke no word of English. "Más fuerza con las piernas" (more impulsion with the legs) and other admonitions rang in our ears.

In the Fall of 1965, Dee started hunting and we both had a grand time. In the Spring of 1967, a new Master was appointed and I was asked to be a Whipper-in, which entailed spending considerable time helping with the training and breeding of the hounds, as well as keeping control of them when we hunted. It was at this time that I bought my big Irish-bred hunter, Jogglebury, on whom I am mounted in the frontispiece. Now I began to appreciate what really went into handling a pack of hounds. I also began to participate fully in the social life associated with the world of foxhunting. One of the great institutions, in addition to balls and other purely social events, were the hunt "breakfasts." After a brisk morning of hunting, some member would have all the other members to breakfast, at around noon usually. The most memorable of these were put on by one couple who would have, *inter alia*, barrels of fresh oysters, with a man to open them and a barman to make black velvets, a heavenly cold drink consisting of equal parts of champagne and Guinness stout. It didn't get any better than that.

At the close of the 1969 season, the Master, who was subject to severe bouts of disequilibrium, had to be replaced and I was asked to be the joint Master together with a wonderful lady, Kerry Glass, who had, along with her husband John, become a great friend of ours. Kerry appointed Dee as a Whipper-in and all of us began the awesome task of renewing the pack, much of which had been stolen, and landowner relations that had been seriously damaged in the previous two years. My responsibilities were to deal with landowners, work with Kerry to plan the runs, and act as Field Master, so as to keep the field out of trouble while taking them over as many fences as possible. Through superhuman efforts by Kerry to reconstitute the pack, we managed to show good sport in 1970 and thereafter.

The Norfolk Hunt Club, which included many people who didn't ride to hounds, and did not necessarily include some who did, provided a remarkable social milieu. In some ways it was snobbish, but in others very egalitarian. Some of our members (like me) weren't rich and weren't in the investment business and some weren't white, or weren't Anglo-Saxon, or weren't Christian. But they were members of the club because they were fun to be with. Another plus for Dee (and myself, too) is that they never wanted to gossip about chemists.

The Saltonstall family were great supporters of the hunt. Mrs. Lewis, the former senator's sister, as well as Richard, her brother, had estates that formed the heart of our country. One of the most delightful members of the club was Mrs. Hilda P. Heffinger, who had gone to dancing school with the senator and invariably referred to him as "the honorable Lev." The senator's daughter, who had inherited his estate, allowed us to use it and supported all sorts of local equestrian activities, though she never rode to hounds herself. Elliot Richardson's family was very supportive and his niece, then in her early teens, was a regular junior rider. In fact, she and I won a trophy as a hunt team, because she was an elegant rider and I managed not to louse things up. Incidentally, Dee and I won the same trophy together another year, following a similar game plan.

Unfortunately for my future as MFH, in the Spring of 1971, I became increasingly fed up with the leftist political atmosphere at MIT, and a Welch Chair at Texas A&M looked too good to refuse. Thus, in February 1972, I said goodbye to one of the most exhilarating periods in my non-professional life and made the trek, as will shortly be described, to start a new life in Texas. I have occasionally attended the annual meeting and ball of the Masters of Foxhounds of America in New York, but I have never ridden to hounds again. After you've been a Master, nothing less will do. I do have the peculiar satisfaction of knowing (or at least believing) that I am the only person in history to be a member of both the NAS and the MFHA (Masters of Fox Hounds of America). It's not a combination one expects to find.

A Visit to Argentina

In September of 1963, I received a letter from Professor Rodolfo Busch, head of the Department of Physical, Inorganic and Analytical Chemistry at the National University of Buenos Aires (NUBA) with the proposal that I should spend anywhere from three to twelve months there as a visiting professor, with the support of the Ford Foundation. The Foundation had, over a period of some years, made a major investment in this department, which entailed funding for instrumentation and stipends to students to obtain Ph.D. degrees, or do postdoctoral work, abroad. About half of them had come to the United States and the rest had chosen the United Kingdom, France, or Germany. As a result, there was now a cadre (ca. ten) of young faculty members, well trained and eager to develop a chemistry department that was world-class in respect to both teaching and research. As the final stage of their support, the Ford Foundation had asked these young people to choose a visiting professor, whom the Foundation would support for an extended visit. I had been their choice.

I should explain that at NUBA there were, in fact, two chemistry departments, one for organic chemistry and one for physical, inorganic, and

Pablo Arcurain (our teacher), a horse we bought, Dee and I in Buenos Aires, 1965.

analytical chemistry. It was the latter that the Ford Foundation had supported and that had chosen to invite me.

It so happened that I received this letter of invitation on the day that Alfred (Alfie) Maddock, a friend of mine and a former classmate of Geoff Wilkinson's at Imperial College, was visiting at MIT as the guest of Charles Coryell and Jack Irvine. The Irvines had arranged a dinner at their house for Alfie and some guests, including Dee and myself. Alfie (who was at Cambridge University) had for some years been a frequent visitor and guest professor all over Latin America, from Mexico to Chile and Argentina, so I was able to get his expert opinion about the wisdom of accepting this offer. I myself had no knowledge at all of the situation at NUBA. Alfie recommended it enthusiastically.

Dee and I were initially in a quandary as to how long to spend and finally opted for nine months, January–September, so that we would not miss the Fall foxhunting season. MIT was very cooperative about giving me leave (which didn't cost them anything) and our girls, Jane and Jennifer, were then only 1½ and 3½, so school was not a problem. The Ford Foundation's offer was extremely generous: round-trip first-class air fare for the family, a house and a car at our disposal, and two trips back to the United States for me so I could do my consulting and talk with my research group.

By the end of 1963, we started making detailed plans with regard to renting our house, arranging for the care of our horses, and learning Spanish. Needless to say, the year 1964 passed in a whirl. It had begun in great joy with the birth of our second daughter, Jane, on January 27. Attending to Jennifer and Jane kept Dee very busy, and I was also beginning to concentrate on how to revise the first edition of *Advanced Inorganic Chemistry*, which I planned to continue working

One of my Argentine memorabilia — a pato ball. This ball has six handles arranged so that no matter how it lands on the ground three handles stick up to enable a horseman to ride by, lean over, and pick it up. The arrangement fascinated me because it has T_h symmetry.

With my daughter Jennifer at a livestock show in Buenos Aires.

on in Buenos Aires. Dee and I, our two daughters, two siamese cats, and lots of baggage arrived in Buenos Aires at 7:00 am on December 30, 1964.

Between July and December of 1964, I submitted 14 manuscripts, including the three leading reports on the preparation, structure, and bonding of the $Re_2X_8^{2-}$ ions and the first structural and bonding characterization of the $Tc_2Cl_8^{3-}$ ion. Thus, it was a very important and exciting time in research and I made careful plans for how it would be carried out in my absence. It did, in fact, go on quite well, and during the period I was away, 13 papers were written and submitted. Considering that in those days fax and e-mail were unknown, I can hardly imagine, in retrospect, how this was possible. On top of this, I completed my contributions to the second edition of *Advanced Inorganic Chemistry*, which was published in 1966.

It should not be thought that all this work prevented me from thoroughly enjoying life in Buenos Aires. At the university itself, I found the people who had invited me to be intellectually outstanding and very stimulating, and my lectures were well received. I progressed slowly in Spanish and toward the end I was able to give a few lectures in Spanish. We lived in a very nice semi-suburb of Buenos Aires where there were no English-speaking neighbors or shop owners, so improving our Spanish was not only a pleasant game but very practical. We also had two maids, one who lived in and helped look after the children and another who came in several days a week to do the cleaning, as well as a gardener who came every week. None of these people spoke a word of English, so here were further opportunities to learn and practice. We also sent our elder daughter, Jennifer, to a local nursery school where she too learned the language. Regrettably, because of lack of opportunity, Dee and Jennifer have lost the skill and I can barely manage, although I have little problem in reading.

Dealing with Spanish (or Castellano, as it is called in Argentina) in Latin America is a multifaceted problem. The language varies considerably from country to country. In Buenos Aires (as well as in neighboring Montevideo), Spanish is spoken with a heavy porteño accent, and there are also many usages found nowhere else. In Peru and Colombia, however, they speak fairly pure Castillian, even including, at times, the lisp. Most troubling sometimes is the way vocabulary varies. This led to a rather amusing linguistic incident toward the end of our stay when we visited Santiago, Chile. Incidentally, on this trip, which involved a nighttime flight over the Andes, we saw an unforgettable sight. It was a perfectly clear night and we saw the moon shining on the snow-covered peak of Aconcagua, the highest mountain in the western hemisphere.

But to get back to the linguistic incident, I delivered my lecture in Santiago in Spanish, and in the course of it, I had frequent occasion to present and describe absorption spectra. I had learned in Buenos Aires that one used the Spanish equivalents of peak and band, *i.e.*, pico and banda, just as in English. I therefore did this in my lecture in Santiago, but it became evident that members of my audience were variously uncomfortable or amused by something I was saying. Immediately following I asked one of my hosts what the problem was. It seems that while banda was a fine word to use, pico, as he put it bluntly, is a slang term for the male sex organ, and is not used in Chile to describe spectral features. How was I to know?

During the Fall of 1964, we had seen a letter to the editor of *The Chronicle of the Horse*, the magazine that covers, *inter alia*, show riding, other competitive activities such as dressage, cross country racing, and especially fox hunting in the United States and Canada. This letter was from a man in Argentina, Alexis Wrangel, who, we found out, was the cultural attaché at the American embassy in Buenos Aires. From his letter, it was clear that he was very involved in equestrian activities there so we sought him out soon after our arrival. He turned out to be the key to much of our enjoyment of our stay.

Alex Wrangel was Baron Wrangel, son of the last commander of the White Russian Army, which was, of course, extinguished by the Bolsheviks. As a young boy, he and his family had escaped *via* China, come to the United States, and he had become an American citizen. He had also spent time in Jordan and several European countries and was an expert on Arabian horses. He was wired in to the equestrian scene in

Buenos Aires and soon got us wired in as well. We became associate members of the Club Hípico Argentina; met all the leading show riders and trainers; purchased saddles, bridles, boots, and horses; and had the horses stabled close to the Club. All this was accomplished in a matter of weeks through the kindness of Alexis. Thus, we had an absolutely wonderful equestrian vacation in Buenos Aires. As mentioned earlier, we had excellent instruction, which improved our riding greatly, and had the pleasure of finding our horses well taken care of and all tacked up and ready to go whenever we came to ride. It was idyllic.

Alexis and his wife Katya became warm friends, who showed us the best restaurants and also introduced us to all sorts of interesting people, including other Russian expatriates. He also gave me the formula for vodka as it was made at the Romanov court (and was still made by him). Alexis was absolutely devoted to equestrian activities and in addition to riding his Arabian stallion, he kept a couple of race horses. Dee and I have never before or since been involved in racing but we enjoyed going to the track whenever Alexis had a horse running, and also joined him several frosty early mornings to watch his horses work out.

Over and above Alexis's many other wonderful qualities was his linguistic ability. I have known many bi- and even polylingual people, but none to match him. On one occasion when I called in on him at the embassy, he sat at his desk and talked to me in English, to his secretary in Spanish, and to his uncle who called on the phone in French. Of course he and Katya spoke Russian (or French) together, and he had some command of both Italian and Arabic. Just when I had thought I knew all about him, we were riding together at Club Hípico one morning when a young rider in uniform approached us and Alexis had a lively conversation with him in German. The young man was the military attaché at the German embassy. I once asked Alexis how he did it and he said, modestly, that it was effortless. As he put it, "If you put me down with the gibbons I would soon be speaking gibbonese."

I think Dee and I both feel that our nine months in Argentina (about which I have only touched on the highlights, omitting such things as a trip to Machu Picchu) was the happiest time in our adult lives. But there is a sad sequel. The people in the university and the Ford Foundation were so pleased with the way my visit worked out that they proposed a plan whereby I would spend six months a year in Buenos Aires for five years. With some fear that he would say no, I asked Jerry Wiesner, then the

MIT Provost, if he would allow this. His reply couldn't have been more generous. He said, "It's up to you; if you think this is a good idea, do it." So plans were made, such that Dee and the girls would live there nine months a year, so that the girls could go to school there, and I would come for six months. We would all be in Massachusetts during the Fall so that we could enjoy the hunting.

This idyllic plan collapsed one fine day the following year when a junta of military officers decided that the civilian government of President Arturo Illia (which was democratically elected and running the country very successfully) was unsatisfactory. They carried out a coup d'etat and installed some general named Juan Onganía as President. The generals and admirals not only knew nothing about how to manage the government in general or the economy in particular — years of economic decline began immediately — but they were fascistic in general and antisemitic in particular. In a fairly short time, practically all of the friends I had made in the university had either quit or been fired and neither I nor the Ford Foundation had any further intention to proceed with our plan.

Although many of the university people who left found other livelihoods in Argentina (generally in private business), many found positions in other countries, mainly Chile and Venezuela. In the Spring of 1967, the Ford Foundation asked me to go down and, as it were, survey the wreckage of what had been their splendid effort in chemistry in Buenos Aires. Thus, I spent about ten days in Buenos Aires and other places such as Bariloche and Bahía Blanca, and then several days each in Santiago and Caracas, finding and talking to the participants in this tragic diaspora. Few of those who left Argentina remained away in the long term. They returned and made other lives for themselves outside the University. Saddest of all, however, was that one of those I had known, Antonio Mesetich, a young physicist, later became one of the "desaparecidos" and, so far as I know, no one ever knew how or why or where.

The only bright spot on my survey trip occurred one morning at the end of April when I came down to breakfast to find a telegram from Bob Alberty, then Dean of Science at MIT, with congratulations on being elected to the National Academy of Sciences.

A Pleasant Sojourn in New York City

Although the city of New York has never been one of my favorite places and I avoid it, especially in recent decades, I have had a few marvelous times there. The only thing that has recently attracted me was visiting my old high school pal Bill Holt, who had a wonderful brownstone a few doors from Bill Cosby's and still owns the magnificent restaurant, The Firebird. Bill Holt was another member of the championship Bartram High debating team and a protegé of Elisabeth Gentieu.

One of my marvelous times in New York was in 1969, when I spent two weeks at Columbia University as the Falk-Plaut Lecturer. Apart from my interaction with a lot of people I liked very much, including Nick Turro, Ron Breslow, and Steve Lippard, on a daily basis, at Columbia I tasted some of the delights of Manhattan. I particularly remember going to that remarkable establishment called Claremont Riding Academy and hiring a horse for a ride in Central Park. I believe this is still possible today. It was a special experience to ride out the front door there and make one's way across the morning traffic of Central Park West into the park. Of the hundreds, perhaps thousands, of hours I have spent with my legs over a horse, I still have a special remembrance of that extraordinary one.

My stay at Columbia was during the early part of December, and New York was grandly gussied up for Christmas. One evening, Steve Lippard, Charles Cantor, and I went downtown for dinner. We went to a then trendy Spanish restaurant called El Faro (the lighthouse) and had a very good meal. However, there, in the midst of Christmas cheer, we looked at the prospects for dessert and, to paraphrase Her Majesty, "we were not amused." Although it was pouring cats and dogs, we decided to run over to Lüchows's (a New York treasure, now gone) for a real German dessert.

Lüchows's was then offering a German Christmas in absolutely grand style, with a classic umpah band going full tilt. We got in, got a table and had a terrific, rich Bavarian style dessert. I might recall that after Dee and I were married we went to Lüchows's every time we were in New York until it closed, sometime in the 1980s, I believe. One day we went there with Peter and Marilyn Verdier and we all decided to start with clams on the half shell and martinis. Both the clams and the martinis were very good, so we ordered another round, and then another. After we had all eaten about two dozen clams apiece, we confounded our waiter by announcing that we couldn't eat any more and wanted the check.

One other memorable incident in December 1969 was a trip late one afternoon from downtown Manhattan to Morningside Heights when I took the wrong train. I realized this only after the routes of the A-train and the B-train had diverged, so I got off on the east side of the park, walked uneventfully through the park, climbed up to Columbia, and dropped in to say hello to Ron Breslow. When I told him what I had just done, he said, in genuine astonishment, "you didn't?" He then informed me that it was not uncommon for white people to be robbed and even murdered there, even in broad daylight. Well, as the saying goes, "It's better to be born lucky than bright."

Fraternizing with one of the natives in Australia, 1966.

In July and August 1966, I did my first circumnavigation of the globe, broken by lengthy stops in New Zealand and Australia. Dee and I recall the initial part of the trip, which took us first to Fiji, as "the long day of three dinners." From Boston to San Francisco we had dinner, and again from San Francisco to Honolulu, and yet again from Honolulu to Suva, Fiji. We spent two very pleasant days in Fiji partially overcoming jet lag and then flew south to Auckland, New Zealand. Then, over ten days we made our way south as far as the glacier on the south island with stops in Wellington to see Neil Curtis, who had been a postdoc in 1963–64 and in Dunedin, the southernmost city in New Zealand. I think it is true that the variation of climate and topography in New Zealand is greater than in any other country in the world, and the top to bottom journey is very interesting. They also drive on the left and I nearly got us killed one day by forgetting that.

We flew over to Melbourne where Ray Martin was our host for a few days. Next stop was Sydney and then Dee went home *via* Mexico, where she discovered that yankees could *NOT* enter Mexico without a visa if they are coming from anywhere else besides the United States. She was unceremoniously put on the next plane out to

Los Angeles and never got to see Mexico City. Meantime I made an extensive tour of Australia including Tasmania, Brisbane, Adelaide, and finally, Perth. From Perth I flew to Europe and later home, thus completing my circumnavigation. While in Sydney, I was presented with the Frank Dwyer medal, which gave me special pleasure because I had known, enjoyed, and admired Frank. In many ways, especially in his speech, he was an Aussie's Aussie.

Calm Before the Storm

One very pleasant interlude just before life at MIT began to give me a sense of despair about the future took place at the Georgia Institute of Technology in Atlanta. I was the Seydal-Woolley Visiting Professor for the Winter quarter in 1968. Dee and I and the girls, who were then only six and four, drove down right after Christmas in two separate vehicles. Dee and Jennifer traveled in one and Jane and I were in our pickup truck pulling a two-horse trailer, since we had arranged to enjoy hunting with the Shakerag Hounds. The members of the Shakerag Club, whose extensive country was located north of Atlanta with the Chattahoochee River running through it, were a wonderfully hospitable and socially delightful group. The same could be said of the

Jennifer and Jane, at ages of about six and four.

Chemistry Department at Georgia Tech. It was with great reluctance that we made our way home around the middle of March.

This was just in time for me to leave on a lecture tour in Ireland, which I began in the west, and included a day I'll never forget hunting with the Galway Blazers, one of the storied hunts of Ireland. The Irish-American from whom I had bought my wonderful hunter Jogglebury had made the arrangements for me in Galway, and the people there saw to it that I was well mounted on a horse that was used "for all the Americans," who rode over jumps with a "forward seat" rather than in the common "feet on the dashboard" style favored by the Irish. This little horse, "The Beatle" was as faithful a jumper as I have ever been on. All I had to do was release him as we approached the jump — ditch, bank, or stone wall — and he sprang over it. The Galway country was glorious on that early Spring day, and their hounds and staff showed sport of the highest calibre. I have always meant to go back, but never have.

Goodbye to MIT

When Texas A&M and the Welch Foundation formally offered me the Robert A. Welch Chair at A&M, my life in Massachusetts had become very enjoyable and satisfying in some respects but very unsatisfactory in others. My social life and family life had never been better. Dee and I had two beautiful and delightful girls who gave us great pleasure, and our involvement in the Norfolk Hunt as already mentioned and other equestrian activities were giving us enormous satisfaction. In 1968 we had completed a beautiful contemporary house on twelve wooded acres, complete with a handsome five-stall barn for our horses. Sherborn and its twin town, Dover, on the other side of the Charles River, were still relatively uncrowded and almost rural. We felt very much in our element. It was also true, however, that there were worrisome signs of impending rapid development. The extension of the Mass Turnpike from Route 128 on into Boston boded ill for the retention of rural character in Wellesley, Framingham, and even the next layer of towns that included Sherborn and Dover. One could already see them becoming bedroom communities for people working in Boston. However, this alone would not — or at least not yet — have made me seriously consider leaving. What did?

Although it seems long ago and far away now, there was enormous civil unrest everywhere and especially in universities in the years 1968–71. Leftist students and a small army of leftist non-students (simply roving anarchists who pretended to be students) were using the excuse of opposition to the Vietnam war to disrupt, demoralize, and to a significant extent, destroy the fabric of rational, civilized academic life. One band of these filthy (literally) nihilists came to MIT, took over the student activities building, and then with the very active assistance of some members of the faculty, broke into the office of the president and a large number of adjacent offices that were part of the upper administration. They stayed for a week or so and no effort was made to remove them. Instead, they were pandered to. When they left, the offices were not only in shreds and tatters, but the occupiers had urinated and defecated everywhere to show their disdain for civilized society.

The president during this period, one Howard Johnson, moved out of the official president's house on the corner of Ames Street and Memorial Drive because he was afraid of "student" violence. In late 1971, this same feckless president canceled an appointment with me, in which he was supposed to have tried to persuade me not to leave MIT, because some of these vile "students" insisted (so he said) on seeing him. Perhaps they were among those who had pillaged his office, or perhaps they were some of those who a little earlier had come into a faculty meeting, looking like utterly degenerate slime balls, and sat hunched over on the floor in the front of the large lecture room where the meeting was being held, while Howard Johnson praised them as, and I quote, "fine young men." Does all this sound unbelievable? Of course, it does — *BUT IT HAPPENED!*

It is my opinion that Howard Johnson in that period was the most spectacular example of moral, physical, and every other kind of cowardice and faint-hearted abdication of adult responsibility that it has ever been my misfortune to encounter.

There were two final episodes in MIT's halfhearted attempt to prevent my leaving. Johnson never again offered to talk to me. He was much too busy running away from moral challenges and pandering to his "fine young men." One thing that was done was to offer me an endowed Chair. However, it quickly became apparent that, to paraphrase what I put in my letter of declination, all this would mean to me was that every time I sent out a letter, my secretary would have to type a certain number of additional letters (28, I believe) to declare that I was the "whozy-what's-it Professor." I wasn't to get a penny for my research program, nor a penny of salary increase, nor one iota of reduction in my teaching load. The Institute planned to keep all the income from the endowment and present me with nothing but the empty honor of being known as the "whozy-what's-it Professor."

Finally, in December of 1971, about six weeks before the effective date of my announced resignation, Jerry Wiesner, who was then Provost, found time in his busy schedule to take me to lunch for a tête-à-tête in which he assured me I was acting rashly and making a great mistake. I told him how disgusted I was with the MIT administration (in which he, as Provost, was number 2) and he explained to me how difficult, even impossible, it would have been for them to have behaved in any other way. I found that sob story amusing, in a pitiable sort of way. It was the final confirmation that I would do well to escape from an atmosphere of moral cowardice that I found intolerable.

I believe that to some extent, maybe to a considerable extent, the powers-that-were at MIT did not ever really think I'd go. The hubris of places like MIT, then and now, is astonishing. No doubt that when I did really go the administrators and many of my colleagues at MIT thought that I was making a foolish mistake which I would soon regret. They couldn't have been more wrong. Never, at any time, have I regretted the move. As the title of Edith Piaf's theme song has it, Je ne regrette rien.

While I did not leave MIT with much good will toward, or from, the institution itself, this should not be allowed to obscure that fact that my years there were not at all unhappy ones. One of the joys of being at MIT was the pleasure I took in most of my colleagues. Not only were the younger ones, John Waugh, Carl Garland, Fred Green (whom I had already met as a fellow grad student), and Herb House, very congenial, but the senior ones were very kind and approachable. My supervisor in teaching the freshman chemistry program, Clark C. Stevenson, was unpretentious and fun, as was his wife Louise. As I have already indicated, Jack Irvine and Charles Coryell were always helpful and encouraging. Among the senior physical chemists were David P. Shoemaker, Walter Stockmayer, and Isadore (Izzy) Amdur. Bag lunches either in Izzy's office or over at the drugstore across from 77 Massachusetts Avenue (the imposing main entrance to MIT) were always fun. Waugh, Garland, and a little bit later, Gordon Hammes and Jim Kinsey, also belonged to this lunch group. James A. Beattie, who was near retirement also joined the lunch group at times. The very distinguished thermodynamicist George Scatchard was quiet and scholarly. On one occasion, he was also very kind and patient in explaining to me that the determination of single ion hydration energies was not simply impossible but unnecessary, within the framework of classical thermodynamics, which was all he cared about.

Two of the physical chemists whom I remember with particular pleasure were John Waugh and Walter Stockmayer. John, in addition to being an excellent researcher (he won the Wolf Prize, along with Herb Gutowsky and Harden McConnell, in 1984) and an enjoyable colleague in the more conventional ways, had a Til Eulenspiegel complex — an irrepressible itch to see if he couldn't get someone's goat. He was often successful. Sometime in the 1960s, stimulated by Art Cope's intense devotion to the American Chemical Society, an organization John professed to disdain (and which he had never joined), he founded the Unamerican Chemical Society (UCS). This "organization" would admit as a member any chemist who did not belong to the ACS; it also had a higher class of members called Fellows, but eligibility for this honor was reserved only for those who had never been members of the ACS. John even listed his own membership in the UCS in his biosketch for some years, although I note that he did not go so far as to record it in his Who's Who entry. All of this nonsense was, of course, in aid of only one thing: getting Art Cope's goat. And it did.

One who defended himself handily against John Waugh's efforts to be irreverent was Willard Stout, the editor of the *Journal of Chemical Physics*. John had examined the list of symbols that might be employed and noted that spades, hearts, diamonds, and clubs were included. He therefore submitted a manuscript in which he used them in his equations, but Stout pointed out to him that symbols in the list were prioritized so that any particular one could be used only if all those above it on the list had already been used. The card suits were at (or close to) the bottom of the list and thus not acceptable in the circumstances. Undaunted by this rebuff, John then submitted a paper entitled (or at least including) the term "proton" enhanced nuclear induction spectroscopy. The idea, I believe, was first to slip this term into the journal and then send a follow-up paper in which, to save space, and very much as was being done with many other cumbersome NMR terms (like nuclear magnetic resonance itself), use the handy, and highly pronounceable, acronym formed by the first letters. Stout headed this one as well off at the pass.

Walter Stockmayer, whom his friends called Stocky, was a leading theoretician, who also did some experimental work. Apart from his research (which earned him a National Medal of Science in 1987), Stocky was a polymath. He not only knew more limericks than anyone else I've met, but he knew some in German and French as well as English.[2] He had been a Rhodes Scholar and played the piano very well. What really distinguished him, however, was that he wore all his brilliance and erudition very lightly and had a wonderful gift for humor and whimsy.

One of the most delightful manifestations of Stocky's whimsy was his invention of Waldemar Silberszyc. As I understand it, this fellow first appeared as the author of a paper in a mock journal, the *Journal of Comical Physics*, back in the early 1940s. He might well have been forgotten ever after except that in 1963 he published a brief note (*Polymer Letters* **1963**, *1*, 577) correcting an error in one of Stocky's papers. Stocky later confirmed that Silberszyc was correct (*J. Polymer Science* **1966**, *4*, 437). Silberszyc, whose entire career was spent at the Northeast Poultry Analysis Institute, Norwich, VT, resurfaced again in 1973 when he pointed out a serious error in an article in *Chemical & Engineering News* (May 21, p 35) concerning the quantity of dog feces produced in the United States. At some point, I was the recipient of a letter from Silberszyc, addressed to a Herr Professor Baumwolle. For some years, until his death in 2004, my written communication with Stocky was indirect, his colleague (whom I always referred to as his "more celebrated colleague") Silberszyc being the intermediary. At a time when I had some Polish postdocs in my lab I was able to address Dr. Silberszyc in his native language and received a very courteous reply, in elaborately formal, almost medieval Polish.

Stocky was never ambitious in political or administrative ways and in fact deplored the need to devote any time to such things. Thus, having also no regard for the shibboleth of being at a world famous place, in 1961 he betook himself to Dartmouth College in rural New Hampshire where he sought a life with less stress and general arglebargle. It was a significant loss to the cultural level of life at MIT.

After the disappearance of the drugstore in a fire and then a few years later Izzy Amdur's death, I frequently lunched at the "chemistry table" at the faculty club. This was not a table officially assigned to chemistry, but had established itself by long custom, as had the one next to it for the economists, where such luminaries of the dismal science as Paul Samuelson and Robert Solow, and a then young and skinny Lester Thurow gathered. Most of those at the chemistry table were organic chemists but there were others such as Dietmar Seyferth who came pretty regularly. Over the

---

[2]I forbear to give examples since the best limericks in French and German have one thing in common with the best limericks in English: They are not suitable for "polite" company.

years regular lunching with physical or organic chemists broadened my awareness of the scope of chemistry.

Among the senior organic chemists I particularly remember are John Sheehan and George Büchi. In the very beginning when I needed secretarial help, I used Charlene Vanelli, who worked for the Coryell-Irvine group, but after my elevation to Assistant Professor, I was given whatever time I could scrounge with a lady who was secretary to both Sheehan and Büchi. I well remember going down to her office one morning in 1957 with several letters and waiting while John Sheehan was putting the final touches, by telephone, on something he had prepared for a *New York Times* reporter who was writing a story on his total synthesis of penicillin. John fed him that remarkable simile that showed up in the front-page story next day, namely, that completing the final steps without rupturing the β-lactam ring was comparable in difficulty to "balancing an anvil on a house of cards." I think, with all due respect, that John overstated his case, but I was, and still am, mightily impressed by his skill as a communicator. As a result of his work, John got a patent on a small but indispensable synthetic step that was enabling for the enormously important later work on semisynthetic penicillins, and he became very rich. MIT also derived an enormous income from the patent and, after John's death (1992), honored him appropriately by creating a John C. Sheehan Chair, first held by Joanne Stubbe and now by Sylvia T. Ceyer.

Some people criticized John for subsequently acting like a nouveau riche in his frank enjoyment of his prosperity. I took quite the opposite view and thought it wonderful that he was so thoroughly enjoying himself. As a scientist, John was perhaps more persistent than brilliant, but he was a charming and fun-loving guy and I enjoyed his company. On several occasions, he treated me to lunch at Joseph's (a gourmet restaurant in Back Bay) and was a generous host.

George Büchi was also a splendid colleague. He was a man of great natural dignity, with a quiet but very warm personality. On several occasions, he gave me very good counsel. I suppose one reason we got along quite well, despite having little overlap in our chemical interests, is that we were both very fond of what the French call *la bonne chère* (gourmet dining, roughly translated) and we were both conservative socially and politically.

George was very much a man of principle. To illustrate, I telephoned him sometime after leaving MIT to point out to him that one of our mutual friends had not yet been elected to the American Academy of Arts and Sciences and to suggest that since he was in closer proximity to that somewhat regional organization, he might be better able than I to promote the case. He said that, unfortunately, in spite of being very sympathetic, he could do nothing because he was not a member himself. When I said I was astonished that he had never been elected, he calmly informed me that he had been elected. However, during the Vietnam war he had resigned because "I found out that the Academy was spending some of my dues" to support political action against the war. George disapproved of this and resigned in protest. I dare say that most people would have felt that the prestige of membership in the Academy would have been worth the price of tolerating such a mild irritation, but not George.

George was Swiss by birth and probably never in his life in the United States had the pleasure of having his name pronounced correctly by an American. My sorrow at hearing of his death, at age 77, in 1998 was tempered by knowing that it had occurred during a walk in the Swiss Alps which he loved very much.

On the whole, collegiality within the MIT Chemistry Department enriched my life there and is dominant in my memory of those seventeen years.

CHAPTER 5

MIT 1961–71: MOSTLY ABOUT SCIENCE

No amount of skillful
invention can replace
the element of imagination.

—*Edward Hopper*

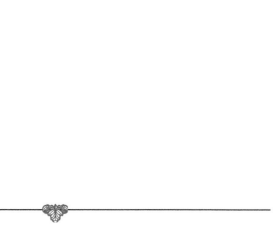

I HAVE ALREADY discussed some but by no means all of the scientific work done at MIT following my being promoted to the rank of Professor in 1961. Now I want to give a more complete account of the main research activities that I pursued in that same decade. I shall try to avoid unnecessary technicalities so that perhaps even chemists who have only a slight acquaintance with my areas of interest may find this chapter readable, although for a non-chemist it may be over the top.

As I look back, I am astonished to see how much work was done and how varied it was. Hard as I work today, I can't get nearly as much done nor keep so many different balls in the air as I did then. There is nothing to match the energy of youth and the creativity of a young mind. To any reader who still has those attributes, I say, give it all you've got now and enjoy it while it lasts.

The Discovery of the Quadruple Bond

Undoubtedly, the most far reaching consequence of finding the $Re_3Cl_{12}{}^{3-}$ structure, which I discussed in the previous chapter, was that it led to the discovery of quadruple bonds, and hence all the metal-metal multiple bond chemistry that followed. This is how it came about. In our further pursuit of rhenium(III) chemistry, as previously outlined, we were annoyed by the effort required to prepare Re^{III} halides from the metal, and I proposed to Dr. Neil Curtis (then a postdoc who later became a chemistry professor at Victoria University, Wellington, New Zealand) that maybe we could find a way, by aqueous chemistry, to go directly from $ReO_4{}^-$, which was the most convenient source of rhenium, to $Re_3Cl_{12}{}^{3-}$. We were also interested in seeing if a mixed metal cluster such as $Re_2OsCl_{12}{}^{3-}$ could be made by a reduction of $ReO_4{}^-$ in the presence of $OsCl_3$ in an aqueous medium.

One problem we faced is that $ReCl_6{}^{2-}$ is a thermodynamic sink for higher valent rhenium in aqueous HCl under reducing conditions. To avoid this, numerous reductants were tested, one of which was hypophosphorous acid. Curtis found that with this as a reducing agent, in concentrated hydrochloric acid, he obtained a beautiful, intensely blue solution. This was a surprise because both $Re_3Cl_{12}{}^{3-}$ and

$ReCl_6^{2-}$ are red or red-brown. From the blue solution he was able to obtain "$CsReCl_4$," but, of course, since it was blue, we knew it wasn't $Cs_3Re_3Cl_{12}$. We did not, however, proceed immediately to determine the crystal structure of this particular substance because of other simultaneous developments in the lab.

By a coincidence, of a sort that seems to occur rather often in research, there was another visiting research associate in the group at the same time, namely, Dr. Brian Johnson (later professor of chemistry, Cambridge University), who had been checking a rather puzzling report from the USSR to the effect that reduction of ReO_4^- in hydrochloric acid by hydrogen gas under pressure gave $[ReCl_6]^{3-}$. This was obviously of concern with respect to Curtis' work, since it suggested that aqueous reduction of ReO_4^- might give (previously unknown) mononuclear Re^{III} chloro complexes. However, Johnson showed that the claim of $[ReCl_6]^{3-}$ salts was erroneous and that the compounds were in fact the familiar $M^I_2ReCl_6$ salts [133].

The same Russian authors had also claimed that there was a dark blue/green product, to which the formula K_2ReCl_4 was assigned. Johnson found that there was indeed such a product and, in view of its apparent similarity to Curtis' new blue "$CsReCl_4$," we immediately wondered if the Soviet chemists had simply got their formula wrong and that they really had "$KReCl_4$." It did not take long to show that this was precisely the case and that the substance had the empirical formula $KReCl_4 \cdot H_2O$. Since it formed better-looking crystals than did the cesium compound (which we showed later to be $Cs_2Re_2Cl_8 \cdot H_2O$) and these nice crystals had a small triclinic unit cell, we considered $KReCl_4 \cdot H_2O$ to be the preferred subject for an X-ray crystallographic study. Charles B. Harris (now Professor of Chemistry, University of California, Berkeley), who was just beginning his doctoral research and had never previously done a crystal structure, began a study of these crystals.

I also examined the Soviet chemical literature more carefully to see if there were any further reports of interest on the chemistry of lower-valent rhenium. I found that between 1952 and 1958 V. G. Tronev and coworkers had published several papers that described an assortment of low-oxidation-state rhenium halide complexes in which the metal oxidation state was proposed to be +2.

Much of the impetus for the Russian investigations was a search for analogies between the chemistry of rhenium and platinum, an approach which, apparently, prejudiced them in favor of the Re^{II} oxidation state. Most of the compounds described in these reports have never been substantiated, but the "K_2ReCl_4" already mentioned and blue-green "$(NH_4)_2ReCl_4$," which was also obtained by the action of hydrogen under pressure upon solutions of NH_4ReO_4 in concentrated hydrochloric acid at 300°C, were further discussed in a later paper and appeared genuine. Other than rhenium and chlorine microanalyses and an occasional oxidation state determination, no further characterizations were described that supported these formulations or any of the numerous other formulas reported. With respect to the oxidation state determinations, which the Russians reported as supporting the oxidation state +2 for rhenium, two points are pertinent. First, the method used has often been found unreliable. Second, however, when this procedure was repeated on one of our own compounds it gave an oxidation number of +2.9 ± 0.2. Presumably, the Russian chemists, having once obtained results that they thought required an oxidation number of +2, adjusted the number of cations, usually by putting in an otherwise unsubstantiated H^+, to get a formula such as M^IHReCl_4, consistent with the analytical data they had.

Except for one more brief report in 1962, describing the formation of crystalline $(C_5H_5NH)HReCl_4$ by hydrogen reduction of a hydrochloric acid solution of the rhenium(IV) complex $ReCl_4(C_5H_5N)_2$ in an autoclave, the early work of Tronev *et al.* was not further examined, either by the authors themselves or by anyone else, until 1963.

While Harris was carrying out his crystallographic study of "$KReCl_4·H_2O$," proceeding rather slowly and deliberately (since he was learning X-ray crystallography as he went), an issue of the *Zhurnal Strukturnoi Khimii* was received at MIT, and I noted that it contained an article dealing with "$(pyH)HReCl_4$." Since I did not read Russian, it was not immediately clear what was being reported, though tables and figures within the article implied that it was reporting a structure determination. Fortunately, Steve Lippard, then a graduate student in the group, had completed a crash course in Russian the previous summer at Harvard University and he was able to enlighten us. The paper (in Lippard's translation, which is substantively identical to the commercial translation that appeared nearly a year later) reported that:

> Eight chlorine atoms constitute a square prism with two rhenium atoms lying within the prism, whereby each rhenium atom is surrounded by four neighboring chlorine atoms situated at the apices of a strongly flattened tetragonal pyramid. The apices of two

such pyramids approach each other generating the prism. In such a structure, each rhenium atom has for its neighbors one rhenium atom at a distance of 2.22 Å and four chlorine atoms at a distance of 2.43 Å. As a result, the dimeric ion $[Re_2Cl_8]^{4-}$ is generated.

With regard to the structural situation of the H atoms present in the formula, the following statements were made:

> The isolated $[Re_2Cl_8]^{4-}$ grouping is bonded ionically to the pyridinium ion $[C_5H_5NH]^+$ carrying a positive charge, and its free hydrogen ion …. The detached free hydrogen ion is identified as situated on a fourfold position, which is electrostatically stable. It may be surmised that four hydrogen atoms are situated between Cl_{II} atoms on centers of symmetry… and serve to bond the $[Re_2Cl_8]^{4-}$ groups even further to each other.

In addition to the completely unprecedented Re-to-Re distance of only 2.22 Å and a puzzling discussion of the structural role of the "hydrogen ions" (also sometimes called "hydrogen atoms"), there had been, according to the experimental section of the paper, severe difficulty with crystal twinning. For all these reasons, we felt that this work was probably in error, possibly because the twinning problem had not, in fact, been successfully handled. Harris therefore hurried to complete his work on "$KReCl_4 \cdot H_2O$."

To our considerable surprise, he found an anion essentially identical in structure to that described by the Soviet workers. There were some small quantitative discrepancies, which we later resolved by re-refinement of the Soviet structure. The structure of the $[Re_2Cl_8]^{2-}$ ion, exactly as found and reported by C. B. Harris in $K_2Re_2Cl_8 \cdot 2H_2O$ [158], is shown in Fig. 5-1.

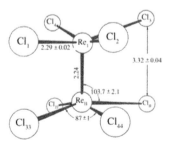

Figure 5-1. The structure of the $[Re_2Cl_8]^{2-}$ ion as originally reported.

While Harris was completing his structural work, several others in the laboratory had also prepared a number of new compounds containing the $[Re_2Cl_8]^{2-}$ ion, using both our method (H_3PO_2 reduction) and the Tronev method (high-pressure H_2 reduction). It was clear that the same products were obtained by both methods (although the former was far more practical), and that the charge on the Re_2Cl_8 unit was indeed 2- and not 4-, as believed by the Soviet workers.

In late 1963, Wilkinson reported that reactions of rhenium(III) chloride with neat carboxylic acids give diamagnetic, orange products with molecular formulas $[ReCl(O_2CR)_2]_2$. It was proposed, by analogy with the known structure of Cu^{II} acetate, that the compounds were molecular, with bridging carboxylato groups and axial chloride ligands, but there was no direct evidence presented for this. Perhaps the most interesting statement in the report of this work is this: "(I)t is, however, not necessary to invoke metal-metal bonding to account for the diamagnetism" of these compounds.

Recognition of the Quadruple Bond. In only one of the Soviet papers discussed so far was anything said about the bonding in the putative Re^{II} compounds, namely, in the structure paper, where the following statement was made:

> It should be noted that the Re–Re distance ≈2.22 Å is less than the Re–Re distance in the metal …. The decrease in the Re–Re distance in this structure, compared with the Re–Re distance in the metal, indicates that the valence electrons of rhenium also take part in the formation of the Re–Re bond. This may explain the diamagnetism of this compound.

The timidity and vagueness of this remark, in the face of the astonishingly short Re–Re distance has never ceased to amaze me. It's as if someone caught in a blazing inferno were to remark that "it's a little warm in here, don't you think?"

It appears that it was not until 1977 that Russian chemists accepted the concept of the quadruple bond. They appear to have remained quite ambivalent for years about the related problems of composition (*i.e.*, the oxidation state of the rhenium and the question of whether hydrogen is present) and bonding, and the discussions in their papers are sometimes confusing. In 1970, there was a paper entitled "Crystal Structure of $Re_2Cl_4[CH_3COO(H)]_2 \cdot (H_2O)_2$ with a Dimeric Complex Ion," in which it was stated that "In the two (α and β) modifications of (pyH)HRe^{II}Br$_4$ the authors found triple ($1\sigma + 2\pi$)Re–Re bonds." The correct formulas and oxidation numbers for at least some of their compounds still appeared to elude them. In the formula used in the title, the appearance of "(H)" is certainly an arresting feature, but what it was meant to imply was left entirely to the reader's imagination, unless it was an attempt to evade what they referred to as "the question of whether acetic acid is found as a neutral molecule or as an acetate ion." The authors described that question as one which "remains unclear."

Thus it was that in 1964 the explanation for the remarkable structure of the [Re$_2$Cl$_8$]$^{2-}$ ion was first put forward in the first article I ever published in *Science* [143]. Prior to this the chemistry of the [Re$_2$Cl$_8$]$^{2-}$ ion had been extensively clarified. We had shown that the ion could be prepared much more conveniently from [ReO$_4$]$^-$ using an open beaker with H$_3$PO$_2$ as the reducing agent, that the analogous bromide could be made, that Re$_2$Cl$_8^{2-}$ reacted with carboxylic acids to give the Wilkinson compounds, and that this reaction is reversible:

$$[\text{Re}_2\text{Cl}_8]^{2-} \underset{\text{excess HCl}}{\overset{\text{excess RCO}_2\text{H}}{\rightleftharpoons}} \text{Re}_2(\text{O}_2\text{CR})_4\text{Cl}_2$$

The existence of a quadruple bond between the rhenium atoms was proposed and explained in the *Science* paper as follows:

> The fact that [Re$_2$Cl$_8$]$^{2-}$ has an eclipsed, rather than a staggered, structure (that is, not the structure to be expected on considering only the effects of repulsions between chlorine atoms) is satisfactorily explained when the Re–Re multiple bonding is examined in detail. To a first approximation, each rhenium atom uses a set of s, p_x, p_y, $d_{x^2-y^2}$ hybrid orbitals to form its four Re–Cl bonds. The remaining valence shell orbitals of each rhenium may then be used for metal-to-metal bonding as follows. (i) On each rhenium d_{z^2}–p_z hybrids overlap to form a very strong σ bond. (ii) The d_{xz}, d_{yz} pair on each rhenium atom can be used to form two fairly strong π-bonds. Neither the σ nor the π bonds impose any restriction on rotation about the Re–Re axis. These three bonding orbitals will be filled by six of the eight Re d electrons. (iii) There remains now, on each rhenium atom, a d_{xy} orbital containing one electron. In

the eclipsed configuration these overlap to a fair extent (about one third as much as one of the π overlaps) to give a δ bond, with the two electrons becoming paired. This bonding scheme is in accord with the measured diamagnetism of the $[Re_2Cl_8]^{2-}$ ion. If, however, the molecule were to have a staggered configuration, the δ bonding would be entirely lost (d_{xy}-d_{xy} overlap would be zero) .... Since the Cl–Cl repulsion energy tending to favor the staggered configuration can be estimated to be only a few kilocalories per mole, the δ-bond energy is decisive and stabilizes the eclipsed configuration. This would appear to be the first quadruple bond to be discovered.

In a full paper [159] that I published shortly thereafter, this proposal was further elaborated and supported with numerical estimates of d-orbital overlaps. It was proposed that Re–Re quadruple bonds also occur in the $Re_2(O_2CR)_4X_2$ molecules. Finally, the correlation of metal-metal distances with bond orders ranging from < 1 to 4 was explicitly discussed, and the concept of an entire gamut of M–M bond orders in an entire field of non-Wernerian compounds was introduced. A broad survey of the field was presented very soon after in a review article [189].

The quadruple-bond chemistry of rhenium was opened up quickly in several papers [157, 172, 190, 205, 217], and before the end of 1966, the first metal-metal triple bond had also been reported [187] in the dirhenium compound $Re_2Cl_5(CH_3SCH_2CH_2SCH_3)_2$, which is obtained from the reaction of the $[Re_2Cl_8]^{2-}$ ion with $CH_3SCH_2CH_2SCH_3$.

Today the concept of quadruple bonds has become commonplace, with hundreds of compounds known to contain them, and the physical and theoretical characterization of them is very comprehensive. However, prior to 1964, quadruple bonds were totally unknown, and the idea even seemed to alarm some organic chemists, who were slow to accept the fact that d-orbitals can do things that s- and p-orbitals cannot. The newness of the concept of a quadruple bond is well illustrated by Linus Pauling's comment (3rd Ed. of *The Nature of the Chemical Bond*, 1960, p 64) that no one had ever presented evidence "justifying the assignment to any molecule of a structure involving a quadruple bond between a pair of atoms." Actually, the notion of quadruple bonds had been broached as early as 1923, when Langmuir had proposed to G. N. Lewis that the structure for nitrogen and carbon monoxide might involve "a quadruple bond such that two atomic kernels lie together inside a single octet," but this possibility (not surprisingly) was quickly eliminated as a realistic description of the bonding in any homonuclear or heteronuclear diatomic molecule formed by nonmetals.

Extension of the concept of quadruple bonding to elements other than rhenium occurred in the following way. In late 1963, a paper had appeared (Eakins, Humphreys and Mellish, *J. Chem. Soc.* **1963**, 6012) in which a series of technetium compounds that, by chemical analysis, were assumed to contain a $Tc_2Cl_8^{3-}$ ion were reported. The authors noted that "the stoichiometry of the $[Tc_2Cl_8]^{3-}$ ion is unusual and it seems to have no analogs." I may have been the only person in the world (except, perhaps, a few people in the USSR) who did not find the stoichiometry entirely surprising because, apart from the charge, it reminded me of $Re_2Cl_8^{2-}$. I immediately, in early 1964, asked a graduate student, Kevin Bratton, to prepare a sample of $(NH_4)_3Tc_2Cl_8 \cdot 2H_2O$ and begin collecting X-ray data on a single crystal. As was typical in those days, the solution and refinement took about six months, so it was not until about September that we had an accurate structure and knew that, as I had suspected, the $Tc_2Cl_8^{3-}$ ion was an exact structural analog of the $Re_2Cl_8^{2-}$ (although it has one more electron), with a Tc–Tc distance of 2.13 ± 0.01 Å.

About the time that we were preparing to publish this result I learned from Mel Churchill, who had been a student of Ron Mason when Ron was at Imperial College, that Ron had solved the structure of $Mo(CH_3CO_2)_2$ and that there were two close Mo atoms and four bridging acetate groups, as in the known copper(II) acetate type of structure, but with a much shorter metal-metal distance. I immediately wrote to Ron (October 21, 1964) and said that if what I had heard was true "such a structure is exactly what I would have expected for the compound since it is isoelectronic with the rhenium(III) chloro acetate type compounds and we have been doing a good bit on the chemistry of these especially in relation to the chemistry of the $Re_2Cl_8^{2-}$ ion. At the same time we are looking into some chemical aspects of the dimolybdenum compound, particularly trying to prepare the $Mo_2Cl_8^{4-}$ ion which would be isoelectronic with the $Re_2Cl_8^{2-}$ ion. I would very much appreciate knowing the latest word on the molybdenum compound." I soon heard from Mason telling me that he had finished the structure of $Mo(CH_3CO_2)_2$ and giving me all the details.

Before proceeding further, it will be appropriate to stop and review some work that had gone on in Wilkinson's laboratory in the period 1958–63. The impetus for his work goes back, at least in part, to the first thing I had done when I started my

Ph.D. research in 1951, namely, to prepare anhydrous lanthanide acetates by refluxing $LnCl_3 \cdot xH_2O$ compounds in acetic acid/acetic anhydride mixtures. My work, as I noted earlier, had produced this result unintentionally, but Geoff had kept it in mind, with the idea that anhydrous carboxylates of transition metals might be prepared similarly and might be useful starting materials for synthesis of other compounds. He was also still interested in the synthesis of metal arene complexes, and these two lines of interest came together in two papers he published in 1959 and 1960. In the first one (Abel, Apar Singh and Wilkinson, *J. Chem. Soc.* **1959**, 3097) he reported reacting $Mo(CO)_6$ with benzoic acid in diglyme and obtaining a compound which analyzed as $Mo(C_6H_5CO_2)_2$. He assumed this to be an arene complex and assigned to it one of the structures shown in Fig. 5-2.

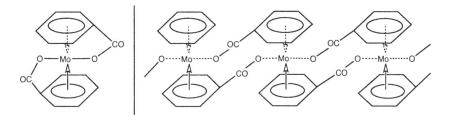

Figure 5-2. Structures initially proposed by Wilkinson for $Mo(C_6H_5CO_2)_2$.

Close on the heels of this, Eric Bannister (who had shortly before been a postdoctoral with me) and Wilkinson reported (*Chem. & Ind.* **1960**, *319*) that acetic, propionic and other (unspecified) aliphatic carboxylic acids also reacted with $Mo(CO)_6$ to give $Mo(RCO_2)_2$ products. As a result, Geoff had to withdraw his proposals for the benzoate structure and suggested a polymeric formula for all such compounds, shown below in Fig. 5-3.

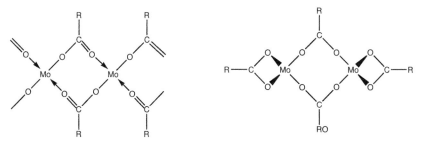

Figure 5-3. Later proposals for structures of all $Mo(RCO_2)_2$ compounds by Wilkinson.

It was not until November of 1963 that Stephenson, Bannister, and Wilkinson submitted another paper on $Mo(RCO_2)_2$ compounds. Here they proposed the binuclear structure shown on the right in Fig. 5-3. They gave three reasons for suggesting such a structure: (1) for two of their compounds, molecular weight measurements in solution corresponded to a binuclear structure; (2) the diamagnetism was "consistent with tetrahedral coordination"; (3) a personal communication from Lawton and Mason concerning the crystallography of the acetate was said to "substantiate the dimeric geometry of the molecule."

Of these three reasons, only the first makes sense. The second flies in the face of the well-known fact that no low-spin tetrahedral complex had ever been identified and it was generally believed to be impossible to make one. Had these compounds been low-spin tetrahedral complexes, that would have been front page news. Finally, if Lawton and Mason could have assured him the compound was binuclear on the basis of crystallography, they would have to have known that the Mo–Mo distance was only about 2.1 Å, which is certainly not implied by the proposed structure.

The remarkable number of completely incorrect structures proposed for the $Mo(RCO_2)_2$ molecules is exceeded, to my recollection, only by the number of incorrect structures suggested during the 1960s for the four-iron ferridoxins. In the $Mo(RCO_2)_2$ case, however, all the incorrect structures came from the same laboratory.

Let me return now to the situation in October-November of 1964. Once I had heard from Mason the details of what I shall now refer to as the $Mo_2(O_2CCH_3)_4$ structure, I wrote him on November 13 to say: "Thanks very much for your letter concerning $[Mo(O_2CCH_3)_2]_2$. Enclosed is a manuscript of a preliminary communication of something related. (This was the $Tc_2Cl_8{}^{3-}$ structure.) How about you putting in a simultaneous note on the $[Mo(O_2CCH_3)_2]_2$ structure? I was intending to send ours to J.A.C.S."

Mason promptly agreed to this and sent his manuscript to me. Thus, both of our notes bear a received date of November 30, 1964, and they appeared back to back in JACS, both starting on page 921 in early 1965. Our note on the $Tc_2Cl_8{}^{3-}$ ion was titled "A Multiple Bond Between Technetium Atoms in an Octachloroditechnetium Ion" and made explicit reference to the following report on the molybdenum molecule. I commented that "It appears that the formation of extremely short, presumably quadruple bonds between d^4 ions of the second- and third-row transition elements may be quite general." This judgement has been abundantly confirmed.

Over the next two years, we extended the chemistry of $Re_2Cl_8{}^{2-}$ and species derived from it in various ways, some of which I have already mentioned. One of the others that had considerable implications for the future was to show that Cl^- ligands could be replaced by phosphines, as in the molecule shown at right, which was the first one to be reported [218, 252], with full structural characterization. We also showed [217] that a thiocyanate, $[Re_2(NCS)_8]^{2-}$, could be made and that both this and $Re_2Cl_8{}^{2-}$ could be reversibly reduced to the 3– ions [219]. This gave credibility to the charge on the $Tc_2Cl_8{}^{3-}$ ion. Flavio Bonati, one of my many excellent Italian postdocs, showed that the action of Cl_2 or Br_2 on

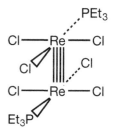

the $Re_2Cl_8{}^{2-}$ and $Re_2Br_8{}^{2-}$ ions led to the corresponding $Re_2X_9{}^-$ ions [222], which have since been shown to have face-sharing bioctahedral structures, as would have been expected. We also did the first X-ray structure of a $Re_2(O_2CR)_4X_2$ compound, namely, the one with $R = C_6H_5$ and $X = Cl$ [243]. Thus, by early 1968 the broad outlines of $Re_2Cl_8{}^{2-}$ chemistry were pretty well drawn.

During 1968, three significant discoveries were made concerning multiple metal-metal bonds. I shall describe these in increasing order of importance — at least as viewed retrospectively. The first was the accurate characterization of a molecule with a structure of the type shown to the left. While this was a foreseeable possibility, and there had been preliminary reports of such species, the first one to be definitively characterized was the compound in which $M = Re$, $X = I$, and $R = C_6H_5$ [268].

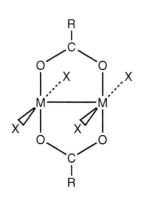

In early 1968, I looked back at a paper Tony Stephenson and Wilkinson had published in late 1966, in which they reported complexes of formula $[Ru_2(O_2CR)_4Cl]$ with $R = Me$, Et, and Pr^n. They had not only confirmed the dinuclear formulation by molecular weight determination but also found that these compounds were highly paramagnetic. They suggested a copper acetate type structure but proposed that "the metal-metal distance in these systems is large enough to prevent direct orbital overlap" between the metal atoms. In view of what we had learned about dirhenium and dimolybdenum systems, I was not inclined to agree with this, and in collaboration with two postdocs, Mike Bennett and Ken Caulton, the compound $Ru_2(O_2CC_3H_7)_4Cl$ was made again and its crystal structure determined [260].

We struck gold. There were, indeed, $Ru_2(O_2CC_3H_7)_4$ units with a copper acetate-like structure, and they were connected in crisscrossing zigzag chains *via* Cl bridges. However, the Ru–Ru distance was definitely *not* "large enough to prevent direct orbital overlap." On the contrary, it was short enough, 2.28 Å, to *demand* direct orbital overlap, that is, to leave no possible doubt that a Ru–Ru bond existed. We also pointed out that the presence of three unpaired electrons could be explained without giving up the idea of metal-metal bonding, although our explanation was flawed in detail and the correct one was not given until eleven years later by one of my former students, Joe Norman. The key point, however, was that we had shown that multiple metal-metal bonding occurs in group 8. This, in turn, gave me the conviction that the field might cover a great many elements, as we know today it does.

I have already mentioned that as early as 1964, that is, as soon as the structure of $Mo_2(O_2CCH_3)_4$ was clear, I had drawn on the analogy to the interconversion of $Re_2Cl_8^{2-}$ and $Re_2(O_2CR)_4Cl$ compounds to think that an $Mo_2Cl_8^{4-}$ ion might be obtained by the reaction

$$Mo_2(O_2CCH_3)_4 + 4HCl + 4Cl^- = Mo_2Cl_8^{4-} + 4CH_3COOH$$

In fact I had even made a few hasty attempts to do this, but without success. Nevertheless, I remained convinced that it could be done but that the right conditions had to be found. It was clear that some patience might be required to do this.

In 1967, I welcomed a new postdoc from Ljubljana, Jugoslavia, named Jurij Brenčič. He had, on his own initiative, obtained support for postdoctoral work in the Untied States from the Samuel Rubin Foundation. I believe this was a relatively small foundation set up by one Samuel Rubin who had made money as an entrepreneur in making innovative batteries. To this day, I know no more about it than that. Brenčič proposed spending his postdoctoral year (which we later got extended to two) in my lab. This turned out to be a great stroke of luck for me.

I proposed to Jurij that he try to find the right conditions for the conversion of molybdenum acetate to $Mo_2Cl_8^{4-}$ and within a few months he had done it. Either he already knew something about crystallography when he came or he picked it up

A picture of me with Jurij Brenčič taken in 1994 when I visited him in Ljubljana.

very quickly, but we soon had the structure of $K_4Mo_2Cl_8 \cdot 2H_2O$, a beautiful red, crystalline solid, and we were able to submit a manuscript by July of 1968 [261]. The existence of the rather stable $Mo_2Cl_8^{4-}$ ion, isoelectronic and isostructural with the $Re_2Cl_8^{2-}$ ion was thus established. The basis of the most voluminous quadruple bond chemistry now known for any one element was now in place.

Brenčič's work continued and he was able to isolate and characterize several more compounds containing the $Mo_2Cl_8^{4-}$ ion [277, 284, 285]. He found that to avoid getting Mo^{III} compounds instead of the desired product it required careful attention to the reaction conditions; acid concentration and temperature were critical. Under one particular set of improper conditions, he found that compounds having, apparently, the formulas $Rb_3Mo_2Cl_8$ or $Cs_3Mo_2Cl_8$ could be obtained, and from an X-ray study of the former, he got a very curious result. As the structure first developed, it seemed to be a face-sharing bioctahedral structure, but it could be refined only by assigning an occupancy factor of only 0.67 ± 0.01 to each of the bridging Cl atoms, as shown at right. The only explanation we could think of was that there were $Cl_3Mo(\mu$-$Cl)_2MoCl_3^{3-}$ ions disordered in such a way that the vacant bridge position was one-third of the time at each possible position, as required by the space group. I didn't really like that structure although the concept of such a disorder was perhaps not unreasonable. For the postulated structure of the anion, whether ordered or not, there should have been one unpaired electron, and a measurement that should have been more carefully done appeared to indicate this. This was not so, but I'll come back to that later.

Jurij Brenčič returned to Ljubljana in 1969 and has since become the professor of inorganic chemistry there. I have been to visit him once — since Slovenia became an independent state. I shall always remember him as one of the most capable, but modest coworkers I have had, and as the man whose patience was responsible for one of the critical breakthroughs that have shaped the course of my research.

Up to the end of 1969, essentially no relevant work on metal-metal bonds had been published elsewhere. I was in the pleasant position of having the field of quadruple bonds almost entirely to myself, a situation that would soon change drastically. I should mention, however, that two East Germans, Albrecht and Stock of Jena, published an extremely brief note in an obscure journal, *Zeits. für Chemie* (**1967**, *7*, 231), in which they reported the first example of what I would later call a *supershort* chromium-chromium quadruple bond, in the compound $Cr_2(C_3H_5)_4$. The structure they presented was extremely rough, showing almost nothing about the allyl ligands, and giving a very imprecise Cr–Cr distance of 1.98 ± 0.06 Å. Two years later, Japanese workers (*Bull. Chem. Soc. Japan*, **1969**, *42*, 545) reported a more complete structure of the same compound showing the arrangement of the allyl ligands and giving a Cr–Cr distance of 1.97 ± 0.01 Å. These workers made no reference to the East German work; in fact this is the only paper I have ever seen that contained no references at all, to anything. In 1971 we showed that an isostructural $Mo_2(C_3H_5)_4$ molecule with an Mo–Mo quadruple bond existed [314]. I will resume the fascinating story of supershort bonds in Chapter 7.

In 1969, a meeting of the International Union of Crystallography was held on the Stony Brook campus of the State University of New York. I was invited to speak and I discussed some of our work on trirhenium clusters, $Re_2Cl_8^{2-}$, and also the newly discovered $Mo_2Cl_8^{4-}$ ion. At the end of my lecture, someone in the audience whom I did not recognize came to the front of the room and announced that he had something to present to me. It turned out that this man was N. A. Porai-Koshits, a well-known Russian crystallographer, from the Kurnakov Institute in Moscow; what he had was

a newly issued USSR postage stamp. The stamp featured a ball and stick model of an $Re_2X_8^{2-}$ ion. I was very surprised and very impressed with this and thanked him warmly, but also managed to put my foot in my mouth. In my hasty examination of the stamp, I took an object that was represented in the background to be a punched card (an IBM card was the common term) and commented on this. With a pained expression he informed me that it was *not* an IBM card but his institute. I apologized as best I could, but felt rather foolish.

During my remaining time at MIT, no other major discovery occurred in the metal-metal bond field, but the groundwork was laid for several future developments. We determined the structures of $Cr_2(O_2CCH_3)_4(H_2O)_2$ and $Rh_2(O_2CCH_3)_4(H_2O)_2$, both of which had previously been done but very inaccurately. In the chromium case, in fact, little was certain except that the structure was qualitatively like that of $Cu_2(O_2CCH_3)_4(H_2O)_2$. In the rhodium case, there was a better structure, but still not a very precise one, so I thought it would be a good idea to get structures that had state-of-the-art accuracy for both of these compounds. We found [291, 313] that the chromium compound had a far shorter Cr–Cr distance, 2.362(1) Å, than had previously been estimated (~2.64 Å being the estimate), and this certainly gave a big boost to the idea that there was a quadruple bond here. The Rh–Rh bond was also found to be shorter than previously believed, namely, 2.386(1) rather than 2.45 Å. These results caused me to become far more interested in Cr–Cr and Rh–Rh bonds than I had been and led directly to extensive work on them a few years later at Texas A&M.

During the last few years that I was at MIT, I had become quite friendly with John Slater and had told him about our short Re–Re, Tc–Tc, and Mo–Mo bonds. I had lamented the fact that, so far as I knew, such systems were completely beyond the reach of the conventional computational procedures available for molecular electronic structures. One day in the Fall of 1971, John came up to my office and announced that he had a technique that might allow quantitative calculations on our compounds. This method was a direct outgrowth of the Muffin-Tin-Xα treatment that he had developed for calculating the band structures of metals. Recently, he and a young coworker, Keith Johnson, had adapted it for molecular problems and a program had been written to implement it.

This was exciting news, but since I was virtually in a transition state between MIT and TAMU, I was not prepared to take direct advantage of it. However, I had a graduate student, Joe Norman, who would finish his experimental work by the end of 1971. Joe had an interest in theory and an NSF predoctoral fellowship that continued until mid-1972. It was therefore arranged that instead of his coming with me to Texas he would stay at MIT and apprentice himself to Keith Johnson. This bore fruit in 1975 when Joe and his first graduate student at the University of Washington, where he had taken an assistant professorship, published calculations on $Mo_2(O_2CH)_4$ and $Mo_2Cl_8^{4-}$. These results quantitatively supported the $\sigma^2\pi^4\delta^2$ description of the quadruple bond that I had presented ten years earlier. Moreover, contour maps of the σ, π and δ orbitals corresponded beautifully to the shapes expected from the $d_{z^2}-d_{z^2}$, $d_{xz}-d_{xz}$, and $d_{xy}-d_{xy}$ overlaps from which the σ, π, and δ molecular orbitals arose. I considered these to be the most beautiful pictures of molecular orbitals I have ever seen, and even though I am only their godfather (along with John Slater) I am reproducing them here (Fig. 5-4). I will pick up the story of metal-metal multiple bonds again, after my move to Texas A&M.

Figure 5-4. The σ, π, and δ orbitals of the quadruple Mo–Mo bond in $Mo_2Cl_8^{4-}$ as calculated by Norman and Kolari (*J. Am. Chem. Soc.* **1975**, *97*, 33).

I might add here a personal word about John Slater, whom I admired very much. He was not an easy man to get to know because he did not seem to welcome close contact. I found him to be a basically friendly person, but too shy to let it show. When you passed him in the hall, he would produce what became known as the Slater millisecond smile, a curious smile-like twitch of the mouth. When there was something serious to discuss, or organize, however, he was a pleasure to deal with. Only once did I have a (transient) problem with him. One morning, shortly after I had arrived in my office, John marched in and immediately made it clear, in his relatively undemonstrative way, that he was feeling outraged. What had happened was that during the preceding night a hose on a condenser in one of my labs had popped off and a flood had resulted. This lab was right over one of the major work rooms next to John's office on the floor below and large stacks of computer output had been turned into incipient papier mâché. It was not easy to explain to this quintessentially theoretical-physics-oriented person that such accidents are not uncommon in a chemistry laboratory, and that when they occur in the middle of the night, there may well be serious damage to those below. He did eventually accept the fact that I was not to blame. I myself have never been in jeopardy of such damage since I have always been on the top floor, both at MIT and at Texas A&M.

When retirement age came for John, MIT maintained its place among the ranks of universities who, in those days, required their most distinguished scholars to abide by the mandatory retirement age, and thus John moved to the University of Florida. This uncivilized treatment was similar to that given to Lars Onsager by Yale. It was incidents like these that I had strongly in mind when, in the early nineties I was on a National Research Council committee to report to congress on the question of doing away with any forced retirement age for faculty.

John Slater was a prodigy, having been elected to the National Academy of Sciences at the age of 31 (younger, even, than Linus Pauling), and his contributions to the theory of chemical bonding and related matters were as important as those of Mulliken, Pauling, or anyone else. However, he lived to the age of 76 without receiving the Nobel Prize. Perhaps he suffered the curse of being interdisciplinary — a physicist in the eyes of the chemists and too chemical for the physicists.

Infrared Spectra of Metal Carbonyls

Not all of my work in the early 1960s stemmed from the advent of X-ray crystallography in my repertoire of tools. I have already described how, as a graduate student, I had developed an interest in the infrared spectra of molecules containing carbon monoxide as a ligand. This began with rationalizing the spectra of molecules such as $(C_5H_5)V(CO)_4$, $(C_5H_5)Mn(CO)_3$, and $(C_5H_5)Co(CO)_2$. In 1961, my interest was rekindled because of the rapid increase in the number of compounds being reported in which one or more of the CO ligands in homoleptic metal carbonyls, particularly the group 6 hexacarbonyls, were substituted by phosphines, arsines, dialkyl sulfides, amines, and other ligands with a lesser capacity than CO to back-accept π electron density from the metal atom.

There were two related questions raised by these compounds: How could their infrared spectra be correlated with their structures? How could the spectra be interpreted to derive information about the relative capacities of the substituents to engage in π back-bonding? These are all large molecules and there are $3n-6$ fundamental vibrations for an n-atomic molecule. To be able to answer the questions in a broadly useful, practical way, the general equations expressing the dependence of the vibrational frequencies on the molecular symmetry, atomic masses, and force constants would have to be drastically simplified. What was needed — if it could be found — was a ruthless approximation. What I did was find one.

The ruthless approximation was to treat the CO stretching modes separately from all other vibrational modes. In this way relatively simple secular equations for the CO stretching frequencies could be written using Wilson's F and G matrix method. In addition, I proposed several sensible but simple ideas about the electronic factors that would come into play, especially in the case of octahedral molecules, where the angles between all pairs of CO groups are either 90° or 180°, or close to these values. The three principal ideas were as follows:

1. All (CO stretching)–(CO stretching) interaction constants in the F matrix should have a positive sign. This was based on an essentially certain argument similar to the well-known explanation of why the interaction constant in CO_2 is positive.

2. The interaction constants between a *trans* pair of CO's should be about twice that between *cis* pairs.

3. All CO-stretching force constants in a substituted molecule should be lower than that in the unsubstituted parent hexacarbonyl molecule, with that for a CO group *trans* to a substituent being lower than that for one *cis*.

By employing some data from the literature but also new data obtained by a postdoctoral coworker, Charles S. Kraihanzel (who later spent his career at Lehigh University), it was found that the factoring out of the CO-stretching vibrations worked quite well and the three supplemental ideas were all validated. This led, in 1962, to the first publication [109], in which compounds with phosphine ligands were treated.

The following year it was shown that molecules with a variety of amine ligands could be equally well analyzed by the same method [116]. The collaboration of Charles Kraihanzel was very important in this work because nineteen compounds had to be made and their spectra recorded. In 1964 I showed [132] that data in the literature for a further eighty-six compounds, containing an enormous variety of ligands could be consistently analyzed and correlated by the methods already developed. Included in this study were five $XMn(CO)_5$ molecules as well as eighty-one derivatives of the group six $M(CO)_6$ molecules.

It was in this third paper that the real chemical impact of the methodology proposed and illustrated in the first two was manifested. For one thing it was shown that the methodology was applicable to compounds with any type of ligand normally encountered in the chemistry of substituted metal carbonyls. More important was the rank ordering of ligands in a semiquantitative way as to their capacity as π-acids. Two especially interesting conclusions of this kind were that PF_3 is an even better π-acid than CO (previously regarded as the ultimate) and that methylformamide and dimethylformamide are actually π-donors.

Over an approximately fifteen-year period, parts 1 and 3 of the papers published in 1962-64 became my two most cited publications with 625 and 380 citations up to the end of 1980, and I was asked to write a commentary on the former as a "Citation Classic" which was published in the May 31, 1982 issue of *Current Contents* (Physical, Chemical, and Earth Sciences).

Fluxional Organometallic Molecules

The earliest trace, so far as I know, of the concept of fluxionality in organometallic chemistry goes back to 1956, when Wilkinson and Piper published a paper (which ceased to exist in 1981, to be replaced by the journal *Polyhedron*). In this paper (*J. Inorg. Nucl. Chem.* **1956**, *3*, 104), they reported the NMR spectra of the compounds $(C_5H_5)_2Hg$ and $(C_5H_5)_2Fe(CO)_2$. For these they expected the structures to be the following:

Since X-ray crystallography was not an available option to verify these structures at that time, they resorted to ^{1}H NMR spectroscopy. They were greatly surprised to find that the mercury compound had a spectrum consisting of but one sharp line, while the iron compound had only two lines, one very sharp, the other not quite so sharp. They proposed an explanation of these entirely unexpected results. To retain their belief in the above structures they postulated that in each case a σ-C_5H_5 (in the notation of the day) ring or rings was rotating rapidly so that the point of attachment of the ring to the metal atom changed from one carbon atom to an adjacent one. Thus, over time, they would all appear equivalent in the ^{1}H NMR spectrum. So far as I am aware there is nothing in the literature to suggest that any attempt was made by anyone (until my work with Alan Davison in 1965) to investigate this remarkable suggestion.

In 1959, several laboratories reported the preparation of the $C_8H_8Fe(CO)_3$ molecule, where C_8H_8 is cyclooctatetraene, and that its ^{1}H NMR spectrum consisted of a single line. Several people, including myself [53], suggested 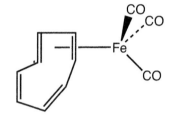 that the iron atom lay over the center of a flat C_8H_8 ring. However, it was soon found that in the crystal the structure was as shown at right. To reconcile this with the solution ^{1}H NMR spectrum, it could either be supposed that the structure changed on going into solution or, as Dickens and Lipscomb (*J. Am. Chem. Soc.* **1961**, *83*, 4862; *J. Chem. Phys.* **1962**, *37*, 2084) proposed, the solid state structure persists but there is a "dynamical effect amounting to permutation of the C atoms of the ring relative to the $Fe(CO)_3$ group." No reference was made to the earlier idea of Piper and Wilkinson.

Not long after this, Bill Doering (Doering and Roth, *Angew. Chem., Int. Ed. Engl.* **1963**, *2*, 115) invented his marvelous molecule, bullvalene, which he predicted would undergo an endless series of Cope rearrangements:

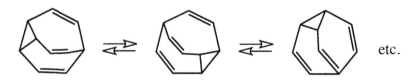

At a given instant, there will be four different types of C–H groups, in a ratio of 3:3:3:1, but as the rearrangements are repeated, all C–H groups will experience all four environments, over and over. It turned out that these rearrangements proceed fast enough at room temperature to cause all hydrogen atoms to appear equivalent in the ^{1}H NMR spectrum, but as the temperature is lowered this line broadens, eventually disappearing into the baseline to be replaced by a complex set of multiplets corresponding to the structure as drawn. This phenomenon wherein site exchange takes place rapidly on the NMR time scale was called fluxionality by Doering and Roth.

It is clear that the choice of the NMR time scale (ca. 10^{-3} sec) in defining a fluxional molecule is arbitrary, and in a very insightful letter to the editor of *Inorganic Chemistry* (**1965**, *4*, 769), Earl Muetterties stressed this point, noted that each sort of physical technique for structure determination has its own time scale, and cited some purely inorganic examples of what he called "stereochemically nonrigid" molecules. His points were important and well made but his term has not stuck and the word "fluxional" has been generally adopted.

It is interesting that neither purely organic molecules nor simple inorganic ones had immediate futures in the field of fluxionality. The scene soon shifted to organometallic and metal carbonyl molecules where rapid growth occurred. I think it is fair to say that the main impetus in the organometallic area was given by work done by Alan Davison and me at MIT in 1965.

As mentioned in the previous chapter, I had arranged already in the Summer of 1964 to spend January to September of 1965 in Buenos Aires. However, in the Fall of 1964, I was invited to give a plenary lecture at the annual Welch Foundation Conference in November 1965, the topic of which was to be organometallic chemistry. I wanted to have something new and exciting to talk about and my mind turned to the previous proposals by Wilkinson and Piper and by Dickens and Lipscomb. In neither case had there been any experimental confirmation. Fortunately, I chose the $(\eta^5\text{-}C_5H_5)\text{-}Fe(CO)_2(\eta^1\text{-}C_5H_5)$ molecule to look at first.

I scoped out the necessary steps. First, an X-ray crystal structure would be required to confirm the proposed structure, which I felt was highly probable but should not be taken for granted without proof. Second, it would be necessary to find out if the proposed rapid rotation of the $\eta^1\text{-}C_5H_5$ ring could be slowed down at low temperature so that the pattern of resonances for that ring would correspond to its instantaneous nature.

There were several problems in embarking on this program, the most obvious of which was that I was going to be absent during the period when it would have to be

Discussing the $(\eta^5\text{-}C_5H_5)Fe(CO)_2(\eta^1\text{-}C_5H_5)$ problem [192] with Steve Lippard, John Bergman, Alan Davison, Mike Bennett, and Jack Faller in 1965.

carried out. It was also a fact that running NMR spectra at low temperatures had rarely been done up to that time, and practical questions like the choice of suitable solvents needed to be addressed. I might also add that whoever was in charge of the NMR spectrometer at MIT (I don't remember who) was strongly opposed to risking the probe by cooling it to –100 °C.

I talked to Alan Davison about this, and it turned out that he too had such a project in the back of his mind, so we agreed to collaborate. I would assign some people in my group (Steve Lippard, Sheila Morehouse, and Mike Bennett) to determine the crystal structure and Alan and his student Jack Faller would have a go at the low temperature NMR spectrum. It all worked out, although it took a little more time than we had expected. However, by the November 1965 Welch Conference I was able to report [188] that the structure was confirmed and that "the simple NMR line due to the 'whizzing' ring broadens and eventually, at ~ –80°, there is a 2,2,1 pattern with chemical shifts which seem to be consistent with one σ-bonded ring."

As above and elsewhere in my Welch lecture, I used the term "whizzing." It was Roland (Rollie) Petit who coined the term "ring whizzers" for molecules of this type including a $C_7H_7Fe(CO)_3^+$ ion that he had reported (*J. Am. Chem. Soc.* **1964**, *86*, 3589) but so far as I know, he never used the term in writing. This colorful term was typical of Rollie who had a wonderful, irreverent sense of humor. His too early death was very sad for all of his friends. He, like Earl Muetterties, just would not give up cigarettes.

I cannot resist one further remark about ring whizzers. I tried very hard to convince my former student Terry Haas (Ph.D. 1963) who was by then on the faculty of Tufts University to take just a little time to make a contribution to the study of one of the further molecules I intended to look at so we could publish at least a note on "The Wonderful Whizzer of Haas" but he wouldn't do it, much to my chagrin.

It was not until March 1966, several months after the Welch conference, that we completed all of the work on $(\eta^5\text{-}C_5H_5)Fe(CO)_2(\eta^1\text{-}C_5H_5)$ and submitted a detailed report to JACS [192]. This paper took a very important step beyond simply showing that the molecule really is a whizzer. The NMR spectra that we reported are shown in Fig. 5-5. The low temperature (limiting) spectrum of the $\sigma\text{-}C_5H_5$ ring broadens and collapses completely (ca. –25 °C) before the single line begins to appear and then narrow. However, there is a feature of this process that is of special interest. As the two multiplet signals for the olefinic hydrogen atoms broaden, they do not broaden at the same rate. This simple fact, which might easily have been overlooked (but wasn't), has far-reaching significance.

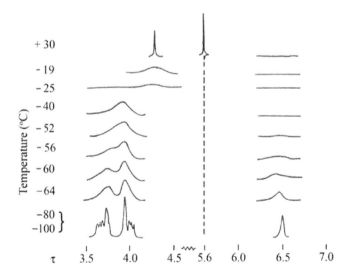

Figure 5-5. The proton NMR spectra (60 MHz) of $(\eta^5\text{-}C_5H_5)Fe(CO)_2(\eta^1\text{-}C_5H_5)$ in CS_2 at various temperatures. The line for the $\eta^5\text{-}C_5H_5$ protons is shown only on the 30 °C spectrum.

As I had noted in my Welch lecture, Wilkinson's suggestion that the Fe–C bond jumped from one position to the next (1,2-shifts) was only one of three possibilities for NMR signal averaging. 1,3-Shifts as well as random shifts would equally well accomplish the final result of making all hydrogen atoms appear equivalent at higher temperatures where the process, whichever one it was, proceeded rapidly.

The importance of the unequal rates of collapse of the two lines in the low-temperature limiting spectrum is that, in principle, it can tell us which of the three pathways is the correct one. The basis for this is the inverse relationship between line width and the time a nucleus remains in a given environment. Obviously, in a non-fluxional molecule, where each nucleus remains in its environment indefinitely, the lines are extremely narrow. If the nucleus can pass from one environment to another where its chemical shift is different, the line will be broader and become more so in proportion to how often it jumps.

For the case of 1,2-shifts, when a jump occurs, as shown below, each of the two ^{1}H nuclei initially in a b or b' position changes to a different kind of environment, that is, one goes from b to c while the other goes from b' to a. On the other hand, of the two ^{1}H nuclei initially in a c or c' position, one changes ($c' \rightarrow b'$) but the other one doesn't ($c \rightarrow c'$). Thus, of the two signals, the one arising from protons in the b environment should broaden twice as fast as the one arising from protons in the c environment. If, however, 1,3-shifts occur, exactly the opposite behavior of the spectrum will result.

1,2-shift   1,3-shift

For a random mixture of both 1,2- and 1,3-shifts, or if the metal were to go to the middle of the ring whence it could go to any new position with the same probability, all ^{1}H nuclei would change environments with equal frequency, and both the b-type and c-type NMR signals would broaden at the same rate. Obviously these last two possibilities are ruled out completely by the observations, and the only acceptable choice is between 1,2- and 1,3-shifts. To make this choice, the assignment of the two multiplets of relative intensity 2 had to be made. The preferred assignment at the time (which has since been further validated) requires the conclusion that 1,2-shifts are occurring.

The importance of this result in this specific case was not the major point. What mattered far more is that this was a paradigm for many other cases. This was the first time to my knowledge (and I have looked and enquired extensively) that line shape analysis had ever been used to answer the question of which among several possible rearrangement pathways is the correct one. Previously the broadening, collapse, and redevelopment of an NMR spectrum had been used only to determine the rate *vs* temperature of a process (whence $\Delta H^{\ddagger}$ and $\Delta S^{\ddagger}$ could be evaluated). I believe the first instance of rate determination was the tautomerization, shown below, of dimethylacetamide by Gutowsky and Holm (*J. Chem. Phys.* **1956**, *25*, 1228); but here, there is only a one-for-one exchange and no question of pathway arises.

$$H_3C \diagdown CH_3{}^a H_3C \diagdown CH_3{}^b$$
$$C{-}N \longrightarrow C{-}N$$
$$O CH_3{}^b O CH_3{}^a$$

Similarly, in 1960 Frederick Jensen and coworkers (*J. Am. Chem. Soc.* **1960**, *82*, 1256) had measured the chair-to-chair inversion rate for cyclohexane, another case where no information on a pathway was sought. In the case of bullvalene, the NMR spectrum was again used to establish rates *vs* temperature. The observations were consistent with the Cope rearrangement pathway, but then the whole concept of the bullvalene molecule had arisen on the *assumption* of such a pathway in the first place. No question about the pathway was asked or answered; rather a single assumption having no plausible alternative was supported.

We next turned to the problem of $(C_8H_8)Fe(CO)_3$, as did several others. For some details of that work, I will refer the interested reader to a review I published in *Accounts of Chemical Research* in 1968 [254]. There were three independent studies that all showed that the molecule was fluxional, as Dickens and Lipscomb had proposed. However, because even at the lowest temperature that could be reached (ca. –150°C) a true limiting spectrum was not obtained [194], the rearrangement pathway remained unknown. It was not until about a year later that we found a way around this difficulty [226].

We worked with the ruthenium analog, η^4-$C_8H_8Ru(CO)_3$, which behaves like its iron analog but with a higher activation energy. Thus, a limiting spectrum was reached at about –130 °C, and by examining all of the line shape changes that occurred on the way down, we were able conclusively to exclude 1,3-, 1,4-, and 1,5-shifts while showing that the observations were fully consistent with 1,2-shifts, as shown in Fig. 5-6.

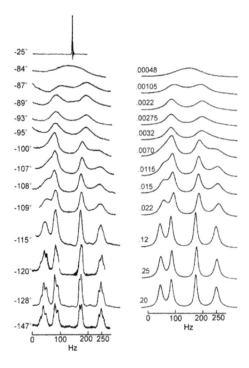

Figure 5-6. The pattern of collapse of the $C_8H_8Ru(CO)_3$ ^{1}H NMR spectrum as observed (left) and computed (right) for a 1,2-shift process.

It was the study of η^4-$C_8H_8Ru(CO)_3$ that set the pattern for most of our future work. In cases where more than two possible pathways had to be considered, the preferred procedure was to compute spectra as a function of exchange rate for each of the candidates and see which set matched the observed spectra. Moreover, by matching each observed spectrum at temperatures $T_1, T_2 \ldots T_n$ with a computed spectrum, at rates of $R_1, R_2 \ldots R_n$ (as shown in Fig. 5-6) we could obtain, from the dependence of rate on temperature, the activation energy of the process. The computation of a spectrum at a given rate was done by using a program (EXCH) that George Whitesides had written, based on theoretical treatments in the literature. It was normally necessary (or at least very helpful) to have the spectrum in the slow exchange limit as a starting point. One of the experimental problems in most of our work was to find solvents that could be used at very low temperatures (ca. –150 °C). It was found that mixtures of chlorofluorocarbons were generally useful, an example being a 1:1 v/v mixture of $CFHCl_2$ and CF_2Cl_2.

The paper we published on η^4-$C_8H_8Ru(CO)_3$ is memorable to me for a personal reason. I had been elected in April of 1967 to the National Academy of Sciences and this work was the first that I had the pleasure of communicating to the *Proceedings of the National Academy of Sciences* under my own aegis.

I was of course keen to extend our new technique to still other cases. I decided to look at what might be (and in the event was) a more complicated case [193]. I had been joined by an excellent Italian postdoc, Alfredo Musco, and he and Jack Faller set about preparing 1,3,5,7-tetramethylcyclooctatetraene (TMCOT) and then its complex $(\eta^6$-TMCOT)Mo(CO)$_3$, with the following structure:

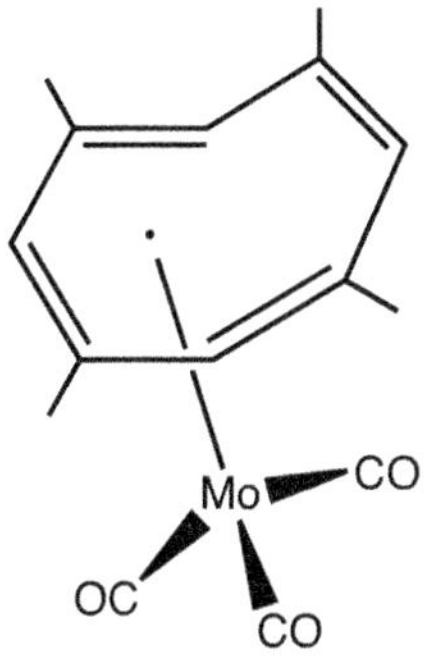

In this molecule, we could follow the broadening and coalescence of both ring proton and methyl proton signals. It was already known from work of Winstein and Kaesz in 1965 that η^6-$C_8H_8Mo(CO)_3$ had a one-line spectrum at room temperature.

The $(\eta^6$-TMCOT)Mo(CO)$_3$ molecule has fascinating behavior. A limiting low-temperature spectrum was reached easily (–10 °C) but as the temperature was raised, two separate stages of fluxional behavior were observed. In the first one, a reversible 1,2-shift occurred but only back and forth between the methyl groups bracketing the two adjacent points of attachment of the Mo atom to the ring. Evidently, a 1,2-shift across a methyl-substituted ring carbon atom was disfavored. In fact, it never does occur since after the 1,2 averaging is completed at about 70 °C the next process that sets in is something that had never been observed before: transannular jumping. We published a very detailed account of studies on the three $(\eta^6\text{-TMCOT})M(CO)_3$ molecules, M = Cr, Mo, and W, about a year later [237].

In view of this fascinating behavior of the $(\eta^6\text{-TMCOT})Mo(CO)_3$ molecules, Musco and I were interested in seeing how the $(\eta^6\text{-TMCOT})Fe(CO)_3$ molecule would behave, but Nature would not let us find out. The attempt to prepare it [238] gave instead only the following complex based on the bicyclic tautomer of TMCOT:

A Digression Concerning Notation. As our investigations of fluxional behavior continued, the accompanying synthetic and X-ray structural work led to the discovery of a great variety of ways in which a polyolefin could be attached to one or more metal atoms. It became very evident that no suitable notation existed to describe these structures unambiguously. I therefore set my mind to the task of devising one, and by early 1968, I had a scheme that seemed generally satisfactory. I then checked with people at Chemical Abstracts and also gave a talk on it at the Spring ACS meeting. People generally liked it and made only a few minor suggestions for change, so in July 1968 I formally proposed it in the form of a Letter to the Editor of JACS [255]. In a relatively short time, it was formally accepted by IUPAC, where the only significant change was to use Greek η (eta) instead of Roman h for hapto. This hapto (η) notation is now universally used and indispensable. I have already illustrated it in a very simple way in distinguishing between the ring binding in $(\eta^4\text{-}C_8H_8)Fe(CO)_3$ and $(\eta^6\text{-}C_8H_8)Mo(CO)_3$, but it can be used in far more complex cases, often in conjunction with the μ_n notation to designate bridging.

In the years 1968–70, we proceeded to investigate more and more complex examples of fluxionality. For example, in the Winter of 1967–68, I had a postdoc named Charles Reich who studied the following molecule [263]:

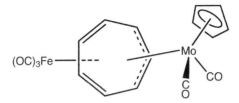

It was found that all seven hydrogen atoms of the large ring became NMR equivalent on going from –70 °C to +50 °C. Line shape analysis strongly favored 1,2-shifts. Since the massive $Fe(CO)_3$ and $(\eta^5\text{-}C_5H_5)Mo(CO)_2$ groups will move very little, we must have a little molecular machine in which the C_7H_7 ring resembles a bicycle chain held in place by the sprocket and the gear wheel on the back axle. There is much hoopla today about molecular machines. This may well be the first ever deliberately made.

Another molecular machine we discovered [262, 272] was a molecular windshield wiper. As shown below, the instantaneous structure of this molecule has the $(OC)_3Fe–Fe(CO)_3$ moiety at a slant relative to the ring, but it flicks back and forth like a wiper blade.

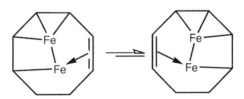

One episode in the work on fluxional organometallic molecules proved to be both entertaining and challenging, but also to have consequences concerning the modes of metal to cyclopentadienyl bonding. This molecule, which in my lab was called the "three-ring circus," was made following the report by Bruce King of a molecule that he proposed to be $(\eta^3\text{-}C_5H_5)(\eta^5\text{-}C_5H_5)Mo(NO)CH_3$, using the eta notation which was only later introduced. It occurred to me that if we replaced the methyl group by an $\eta^1\text{-}C_5H_5$ group, we would have a molecule with three different kinds of cyclopentadienyl groups attached to the same metal atom. If fluxionality also occurred, we would have three rings, all going around in circles: hence the nickname.

It should be stressed that we, and King before us, had decided on the η^5, η^3 assignment based on the usual, but somewhat naïve rationale that (1) the total number of electrons donated by two C_5H_5 rings to an Mo(NO)R fragment had to be 8, so as to satisfy the 18-electron rule, and (2) that a C_5H_5 ring that was neither η^5 nor η^1, would have to be, at least formally, η^3.

Peter Legzdins, a Ph.D. student, quickly accomplished the preparation of the putative η^5, η^3, η^1 compound and then showed that it displays two-phase fluxional behavior [256]. At –110 °C, the ^{1}H NMR spectrum showed that there is a structure in which there are three different kinds of rings, one of which is certainly an η^1-C_5H_5 ring in an environment in which free rotation about the Mo–C bond is restricted, so that the two edges are not equivalent. In addition to the signals assigned to the η^1 ring, there were two singlets, separated by about 1.1 ppm (66 Hz), which could be assigned to each of the two other rings. We had to presume that even at –110 °C the η^3 ring was still "whizzing" fast enough to average out the resonances of its five protons.

In the temperature range –110 to –50 °C, the two edges of the η^1-C_5H_5 ring become equivalent and, simultaneously, the two singlets for the other rings collapse and coalesce. Our interpretation was that the η^5 and η^3 rings were interconverting, which would necessarily cause the two edges of the η^1 ring to become equivalent. In the range –50 to +15 °C, all resonances collapse and coalesce into one. Thus, we have complete, rapid interconversion of all the rings.

As to how the presumed η^5 and η^3 rings became NMR equivalent we saw two possibilities. (1) The NO group participates. It temporarily becomes a bent NO which is a one electron donor, thus causing the η^3 ring to become an η^5 ring to keep the total electron count at 18. When the NO group goes back to being a linear three-electron donor, either of the η^5 rings has the same probability of resuming η^3 status and thus η^5/η^3 scrambling is accomplished. (2) The NO group remains unchanged and the η^5 and η^3 rings both move, so that both adopt the same kind of irregular relationship, neither η^5 nor η^3 but something in between, to the Mo atom. From this transition state (or possibly an intermediate) the resumption of the presumably more stable η^5, η^3 status quo ante could take place in two equally probable ways.

The idea very naturally occurred to me that before considering this project complete it would be nice to determine the crystal structure of the molecule and actually verify the assumed η^5, η^3, η^1 nature of the molecule. In fact, the structure we determined [267] was a complete surprise! It *falsified* the assumed η^5,η^3,η^1 structure. There is an η^1 ring, exactly as expected, but the other two rings have, within experimental error, exactly the same relationship to the metal atom; this is shown in Fig. 5-7. This was certainly the first time and (as far as I know, to this day) the last that a crystal structure did not appear to jibe with the implications of the low-temperature limiting NMR spectrum. This was very disconcerting.

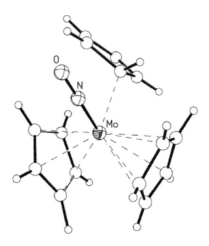

Figure 5-7. The structure of the $(C_5H_5)_3$MoNO molecule.

We considered the possibility that perhaps, as a strange coincidental result of packing forces, the second transitional structure mentioned above had been trapped and stabilized in the crystal. If this were so, then surely a different compound, having different intermolecular forces, could not possibly show the same structure. We therefore determined the structure [323] of the purported $(\eta^5-C_5H_5)(\eta^3-C_5H_5)$Mo(NO)$CH_3$. It showed exactly the same virtual equivalence of the two rings, each in a relationship less close than η^5 but closer than η^3 to the metal atom.

After a little thought, I realized that it was possible to reconcile the seemingly incompatible NMR and crystallographic results. If we look at the $(C_5H_5)_3$MoNO molecule from the point of view shown in Fig. 5-8, we can see how. Two of the C_5H_5 rings differ because of their different relationships to the $\eta^1-C_5H_5$ ring, not because they are η^5 and η^3, or in any other way different in their relationship to the metal atom. At low temperature rotation, about the Mo–C bond is so slow that these two C_5H_5 rings, while rotating individually fast enough to give only one signal each, have different chemical shifts; at the same time, the two sides of the η^1-C_5H_5 ring are non-equivalent.

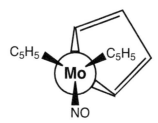

Figure 5-8. A schematic view of the $(C_5H_5)_3$MoNO molecule.

All that is required to eliminate both of these non-equivalences simultaneously (as observed) is for rotation about the Mo–C bond to become rapid. It is also clear, from this viewpoint, why the $(C_5H_5)_2$Mo(NO)CH_3 and $(C_5H_5)_2$Mo(NO)I molecules *never* show inequivalence of the C_5H_5 rings [305]. The final, and most important, result of all of our work on this "three-ring circus" was that the idea of an η^3-C_5H_5 ring, at least in this and related molecules, is a myth.

In 1970 we studied a molecule expected to have the structure shown in Fig. 5-9, with two η^1-C_5H_5 rings. This structure was confirmed in the solid state [304] and also at very low temperature in solution [303]. At –25°C, the η^1-C_5H_5 rings are already whizzing rapidly and give a sharp single line, but what occurs above that temperature is quite remarkable. The η^1-C_5H_5 and η^5-C_5H_5 rings begin to exchange roles and the signals collapse and reappear as a single line for all 20 hydrogen atoms in the molecule. There is no hint of this ring exchange in $(\eta^5$-$C_5H_5)$Fe(CO)$_2(\eta^1$-$C_5H_5)$, in which the metal atom has an 18-electron configuration. It was our hypothesis that this is possible

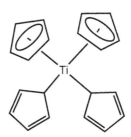

Figure 5-9. The structure of the $(C_5H_5)_4$Ti molecule.

in the titanium compound because with only a 16-electron configuration there is an unoccupied metal atom orbital that allows the formation of a transitional structure. To confirm our hypothesis we showed that in $(\eta^5-C_5H_5)_2Mo(\eta^1-C_5H_5)_2$, which has the same composition as the titanium compound but also has an 18-electron configuration, no ring exchange occurs [305].

I will leave the subject of fluxional organometallic molecules now, although I have omitted a number of other molecules we studied. I have always found fluxional molecules to be fascinating because they undergo rapid change — and yet they do not change. The well-known French adage, plus ca change, plus c'est toujours la même chose, describes them admirably [246].

An Enzyme Structure — *Staph* Nuclease

As I became more involved in X-ray crystallography, I began to think about going beyond the relatively routine practice of solving the structures of medium size inorganic compounds and trying something at the frontiers of the subject. As I saw it, there were two challenges that might interest me. One was to "go small" and try to determine experimental electron density maps of very small molecules. This tied in with my lifelong interest in electronic structure, but as I looked into this more closely, I became disenchanted. It did not seem to me that even in the most favorable cases — very small molecules consisting entirely of light atoms — the accuracy could be high enough to tell you something you didn't already expect. Moreover, the bonding questions that interested me all tended to involve transition metal compounds, and particularly those with very heavy metal atoms, where there was (and still is) no hope of obtaining meaningful electron density results.

I then considered "going big," that is looking at a protein of some sort. My first opportunity to do that came a bit too soon and I turned it down. Bert Vallee, at Harvard Medical School, was deeply involved in chemically characterizing the enzyme bovine carboxypeptidase, and suggested this as a project, sometime in the early 1960s. This is a moderately large protein (MW 37,000 D) and I didn't feel I could handle it. I knew, however, that Bill Lipscomb had been thinking about tackling a protein, so I arranged a dinner at our on-campus apartment at MIT that comprised the Vallees, the Lipscombs, and the Stockmayers. I encouraged Bill and Bert to talk and this led to the start of Lipscomb's work on carboxypeptidase. The collaboration did not last, but

Ted Hazen in 1980.

Lipscomb continued and eventually carried that structure to atomic resolution (1967).

In the Fall of 1962, Ted Hazen started an NIH postdoctoral fellowship with me. Ted had just completed a Ph.D. at Harvard with J. L. Oncley in biochemistry, and Oncley had directed him to me (with an NIH fellowship) so he could learn more about transition metal chemistry and its possible role in biochemistry. Had all gone as intended, I might have become one of the early bioinorganic chemists, but that wasn't to be. Again, Alex Rich had a pivotal influence on my career.

At that time, Christian B. Anfinsen was at the Harvard Medical School and he was very keenly interested in an enzyme produced by *Staphylococcus aureus*, which was an extracellular nuclease that had several unusually attractive features. From a purely chemical point of view, it was interesting because it is both a DNase and an RNase. From a structural point of view, it was exceptional because its tertiary structure did not depend on the presence of any disulfide linkages, but only on hydrogen bonding and other non-covalent interactions. Thus, it provided an unprecedented opportunity to study the relationship between the sequence (primary structure), the regions of secondary structure (α-helix, β-sheet), and finally, the overall tertiary structure.

From a practical point of view, Anfinsen's team was now skilled at growing certain *Staph* strains and at purifying the enzyme, an absolute necessity if it were to be crystallized. Moreover, the enzyme, like baby bear's porridge, was "just right" in terms of size. With a molecular weight of about 16,000 (149 amino acid residues), it is large enough to be typical in terms of its tertiary structure but small enough to be amenable to the methodologies then available for the solution and refinement of its crystal structure. A final point was that Anfinsen had already determined the sequence, which was also a necessity, if the electron density map was to be interpreted.

Chris Anfinsen proposed the problem to Alex, but Alex had already embarked on one of the earliest attempts to do a *t*-RNA structure, which he, Sung-Ho Kim, and other coworkers later completed. He therefore referred Chris to me, because he knew

I might be interested. Suddenly, for me all the pieces fell into place. I proposed to Ted Hazen that doing an enzyme structure would be more exciting than what he was already doing and he readily agreed. Chris promised full cooperation in supplying material and advice. I also persuaded David Richardson, a graduate student who had entered in the Fall of 1962 and had already begun thesis research by doing a small molecule crystal structure, to join in the effort.

Thus, sometime in the Spring or Summer of 1963, Hazen, Richardson, and I went to work. The first job, and one where Ted's familiarity with practical protein chemistry was invaluable, was to grow crystals. Following that, it was necessary to find out how to make crystals with heavy atoms isomorphously introduced. We also, of course, had to set up a laboratory properly equipped for this sort of work, to obtain NIH funding, and to purchase and set up a new diffractometer. Thus, it was not until almost three years later that we were able to submit our first manuscript, to the *Journal of Biological Chemistry*, reporting how crystals could be grown and that we had reached the point of beginning extensive data collection [196]. We knew by then that *Staph* was going to be a winner, because the crystals diffracted well beyond the $\sin2\theta$ values needed for 2 Å

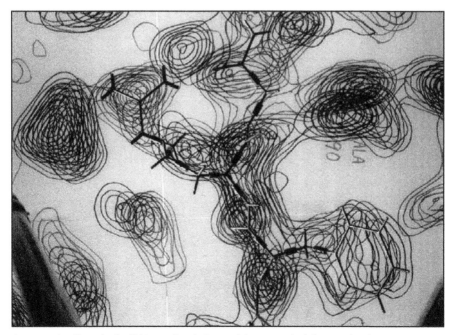

An example of how, in the 1960s, we fitted wire models to a stack of electron density maps on a light box. Computer fitting was still a long way in the future.

resolution. This meant that we could expect, ultimately, to achieve atomic resolution. In the end (1974), we had collected and used data to 1.5 Å resolution, and knew there was still more available (to ca. 1.4 Å).

From there things moved a bit more quickly and after three more years (May 1969), we had a structure at 4 Å resolution, which we published in *Proceedings of the National Academy of Sciences* [279]. In this PNAS paper, we were able to describe the shape of the molecule (very globular), to recognize secondary structure elements (about 15% α-helix and a nice three-strand slab of β-sheet), and to find the active site and a calcium ion therein.

Initially we employed a conventional approach, if anything in a field then so new could have yet been called conventional. We grew crystals of the enzyme and then soaked these crystals in solutions containing $PtCl_4^{2-}$, UO_2^{2+}, or *p*-acetoxymercurianiline — that is, species that were expected to go into specific positions in the crystal without making any significant change in the structure other than introducing a strong scattering center. Theory showed that at least two such isomorphous substitutions are needed to fix the phase angles of the reflections and thus allow the electron density to be mapped, and more are desirable. This method is known as multiple isomorphous replacement.

However, in theory there is another way to proceed that requires only one heavy atom derivative, provided that anomalous dispersion by the atom is large enough

Working out the structure of staph nuclease in 1967-68. Jim Bier, me, Ted Hazen, Dave Richardson, Jane Richardson, and Arthur Arnone.

to measurably affect the intensities. In practice, this possibility had not yet been employed successfully, but we were able to make it work. It was known that for the nuclease there was a strongly bound competitive inhibitor, thymidine-3',5'-diphosphate (Fig. 5-10a), and we found that an enzyme-inhibitor-Ca^{2+} combination could form beautiful crystals, again with scattering to high resolution. Our idea was to employ 5-iododeoxyuridine-3',5'-diphosphate (Fig. 5-10b), which is the same molecule except

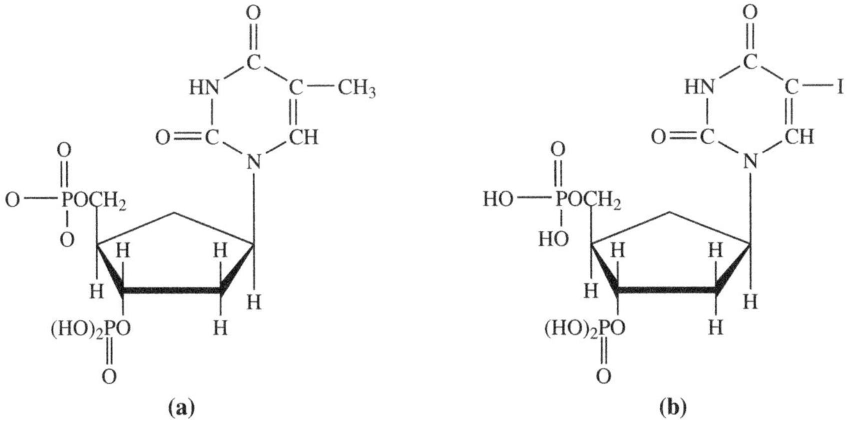

Figure 5-10. (a) A natural inhibitor of *Staph* nuclease. (b) An isosteric dummy inhibitor.

for the replacement of the thymine base by 5-iodouracil. Sterically, the difference between CH_3 and I is negligible, but a group of (total) atomic number 15 (CH_3) is replaced by one of atomic number 53, a net difference of 38. This corresponds to a very modest "heavy" atom (comparable to an Sr atom) as compared to the typical heavy atom Pt with atomic number 78, but it was sufficient and had the advantage that the precise replacement of CH_3 by I results in perfectly isomorphous crystals. The final point is that the iodine atom displays a relatively large anomalous scattering for copper Kα X-rays. Thus, from only two (instead of at least three) isomorphous crystals, the phase problem could be solved. In fact, however, we also availed ourselves of the chance to replace the Ca^{2+} ion in the enzyme/inhibitor complex by Ba^{2+} (a 36 electron change).

We convinced ourselves that the overall structure of the nuclease was essentially the same in both the native (uninhibited) form and in the enzyme/inhibitor/Ca^{2+} complex and then concentrated on the latter for two reasons. It gave better data to higher resolution, but, more important, on the reasonable assumption that the

3',5'-diphosphate nucleosides inhibit by occupying the active site in the same way as would a DNA chain, we could go directly to the question of how the enzyme binds and activates its substrate. This, after all, is the main reason (though not the only one) for determining an enzyme structure.

It was during the Summer of 1969 that I believe we had what we thought was the correct assignment of the polypeptide chain to the experimental electron density map. By mid-1970, we had been able to refine the structure to the extent that there was no doubt at all that it was correct. Details of the α-helix portions looked just as they were expected to and a three-strand β-sheet looked very beautiful.

We now moved fast, adding new coworkers (see later) and by the early Fall of 1971 we were able to publish a high-resolution (2 Å) structure [299] that traced the entire peptide chain and revealed in fair detail how the inhibitor was bound *via* a Ca^{2+} ion and several hydrogen bonds. This led also to publication in Boyer's *The Enzymes* [302] and in a symposium organized by Jim Watson, where our article [319] together with one by Chris Anfinsen, gave a broad overview of the crystal and solution studies.

In recent discussions with Dave and Jane Richardson, we have come to the conclusion that the *Staph* nuclease structure was the tenth protein structure to reach a high degree of resolution. The first was Myoglobin in 1960 which Kendrew *et al.* described at 2 Å resolution. There followed lysozyme (Phillips *et al.*, 1965), ribonuclease (Harker *et al.*, 1967), and then others at resolutions of 2 to 3.5 Å. Finally, our 2.0 Å structure led to a publication by Arnone, Bier, Cotton, Day, Hazen, Richardson, Richardson, and Yonath [299] which appeared at about the same time as a 2.8 Å structure of cytochrome C by Dickerson *et al.*

For the denouement of the *Staph* nuclease story, it is necessary to go on to Texas A&M, which, for the sake of continuity, I shall do now. When I moved to A&M, Ted Hazen came with me, with a faculty appointment. There we recruited an entirely new crew of coworkers and built an entirely new lab. Our major goals were to achieve even better resolution (1.5 Å) and to formulate a mechanism for enzyme action. Getting to higher resolution was largely a matter of sweating out the task of collecting and refining more data. Speculating about the mechanism was the fun part. Our principal coworkers in these tasks were Phillip Tucker and Marge Legg, graduate students, and Douglas M. Collins, a postdoc from Lynn Hoard's laboratory at Cornell.

Over the years 1972–79, we accomplished these tasks. In 1978 and 1979, Ted Hazen, Phil Tucker, and I published four papers [573–76] in *Molecular and Cell Biochemistry* in which we presented some new data but principally made an effort to review and integrate all that was known about *Staph* nuclease. In this effort, it was Phil Tucker who did all the heavy lifting. In part III of this series we first presented a detailed mechanistic proposal for the action of the enzyme. In our final publication [598] on *Staph* nuclease, Ted and I and Marge Legg said our final word on the mechanism.

In the years since there have been a number of searching tests of this mechanism by enzymologists employing many sophisticated techniques of physical organic chemistry, and apart from one very minor modification, the mechanism that we proposed has stood up. In the course of many subsequent studies, in which site-specific amino acid replacements have been made, *Staph* nuclease has been a valuable workhorse for testing basic concepts of static tertiary structure, the dynamics of protein folding and unfolding, and other things that enzymologists are interested in. One of the most recent of these studies by Peter Schultz *et al.* appeared in *Angew. Chem.* in 1995, where a "tapeworm" drawing of the structure, similar to that in Fig. 5.11, appeared on the cover of that issue. I myself think that what Ted Hazen and I and our many coworkers

Figure 5-11. A schematic drawing of the peptide chain in *staph* nuclease as drawn by Jane Richardson in 1978 [575].

produced is one of the early landmarks in structural enzymology. One is reminded that the moving finger writes, and having writ, moves on.

It could be asked of me, why, after such a successful maiden voyage into the field of protein crystallography, have I never done anything more. The answer is clear from three considerations: (1) I have never wanted to be less than the leader in any field I go into; (2) I have abiding and absorbing interests in inorganic chemistry; and (3) there are only twenty-four hours in a day. To be more specific, it became obvious by the early seventies that protein crystallography had proved to be very do-able and that traffic would soon get very heavy. Moreover, to work meaningfully at the cutting edge of the field, one would have to stay up with the already huge literature in enzymology as well as with the rapidly evolving experimental and computational techniques for the X-ray work *per se*. In short, leadership in the field would require full time effort. I was not prepared to abandon all my work in inorganic chemistry in order to do this, so I decided to make a clean break and stay with my first love, inorganic chemistry. But my excursion into biochemistry was a great ride while it lasted.

Before entirely leaving this subject, I want to talk a little more about my coworkers in the *Staph* nuclease structure. During the period 1966–69, several more people joined the group, Arthur Arnone and James Bier, Ph.D. candidates, and Jane Richardson, David's wife, and Ada Yonath, an Israeli who has since become famous for her brilliant work on ribosomes.

Arthur Arnone was a truly exceptional graduate student. He arrived in September of 1966 with a Master's degree, by way of which he had already learned some X-ray crystallography. He took to protein crystallography like a duck to water and made major contributions to our work. I found him so impressive I recommended him to Max Perutz for postdoctoral work. He went to Max's lab on an NIH postdoctoral fellowship for two years, during which time he solved the longstanding and important problem of how 2,3-diphosphoglycerate associates with deoxyhemoglobin. He then joined the biochemistry faculty at the University of Iowa where he is a professor today.

Others of note who joined the team during the academic year were Victor W. Day, who went on to an academic career at the University of Nebraska, and Sine Larsen, who is now a professor in the University of Copenhagen.

In some ways, however, the most extraordinary member of the MIT group was Jane S. Richardson. At some point, in about 1967 or 1968, Ted Hazen and I realized that we needed and could afford another coworker, one to do a number of non-Ph.D. level tasks. After several months of not finding anyone both suitable and available, Dave Richardson made, with some hesitation, the suggestion that perhaps his wife Jane could be useful. Jane had a bachelor's degree from Swarthmore and two master's degrees (one in education) and she was then teaching in a high school in one of the Newtons. She was not enjoying teaching high school and we were really in need, so she came and joined our research group. Within a year, it was evident that this modest young woman was simply brilliant. We couldn't have done better. I've had the pleasure of working with quite a few brilliant people — many in fact — but none with more native intelligence than Jane.

After following his Ph.D. work by two more years with me as a postdoc, Dave Richardson went to Duke University in 1969 where he subsequently rose through the ranks to Full Professor. During this period, Jane continued to work with him in various non-faculty positions. She soon began to acquire a reputation of her own as an imaginative analyst and a correlator of protein secondary and tertiary structures. I was able to see that she was nominated for a MacArthur award, which she received sometime in the 1980s. This awoke Duke to an awareness that they had a major scientist in their midst, but not on their faculty. However, their response was limited to giving her the title Adjunct Professor — which many consider to translate as "not a real professor." Finally, after she was elected to the National Academy of Sciences in 1991, Duke had to, and did, make her a *real* professor. She now holds the title of James B. Duke Professor. Other than R. V. Pound at Harvard and Lars Onsager at Yale, I have never heard of anyone without a Ph.D. becoming a full professor in a major university and a member of the NAS. I think Jane's theme song could appropriately be, "I did it my way."

Two key coworkers at Texas A&M were Philip W. Tucker and Margaret J. Legg. Phil Tucker was not one who made a first impression as a budding academic scientist. He was a native Texan who wore an abundant beard, rode a Harley-Davidson, and played (I forget what instrument) in a rock band. Nonetheless, he was a smart, hardworking student who could write. He has gone on to have a highly successful career and is now a professor of microbiology at the University of Texas, working on mechanisms of gene expression. Marge Legg was a very quiet, unassertive person who proved to be very smart and is now a professor at St. Ambrose University in Davenport, Iowa.

CHAPTER 6

YEE HA!
OFF TO TEXAS

MY MOVE TO Texas was not specially complex or difficult, but did have its moments. The very first step predated my own actual transfer. I had heard from others that a move typically sets back one's research by a year, and I resolved that with good planning that was unnecessary. In early 1971, I began inquiring for a postdoc in the field of X-ray crystallography who could go down a semester ahead of me and set up my X-ray lab as well as other things so that I could, as the saying goes, hit the ground running when I arrived. This worked out to perfection, thanks to Bertram A. Frenz, who finished his Ph.D. at Northwestern in the Summer of 1971 and went to College Station in September. An X-ray diffractometer was delivered concomitantly, and by November, Bert notified me that it was up and running. I sent several crystalline samples down to Bert from MIT and thus even before I arrived we had several structures ready to publish from A&M.

Bert had also been looking over the crop of new TAMU graduate students and had one lined up for me, Jan Troup. Actually Jan had come to A&M already more or less lined up for me. Jan had done an M.S. at Ohio University in Athens, Ohio, with Abe Clearfield and was already a very competent crystallographer. The most important thing that Bert did in regard with students was to check some others out and warn me off of them.

Jan Troup could not continue for the Ph.D. at Ohio University and Abe, who had been a friend since my undergraduate days at Temple

With my first Texas A&M graduate student , Jan Troup, May 1973.

My daughter Jane as a high school student.

University, knew I was to arrive at A&M and had advised Jan to enroll there for the Ph.D. Little did Abe or I think then that a few years later we would be colleagues. Jan proved to be an excellent student and I am very grateful to Abe for sending him my way.

Another thing that helped me to get into top gear quickly on arrival in Texas was that several students came with me from MIT. These were students who still had a year or more to go, namely Richard D. (Rick) Adams, Alan White, Charles M. (Chuck) Lukehart, and Zoltan Mester. In addition to Bert Frenz, I also had other postdocs, Len Kruczynski, who came with me from MIT, and Tom Webb, who started shortly after my arrival. In the meantime, several people were still working, during the Spring semester, at MIT. The protein structure project also moved smoothly. Ted Hazen had been given a tenured position at A&M, and we soon had the diffractometer moved and very quickly picked up where we left off. There was no dip in my rate of publication as a result of moving.

On the domestic side, the move was more of an adventure. It was not a particularly propitious time for our trip. In mid-January of 1972, bitter winter storms were sweeping across the northern part of the country, from the midwest to New England. We loaded the back of our Ford pickup truck with numerous suitcases and trunks, as well as crates containing one basset hound and five foxhounds — breeding stock for the foxhunt we planned to start on the ranch we intended to buy — and a couple of siamese cats. In the front were Dee and I and our daughters, then aged 10 and 8. We were highly compressed, but only as far as Philadelphia. There we stopped for a day with my mother and left the girls with her. They were to join us later after we had settled down in College Station. We then headed west to Rock Island, Illinois, where Dee's parents lived. For a couple of days we unwound and prepared for the rest of the journey.

As we left Rock Island, the weather was bitterly cold with snow and ice on the ground. Somewhere in southern Iowa, on the first day, the truck began to malfunction and we barely managed to coast into a filling station, feeling very concerned because of all the animals. It was in a state of very low morale that I told the proprietor my problem, but he replied cheerfully, "Don't worry, you just need some heat." I watched in a state of utter incomprehension while he brought to the counter a pint can bearing the word HEAT. HEAT turned out to be essentially methanol, and he explained to me that our problem was, in his opinion, nothing more than ice in the fuel line. He was right. The can of HEAT did the trick and after purchasing several more to take along, we were on our way again. By the evening we were through Missouri and into Oklahoma.

We now had the problem of spending the night in a motel. Why a problem? Because we had eight animals who also needed somewhere to spend the cold night. I checked into a motel and, with as straight a face as I could manage, I said I wanted two rooms, one for my wife and myself and one for "the little ones." I am afraid I did not explain that the little ones in question went about on four feet. All went smoothly, except that while we were outside feeding, watering, and exercising the foxhounds, our pet basset, Humphry, ate approximately 120 Reese's peanut butter cups, most of them paper and all. His cast iron stomach, as it always had, proved equal to the situation and he was his usual perky self in the morning, when we made an early start on the last leg of the journey.

As a result of contacts made through the Pony Clubs of America we had been in touch with a very kind person, Col. Sid Loveless, who had dealt with a realtor on our behalf and had found us a small country house with some acreage, suitable for hounds and the three horses who were scheduled to arrive shortly after us. He even kept the horses for us until I had time to build accommodations for them.

We arrived in College Station around noon, had lunch with the realtor, and she then led us about ten miles out of town to our new, but temporary, home. By June we had sold our first home in Texas, purchased our present ranch, and with the ranch we also bought fifty head of cattle. Thus, we were immediately embarked upon on-the-job training in cattle ranching. We learned two things quickly: that horses trained for foxhunting and show jumping cannot be used to round up cows and that we needed a ranch manager. In due course, we sorted all this out.

In the meantime, we had arranged to have more foxhounds shipped down and we had to get a kennel built. It was our intention to establish a hunt. Unfortunately, this never worked out, although we made a valiant effort, and we eventually gave our

According to an old Texas adage, nothing so improves the profitability of a ranch as a few oil wells. This was the entrance to our first well.

foxhounds to local night hunters. We did, however, meet a lot of very nice people in the horse world here and had a good time riding in local horse shows. Today, our horse activities are all vicarious, through our daughter Jennifer, who runs the ranch, breeds horses, and competes in dressage. She's a better rider than we ever were and because of her we are still in touch with activities that used to be such a major part of our lives.

Several times during our early days in Texas we heard it said that to be (or become) a true Texan one had to do one of the following: (a) drive a pickup, (b) have cows, (c) have an oil well. Right from the beginning we met the first test because we arrived in a pickup. By May of 1972 we had further solidified our standing by meeting the second criterion. The third one took us a decade longer, but we got our first oil well in 1982. We still have one operating well (not the one shown) but are pooled into a multi-well field.

Settling into the laboratory also went fairly smoothly. I was temporarily put into an office and labs in a building that had been opened in 1958, while a newer one was being completed. The labs were a bit run down, but then I was used to that from MIT. I was given a modest-size office next to Minoru Tsutsui's with an attached secretarial office. My first major task was to find a secretary. The department had a list of three for me to interview.

The first candidate was an intelligent, well-mannered, and evidently competent young woman, obviously not a local. On learning that her husband was a graduate student, I decided she would not do. I had no intention of having to find another secretary in a few years. The next candidate never had a chance from the moment I set eyes on her. She was a young, gorgeous Eurasian woman. I got rid of her politely but perfunctorily, since I had no intention of having an outer office full of young men wasting her time and theirs chatting her up.

Fortunately, the third candidate was not strike three but a home run. Mrs. Irene Casimiro was a forty-something married woman, with children who were already teenagers. Moreover, she had grown up in the community, had been employed as a secretary elsewhere in the university, and could type like a demon. I hired her on the spot, and she stayed with me for twenty years, during which she must have typed the manuscripts for about 850 papers as well as those of six books, and indenumerable letters. Word processors and computers, which she did not like, finally drove her into slightly early retirement.

The students who came with me from MIT soon arrived and settled, temporarily, into labs close to my office. All of them subsequently received their Ph.D. degrees from MIT in the years 1972–74. Chuck Lukehart stayed for an additional year as a postdoc.

By about July, we were all able to move into my brand new and permanent quarters in what is now known as the 1972 building. This very nice building was built under the direction of Arthur Martell and I think he did an excellent job in controlling the design.

A few words about Arthur Martell as a department head are appropriate here, because his talent for the job was very beneficial to me. I had known Arthur since my MIT days, when he was head of the chemistry department at Clark University in Worcester, Massachusetts, about thirty miles west of Cambridge. We met often and became friends because he attended MIT-Harvard seminars and colloquia and frequently used the MIT library.

Arthur enjoyed running a department, but he wanted to build and improve, and when scope for doing that at Clark was not on the cards, he found a better opportunity in 1961

With Arthur Martell in 1980 when I received the Gibbs medal in Chicago.

at the Illinois Institute of Technology (IIT) in Chicago. Here he laid into his work energetically and soon made great improvements.

Things weren't moving as fast as Arthur wanted, so he looked around for another opportunity and found one at TAMU. Thus he had arrived there in 1966 (bringing with him my old pal from graduate school Barry Shapiro as Associate Head) with a mandate to convert a low rated, service-oriented department into one with as much pizzaz as possible. This was Arthur's meat, and he again laid into the job with alacrity, this time with continuing support from the administration.

By the time I arrived, the department was definitely a research oriented one, and while it certainly wasn't MIT, it was a reasonable place to be. It had one great advantage for me, namely, that of my being a very big fish in a little pond, and having Arthur so regard me. Arthur was more than a capable administrator, he was a gifted one, and he was assisted by a business manager named Pete Rodriguez, who was also a man of rare talent in making sure that rules, regulations, and state bureaucracy did not interfere with doing what was in the best interest of the department. Arthur and Pete were maestros of manipulation, and there was no bureaucracy that could thwart them when they were determined to do something.

Making a point with the staph nuclease model, 1973.

For example, shortly after I arrived, I got a yen to have a nice model of *Staph* nuclease built by a company in Edinburgh, Scotland. The cost was, I believe, about $800, and I asked Arthur if he could provide the money, which was not a trivial sum in those days. He said he would think about it, and within a week he gave me the OK to order it. At that time, TAMU was preparing to celebrate its 100th anniversary, and a large fund had been created, the Campus Beautification Fund. Arthur considered that purchasing this model could be construed as beautifying the campus, and of course I agreed.

In general, Arthur ran the department as a benevolent dictator, or, if you prefer, with an iron fist in a velvet glove. It was the best form of administration at the time because Arthur had generally good judgement and an avuncular manner. For the rest of his tenure as head (until 1980), the department ran efficiently and made rapid progress. Arthur was by no means averse to seeking and taking advice, but he chose his advisors to suit himself, and not according to any pseudodemocratic formula. He and his handpicked "kitchen cabinet" would meet for lunch every month or two and discuss plans for the development of the department and Arthur would handle the execution. In recent years, the running of the department as well as the whole university has grown incredibly complex and I often think nostalgically back to the simpler and happier days of King Arthur.

Within a few months I had reestablished a working atmosphere and research began to go very well again. The nuclease work, which was guided on a day-to-day basis by Ted Hazen, quickly picked up where we had left off at MIT. Our task at this point was to keep pushing to higher resolution (which we knew was attainable) so that we could obtain a sufficiently accurate and detailed picture of how the inhibitor was bound at the active site to allow realistic speculation about the mechanism. I have already explained how this phase of the work proceeded in the previous chapter.

Work along inorganic and organometallic lines was mainly in the hands of Adams, Lukehart and my newly acquired student, Jan Troup, with Bert Frenz overseeing the crystallographic side of things. With Bert's supervision, all of these students soon became very capable crystallographers. By 1973, I had picked up some TAMU students. I have already mentioned Phil Tucker, who worked on the nuclease. In addition, I soon had Doug Hunter, Larry Shive, Anna Stanislowski, and Carlos Murillo. In this same period (1972-73), some very good inorganic postdocs joined the group: Francis Carré from Montpellier, Pascual Lahuerta from Zaragoza, and Erik Pedersen (now deceased) from Copenhagen. By the end of 1973, things were moving at a pretty good clip.

The Discovery of Agostic Interactions

I want to get back now to discussing some research topics. One of the very important ones, in terms of its impact on organometallic chemistry, even though only a few of my coworkers and publications are involved, is the question of how X–H bonds, particularly C–H bonds, may interact with metal atoms that would be electronically unsaturated without such interactions.

The story begins just before I left MIT. Toward the end of 1970, Jerry Trofimenko of DuPont reported (*Inorg. Chem.*, **1970**, *9*, 2493) that he had made a number of compounds that had a non-inert-gas configuration. For one of these compounds, he proposed structure (a) in Fig. 6-1, although noting that the 18-electron structures (b) and (c) might also be considered. I think his preference for (a) was based on the fact

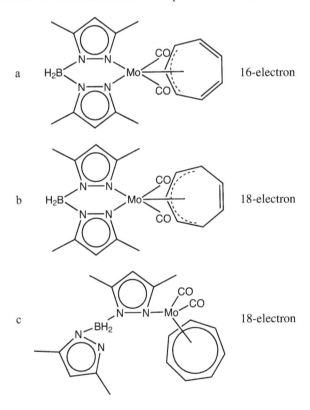

Figure 6-1. The three sturctures initially suggested for a $(H_2Bpz_2)Mo(CO)_2C_7H_7$ molecule.

that he had made analogs in which only simple allyl groups were present, and in these compounds, structures similar to (b) and (c) would be impossible. I was skeptical of the idea of 16-electron compounds of this type and so we undertook an X-ray crystal structure determination on this compound [334, 342]. As a result, we showed that the actual structure was none of those in Fig. 6-1, but instead the one shown in Fig. 6-2. The data set was a very good one and the structure behaved so well that all hydrogen atoms were refined isotropically and independently.

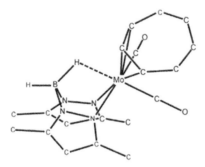

Figure 6-2. The actual structure found for the $(H_2Bpz_2)Mo(CO)_2C_7H_7$ compound.

The structure does show a η^3–C_7H_7 ring, as proposed by Trofimenko, but there is not a 16-electron configuration. It can be seen that the conformation of the pyrazolylborate has an unprecedented severely puckered boat configuration. This is what I said about the structure: "The reason for this puckering is that one of the hydrogen atoms on the boron atom approach[es] the molybdenum atom closely, forming a B−H−Mo 3-center, 2-electron bond. Thus, the molybdenum atom can be considered to achieve an effective 18-electron configuration when the two electrons in the bonding orbital that extends over the B−H−Mo group are considered." And later in the paper "the central Mo atom achieves an effective 18-electron configuration by virtue of its share in the electron pair occupying the bonding orbital spread over the three-electron system B−H−Mo." In this compound we found and explained the *very first* example of what was, more than a decade later, called an *agostic* interaction. This interaction was fully displayed by the experimental data and was fully explained and accounted for in terms of the electronic structure [334,342].

With the "agostic" B−H case before me, I wondered what would be the result of substituting a BR_2 group for the BH_2 group in the pyrazolyl ligand, while also having a simple allyl group which could not possibly be more than a 3-electron donor. I discussed this with Jerry Trofimenko and he came up with just the compound that was needed, and in the Spring of 1973 we determined its structure. The result was the first example of an agostic C–H bond [378], as shown in Fig. 6-3. In our paper, we pointed out that "there is an interaction between an aliphatic C–H bond and a metal atom" and described this interaction as "a three-center, two-electron bond encompassing the C···H···Mo atoms."

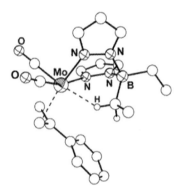

Figure 6-3. The structure of $(Et_2Bpz_2)Mo(CO_2)(2-C_6H_5allyl)$.

Almost immediately after that we showed that such an interaction is a strong one [401]. This was done by using NMR spectroscopy to measure the activation energies for the alternation of the two α-hydrogen atoms on the endo CH_2 group in the bonding position as well as several other fluxional processes in the molecule. These results were used to estimate that "17-20 kcal mol^{-1} ... probably approximates to the strength of the C–H···Mo interaction."

Finally, the structure of the (diethyldi-1-pyrazolylborato)Mo(CO)$_2$C$_7$H$_7$ molecule was examined [385] to answer the following question: "Can a CH–to–metal interaction of the type found in [the analogous compound with a 2-phenylallyl group] ... be strong enough to compete with the far better established process of more extended interaction of a metal atom with a cyclic polyolefinic ligand?" The answer, which may be seen in Fig. 6-4, is "that the C$_7$H$_7$ ring is [only] a three-electron donor [and] there is a C–H–Mo interaction. Such a C–H–Mo ... interaction is structurally and thermodynamically competitive with [more extended] olefin-metal bonding."

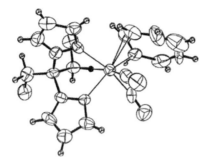

Figure 6-4. The structure of $(Et_2Bpz_2)(Mo(CO)_2C_7H_7$

It was not until several years later that other examples of C−H⋯M bonding began to appear in the literature. The first one was in $\{Fe[P(OMe)_3]_3(\eta^3-C_8H_{13})\}(BF_4)$, which was obtainable in the form of large enough crystals to permit collection of neutron diffraction data (*J. Am. Chem. Soc.* **1980**, *102*, 981). This revealed a situation identical in two essential ways to the one I had described six years earlier: (1) a metal atom that would have a 16-electron configuration without the C−H⋯M interaction; (2) significant distortion of the ligand owing to the strength of the C−H⋯M interaction. There was a small difference in the sense that the C−H⋯Fe interaction was somewhat more toward the side-on type while the C−H⋯Mo had been more toward the linear type. Many other genuine examples (as well as chimerical ones, such as $TiCl_3CH_3$) have since been found, and in a review in 1983 (*J. Organomet. Chem.* **1983**, *250*, 395), the term "agostic" was proposed for this interesting and important phenomenon which I am proud to have discovered some years before others began to recognize it.

There is one recent piece of work done in my laboratory that I would like to mention because it shows that the very common practice of inferring a reaction mechanism from a feature seen in a crystal structure can be unreliable. In 1986 the molecule shown in Fig. 6-5 was reported.[*J. Chem. Soc., Dalton Trans.* **1986**, 1629] This molecule provides an excellent example of an agostic interaction; the ethyl group is highly distorted ($\angle$ Ti−C−C = 84°) so that a β-hydrogen atom can approach the titanium atom closely. The posture of the ethyl group seems to suggest that, as the authors themselves put it, it "models a stage about half-way along the reaction coordinate for a β-elimination reaction to form the titanium-ethylene-hydride complex $[TiH(dmpe)(\eta-C_2H_4)Cl_3]$." As they also pointed out, the latter compound would certainly be unstable. Since the

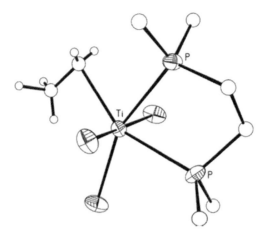

Figure 6-5. The TiCl$_3$(dmpe)C$_2$H$_5$ molecule.

parent compound is itself thermally unstable (decomposing at ca. 37 °C in solution), it would seem likely that ethylene would be eliminated easily from TiCl$_3$(dpme)C$_2$H$_5$, and ultimately the dinuclear species Cl$_2$(dmpe)Ti(μ–Cl)$_2$TiCl$_2$(dmpe) would be formed. In 1997 we reported the isolation and characterization of this latter species [1391]. However, in spite of the suggestion that seems implicit in the structure shown in Fig. 6-5, we found [1373] that decomposition actually occurs by a Ti–C bond heterolysis, giving ethyl radicals which then form C$_2$H$_4$, C$_2$H$_6$ and C$_4$H$_{10}$ in the expected ratio.

More Metal—Metal Multiple Bonds

There had been a short pause in our work on M–M multiple bonds and by the time it was resumed in late 1972 at TAMU, others had begun to make interesting contributions to the subject. Henry Taube had reported that the Mo$_2$Cl$_8{}^{4-}$ ion in aqueous solution could be converted to Mo$_2$(H$_2$O)$_n{}^{4+}$, [Mo$_2$en$_4$]Cl$_4$, and [Mo$_2$(SO$_4$)$_4$]$^{4-}$, and Joe San Filippo then at Rutgers had reported the first Mo$_2$Cl$_4$(PR$_3$)$_4$ compounds, and a number of other interesting observations. There was no X-ray crystallographic support for any of these species, however. In addition, in 1971 Wilkinson had reported the preparation of the triply bonded compound Mo$_2$(CH$_2$SiMe$_3$)$_6$ along with a crude supporting crystal structure.

In our first effort to obtain structural data on Taube's [Mo$_2$(SO$_4$)$_4$]$^{4-}$ ion, Tom Webb (a postdoc who is now at Auburn University) not only obtained the structure of that anion (the first [M$_2$(SO$_4$)$_4$]$^{n-}$ ion to be structurally characterized) but also found that it was easily oxidized to the $[Mo_2(SO_4)_4]^{3-}$ ion which has a metal-metal bond of fractional order, 3.5 [420]. Tom stayed for two years and did a great deal more important work on M–M multiple bonds. At about the same time Erik Pedersen came for the year 1973–74 from the Technical University of Denmark, where he was a lecturer. He was a very talented and experienced physical chemist and his work led to five papers in which electrochemical, EPR, and magnetic susceptibility measurements were all used to bring a great deal of new insight concerning the electronic structures of Re_2^{5+}, Mo_2^{5+}, and Ru_2^{5+} compounds. I was very sad to hear that this quietly brilliant man died prematurely in 1991. The last of his papers [423] cites the preceding ones.

In 1975, the curious story of what had initially been presumed to be the $Mo_2Cl_8^{3-}$ ion was finally resolved. As already noted (see page 137), a compound that appeared to contain this ion was first made by Jurij Brenčič at MIT in 1969 and characterized crystallographically as resembling a confacial bioctahedron with one bridging chloride ion randomly missing. According to this formulation, there would have to be an odd number of electrons and this anion would therefore have to be paramagnetic. A magnetic susceptibility measurement made at that time appeared to be consistent with the presence of one unpaired electron. At Texas A&M, in 1972, one of my students, Zoltan Mester, examined the bromo analog, $Mo_2Br_8^{3-}$, and found that in the compound $Cs_3Mo_2Br_8$ there was a very similar structure, which we published [366]. Only after this publication did it occur to me to ask Zoltan to measure the magnetic susceptibility of this compound. He found it to be diamagnetic, with about the right magnitude to correspond with its composition. This was a shock to say the least, and I immediately asked him to remeasure the susceptibility of $Rb_3Mo_2Cl_8$ using a carefully purified, crystalline sample. In contrast to the previous results, which must have been obtained on an impure sample, $Rb_3Mo_2Cl_8$ was now also found to be diamagnetic.

Several possible ways to resolve this inconsistency suggested themselves, but the most plausible one was that in addition to two bridging halogen atoms, there must be a bridging hydrogen atom. This, of course, would never have shown up in the X-ray studies. The challenge, however, was to obtain proof of this. At this point another postdoc, Dr. Barbara Kalbacher, took over [460]. She first showed that no gas was either taken up (O_2) or released (H_2) in the reaction that led to these products. Thus, we concluded that their formation could be attributed to the following reaction:

$$Mo_2(O_2CCH_3)_4 + 5HX + 3M^+ + 3X^- \rightarrow M_3Mo_2X_8H + 4CH_3CO_2H$$

We then did several experiments to obtain evidence that the μ–H atom is really there. The first was to carry out the reaction in water labeled with tritium and show that 1± 0.2 atoms of T were present in each labeled molecule so formed. We also used deuterium labeling experiments; the shift of a band in the infrared spectrum of the $Mo_2X_8H^{3-}$ ion at 1245 cm^{-1} to 915 cm^{-1} in the deuterium analog provided more support for the presence of the μ–H atom.

What was most needed to provide definitive structural data on the $[Mo_2Cl_8H]^{3-}$ ion was a compound in which there was no disorder about the 3-fold axis, and Avi Bino was the first to provide one [590]. This and several more compounds he made allowed us to identify and refine the position of the μ-H atom [720]. Finally, in 1984 we obtained a neutron diffraction structure of $(NMe_4)_3 \cdot Mo_2Cl_8H$ which was the final definitive result [775].

Collaboration with Malcolm Chisholm

In September of 1971, I received a letter from a young Englishman who had earlier done his Ph.D. with my old friend Don Bradley at Queen Mary College and was then

doing postdoctoral work with Howard Clark at the University of Western Ontario. He expressed an interest in joining me for more postdoctoral work and arrangements were soon made for him to start the following September. However, in the early Spring of 1972, he informed me that he urgently needed to see me to discuss an unexpected development and that he intended to fly down. When he came, soon thereafter, he explained that he had been offered an assistant professorship at Princeton University, to begin the following

Malcolm Chisholm making a point.

September, and he was concerned about the conflict between this and his agreement to come to work with me. This, I told him, was a no-brainer; the offer from Princeton was one I couldn't imagine his not accepting.

During his brief visit, it became obvious to me that Malcolm Chisholm was an exceptional person in terms of talent, integrity, and personality. It also turned out that one of the reasons he had been interested in working with me was that back when he was in Bradley's lab he had made some dinuclear dimethylamido compounds of molybdenum and tungsten which he thought might contain metal-metal bonds. So we agreed that after he got settled at Princeton, he would make them again and we would try to establish their structures. Thus began a cherished friendship not to mention a series of collaborative publications now numbering in the forties.

The first of these publications [391], in 1974, reported the structure of the triply-bonded $(Me_2N)_3Mo\equiv Mo(NMe_2)_3$ molecule, as shown in Fig. 6-6, and the next year we reported the tungsten homolog. Thus, it became clear that Wilkinson's isolated report in 1971 was just the tip of an iceberg.

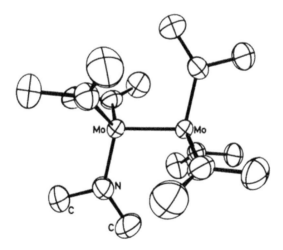

Figure 6-6. The $Mo_2(NMe_2)_6$ molecule as first reported in 1974 [391].

As Malcolm enlarged his band of students, this area expanded very rapidly, driven mainly by Malcolm's creativity. Two more of our important publications from that period were a combined theoretical (by George Stanley and Barbara Kalbacher at A&M) and experimental PES (by Jennifer Green at Oxford) study of the triple bonds in $X_3Mo\equiv MoX_3$ compounds (X = RO, R_2N, R) [506] and a review of metal-metal triple bonds in *Accounts of Chemical Research* [558] by Malcolm and myself. Fig. 6-7 shows how the combination of SCF-Xα-SW calculations and the PES spectrum of a $(RO)_3Mo\equiv Mo(OR)_3$ compound made a compelling case for the $Mo\equiv Mo$ bond.

One might have supposed that Malcolm's future at Princeton would have been assured, but a complication arose. Jack Norton (today at Columbia University) was also an assistant professor and Malcolm became convinced that Princeton intended to keep only one of them. Considering this to be an unsavory situation, he resolved to look elsewhere and soon had a tenure offer from the University of Indiana. Had Princeton been willing to assure him that he and Jack could both earn tenure there, Malcolm might still be there today. Ironically, a couple of years after Malcolm left, Princeton decided not to offer tenure to Jack Norton.

These two misguided decisions had the effect of putting me, briefly, into an unaccustomed relationship with Earl Muetterties. Both of us had been advising Princeton, for several years before Malcolm left, that they should make it clear that in due course both Malcolm and Jack would be eligible for tenure. We were both

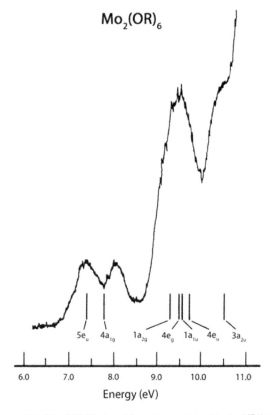

Figure 6-7. The measured PES of $Mo_2(OCH_2CMe_3)_6$ and the positions calculated by the SCF-Xα-SW method. The $5e_u$ and $4a_{1g}$ features correspond to the π and σ components of the Mo$\equiv$Mo triple bond.

Dee and I with Joanne and Earl Muetterties in Venice, ca. 1975.

very annoyed when their mishandling of the situation caused Malcolm to leave. Earl was the first to learn of their foolish decision not to keep Jack and Earl phoned me in a state of high dudgeon, proposing that we both "raise cain" with the powers that be at Princeton. My own reaction was something like "to hell with a place that piles one stupid move on top of another" and I proceeded to talk Earl around to agreeing that "raising cain" really wouldn't do any good to anyone. It was quite a role reversal because normally, over many years, it had always been the calm, rational Earl who had had to persuade me to calm down about something.

The final episode of Princeton's misbegotten effort to establish an inorganic division began with a letter sent to me about a year after Jack Norton had departed for Colorado. I believe the letter came from a Dean who first informed me of Princeton's keen and sincere desire to hire an outstanding young inorganic chemist and then asked for my help in doing so. My reply was that given Princeton's very recent absurd performance in this regard, in the face of the advice they had received from two of the leading inorganic chemists in the world, my attitude was (and I so expressed it): "I don't care if you hire Mickey Mouse, as long as you don't bother me about it."

The Christmas/New Year break in the winter of 1977/78 is one I still remember with a glow of satisfaction. Until then no set of homologous M–M multiply bonded compounds in the same group had been made. Two postdocs, Ron Niswander and Phil Fanwick, were, like me, working during the week between Christmas and New Year's day, and Ron was trying the reaction of the group 6 hexacarbonyls with 6-methyl-2-hydroxypyridine, in the hope that at least one or two of the following reactions would work:

To our delight, all three of them worked [546], giving the three isomorphous crystalline products, whose structures were determined by Phil Fanwick, with help from Janine Sekutowski. The fact that the reaction with $W(CO)_6$ worked was particularly important, because prior to that there had been no reported paddlewheel molecule containing a W–W quadruple bond. In the next few years, work by Tim Felthouse (another of the outstanding postdocs I have had) and others greatly enlarged the range of quadruply bonded W_2 compounds.

A few more words about Phil Fanwick are in order. Phil did his Ph.D. with my good friend Don S. Martin at Iowa State and came to me with the direct purpose of getting my lab into the realm of polarized single crystal spectroscopy. This he did superbly, but he eventually turned into a crystallographer, and is now the departmental crystallographer at Purdue University.

The Rise and Decline of the Crystal Structure Industry

In the early 1970s, awareness that crystal structures could be critical to the solution of many research problems was growing rapidly, but access to structural results remained very limited. The only way for a non-crystallographer to get a crystal structure done was to persuade a friend or colleague in a university who was a crystallographer to do it for him. Researchers in industrial laboratories were almost never able to do that for security reasons. It was several of my early coworkers at Texas A&M who totally changed this picture.

My first A&M Ph.D. was Jan Troup, as I have already mentioned (see page 167). He also was one of my all-time best, well up to the standard I was used to at MIT. His

Ph.D. work led to about twenty publications, in all of which crystallography featured prominently, although Jan was also a very capable preparative chemist. One of the first of the publications based on his work was a Letter to the Editor of the *Journal of the American Chemical Society* in 1973 [355]. Here we not only reported the results of four crystal structure determinations, but emphasized the speed, efficiency, and low cost with which these had been done. The real message was not the results themselves but the idea that X-ray crystallography had reached the point where it could be a common rather than an uncommon tool. It was I who had written this note in this particular way, but it was Jan who was to translate the message into an enterprise.

Jan, a born entrepreneur, got to thinking about whether there might be a business opportunity here and talked to Bert Frenz and Chuck Lukehart about it. The three of them did some very serious thinking, some cost analysis, and finally sent out a fairly large number of letters to both academic people and industrial laboratories to see who might be interested in using a confidential fee-for-service X-ray crystallography facility. The response to this market research was encouraging, and in late 1973, they issued the *Prospectus* of the *Molecular Structure Corporation*. With my cooperation, they did a few structures on my equipment and reimbursed me at some rate we had agreed on for the use of the facilities.

It soon became evident that this activity might have a real future, and the volume of business quickly grew to the point that they needed their own diffractometer. They applied to a local bank for a loan, but were greeted with cautious incredulity and reluctance, which was not surprising. How else could you expect a banker in a small town in east central Texas to respond to a proposal by three young men with no business background, no credit rating, no security, to establish a type of business that was without precedent, doing something that was completely incomprehensible to a banker, but required a substantial amount of venture capital? I stepped in and convinced the head of the bank, whom I knew, that the proposal was sound and the young men highly capable and motivated. What was probably more to the point, I promised that if the business went bust I would pay off the loan and take the diffractometer and computer off their hands. I was not sure how I would manage to do that, but then I was confident that they weren't going to fail.

In a very short time, the Molecular Structure Corporation (MSC) was housed in a tiny commercial property just across Texas Avenue from the main entrance to the

University. Their reputation and their volume of business, especially from industry, continued to grow rapidly and after only a year or so they moved to much larger quarters in south College Station. In the meantime, Chuck Lukehart got a faculty position at Vanderbilt University and had to drop out. Mike Extine, a Ph.D. of Malcolm Chisholm's from Princeton, had been a postdoc with me, 1976-78, and had become a very skilled crystallographer. He joined the company and was instrumental in writing software for the Texray package, while, in the meantime, Bert Frenz had decided to leave and start a computer retailing business of his own.

The Texray package was a complete software system for structure solution, refinement, and graphics, and MSC also began marketing Rigaku diffractometers. Finally, Jan Troup decided to move the whole corporation to the Woodlands (a high-tech commercial and industrial park) just north of the Houston Intercontinental Airport, at which time Mike Extine decided he would go home to the northwest. I won't continue with the story of the MSC because it would take me too far afield, but it was one of the most unusual and interesting developments I have been connected with.

A few years later, Victor W. Day, a former member of the *Staph* nuclease team, who had joined the faculty at the University of Nebraska, started a company called Crystalytics, which he and his wife Cindy still run. There is still, obviously, a call for crystal structures on an *ad hoc* basis, particularly by smaller industrial laboratories, but faster diffractometers, supersmart software, and the new profession of staff crystallographer have considerably curtailed the market. Crystalytics is still in business, but the MSC, now principally a purveyor of equipment, is a wholly owned subsidiary of Rigaku and Jan Troup has other interests such as designing specialty equipment for scuba diving.

My First Visit to Israel and the Chemistry It Led To

During 1975, I was invited by the chemistry department of the Technion, in Haifa, Israel, to come for a lengthy stay as a visiting professor. I think they would have liked me to stay for as long as a semester, but it was finally agreed that a visit of three weeks would be mutually satisfactory. Thus I spent the first three weeks of May, 1976, in Haifa. During this period, I gave seven lectures at the Technion as well as one at Ben Gurion University in Beersheva and one at the Weizmann Institute in Tel Aviv.

On my arrival in Israel, Zvi Dori met me at the Lod airport, just east of Tel Aviv, in his Volvo station wagon and we set off for Haifa. We had not left the environs of Tel Aviv before it became clear that the car was in dire straits and was not going to get us to Haifa. As he pulled off into the breakdown lane, Zvi said, wryly, "Welcome to the holy land." I do not recall how he did it, because cell phones did not then exist, but he soon contacted a friend in Tel Aviv, and it was not long before we were sitting in his friend's apartment having a cup of coffee. This friend was someone he had known during his graduate student days in New York. The man was a specialist in automobile repairs, who

Zvi Dori at the entrance to the chemistry building at the Technion (1975).

had run a garage in the Bronx and was now in the same business in Tel Aviv. What an example of knowing the right person, in the right place, at the right time! Zvi's friend lent us a car and undertook to get Zvi's repaired.

I stayed in a kind of hotel-cum-boarding house where good Israeli breakfasts and laundry service were available. Most days I walked the two to three miles to the campus, but sometimes I took a bus. Lunch was generally had at a dining room at the student center and for evening meals I went to restaurants closer to the center of town. I frequently went to the restaurant at the Dan hotel, which was a first-class place.

Besides my activities at the Technion, I was taken by my hosts, especially Zvi Dori, as well as by Dan Meyerstein in Beersheva, to see nearly all of the country, from the Golan Heights to the Negev. Among the many wonderful things I saw were the ruins at Acre, the Galilee region, a kibbutz, Jerusalem, the Dead Sea, Masada, and the caves where the Dead Sea Scrolls had been discovered, not too many years before. There wasn't a dull moment the entire time.

The chemistry department at the Technion was warm, convivial, and very stimulating. The faculty covered the whole range of chemistry with people of the highest quality. Those I remember very warmly were Magda Ariel (electrochemistry), Michael Cais (organometallic), Mordechai Folman (physical), Amitai Halevy

Zvi Dori, Avi Bino, and I when we were working together at A&M. Avi is holding a model of one of the Mo_3 or W_3 molecules we worked on.

(physical), Frank Herbstein (crystallography), Ruben Paunz (theory), and Mordechai Ruben (organic). My very special memories are of Zvi Dori and David Ginsburg, of whom I shall say more.

I believe it was Zvi Dori, with perhaps an assist from Mike Cais, who was primarily responsible for my invitation. Zvi and I barely knew each other prior to my visit to Israel, but a close friendship rapidly developed, and exists to this day, although we rarely see each other. Zvi is an extraordinary person. After graduating from the Technion in 1956, he served in the Israeli merchant marine between the ages of 22 and 27, but then decided to become a marine engineer. He applied to MIT but they advised him to take a year of university work somewhere else to see if he was still academically qualified. The world thus lost a great marine engineer.

He enrolled at Columbia University for a one-year general studies program, during which he took a chemistry course taught by Harry B. Gray. He decided he would really like to be a chemist and stayed at Columbia where he completed a Ph.D. under Gray's supervision in 1966. After five years as an Assistant Professor at Temple University, he returned to the Technion in 1971 as an Associate Professor. Zvi's decision to become a chemist was a good one. It turned out that he is exceptionally gifted in preparative chemistry.

I should also mention that Zvi's father, Yaakov Dori, was commander of the Jewish Underground Army (1946-49), then the first Chief of Staff of the Israeli Army (1949-54), and then President of the Technion (1954-68).

At the time I went to Israel, two related questions had been on my mind (though I had not yet done anything about them) since the original report by Wilkinson (1959) that $Mo_2(O_2CCH_3)_4$ was formed in about 20% yield by refluxing a suspension of $Mo(CO)_6$ in a mixture of acetic acid and acetic anhydride. The first question was: "What happens to the other 80% of the molybdenum?" It was known that all of the carbonyl groups were lost, but beyond that, the question of what was in the green-brown, oily supernatant liquid from which the $Mo_2(O_2CCH_3)_4$ crystallized was unanswered. The second question was "Why does a similar procedure with $W(CO)_6$ yield no $W_2(O_2CCH_3)_4$, giving only a brown oily liquid?"

I discovered early in my stay that the same questions had concerned Zvi. Moreover, he and two chemists at the Hebrew University in Jerusalem, Prof. Michael Ardon and his student Avi Bino, had already undertaken experiments to find out, and were just about to submit a report to *JACS*, which appeared in October (1976, *98*, 7093). By diluting the green-brown liquid with water and passing that through a cation exchange column, they had isolated a green-brown cationic species that appeared to have a charge of +1. The green-brown eluate solution was oxidized by permanganate to give a red solution from which a red crystalline material was isolated after again using a cation exchange column and eluting with triflic acid. In this red, crystalline product were trinuclear cations for which they reported the structure shown in Fig. 6-8.

The cation shown in Fig. 6-8 was an eye-opener in several respects. Here was the first example in an aqueous solution of a trinuclear molybdenum cluster. Another important point in the Communication was that "Preliminary experiments indicate that tungsten carbonyl reacts with

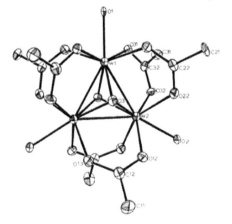

Figure 6-8. The structure of what was thought at first to be $[Mo_3(O_2CCH_3)_6(CH_3CH_2O)_2(H_2O)_3]^{2+}$ as originally reported by A. Bino, M. Ardon, I. Maor, M. Kaftory, and Z. Dori, *J. Am. Chem. Soc.* **1976**, *98*, 7093.

acetic acid to form species which may be related to those obtained with Mo(CO)$_6$."
However, as a thoughtful examination of Fig. 6-8 will show, this structure and the
correct composition of the cation are problematic. The problems were later identified
and corrected by Dori, Bino, and me, as I shall explain below.

Before turning to that I want to say a few words about David Ginsburg, whom I
met for the first time on this visit to Israel. He was a man with an enormous range of
talents, including that of being a warm, perceptive, and sympathetic human being.
He had been born and educated in New York, and retained an unmistakable accent,
but spent his entire career at the Technion. He was an outstanding organic chemist,
and also deeply devoted to the Technion. He had at one time or another in his career
held almost every important post and might well have been Yaakov Dori's successor
as president, but he declined to be considered. It was David who created the famous
"Wall of Fame" in the front of the big chemistry lecture hall at the Technion (described
in an article by Mordechai Rubin in *The Chemical Intelligencer*, **1997**, *3 [3]*, 44.)

While I was still in Israel in May of 1976, Zvi Dori and I became convinced that
the discovery he and his coworkers had made had really opened up a new field. Since
Zvi was due for a sabbatical in the near future, we decided that it would be an excellent
plan for him to spend it with me and the very day I got back to A&M I talked to
Arthur Martell about having Zvi appointed a Visiting Professor and providing him
with some money to supplement his somewhat inadequate sabbatical pay in return for
his giving a few lectures a week in freshman chemistry. In this way, he and his family
(wife, Marilyn, and three boys) arrived here in August, 1977. I still remember that
early on the day he arrived he called me and I asked where are you now? He replied
"Palestine," and we both laughed. He was in Palestine, Texas, some one hundred miles
from College Station.

Before dealing with the chemistry we did while Zvi was here in College Station,
I must say something about his coauthors from Jerusalem on that paper published
in October of 1976, Professor Michael Ardon and his graduate student Avi Bino,
who later (1977) completed his Ph.D. Mike Ardon was an excellent solution chemist
and he and his wife Ora later became good friends of mine. Mike died in 2006. Avi
came to me as a postdoc from January 1978 to the end of May 1979 and thus his stay
overlapped Zvi's during the first half of 1978. The three of us accomplished a great deal
during and after that period. Our work was expedited by the fact that Avi had spent

a short period in Germany where he had already acquired some working knowledge of X-ray crystallography. With these two collaborators one of the most important and satisfying chapters in my research biography opened [547]. And it continued for a decade, our last joint paper in the area [1031] appearing in 1988.

One of the first things we did was actually not a direct outgrowth of the earlier work on the reaction product of $Mo(CO)_6$ with acetic acid. Instead it derived from some earlier work done by Mike Ardon which in turn derived from work by Souchay in Paris on the nature of Mo^{IV} in aqueous solution. It had been shown that the aqua ion was quite stable, but not mononuclear. Solution studies had suggested at least a binuclear species such as $[(H_2O)_4Mo(\mu\text{-}O)_2Mo(H_2O)_4]^{4+}$. We decided to see what it might be possible to isolate from solution and then structurally characterize by X-ray crystallography.

We prepared a solution of Mo^{IV} according to a method published by Ardon, absorbed this ion on a cation exchange column and tried various reagents to elute it. We found that oxalic acid worked very well and that upon adding CsCl to the eluate, beautiful red-purple crystals were formed. The structure we found, Fig. 6-9, was one of the most important I have ever had the thrill of seeing [547]. Here we have a tight Mo_3O_4 core with an octahedron being completed around each molybdenum atom by two oxygen atoms of an oxalate ion and one from a water molecule. The Mo–Mo distances are 2.49 Å — clearly a bonded distance — and there are six electrons available to form three Mo–Mo single bonds.

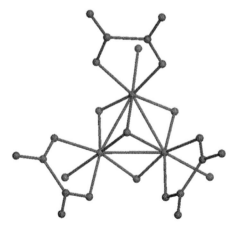

Figure 6-9. The structure of $[Mo_3O_4(C_2O_4)_3(H_2O)_3]^{2-}$.

A structure of this M_3X_{13} type, shown in Fig. 6-10, was already very familiar to me. It had been seen before in condensed phases where the nine outer X atoms were shared with adjacent units, but it had never been found as a discrete unit. In fact, one of the condensed phases was the $Zn_2Mo_3O_8$ compound to which I had previously [142] applied a molecular orbital treatment, and showed that the two lowest metal-based molecular orbitals, *a* and *e*, were occupied so as to give a net average Mo–Mo bond order of one.

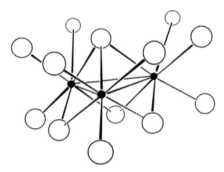

Figure 6-10. The M_3X_{13}-type structure.

Another thing that Bino, Dori, and I did very early in our collaboration was to reexamine [654, 690] the alleged $[Mo_3(O_2CCH_3)_6(OC_2H_5)_2(H_2O)_3]^{2+}$ ion (Fig. 6-8). Several features of the reported structure are extremely suspicious, but were neglected because of the great interest caused by the triangular structure, which *grosso modo* had to be right. The problems are as follows: (1) the β-carbon atoms of the purported OCH_2CH_3 groups are not seen, even fractionally, and (2) the displacement ellipsoids of the presumed oxygen atoms are entirely too large. The presence of CH_3CH_2O groups had been postulated only because they seemed to be the most chemically reasonable thing to arise from CH_3CO_2H molecules.

However, very careful microanalyses and density measurements eliminated CH_3CH_2O and also CH_3O, but agreed with CH_3C, also a possible, though far less expected, derivative of CH_3CO_2H. When CH_3C, a triply bridging ethylidyne group, was used in refining the crystal structure, all glitches vanished and there could be no question that this was correct. This was the first example and is still the only one (except for closely related species with one O atom cap and only one

CH_3C cap) of an ethylidyne compound that is stable to water, aerated water, and even to aqueous permanganate.

With the change from monovalent RO caps to trivalent CH_3C caps, of course, the electronic structure of the compound had to be reformulated. There were now only four cluster electrons, giving an a^2e^2 configuration that is paramagnetic and corresponds to Mo–Mo bonds of order $^2/_3$. In the course of the work leading to the reformulation of $[Mo_3(O_2CCH_3)_6(CH_3C)_2(H_2O)_3]^{2+}$, we made two more important discoveries. We found that by variations in the reaction and workup conditions, we could obtain a very stable species $[Mo_3(O_2CCH_3)_6(\mu_3\text{-}O)_2(H_2O)_3]^{2+}$ as well as the $[Mo_3(O_2CCH_3)_6(\mu_3\text{-}O)(\mu_3\text{-}CH_3C)(H_2O)_3]^+$ ion, that is, a species with two oxygen caps and also one with one oxygen cap and one ethylidyne cap. In these ions, there are six cluster electrons and they are electron-precise for three Mo–Mo single bonds. Accordingly, the Mo–Mo distances, ca. 2.76 Å, are appreciably shorter than those in the $[Mo_3(O_2CCH_3)_6(\mu\text{-}CH_3C)_2(H_2O)_3]^{2+}$ ion, ca. 2.89 Å, where the Mo–Mo bond orders are only $^2/_3$. We also isolated and obtained the structure of the $[Mo_3(O_2CCH_3)_6(\mu\text{-}CH_3C)_2(H_2O)_3]^+$ ion, which should have a bond order of $^5/_6$, and found the Mo–Mo distance, 2.82 Å, to be half-way between the two just mentioned. Clearly, things were working out in a very satisfactory way, and the question of what happens to some 80-85% of the molybdenum when $Mo(CO)_6$ reacts with acetic acid was definitively settled and a new area of molybdenum chemistry was opened up. In summary, a mixture of the three species, shown schematically in Fig. 6-11, is where 80-85% of the molybdenum ends up in the reaction from which $Mo_2(O_2CCH_3)_4$ was first obtained by Wilkinson.

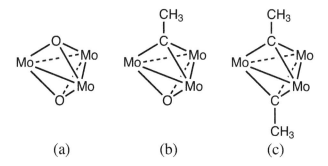

(a)   (b)   (c)

Figure 6-11. The three types of bicapped trimolybdenum clusters that arise in the reaction of $Mo(CO)_6$ with AcOH/ Ac_2O. There are two $\mu\text{-}CH_3CO_2$ groups on each Mo–Mo edge.

We subsequently discovered other ways to make bicapped Mo_3 clusters [819]. One of the new ones was $[Mo_3(O_2CC_6H_5)_6(\mu_3\text{-}O)_2(H_2O)_3]^{2+}$, that is, the benzoate analog of the previous acetate compounds. The amazing thing about this species is that it can be fully and quantitatively nitrated, at one meta position on each of the six phenyl groups, using a mixture of concentrated nitric and sulfuric acids without *any* observed decomposition. The stability of the Mo_3O_2 core is truly astonishing.

We also extended our work to analogous tungsten chemistry. Several others, Steve Koch, Michelle Millar, and Janine Sekutowski, also made important contributions [564]. When $W(CO)_6$ reacts with $CH_3CO_2H/(CH_3CO)_2O$ all of the metal atoms form trinuclear products and no $W_2(O_2CCH_3)_4$ is obtained at all, although this and other $W_2(O_2CR)_4$ compounds exist and can be made in other ways. The first step in this direction was taken by Avi Bino before our collaboration began, although the preparation and characterization of $[W_3(O_2CCH_3)_6(\mu_3\text{-}O)_2(H_2O)_3]^{2+}$ and a related species were first reported by the three of us (plus other coauthors) in 1978 [564]. In later publications, the tritungsten work was extended and we were able to obtain the species with one μ_3-O cap and one μ_3-CH_3C cap [888].

I want to return now to the fascinating story that grew out of our discovery of the structure of the $[Mo_3O_4(C_2O_4)_3(H_2O)_3]^{2-}$ ion which can be isolated from a solution of the aquo Mo^{IV} ion as prepared by Ardon's method. We commented in our first report on the difference between this trinuclear ion and the binuclear $[(H_2O)_4Mo(\mu\text{-}O)_2Mo(H_2O)_4]^{4+}$ ion that Ardon earlier believed to be present in solution. We were not sure what to make of this purported difference and, in fact, it caused us a lot of concern. It might also be mentioned that in the original report of the aquo Mo^{IV} ion by Souchay in 1966, it was said to be mononuclear, and Geoff Sykes (*J. Am. Chem. Soc.* **1975**, *97*, 589) had reported kinetic data which he said supported this. Three different forms of the "same" species seemed like at least one, and probably two, too many.

Avi, Zvi, and I decided to try again to see what could be fished out of the Ardon solution of Mo^{IV}. This time we eluted the cation exchange column with acetic acid and then added the disodium salt of EDTA to the eluate. After a few days we were rewarded with beautiful red crystals which X-ray crystallography showed to contain a central Mo_3O_4 cluster with a tridentate EDTA ion coordinated to each Mo atom [594]. The charge corresponded to Mo^{IV} and the dimensions of the Mo_3O_4 unit were essentially the same as those in the $[Mo_3O_4(C_2O_4)_3(H_2O)_3]^{2-}$ ion. Our doubts about the existence of mononuclear and binuclear formulations were strengthened, since for neither of these was there any direct support. We stopped short of explicitly questioning the binuclear formulation in solution because of our respect for the work of Mike Ardon, but the obvious implication of our work was that Mo^{IV} in non-complexing aqueous solution is $[Mo_3O_4(H_2O)_9]^{4+}$.

Since we were very much occupied with the study of the $[Mo_3(O_2CR)_6(\mu\text{-}X)_2\text{-}(H_2O)_3]^{n+}$ species and their tungsten analogs we postponed more work on the $Mo_3O_4^{4+}$ species, but others soon worked on them. The first was Kent Murmann (*J. Am Chem. Soc.* **1979**, *102*, 3984) who showed that four oxygen atoms per three molybdenum atoms are very slow to exchange. Murmann explicitly proposed that the aquo Mo^{IV} ion is hydrated $Mo_3O_4^{4+}$. The charge of 4+ agrees with Ardon's conclusion as to the total charge; it appears that Ardon's conclusion that this charge is distributed over only two rather than three Mo atoms resulted from an error in measuring the loading of the resin, and that his work does not actually support a binuclear species.

The concluding chapter of this story was published in 1989 by Richens, Merbach, Chapius, and coworkers (*Inorg. Chem.* **1989**, *28*, 1394) who not only did further studies of oxygen isotope exchange in solution but actually isolated the $[Mo_3O_4(H_2O)_6]^{4+}$ ion itself in a crystalline compound. Moreover, several subsequent studies have shown that the trinuclear species accounts for 100% (within experimental error) of all the molybdenum in an aqueous solution of uncomplexed Mo^{IV}. Avi, Zvi, and I are very pleased to have initiated this very satisfying story and to have had the correct answer right from the beginning.

Later on Zvi and I, with coworkers Willi Schwotzer, a postdoc from Switzerland, and Rosa Llusar, a graduate student from Spain, established the existence of the $Mo_3S_4^{4+}$ aquo ion, by again using the oxalate ion to form an isolable derivative [885]. We also showed that a cuboidal $Mo_4S_4^{6+}$ aquo ion exists and characterized the derivative $[Mo_4S_4(NCS)_{12}]^{6-}$ [886].

I mentioned earlier that while Avi Bino was here at A&M, and for a few years thereafter he worked in areas other than the Mo_3 and W_3 compounds, and a few words about those as well as about Avi himself are in order. Avi is a very gifted experimentalist. I have often described Avi as a man who never did a negative experiment. What I mean is that after trying something that we had decided was an interesting idea he never reported simply, "It didn't work." First of all, for Avi, most things *did* work. But

Michelle Millar and Steve Koch in 1980.

when he didn't get the desired results, his report was always something like, "I didn't get any of X, but I got the following interesting result." Avi was the man whose work made my day, many times. Among Avi's other important contributions were the facile synthesis of the $[Mo_2(HPO_4)_4]^{2-}$ ion, which has a very stable triple bond between the metal atoms [614].

Two more of the many outstanding postdocs I had during the 1970s were Stephen Koch and Michelle Millar. They were (and are) not only excellent chemists but they became very good chemists but they became very good friends. They both worked, either separately or together, on some of the most difficult chemistry being done in my lab at the time. However, I shall recall here only one project that they both worked on, because it has proved to be a milestone in the metal-metal multiple bond field. They prepared the first dichromium compound with a quadruple Cr–Cr bond shorter by about 0.1 Å than any M–M bond length ever before reported in a stable compound, 1.85 Å [513]. Indeed when scaled to the inherent sizes of the atoms it was even shorter than the $N\equiv N$ bond in dinitrogen or the $C\equiv C$ bond in acetylene. The molecule containing this prodigiously short bond is shown in Fig. 6-12. It should be noted, however, that even though it is prodigiously short, this Cr–Cr quadruple bond is not prodigiously strong, because of major inner shell repulsive energies that arise when relatively large atoms come so close together.

Michelle and Steve followed up this work by making other Cr_2 paddlewheel compounds with bridging ligands of similar character and a year later made a compound with a Cr–Cr distance of only 1.828(2) Å, which remains to this day the shortest known metal-metal bond [548].

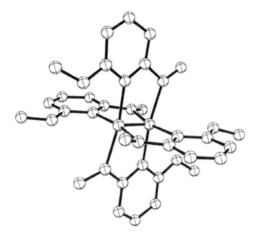

Figure 6-12. The structure of the $Cr_2[2,6-(MeO)_2C_6H_3]_4$ molecule, discovered by Steve Koch and Michelle Millar.

My Adventure in Iran

In March of 1975, I made an unforgettable trip to Iran, then ruled by Shah Reza Pahlavi. The Iranian Chemical Society organized their annual meeting that year as an international one and invited as principal speakers Linus Pauling, Robert S. Mulliken, Cyril A. Grob (a physical organic chemist from Switzerland), C. S. "Speed" Marvel, and me, as well as a few other non-Iranians to give sectional lectures. The meeting was held in Shiraz at the Pahlavi University which had been created a few years earlier to provide Western style education entirely in English. This was a key part of the Shah's effort to modernize and Westernize the country.

I traveled there *via* Italy, where I made some visits, and arrived at a hotel in Tehran a little after midnight, two days before the start of the conference. I was informed that a car would come at seven the next morning to take me to the airport for a flight to Shiraz. So, at about a quarter to seven I groped my way into the lobby. There I found Robert Mulliken, looking and feeling even more ragged that I did. He had arrived at about 3 am. We were taken to the airport, which was very unpleasantly jammed with a chaotic crowd. After much bumping and buffeting, we found ourselves on a plane to Shiraz. I think we both dozed off en route.

The reception in Shiraz was quiet and efficient, and we were soon checking into the Hotel Cyrus, a brand new, grandiose addition to the Intercontinental chain of luxury hotels. This was a very welcome sight and Robert and I agreed that the best way to pass the day until the opening reception in the evening would be to catch up on sleep.

As we headed for the elevators, we passed a pleasant little bar, and almost in unison said, "a nice martini would hit the spot." A young barman, in crisp attire, awaited us, but he lacked one thing — knowledge of English. As we were attempting to communicate our needs to him, we arrived at the decision that quadruple martinis would best provide the solace we craved. It turned out that the barman did understand what a martini was, but we could not get through to him regarding the volume. Pantomime did it. I slowly and deliberately pointed to eight stemmed martini glasses on the shelf behind him and indicated that they should be lined up on the bar. Then, when I pointed at each one, saying "martini, martini, martini, etc.," he made eight martinis.

Similarly, he was given to understand that four were to be poured into one large glass containing ice, and the other four into another. He executed all this with a bemused, not to say amazed, expression. Thus, Robert and I were soon comfortably seated in upholstered chairs luxuriating in the nearest thing to the balm of the gods we could have desired.

The next morning, the first plenary session began with an address by Linus entitled Molecules in Relation to Health and Disease. It was devoted to sermonizing on the evil of smoking, the need to avoid excess sugar, and the panacea-like virtues of vitamin C. As to smoking, the take-home message, proclaimed with arm raised and index finger pointing to heaven, like Rodin's statue of John the Baptist, "Every cigarette you smoke takes twelve minutes off your life."

In the afternoon, Robert presented his plenary lecture. A more striking contrast to Linus could not be imagined. The title was The Progress of Molecular Electronic Structure Theory, and he began by saying he would begin with the Coolidge-James calculation (1933) and trace the growth of the field up to current work in his own group. After about twenty minutes, Robert, who was somewhere in his late seventies, was seated on a stool behind a fairly large and ornate lectern, with his chin down, eyes on his notes, and, as usual, speaking in a barely audible voice. It was a lecture well within the established Mulliken-esque parameters. All was well as far as I could see. However, the President of the Iranian Chemical Society had steadily become more agitated and finally asked me if "Professor Mulliken is ill." I assured him that I was familiar with Professor Mulliken's lecturing style and saw no cause for alarm. He gave me a look of utter disbelief, and ten or so minutes later told me he was sure "Professor Mulliken must be ill." I again tried to reassure him.

Anyway, when Robert eventually reached the end of his allotted time, he was pointing out that the postwar advent of computers was beginning to open new horizons for computational chemists. As this juncture, he made the observation that he was "running a little behind" and would have to fast forward to a few current results. About a quarter of an hour later, the lecture ended, to enthusiastic applause, and Robert came down with a firm step and shook hands with the President, myself, Linus, and others. The President appeared mightily relieved.

The following day we had a wonderful tour of the ruins of Persepolis, which is about 30 miles away from Shiraz. While I can usually take ruins or leave them alone, Persepolis is something I'll never forget. It is the ne plus ultra of ruins, at least in my experience. The civilizations that existed there before Alexander the Great laid waste to it must have given the word "fabulous" real meaning. The following evening there was an excellent son et lumiére presentation at Persepolis, but it couldn't top the experience of walking around in broad daylight and examining the remains at leisure.

The last day of the conference proved to be more memorable than I would ever have expected it to be. At breakfast, we plenary lecturers were told that immediately following a lecture by Gil Haight on teaching freshman chemistry, we would all be taken to the government broadcasting center to tape a panel discussion about the importance of scientific research for Iran.

From about 11 am to noon, we met members of the staff there and were then taken to the TV studio — which was devoid of furniture. Our hosts seemed surprised by this and sent out for help. After about three quarters of an hour, a couple of long tables, some chairs, and some microphones were in place, and we all sat down. Another prolonged period of inactivity then ensued. Of course, we were all getting quite hungry, but no relief on that score seemed likely. We were not a happy group. By the way, the most conspicuous thing in the studio was a large NO SMOKING notice.

Finally, Robert Mulliken could stand no more, pulled out his pack of cigarettes, and enunciated, very clearly, "If Linus doesn't mind, I'm going to take twelve minutes off my life." To which Linus replied, "well, that's entirely up to you, Robert." After a few obviously very satisfying puffs, Robert, who was clearly not enjoying our day, said directly to Linus, "I have never liked the title of your book."

"What's wrong with the title of my book?"

"Well," said Robert, "there is no such thing as *the* nature of *the* chemical bond. It's too simple."

To this, Linus, who was by now nettled, replied, "Well, Robert, you can't put everything into a title."

After another puff, Robert quietly repeated, "I have never liked the title of your book."

This curious exchange was abruptly ended by the entrance of an Iranian broadcasting crew, and the panel discussion soon got underway. The reader may draw his own conclusions about the feelings underlying the incident — much deeper than the proximate ones of hunger and annoyance with the Iranians. I believe it reflected the great difference in the research styles of these two formidable theoreticians. Pauling was always the seeker for a broad, simple, but sometimes too facile generalization, whereas Mulliken wanted to probe deeply into all the intricacies of a question. I have often wondered if Robert's famous observation that "the chemical bond (notice the *the*) is not so simple as some people seem to think" didn't intend "some people" to mean Pauling.

The daytime visit to Persepolis was our first venture out into the streets. It soon became evident that many an Irani who not long before had been riding a donkey as fast as he could make the donkey go, on whichever side of the street he (or the donkey) preferred, was now behind the wheel of a car, proceeding in exactly the same two ways. It was really frightening until we got out of town, where there was no traffic.

Another experience that I remember very clearly was having lunch at a neighborhood restaurant as the guest of one of the young faculty members. It was traditional for entire families to eat their midday meal together and the restaurant had a very warm family-like atmosphere. My host, who was certainly not against progress, pointed out to me that with the rise of Western-style factories and other businesses, men could no longer come home for the midday meal (which was the main one) and that many families were very upset by this. He said to me that the Shah, in his haste to "make progress," was going too fast for most of his people and that while they were generally supportive of the goal, they were not in favor of the pace. He said he thought that if the Shah didn't slow down there would be trouble. Only a few years later, it became clear how prescient this young man was.

A day before I left Shiraz, I asked a man named Mostashari, a Kurd, if I could be reimbursed for my travel expenses before leaving. I asked this man because, although he seemed to be despised by the Iranians who were nominally in charge, he had impressed me as the only one who seemed to know what he was doing, and seemed to do all the real work. He said he would arrange things and soon came back to me to say that we could go immediately to a bank and get the money, about $2,500. The bank had it ready in actual U.S. currency, but I was reluctant to carry this with me (the equivalent of perhaps $12,000 today) in cold cash. I therefore proposed to buy traveler's checks.

This was another problem. The bank had no denominations greater than $20, but, fortunately, they had a lot of them. I made a quick estimate of how many there would be for me to sign there, and then again to cash them, and decided that instead of employing my usual signature on checks and other fiduciary instruments, Frank Albert Cotton, I had better use F. A. Cotton. It still took me over a half-hour to sign about 150 checks. When I brought them to my bank here in College Station, it took another half-hour or so, and I am sure that if the people at the bank had not known me, they would have suspected me of being some shady character trying to launder dirty money.

The French Connection(s)

It was shortly after going to Texas that I began to develop what turned into a lifelong love of France, its culture, and its language. In 1973, I was invited to present a lecture at the annual meeting of the Société Chimique de France in Marseille. Something possessed me to decide that I would give my lecture in French. I soon found, however, that preparing such a lecture was a lot more difficult than I had expected, so I turned to my friend René Hugel (a former postdoc and at that time a professor at the University of Reims) to rescue me. He prepared a French translation of my manuscript [369] from which I teased out a briefer version that I presented orally. I had to work very hard on my pronunciation which, apparently, was at least understandable. There has been improvement since.

With French friends in Paris, 1996. Edmond Samuel, Anne Vessiéres, FAC, Danielle Che, Michel Che, and Gérard Jaouen.

About a year later, I was invited to spend five weeks as a visiting lecturer in Strasbourg. After doing a little research on this city, and taking account of the fact that my hosts would be Jean-Marie Lehn and Raymond Weiss, I found this invitation irresistible, and scheduled my visit for April 2 to May 31 of 1975. I applied for and received funds from NATO to help with travel and living expenses, and arrangements were made for me to stay in the Hotel des Princes, rue des Princes, about a 15-minute walk from the chemistry building, and not too far from the old city. And so it happened that only three weeks after my return from Shiraz I left again for five weeks in France.

L'Université Louis Pasteur, in Strasbourg, had established an annual lectureship honoring the famous Alsatian chemist Joseph-Achille Le Bel and in 1974 the inaugural lecturer had been Konrad Block. I'm not sure about the proportions of shock and pleasure that were mixed in the reaction of my hosts when I first informed them, as I began my first lecture, that I intended to give all of my lectures in French. While the French are always pleased when a foreigner shows respect for their language, against this had to be weighed the fact that in 1975 my French was exiguous and my hosts knew that. For them to listen to me must have been somewhat trying, but not nearly so trying as the preparation of these lectures was for me. I probably spent about 8 hours laboriously writing out each of the eight lectures, looking up about every second noun in the dictionary to check its gender.

As noted, Jean-Marie Lehn and Raymond Weiss were my cohosts but when it turned out that Jean-Marie had to be away for a fair part of the time I chose, lecturing somewhere else, it was Raymond who did most of the hosting. Raymond and I had a natural rapport because of three mutual interests: inorganic chemistry, crystallography, and horses. His office and labs, as well as my office, were on the seventh floor, which would ordinarily have meant frequent use of the typically small, slow European elevator. I took to walking up the seven flights each morning and this eccentric behavior was much talked about. In fact, the Le Bel lecturer the next year, Paul D. Bartlett, heard about it and asked me later if what he had heard was true. I told him it was only part of the whole story, because I had also whistled the Marseillaise all the way — which was not true, but I think he believed me.

It was during my visit in 1975 that the Université Louis Pasteur was considering the possibility of offering a professorial position to John Osborn, who was not awarded tenure as per standard Harvard practice at that time for entry-level appointees. I was able to play a small role in convincing both John and the university that they would suit each other well, and things turned out just that way. John really enjoyed life in Europe and he had a very stimulating effect on the chemistry department there.

My "on-the-job training" in French had a lasting effect in that I have continued to read French and I visit France often. I have given several dozen lectures there since Strasbourg and all of them have been in French. In Nice I later developed a collaboration with Jean Riess, who was then the professor of inorganic chemistry, but was already shifting his main interest to the use of perfluoro compounds as blood substitutes, a field in which he subsequently became a major player.

My French connections were reinforced in 1990 when on a visit to Paris I first met Gérard Jaouen, who is a professor at the École Nationale Supérieure de Chimie de Paris (ENSCP). He and his wife Anne, a biochemist, have since become friends that Dee and I visit at least once a year. Gérard is a true expert on French cuisine, French wines and the restaurants of Paris. Each year the four of us dine in one — or often two — of the two- and three-star restaurants. In several cases, owing to Gérard's expertise, we have followed restaurants from their beginning right up to three-star status, and in one case back down again. In most cases, we have felt that as they rose from one-star to three-star status, they have lost a bit of their zest and freshness. I cannot overstate my appreciation for Gérard's contribution to my education in gastronomy and vinology.

There is not a city in the world, not even Philadelphia where I was born, and still visit frequently, or London, which has in the era of "Tony Blair's Britain" declined precipitously in its cultural level, where I feel so much at home as I do in Paris. The variety and beauty as well as the vitality of Paris refresh me every time I go there. What urban view can match, in every direction, the one that is seen from the middle of the Pont des Arts?

Two other friends that have added enormously to my enjoyment of Paris over the years are Yves Jeannin and Michel Che. Yves retired several years ago and I have not seen much of him lately, but his insight and humor were always welcome. As a one-word description of Yves: sophisticated. Fortunately, I still see Michel Che and his wife Danielle frequently. Michel, with his exceptional generosity, has made major contributions to my familiarity with Paris. He and I, Dee, and Danielle have had some wonderful evenings together that I would never have experienced without his unmatched hospitality and imagination.

Finally, I am deeply grateful to the entire French Academic community for signal honors, such as the Société Chimique de France for the Lavoisier Medal, the Académie des Sciences for electing me a foreign associate, and four universities (Strasbourg, Bordeaux, Rennes and Paris) for making me one of their honorary alumni.

A Meeting in Southern Bavaria

I would venture to say that of all the meetings of inorganic chemists held in the last forty years, the one best remembered by those who were there was "the Ettal Meeting" in July of 1974. This was officially called a European Conference on Inorganic Chemistry, but in fact many Americans were present as well and some were speakers. A main theme of the meeting was chemistry of metal carbonyls and we celebrated the eightieth birthday of Walter Hieber (1895-1976), the father of metal carbonyl chemistry.

The village of Ettal is very small but beautifully situated in the Bavarian Alps. Its most imposing feature is a Benedictine monastery, and it was in one of these buildings that the lectures were held. During one of the midday breaks, a photograph of the attendees in front of the beautiful church of the Benedictines was taken, and I present it here for the historical record. I know of no single picture that contains so many

1. Unnamed
2. Erwin Weiss
3. R. Jira
4. Michael Lappert
5. Niels Wiberg
6. Herbert Kaesz
7. Larry Dahl
8. H. J. Breunig
9. Fred Basolo
10. Gerhard Herberich
11. F. A. Cotton
12. Fausto Calderazzo
13. G. Wilkinson
14. E. O. Fischer
15. Wolfgang Beck
16. Joseph Chatt
17. Walter Hieber
18. Gunther Wilke
19. George Bor
20. Heinz Nöth
21. Helmut Behrens
22. Jon McCleverty
23. Dianne McCleverty
24. Robin Whyman
25. Anthony Poe

26. Bruce King
27. Paolo Chini
28. Jack Lewis
29. Herbert Schumann
30. Henri Brunner
31. F. J. Muller
32. Franz Kummer
33. Reinhardt Schmutzler
34. Andrew Wojcicki
35. Walter Hübel
36. Richard F. Heck
37. J. Ellermann
38. Ronald Mason
39. E. Lindner
40. J. Grobe
41. Jorn Mueller
42. Lamberto Malatesta
43. Umberto Belluco
44. Ugo Croatto
45. Edward Abel
46. Piero Pino
47. Joe Connor
48. Marvin Rausch
49. Meriel Conner
50. G. Intille

51. Raymond Weiss
52. Gordon Stone
53. M. Bigorne
54. James J. Turner
55. H. Krauss
56. Brian Johnson
57. Th. Kruck
58. Rene Poilblanc
59. Elizabeth Fehlhammer
60. Heinrich Vahrenkamp
61. Kees Vrieze
62. Bernard Shaw
63. Cornelius Kreiter
64. Peter Pauson
65. Judith Stone
66. K. Öfele
67. Sergey Gubin
68. Duward Shriver
69. Helmut Werner
70. Robert Angelici
71. Klaus Jonas
72. S. F. A. Kettle
73. M. L. H. Green
74. Marla Wojcicki
75. Max Herberhold

Others present at the meeting
but not identifiable in the
photograph:
Dr. R. Anmann
Prof. E. Bayer
Dr. M. A. Bennett
Dr. F. Boschke
Dr. J. Falbe
Dr. W. P. Fehlhammer
Prof. H. P. Fritz
Dr. G. Huttner
Dr. A. E. Koerner von Gustorf
Prof. L. Marko
Prof. D. W. Meek
Prof. R. Nast
Dr. B. H. Robinson
Dr. S. D. Robinson
Prof. F. Seel
Dr. D. Sellmann
Dr. O. Stelzer
Prof. W. Strohmeier
Prof. H. Tom Dieck
Prof. R. Ugo

of the most prominent inorganic chemists of that era. I also provide a key which identifies most of them.

Since Ettal was so small, social life during the meeting was intramural, and culminated in a conference dinner with glorious Bavarian Speise und Bier, and an energetic band of authentic Bavarian brass musicians. The entire evening was one extended high point, but there were two supermaxima in my memory. One is the well-known pas de deux by Wilkinson and Fischer, arm in arm as the band played and everyone cheered. The sight of these two long-time rivals enjoying each other's company delighted everyone. Since Gordon Stone has already presented a picture of this in his autobiography, *Leaving No Stone Unturned*, I shall not do so here.

The other special occurrence was known to very few people, perhaps to only three: E. O. Fischer, Kees Vrieze's gorgeous and vivacious blond wife, Rita, and me. It is, and was then, known that E. O. was far from being a "ladies man." At that time E. O. was 57 years old and had not married. As I was chatting with Rita, I got a mischievous idea and said to Rita "Why don't you go over, grab E. O., and waltz with him." "Wow," said Rita, "that's a great idea," and off she went. She took E. O. right out onto the floor and off they went — dancing right down the center of the room and out the far door. Rita later told me E. O. knew very well how to waltz and that he had simply taken charge and headed for the door. This was quite a feat, because Rita was no easy person to take charge of.

The Ettal meeting will always stand out in my mind as the best scientific meeting I have ever attended. This was due in part to the sense we had that "everybody who is anybody" was there. It was also because in 1974 the prevailing attitude among academic scientists was that good ideas for basic research had a very good chance of garnering financial support for the simple reason that they were good scientific ideas. Today, of course, in what Ziman has called the "post-academic era" of science, it is almost only ideas that promise (or are advertised as promising) to have financially rewarding applications that garner support. The attitude among young and mid-career scientists today is one of grim competitiveness, and in all too many of my young colleagues, quiet desperation about getting funded. The joy of doing research because exploring ideas is fun seems to me to be a thing my young colleagues have rarely if ever experienced. There is only the grinding burden of endless proposal writing.

In addition to the travels that I have already recounted in some detail, I made many other trips during the 1970s that I can mention only briefly or not at all. In September of 1973, the Climax Molybdenum company sponsored in Reading (UK) the first of a series of conferences on

Walter Hieber and E. O. Fischer at the Ettal Meeting, 1974.

the chemistry and uses of molybdenum. Subsequent ones were held in Oxford (1976) and Ann Arbor, Mi. (1979) and I presented lectures [397, 499, 619] at all of them. Unfortunately, these very enjoyable and useful meetings did not continue.

In February of 1974, I was the Centenary lecturer in inorganic chemistry of The Chemical Society (since renamed The Royal Society of Chemistry) and enjoyed a very well-managed trip to about eight universities to present a lecture entitled "Quadruple Bonds and other Multiple Metal to Metal Bonds," which was published [433].

In 1975 I was invited by Michael Dewar to give a series of five lectures, which I spread over two weeks in November, at the University of Texas in Austin. I enjoyed this very much, especially as an opportunity to get to know many members of that department better. I had known Michael since the late 1950s when he was department head at Queen Mary College, London, and Geoff Wilkinson and I twice went out there to seek his help with theoretical problems. Michael and his wife Mary were both strong and highly articulate characters, and an evening spent at their home for dinner during that period is one I haven't forgotten. Their political outlook was a remarkable iconoclastic mix of the very liberal and the very conservative. While Michael was forthright but gentle when he disagreed with you, Mary was a tiger. She was a scholar in pre-Elizabethan history, and she scathingly labeled the UT history department, which had a distinctly modern cast, "the department of current events." I don't think I have ever spent an evening with only two other people that had more good conversation and more good red wine (Michael held white wine in very low esteem) in my life.

An important byproduct of my visit to Austin was the initiation of the Frontiers in Chemistry lecture series that we still have today at A&M. At Austin I formed the

impression that they had annually budgeted money for a continuing series of lectures. I proposed to Arthur Martell that we should do the same by putting a line item into the budget for a professor to teach a course called Frontiers in Chemistry but instead of hiring a permanent professor we bring in five different

With Linus Pauling in 1976 when I received the Pauling Medal. He usually attended the award ceremonies.

people per semester, each of whom would give five lectures and have three hours of informal give and take with our graduate students. Arthur went for the idea and we soon had the program running smoothly. I think everyone agrees that it has been a great benefit to our department. Stability is assured because it is budgeted. Ironically, about a year after we got our Frontiers in Chemistry program going, I found out that there was no such program at UT. My visit was simply what Michael would have called "a one off."

The decade of the 1970s also brought some very much appreciated recognition. There was first (1974) the ACS Award for Distinguished Service in the Advancement of Inorganic Chemistry which complemented the ACS Award in Inorganic Chemistry that I had received in 1962, and made me the first person to receive both. I also received the Nichols (1975), Pauling (1976), and Kirkwood (1978) gold medals from ACS sections, and the ACS Southwest Regional Award (1977). I was elected to the Royal Danish Academy of Sciences and Letters (1975) and to the Göttingen Academy (1979).

The fifth decade of my life was surely one of the most — if not the most — exciting and successful of my life. I doubt if I fully appreciated this at the time, but I do now.

CHAPTER 7

GOOD TIMES
IN THE 1980s

LIFE IS A FEAST BUT MOST
POOR SUCKERS ARE
STARVING TO DEATH

—Auntie Mame (Patrick Dennis)

THE PACE OF research, and life in general, continued briskly from the 1970s into the 1980s. Several lines of research that I described in the previous chapter, particularly trinuclear molybdenum and tungsten chemistry and several phases of metal-metal multiple bond chemistry, have already been followed into the 1980s.

A large amount of research was published in the decade of the 1980s: about 450 papers altogether, with my lifetime annual high of 59 in 1984. One reason for this high output was that many new avenues had been opened in the late 1970s and natural forward progress along those lines was very productive. Another factor was that beginning in the late 1970s and continuing throughout the 1980s I had a stream of very talented, hardworking postdoctorals. I shall mention some of the work that some of them did in the next few pages but space will not permit discussion of all of it; therefore I shall first cite many by name and the years they were in the lab: Wolfgang Kaim (78–79), Bill Ilsley (79–80), Tim Felthouse (78–80), Willi Schwotzer (81–86), Graham Mott (80–82), Wieslaw Roth (81–88), Milagros Tomás (81–82), Larry Falvello (81–91), Austin Reid (82–85), Akhil Chakravarty (82–85), Stan Duraj (83–84), Derek Tocher (83–84), Piotr Kibala (85–90), Marek Matusz (85–88), Rinaldo Poli (85–87), Susumu Kitagawa (86–87), Xuejun Feng (86–98), K. Vidyasagar (86–90), Maoyu Shang (87–90), Andrew Price (88–90).

By late 1986, my total of research publications was into the 900s, and Rick Adams, foreseeing that there would likely be a paper number 1000 in early 1988, decided that this event should be celebrated. He also knew that I would be Chairman of the Gordon Research Conference (GRC) in inorganic chemistry in 1987. He suggested that the Friday evening following the close of the GRC would be a very opportune time and he asked Steve Lippard to book the American Academy of Arts and Sciences building in the Nortons Woods section of Cambridge as the venue. Steve was able to do this, and so, on August 7, 1987 a memorable reception and dinner for about seventy were held there. A large number of former students and postdocs came, as well as many friends, including several from Europe. I facilitated this by inviting quite a few of my European friends to be speakers at the GRC. Some of the former coworkers who came

---

A few of those at the 1000th publication party. I am holding the handsome trophy presented to me by Rick Adams. Others from left to right: Alan Cowley, Jack Lewis, Steve Lippard, Tobin Marks, and John Fackler.

were people I hadn't seen for two or even three decades, and in a couple of cases, I was hard-pressed to recognize them at first sight.

To mark the occasion, Rick created a handsome trophy that consisted of a beautiful gold-plated model of an $M_2Cl_8^{n-}$ anion mounted on a wood base. This model has been on display in my outer office ever since. I include here just two of the hundreds of pictures taken by George Stanley, one showing me holding the trophy and just a few of the friends who were present, and another of Malcolm Chisholm and Walter Klemperer.

I enjoyed this party more than any birthday party because it marked an accomplishment.

Malcolm Chisholm and Walter Klemperer enjoying the 1000th paper party.

The four coauthors of my 1000th paper at the time it was published. Rinaldo Poli, myself, Marek Matusz, and Xuejun Feng.

Reaching an nth birthday merely means having been lucky enough to stay alive and out of jail for n years. I am reminded of how Mark Twain characterized one of his later birthdays: "my nth escape from the gallows."

It was not until early 1988 that paper 1000 appeared in *JACS* [1000], although it had been submitted in June 1987. This paper was the first report of paddlewheel complexes of nickel and palladium that could be said with certainty to contain Ni–Ni and Pd–Pd bonds, albeit only of bond order ½. My coworkers were Rinaldo Poli and Marek Matusz who did the experimental work and Xuejun Feng who made detailed calculations of the electronic structures. The neutral paddlewheel molecules, $M_2(ArNC(H)NAr)_4$, where Ar was p-tolyl, are not expected to have M–M bonds, because there are enough electrons (8 pairs) present to fill all the bonding ($\sigma\pi\delta$) and all the antibonding ($\delta^*\pi^*\sigma^*$) orbitals. Our idea was to remove at least one antibonding electron thus creating an M–M bond of order ½. We were successful in doing this, but Feng's calculations showed that it was very uncertain what orbital was the half-filled one. To this day, M_2 paddlewheel compounds with 7–9 electrons pose serious challenges to theory.

Before going on, I want to emphasize that I do not consider that publishing 1000 papers is an unqualified hallmark of research distinction — although it has rarely been done and isn't exactly easy. It was never a goal — nor was any other number, such as the nearly 1600 that I have now published. The large numbers reflect my attitude toward publishing research. I believe that results that may be of value to others should be published; it is not necessary that they be earth-shaking. I also believe that research results should be published in coherent, digestable portions, although many of my papers are, in fact, rather large. There is also the point that one's younger coworkers must have publications to advance their careers, and I have had several hundred coworkers, most of whom have been very capable and diligent.

A final point concerning publication concerns the sequence of authors' names. I know of three ways to deal with this. One is to do it on a case-by-case basis, whereby there are often time-consuming, disputatious discussions. This is the system I used for a few years but got fed up with. Another procedure is to consistently put the senior author's name first (or perhaps last) but this usually leaves the old invidious question of ordering the rest, since two-author papers are rare. Third, is the alphabetical system, to which I turned at a very early stage. Once it becomes known that this is always done in ones laboratory, problems almost never arise. Also, as a related rule, either someone has done enough to be a coauthor in alphabetical order or that person is simply mentioned in the acknowledgments.

I should also mention that in 1982 Dick Walton and I published the first edition of our book *Multiple Bonds Between Metal Atoms*. At that time, approximately 500

First and third editions of Multiple Bonds Between Metal Atoms.

compounds, with the metals Cr, Mo, W, Tc, Re, Ru, Os, Rh, and Pt, were known. In the third edition of this book, published in 2005, there are more than 4000 such compounds, and the only transition metals that are not yet included are Ta and Mn. Writing the first edition was not laborious because of the relatively small literature and the fact that Dick and I had probably read all of it more than once (and written much of it ourselves). By the time of the second edition, published in 1993, comprehensive coverage, which was the raison d'etre for the book, had become a much bigger job, but we labored mightily and got it done. Ten years later when we contemplated a third edition, we reluctantly recognized that multiauthorship was an idea whose time had come. We asked Carlos Murillo to join us as an editor, and while we three wrote nearly two-thirds of the book, we had eleven coauthors provide the other third.

My research decade opened with a satisfying discovery, namely, the first compound containing a triple (or other multiple) bond between osmium atoms [648]. The compound, which was rationally designed, is shown in Fig. 7-1. As so often happens in science, the first step in a certain direction may have been a long time coming, but it was quickly followed up. Thus, many other compounds [844] with Os≡Os bonds were soon reported, and today there are something approaching 100.

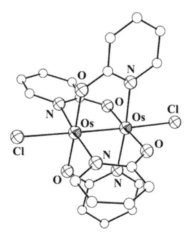

Figure 7-1. The structure of $Os_2(hp)_4Cl_2$, the first compound (1980) with an Os≡Os triple bond.

Another early decadal success came in 1982, in collaboration with Dick Schrock. We were able to isolate and structurally define the $W_2Cl_8^{4-}$ ion as the sodium salt [726]. This anion is exceedingly easily oxidized, unlike the $Mo_2Cl_8^{4-}$ and $Re_2Cl_8^{2-}$

ions which can be handled in air, and success was due to exceptional lab skills by one of Schrock's coworkers and my postdoc Graham Mott. At around the same time, Al Sattelberger had made the first $W_2(O_2CR)_4$ compounds, and $W_2Cl_4(PR_3)_4$ compounds had also been made, some in Schrock's lab and some in mine, mainly by Tim Felthouse. I was very pleased, however, when one of my talented graduate students, Wenning Wang from Taiwan, was able to make $W_2(O_2CC_6H_5)_4$ (Fig. 7-2) very straightforwardly as a beautiful crystalline product and determine its structure [725]. Thus, in a short period of time, quadruply bonded ditungsten compounds went from being virtually unknown to being a numerous and well-characterized group. The key was in recognizing that many of them are hypersensitive to air.

In 1983, with Bruce Bursten, Phil Fanwick, and George Stanley, a combined theoretical and single crystal polarized electronic spectral study of the $Re_2Cl_8^{2-}$ ion was carried out [747]. This led us to make some revisions in earlier assignments and became the definitive spectroscopic study of this iconic ion.

In 1984, Dick Walton and I, and several coworkers, did a very important study of how M–M bond lengths change on removing electrons from the δ orbital or adding them to the δ* orbital. Very naively, one might suppose that there would be successive increases either way, and by about the same magnitude for adding δ* electrons as for

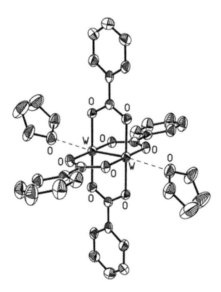

Figure 7-2. The structure of $W_2(O_2CC_6H_5)_4$·2THF. The W–O(THF) separations are ca. 2.6 Å.

removing δ electrons. However, some indications had appeared among dimolybdenum compounds that this seemingly obvious picture was too simplistic, so Dick and I designed a series of compounds to test it. The compounds were $Re_2Cl_4(PMe_2Ph)_4$, $Re_2Cl_4(PMe_2Ph)_4{}^+PF_6{}^-$, and $Re_2Cl_4(PMe_2Ph)_4{}^{2+}(PF_6{}^-)_2$. In that order we are removing δ* electrons to go from a $\sigma^2\pi^4\delta^2\delta^{*2}$ to a $\sigma^2\pi^4\delta^2\delta^*$ to a $\sigma^2\pi^4\delta^2$ configuration. The Re–Re bond lengths changed from 2.241(1) Å to 2.218(1) Å to 2.215(2) Å. This series does not display either the magnitudes or the uniformity expected from the simple model.

We then realized that as δ (or δ*) electrons are removed, the effective nuclear charge felt by the σ and π electrons increases, causing these orbital overlaps to decrease and the strengths of these bonds to decrease. Since the σ and π bonds are far stronger than δ bonds, even small fractional changes in the former may be as important as gain or loss of δ (or δ*) electrons, either enhancing or diminishing the changes in δ bond order. In the specific case cited above, removal of one δ* electron only slightly decreases the Re–Re distance because of the countervailing effect of increased charge, and when the second δ* electron is removed, there is no significant decrease at all in the Re–Re distance. Effects of this sort have subsequently been seen many times, but as a result of this study of the three rhenium compounds, they ceased to be surprising.

In 1984, thanks to the creative work of Akhil Chakravarty and Derek Tocher, a discovery that has subsequently spawned a plethora of important papers was made [789]. The two reactions they discovered are the following orthometallations:

$$Rh_2(O_2CCH_3)_4 + 2PPh_3 \longrightarrow 2CH_3CO_2H +$$

$$Os_2(O_2CCH_3)_4Cl_2 + 2PPh_3 \longrightarrow 2CH_3CO_2H + Os_2(OCCH_3)_2(C_6H_4PPh_2)Cl_2$$

In each case, orthometallation of one phenyl ring occurs on each of two PPh_3 molecules to convert each PPh_3 to a three-atom bridge. No other dimetal units have been observed to do this. The product molecules are chiral because of the head-to-tail arrangement of the two $Ph_2PC_6H_4$ bridges. In the dirhodium case, this initial observation has been brilliantly followed up by Pascual Lahuerta in Valencia, leading to procedures for resolving the enantiomeric products and exploring their potential for asymmetric catalysis of the various methylene-type reactions that are now so well known in organic chemistry.

Two other postdocs whose presence in the decade of the 1980s I must mention are Austin Reid and Rudy Luck. Both of these young men are memorable to me for their larger than life personalities as much as for their work. Each one made contributions to a range of problems. Austin, who has since become a very successful manager at DuPont, devised an extremely convenient and efficient synthesis of $Ru_3(CO)_{12}$ from $Ru_2(O_2CCH_3)_4Cl$ [859]. Rudy, who has been at Michigan Technological University for some years, did some excellent work on compounds containing H_2 ligands.

In 1985, the first experimental results were obtained concerning the dependence of the δ bond strength on the torsion angle about the bond. When the two ends of an L_4M-ML_4 species are perfectly eclipsed, (D_{4h} symmetry) the δ overlap and hence the δ bond strength are maximal; this is straightforward and incontrovertible. As the torsion angle, χ, increases, the δ/δ* separation must decrease, as the δ overlap decreases (linearly with cos 2χ). It might have been (and at that time generally was) simplistically expected that the δ bond strength and the energy of the δ → δ* spectral transition would decrease to zero while the M−M bond length would increase to some limiting value corresponding to retention of the σ and two π bonds but complete loss of the δ bond. The results we obtained showed that while the latter is true [889] as may be seen in Fig. 7-3, the former is not [835], as shown in Fig. 7-4; both figures are slightly simplified from those in the original papers.

Our ability to make these observations goes back to work that originated in Dick Walton's laboratory. With some collaboration from us to get the crystallography done, he discovered that compounds of the general formula $M_2X_4(PP)_2$, where M=Mo or Re, X is Cl or Br, and PP represents a biphosphine such as $R_2P(CH_2)_nPR_2$, may be obtained in one or both of the two isomeric forms shown in Fig. 7-5. Moreover, in many β isomers, the conformational requirements of the rings which are fused along the M−M bond give rise to torsion angles, sometimes quite large, about the bond. The net result is that by determining the structures and spectra of enough different $Mo_2X_4(PP)_2$ compounds we were able to measure δ → δ* transition energies and Mo−Mo bond lengths over a range of torsion angles from 0° to 40°.

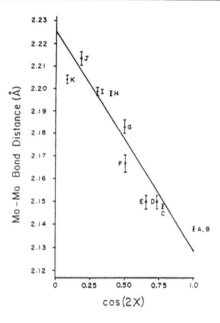

Figure 7-3. Plot of Mo−Mo distances *vs* cos 2χ in compounds of the type Mo$_2$Cl$_4$(PP)$_2$.

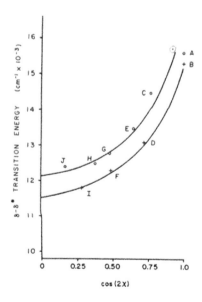

Figure 7-4. Plot of δ → δ* transition energy *vs* cos (2χ) for 10 Mo$_2$X$_4$(LL)$_2$ compounds (o, X = Cl; +, X = Br).

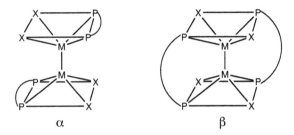

Figure 7-5. The two isomeric forms of $M_2X_4(PP)_2$ molecules.

The non-linearity of the plot of $\delta \rightarrow \delta^*$ transition energy *vs* χ, with a lower limit of 11,000 to 12,000 cm^{-1}, showed, and quantified, the importance of exchange energy in the two-electron δ bond. The whole question of how the δ bond behaves as a function of χ was later discussed in detail by Dan Nocera and me [1450].

Two more significant firsts in the preparation of new M–M bonds were achieved in 1987 by Rinaldo Poli. Compounds of metals in the first transition series were almost entirely unknown except, of course, for the abundance of quadruply bonded Cr_2^{4+} compounds. It seemed to me that cobalt might by worth a try, and I asked Rinaldo to see what he could do. It was clear from the literature that discrete $Co_2(O_2CR)_4$ molecules, homologous to the large class of $Rh_2(O_2CR)_4$ compounds, did not exist, and to this day none have been made. Rinaldo's efforts with formamidinate ligands were also unsuccessful, although today we know how to make $Co_2(ArNC(R)NCAr)_4$ compounds. Success was achieved in 1987 [989] with the synthesis of $Co_2(p\text{-tol}N_3p\text{-}tol)_4$, which has a short, strong Co–Co single bond (2.265 Å).

At the same time Rinaldo was working on making a dicobalt homolog to the numerous Rh_2^{4+} compounds, I asked him to try for a diiridium homolog. This also proved challenging, but he succeeded again, and in 1987, we also reported [969] the first paddlewheel molecule with an Ir–Ir bond (2.524 Å), namely, $Ir_2(p\text{-tol}NC(H)Np\text{-}tol)_4$. I must close with the observation that Poli was certainly one of my all-time most talented postdocs and his subsequent career has been highly successful.

In the realm of extracurricular activities, the decade of the 1980s got off to an interesting start. In 1980, Dee and I went to England for a joint meeting of the British and American Carriage Societies, which included a really grand tour of the finest carriage collections in the south of England, including the one in Buckingham Palace. We also enjoyed a gala banquet in the Royal Mews, hosted by Princess Anne, the royal patron of the British Carriage Association. This trip was bookended by flights on the Concorde. It was neither the first nor the last time I used the Concorde. It had its limitations, but it was the best way there ever has been to go to and/or from Europe. In those days, the price was reasonable, namely, about 20% higher than first class on a 747.

On the return to New York, I received an honorary doctorate from Columbia University, and the following week, the Gibbs Medal from the Chicago ACS section. Both of these were occasions that still glow in my memory. They made me feel that being 50 years old was not such a bad thing after all.

In July of 1980, I participated in a conference organized by Achim Müller at Bielefeld University, and entitled "Transition Metal Chemistry: Current Problems and the Biological as Well as Catalytical Relevance." This was an especially enjoyable meeting because of the number and high quality of the speakers, but also because of Achim Müller's provision of very warm hospitality. He included visits to the charming town, once a principality, of Detmold, where Brahms had lived for a while, and took us to see the Herrmann Denkmal, an enormous statue marking the battle in which the teutons, led by Arminium (Herrmann) drove the Romans from the teutonic homeland in AD 9.

In the Summer of 1980, following the conference organized by Achim Müller, I also made my one and only trip to Africa, thanks to Jan Boeyens who later spent January to July, 1984, as a senior guest in my laboratory. He arranged an itinerary that took me through much of South Africa from Cape Town and nearby Stellenbosch to Pretoria, and Johannesburg with calls in Port Elisabeth, Durban, and the Kruger wildlife reserve. This vast and diverse country was fascinating. The temperate climate around Cape Town, the scenic landscape, including a beautiful sea coast, and the extensive vineyards and farms make the southwest part of the country exceptionally attractive. The other highlight of the trip was three days in the Kruger game park. From charming little dikdiks to massive elephants, the range of animals to be seen in a virtually unspoiled environment was gripping. The sight of a lioness calmly eating a freshly killed zebra only forty yards from the road, a large group of hippos splashing and sloshing in a muddy river, or a group of elephants methodically pulling trees apart for lunch are vivid tableaus that one does not forget.

Speakers at the Bielefeld meeting, 1980.

1. A. Müller	6. H. Vahrenkamp	11. W. M. Beck	16. E. Fluck
2. J. Chatt,	7. W. Siebert	12. W. A. Herrmann	17. M. F. Lappert
3. P. L. Timms	8. H. Brunner	13. O. Smrekar	18. A. X. Trautwein
4. D. R. Williams	9. H. Schmidbauer	14. M. F. Hawthorne	19. D. Coucouvanis
5. F. A. Cotton	10. J. J. Turner	15. B. L. Vallee	20. B. Krebs

In September of 1981, I made my first ever visits to Rome and to Sicily. The Italian Chemical Society, of which Lamberto Malatesta was then president, held a congresso in Catania and I was invited to present a plenary lecture and also to be made an honorary member of the Society. The meeting was an excellent one and my induction (along with Fred Basolo) as an honorary member of the Italian Chemical Society was graciously done.

The year 1982 saw me take two more particularly memorable trips to Europe. In October, I was the Nyholm Lecturer of the Royal Society of Chemistry. My itinerary began in Cambridge and proceeded through Leicester, London, Edinburgh, Liverpool,

Myself, Lamberto Malatesta, and Jack Lewis in Catania, 1981.

Swansea, and Bristol, over a period of two weeks. While each place had its points, I found the visit to Swansea in Wales especially enjoyable. I was put up in a hotel perched above a rocky cove with a spectacular view of waves lashing the shore. My hosts kept

apologizing for the weather and I kept telling them that it was precisely the stormy weather that made my visit so wonderful.

In December of 1982, I went back to Strasbourg to receive an honorary doctorate from the Université Louis Pasteur. Once again, the keenest pleasure came from seeing the city and my many friends there again. On my last night, there was a very late-running and well-lubricated party at Jean-Marie Lehn's place and I arrived back at my hotel decidedly under "the affluence of incohol." Deep sleep

With Raymond Weiss, Jacqueline Weiss, John Osborn, and Jean-Marie Lehn, 1982, when I received a D.Sc. from the Université Louis Pasteur.

quickly ensued and lasted until I was jarred awake by the telephone. I had ordered a cab for 6 am to get me to the airport for an early flight to Paris, and a cheerful female voice on the phone informed me that "votre taxi vous attend." I am sure I set my own personal best record in getting dressed fast that morning, and fortunately my bag had been packed the afternoon before. I made my plane and then slept all the way from Strasbourg to Orly, where I made my connection to Houston.

A Fiasco of My Own Making

In early 1982, my friend George Pimentel and I were nominated by the Council of the American Chemical Society to be candidates for the Presidency. There was then and continues to be, I believe, an unwritten tradition that the Presidency should alternate between academics and non-academics (essentially people in industry and government agencies). Two years before George and I were nominated, that hallowed tradition had been violated by the nomination, by petition, of Fred Basolo, an academic, even though the council, had, pro forma, nominated two men from industry. Some of Fred's friends, including people in industry, had considered the official nominees unsatisfactory and asked if Fred would consent to run as a petition candidate if they got him nominated in that way. (By the way, it is easy to do, since only 500 signatures are required.) Fred agreed and he won the election in November of 1981. Thus, he was to be (and did become) the president in 1983, when according to what a number of ACS members considered the ordained order of things, there should have been a non-academic president.

George Pimentel and I recognized that, because this "perversion" was so resented in some quarters, there was bound to be a petition to put an industrial chemist on the 1982 ballot, and we reckoned that such a candidate would most likely win because he would get strong support from those who had been displeased (to put it mildly) by Basolo's victory, and the academic vote would split between him and me. In other words, the 1982 petition candidate would do to George and me just what Basolo's petition candidacy had done to the two candidates selected in 1981 by the Council.

We agreed immediately that one of us would have to drop out if this were to be avoided. I was quite willing to be the one, but George made the case that he should withdraw because he was way behind on producing what later became known familiarly as the Pimentel report for the National Academy of Sciences. That report, *Opportunities in Chemistry,* was published in 1985. George said that if he went through the three-year cycle on the ACS Board (President-elect, President, Past President) conscientiously (and George was incapable of not doing everything he did conscientiously) the report would not get done for another four years. Thus, we agreed that he would drop out if, and as soon as, an industrial candidate was nominated by petition. Some time in August, as we had fully expected, that happened.

My campaign against the petition candidate turned out to be the biggest botch of my life. In early September, I sent out a letter to about 200 friends in the ACS, a group that extended well beyond my close circle of friends (a practice that was then and still is common) stating why, in my opinion, they should vote for me and encourage others to vote for me. However, I put one paragraph into that letter that showed a wretched lack of political savvy. I discoursed on my opponent's inadequacies and referred to him as a mid-level industrial chemist. There is no doubt that each adjective in that statement was correct, but the juxtaposition proved fatal — to me. Supporters of my opponent, a great many of whom were smarting from the way Fred Basolo had, in their view, euchred them out of "their turn" at the presidency, eagerly grasped the loaded gun I had handed them and took aim. A political frenzy set in. Soon letters — probably 200,000 — were flying around accusing me of having called all industrial chemists second-rate chemists. One backer of my opponent, Al Zettlemoyer (known wide and far as Oily Al because of his technical specialization) was at the time a member of the ACS Board and was so unscrupulous as to write an anti-Cotton letter on ACS letterhead. I am unaware (and would hope it can't be so) that any other officer of the ACS has ever so misused his office.

I lost the election by a margin of about 6000 votes, but the total vote was greater than that in any year from 1972 to 1997, and my votes outnumbered those garnered by many winners, because the electorate was no doubt more polarized and activated than in most, if not all, other elections in this period. My friend Bob Parry, himself a distinguished former ACS President, told me afterward, that if I had simply gone to Pago Pago right after George Pimentel withdrew and refrained from any campaigning I would have won. Another good friend of mine from graduate school sent me a very succinct letter, "Dear Al, you blew it and I'm sorry." I am not sure that even if I had followed Parry's *ex post facto* advice I could have done more than narrow my opponent's margin of victory. There were ACS members who were hyperventilating for revenge, and prepared to go all out to get it. But I didn't have to be stupid enough to lend them a hand.

The National Medal of Science

The awarding of this medal had begun in 1962 with only one that year to the famous Cal Tech aerodynamicist Theodor von Karman. In the years that followed, the awards were characterized by a surprising inconsistency — or maybe not so surprising, since there seemed to be virtually no political mileage in them for the White House, which was solely in charge. There have been as many as twenty in some years and there were none at all seven times between 1971 and 1985.

Ronald Reagan's first term started in 1981 and the numbers seemed likely to continue to go up and down erratically. Early in his term, Mr. Reagan appointed George A. Keyworth, II, as his science advisor. Keyworth was a weapons physicist from Los Alamos (who had been recommended for the job by the president's friend Edward Teller) essentially unknown to anyone outside LANL, and many scientists, myself included, were not pleased. We thought a less obscure, more mature scientist, in the mold of Kistiakowsky or Wiesner should been chosen. I had spoken out publicly several times.

It was against this background that I had a somewhat paranoid feeling when I answered my phone one Spring day in 1983 and heard a voice say: This is the White House calling; please hold on for Dr. Keyworth. In a flash the thought came into my head: He's heard about the critical things I've said and is calling to roast me. However, almost immediately a friendly male voice said: "Dr. Cotton, this is George Keyworth and I have some good news for you. You have been selected to receive the National Medal of Science. Congratulations." After he had given me a few more details we had a very nice chat, because we had a mutual friend, Sir Ronald Mason, who then held an equivalent position (Scientific Advisor to the Ministry of Defence) in the UK. Keyworth and Mason had met a number of times and enjoyed each other's company. In fact, Mason had been trying to persuade me that Keyworth was a good guy and that I should lay off the criticism.

The whole procedure for presentation of the medal was first class. There was a gala dinner in the Great Hall of the National Academy of Sciences and the next morning the medals were presented by Mr. Reagan in the White House. Among the other

Receiving the National Medal of Science from President Reagan.

recipients that year were my admired colleagues Charles Townes, Gilbert Stork, Mildred Cohn, Phil Anderson, and Edward Teller.

A couple of funny things happened in that process. Gilbert Stork sat next to me in the front row. My mother sat behind me and kept trying to photograph the President. The net result of her efforts, as we later saw, were about ten of the finest pictures ever made of the back of Gilbert Stork's head. Dee was holding fire with her camera until I was actually up on the platform with the president. There was, of course, an official photographer, but he took routine shots of the handshake with both parties in profile. As I was receiving the medal, I noticed that Dee seemed to be having trouble unlimbering our new camera, with which she was not fully familiar, and I said to Mr. Reagan, "I hope we can hold on here a moment until my wife gets a shot." He said sure and we both turned to look at her. Thus, she got a better shot than the official one, with both of us smiling right into the lens.

A major part of my own enjoyment of the occasion was the pleasure it gave my mother, then 82 years old. No other honors I had received carried the impact for her that this did. She had always been a staunch conservative and Ronald Reagan was a hero to her. She also loved every minute of her trip to Washington. At one point, I took her out to see the marvelous Einstein statue in front of the academy building, and as

With Dee and my mother in front of the Einstein statue (1983). Photo taken by Frank Press.

luck would have it, Frank Press came along. I asked him to take some pictures, and one of those he took is included here.

In the Spring of 1983, I enjoyed a superb trip to Spain and Italy. It began with a weeklong visit to Valencia, where Pascual Lahuerta, the top professor (catedrático) in inorganic chemistry, had arranged for me to receive an honorary degree. Pascual had been a postdoctoral with me in the early 1970s. My week in Valencia also included a two day side trip to Barcelona. The grandeur of Barcelona, the capital of Catalonia, reminded me of some parts of Buenos Aires, while the elegance of Valencia was very reminiscent of other parts. I later (2000) got to know the smaller city of Castellón that lies between Valencia and Barcelona, because my former student, Rosa Llusar, is catedrática at Universitat Jaumé I in Castellón. Rosa accomplished the remarkable feat of getting two Ph.D. degrees for the work she did with me. She obtained her doctorate from Valencia (1987) and then another from A&M (1988). Needless to say, the work in her Tesis Doctoral does not overlap that in her A&M dissertation. Rosa is a very exceptional person; I think two doctorates in five years attests to that.

According to the custom in Spanish universities, I was expected to present an address of some length at the honorary degree ceremony in Valencia, and I decided to try to revive my moribund Spanish by presenting it in that language. Of course, there was no possibility of doing so extempore, so I prepared a manuscript with much help from Spanish-speaking friends and read it in a decent accent.

From Valencia I proceeded *via* Madrid and Rome to Milan and finally to Florence. The stop in Milan was to visit my friend Massimo Simonetta who only a few years later died of a rare and incurable blood disease he had contracted on a trip to Brazil. Massimo was one of the best chemical theoreticians in Italy and one of the kindest and most cultivated of men. One of the highlights of that visit was an evening at La Scala with him. I never saw him again, but I still remember his wise, cherubic face and his gentle manners.

The final goal of my visit to Italy was to deliver a lecture in Florence in honor of Paolo Chini, who had died a few years earlier. Paolo and Larry Dahl were the two patriarchs of metal atom cluster chemistry. Paolo was one of three friends I have lost prematurely because they could not stop smoking, the others being Rollie Pettit and Earl Muetterties. The Chini lecture continues to be an annual event in Florence. I think it would be appropriate for Wisconsin to set up a lectureship to honor Larry Dahl, especially while he is still alive to enjoy it.

A photograph taken in 1985 of myself with Mercedes Sanaú and her husband Pascual Lahuerta when they were working in my lab.

In June 1983, I was privileged to receive an honorary doctorate from the Technion-Israel Institute of Technology in Haifa. I have already related my visit there in 1975 and the rich harvest of research that resulted therefrom. Only three chemists had previously received honorary doctorates, Bob Woodward, Harold Urey, and Herman Mark. I was one of a group of eight in all fields that year and it gave me the opportunity to meet the legendary mathematician Paul Erdös. He is the most prominent person in a picture of the eight of us sitting on stage because he was wearing his best white sneakers below his black academic regalia. I was also asked to give an address on behalf of the group of recipients and in so doing I tried to emphasize the important role the Technion had already played in applied science and technology as well as its rising prominence in basic research. Needless to say, the joy of the occasion was enhanced by seeing again the many friends I had made at the Technion eight years earlier.

In 1983, an "out of the blue" honor came to me. I was invited to receive an honorary doctorate from one of America's most distinguished liberal arts colleges, Kenyon College. I asked my former postdoc, Bruce Bursten, who was then a young faculty member at Ohio State, if he would consent to meet me at the Columbus airport and come with me to Gambier, Ohio, for the two days I was to be there. Bruce was generous enough to do this. During the first day there, when Bruce and I had a little time by ourselves, I told him that I had become more and more curious about why I had been chosen for this honor at a place where I knew no one personally. I asked him if he would, in his own conversations with some members of their faculty, try to find out. Later that evening he reported to me, "They just think your work is terrific and they wanted to recognize you." I was truly surprised and still am.

During the academic year 1985–86, I was the Alexander Todd (visiting) professor at Cambridge University. I spent about a month there in the Autumn and again in the Spring, staying both times in a flat in Robinson College. This was my first opportunity to get a long-term exposure to Cambridge University, and I enjoyed it thoroughly. My only duties were to give a few lectures, so I had plenty of time to get acquainted with the university, many of the colleges, and the rest of the city. I also had time to do a lot of reading I would not have done at home and to spend a number of days and evenings in London. In that period, music and theater in London were flourishing and I soaked up as much as I could. Dee came over for a fair fraction of the time I was there, and it was a kind of second honeymoon. The curse of e-mail had not yet been invented, and

A dinner with Jack Lewis, Cyndy Chisholm, Dee, Freddie Lewis, and Malcolm Chisholm in Cambridge in 1985 or 1986.

so I was not pestered as one would be today by all the things one has left home to get away from.

While I accept the fact that e-mail has become a useful tool and there is not enough of the luddite in me to say I'd rather it didn't exist, it is true that it can be a pestilential nuisance. E. B. White characterized television as an invention that allows a person sitting in one room to observe the nonsense going on in another room. In a similarly cynical vein, I think of e-mail as an invention that allows people and goings on that one would rather not know or hear about to intrude unbidden into one's life.

In June of 1986, I made a very special visit to Cambridge to receive an honorary doctorate. The ceremony itself was conducted with dignified but elaborate pomp. For each of the seven recipients, the University Orator wrote, and delivered, a Latin encomium. For one of my cohonorees, Dame Cicely Saunders, the founder of the hospice movement, the Orator presumably had the necessary vocabulary on tap, but I think he must have found that necessity was the mother of invention in producing a description in Latin of my research on multiple metal-metal bonds, fluxional molecules and X-ray crystallography.

Following the award ceremony in the Senate House, we repaired to Pembroke College, only a short walk away, where libations and then luncheon were served. At the luncheon, I had the pleasure of sitting beside Prince Philip, the Chancellor of the University at the time. We chatted on three main subjects. One was his preference for beer rather than wine at lunch. This preference had not been anticipated, so he beckoned to a waitress and asked her to bring him a pitcher of beer and a glass. This was quickly done and after she filled the glass he put the pitcher under the table by his feet, remarking that he didn't want to encourage others to make extra work for her.

I then told him that Dee and I had been among the spectators when he won the four-in-hand driving competition the previous year in Great Windsor Park, and he obviously found it enjoyable to talk about driving and about equestrian activities in general. Finally I essayed a subject that didn't go well. I knew that he had long taken an interest in science and engineering, so I put in a plug for basic research as the way to generate profitable technology for the future. His blunt response was that he used to believe that but he didn't any more. So we went back to beer, horses, and other noncontroversial topics. Within five minutes after he left the college we heard his helicopter overhead.

The National Science Board

It was while I was in my flat in Robinson College that I again received a telephone call from the White House. The message this time was that President Reagan proposed to nominate me for a six-year term on the National Science Board (NSB) if I could assure them that I would be willing to do it. I assured whoever it was that I would accept the appointment, and I soon heard from my mother, who was still in Philadelphia, and from several other people, that FBI men had been asking about me. Finally, one of the local FBI men came to see me and toward the end he said he had one more question which he hated to ask but was required to. I assured him that I had no skeletons in my closet and invited him to proceed. "Well," he said, "are you or have you ever been a homosexual?" He seemed very relieved when I said no. I was soon invited to come to Washington to be officially inducted and attend my first meeting, which I did in August of 1986.

As a young faculty member at MIT, I had once raised a question about why the NSF allowed what I thought was a too large portion of my grant to go to the Institute for overhead instead of to me to meet direct research expenses. I remember some senior colleague telling me that this was entirely in accord with policy set by the NSB, the existence of which was news to me. My next, very naïve, question was: Why does this board show such (in my opinion) disproportionately overgenerous solicitude for university administrations? To which my older and wiser interlocutor said something like: If you will look at who is on the board, you will see that the vast majority are university presidents.

Ever since then I had had a secret desire to get on the board myself. Of course, as might be expected, anyone in his thirties would have a long wait. But as the old adage goes, all things come to him who waits.

Thus it was that in about 1983 I wrote to Senator Phil Gramm and told him straight out that I would appreciate it if he would try to get me on the NSB. I got a polite reply that he would do so. In early 1984, I read in our local newspaper that my friend Perry Adkisson, who was then Vice Chancellor for Agriculture (and later System Chancellor) in the A&M University System, had been appointed to the board. When I asked Perry about it he told me he had been after the senator for several years to try to get him on. At least it was now proven that Phil had the requisite pull in the Reagan White House, but I decided that, at least pro tem, I would just bide my time.

Around that period I had gotten into the habit of flying by a chartered airplane service between College Station and Houston airport. This meant debarking in Houston at, or leaving Houston from, a private terminal on the fringe of the airport. One fine day in 1985, returning from a trip, I rang for the little bus to pick me up at one of the commercial terminals. Upon boarding the bus, I found three people already on it: Phil Gramm and two assistants. Phil was going to fly out on a private plane to give a speech somewhere else in Texas. After we exchanged a few pleasantries, I decided to take the opportunity fate had handed me, so I reminded Phil that he had got Perry on the board, but not me. He immediately said to one of his assistants: "Make a note that we are going to get Al Cotton on the NSB." I thanked him, but said I thought it hardly likely he could get two people from the same university on the board at the same time. To this he replied that I might be surprised to see what he could do. And I was.

It took me about a year to get the hang of things on the NSB. The board will work only as well as the director of the NSF and its chairman allow. During my twelve years on it (I was reappointed in 1992 by President Bush), there were several directors beginning with Erich Bloch, a bachelor's degree engineer, retired from IBM, where he had successfully led the development of a new IBM computer system. Eric had little or no sympathy for the idea of individual academic investigators, and he seemed to lack any sense of the scholarly ethos of a research university. In my opinion, he started the trend, which has gained momentum to this day, of putting scientists into the Russian Communists' idea of "collective farms." While Eric and I got along well in a personal sense, his mindset was anathema to everything I believe in. At the same time, another ex-industrial scientist, Roland W. Schmitt, was chairman of the board. Eric was succeeded by Walter Massey (1990–93), who was not from industry and whose strengths were more administrative than in research. The thrust toward collectivization continued. Finally in 1993 Neal Lane, from Rice University, became Director, and the trend that the NSF administrators should offer a lot of "guidance" on what were the "important" areas of research continued and added to the misery for people like me.

Thus, in my twelve years on the board I felt frustrated and helpless most of the time, and often asked myself during my second term why I had been so foolish as to accept reappointment. I also discovered that the NSF was by no means beyond the influence of politics. Of course, when I, who was then a registered Republican, got appointed because of the influence of a leading republican senator with a Republican President, it did not occur to me that political influence was not necessarily a good idea. However, it became apparent to me, over the years, that each president's nominations to the board were carefully screened to be sure the appointee belonged to the correct party. When President Clinton's first appointees arrived, any possible doubt of this was erased.

With Marye Anne Fox (right) and my graduate student Bo Hong in about 1995.

The best thing that being on the NSB did for me was to give me the chance to really get to know Marye Anne Fox. She soon became, and remains, one of my most valued friends. In her progression from Vice President for Research at UT-Austin to Chancellor of the North Carolina State University in Raleigh, to Chancellor of the University of California at San Diego, she has gone from strength to strength. She has a wonderful "take-no-prisoners" approach to reforming sloppy operations that is as efficient as it is rare. I have never seen a better example of how an iron hand in a velvet glove is a superb administrative tool. Before closing this account of my adventures and misadventures on the NSB, I should mention two other friendships that I owe to having been on the board, namely, Bernie Burke, an MIT astrophysicist, and Frank Rhodes, emeritus president of Cornell University.

The Superconducting Supercollider

In October of 1986, I became involved in what promised to be, but never became, the grandest research project in the physical sciences since the Manhattan Project during World War II. Beginning in the early 1980s, high-energy physicists were keenly interested in studying the products of collisions between extremely high energy beams of particles. Theoreticians had made several proposals (such as the famous — and still unverified — Higgs boson) that could only be tested by working with an entirely new generation of two-beam accelerators. Plans to build one (ISABEL) at Brookhaven National Laboratory had been abandoned in 1982 and CERN had risen to the top in accelerator physics, much to the discomfiture of Americans. There was vigorous discussion all around the country, and definitely here in Texas, as to how we could vault back into the lead. An advisory group to DOE called HEPAP (high-energy physics advisory panel) was formed.

Here at A&M we had a young physics professor named Peter MacIntyre who got heavily involved in the questions of design and location of a new collider. John Fackler was then Dean of Science and he strongly promoted the involvement of A&M. Hans Mark, then Chancellor of the University of Texas System and a physicist himself, also became a strong advocate of building what Shelly Glashow proposed to call the Texatron. The general idea was to use a huge underground ring, fifty or even a hundred miles in circumference, to accelerate two beams (possibly of protons and antiprotons) to 20 TeV each, and have them collide. This would require not only a lot of civil engineering but a lot of innovative superconducting magnet design. The project became known as the superconducting super collider (the SSC), and DOE organized a schedule for competition among the states as to where the SSC would be built.

In mid-1986, the Governor of Texas, Mark White, created a Texas National Research Laboratory Commission (TNRLC) to see that Texas prepared and submitted a site proposal. Peter Flawn, emeritus president of the University of Texas, was appointed Chairman, and among the other eight members was myself, appointed October 23, 1986. I think Peter Flawn may have suggested me because he liked to think of me as his successor on the NSB. Among the more obvious luminaries on the Commission were Charles Duncan and Morton Meyerson. We began work early in November.

Our first task was to decide which of the competing sites in Texas we would select, and we chose one centered on the small town of Waxahachie, about 25 miles south of Dallas. The DFW airport was thus about 30 miles away, several interstate roads were nearby, and the main campuses of UT and A&M were within commutable distance. Geologically, the entire site was Austin chalk, an absolutely ideal material for tunneling, and there were no adverse environmental, climatic, or political factors.

At our first meeting we were joined by Lt. Governor Bill Hobby and with his help got a $1B bond issue placed on the ballot. Detailed site planning went smoothly, the bond issue got voter approval, and we came up with a really beautiful proposal. To make a long story short, in the Summer of 1988, our proposal won the nationwide competition. That, sad to say, is all the good news there is on this subject.

With our proposal already submitted by the end of 1987, the political powers that ran Texas decided that they didn't need academics or technical people on the TNRLC any more. Several of us were asked to resign (which I did in January 1988) so we could be replaced by political heavyweights, supposedly needed to see that Texas prevailed in the site competition. Some other Texas scientists (*e.g.*, Norm Hackerman, Neal Lane, and Perry Adkisson) and I were appointed to a new body, an Advisory Council to the TNRLC, which started work on March 1, 1988. This advisory council met a few times, but was not asked for much advice that I can recall.

In the middle of 1992, the U.S. congress pulled the plug on the SSC. I have heard more than one theory about why this happened, and there may, indeed, be several reasons, but in my view, the chief reason was that congress realized, finally, that after four years the SSC was being mismanaged and wasn't getting anywhere. There were major engineering and manufacturing problems concerning the superconducting magnets (of which several thousand would be required) that were still unsolved four years after the project started. Congress decided to stop throwing good money after bad. Sad end of story.

In April of 1989, Dee and I made a visit to Germany where I had the privilege of receiving an "Ehrendoctor" degree from the Johann Wolfgang Goethe-Universität in Frankfurt. I owed this to the good offices of my friend Hans Bock. The pleasure of the occasion was heightened by the uniquely "Bockian" hospitality provided by Hans and his wife Luise. Hans and Luise are an exemplary case of how complementary personalities can be the essence of a good marriage. Hans was the most exuberant of Bavarian men and Luise a perfectly beautiful, intelligent, and sensitive but quiet woman. I found both to be equally charming and lovable.

In August of that same year, 1989, just before embarking on my sabbatical year in England, I attended the Summer graduation at the University of South Carolina, where Rick Adams had done what it took to set up an honorary degree for me. While I don't know this with any certainty, the fact that the then president of USC, Jim Holderman, and I had become friends as members of the NSB may have given Rick's effort a boast. Jim and his wife and the staff of the president's office did a superb job of making the whole occasion a first-class event. Little did I suspect it then, but that was to be the last time I would see Jim Holderman. Jim was as politically ambitious as he was urbane and

Sailing on Lake Winnipesaukee at the Gordon Conference in 1985, with Stanley Kirschner and Dick Walton.

likeable, but evidently his ambition led him to overstep some ethical bounds, and he was forced to resign his position at USC. I liked him a lot and was very sorry to hear that he had stumbled. I don't know what has happened to him since.

A few other satisfying things that came my way during the 1980s were that I was chairman of the 1987 Gordon Research Conference in inorganic chemistry, became Chairman of the Class of Physical Sciences of the NAS (1985–88), was elected an honorary member of both the Indian Academy of Sciences and the Indian National Science Academy. These last two honors were probably promoted by my friend Ram Rao. I also received the T. W. Richards Medal from the Northeastern Section of the ACS in 1986. This had a special resonance for me since I had attended several presentations of this award to others over my twenty-one years at Harvard and MIT.

The decade of the 1980s (which was the decade of my own 50s) was one of consolidation and fulfillment. My capacity to plan and direct research, coupled with good fortune in attracting coworkers of the highest caliber, made it extremely productive. At the same time, I think my creativity began to diminish. I still had a plethora of good ideas, but I think really fresh insights came to me less often. C'est la vie.

CHAPTER 8

FROM 1990 TO THE
END OF THE MILLENNIUM

GROWING OLD IS NO MORE
THAN A BAD HABIT WHICH A
BUSY MAN HAS NO TIME TO FORM.

—André Maurois

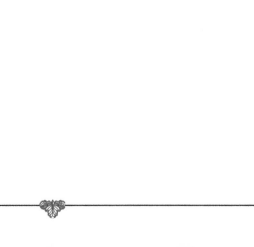

THE DECADE OF the 1990s was a very productive one for research, with 350 research papers published, 23 Ph.D.s completed, and a 2nd edition of *Multiple Bonds Between Metal Atoms* published in 1993. This large volume of work can only be summarized here very selectively, and even then without too much detail. For example, I was able to enlarge the scope of technetium M–M bond chemistry [1300] by collaborating with Al Sattelberger to use the laboratory facilities at Los Alamos to prepare compounds, the structures of which were then done at A&M using just a few crystals. A very good postdoc, Steve Haefner traveled to Los Alamos for short concentrated work periods.

Prior to 1993, there had been no reported V≡V or Nb≡Nb compounds, but in 1992 Carlos Murillo succeeded in making $V_2[(p\text{-tol})NCHN(p\text{-tol})]_4$, with the very short triple bond length of 1.978(2) Å [1180, 1218]. Although much work which did not afford products with a V_2^{4+} core had been done on dinuclear vanadium compounds, the actual preparation of this compound proved to be fairly easy once the choice of a formamidinate ligand was made; reduction of V^{III} to V^{II} is not especially difficult. A naïve effort in 1997 to make a Nb_2^{4+} analog by simply using a formamidinate ligand failed, but that failure had very interesting consequences.

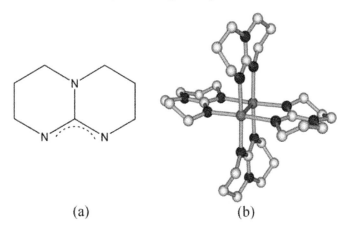

(a)   (b)

Figure 8-1. (a) The hpp ligand. (b) A paddlewheel compound of the type $M_2(hpp)_4$.

Under the exceptionally strong reducing conditions required to put niobium into the oxidation state II, formamidinate ligands did not remain intact. We then turned to the hpp ligand for no more sophisticated reason than the hope that its bicyclic structure would make it sufficiently robust to survive where formamidinate ligands did not. We were delighted when this simple objective was accomplished, in 1997, with the preparation of $Nb_2(hpp)_4$ [1336]. We also used our new hpp bridge at the other end of the d-block by showing that it allowed the preparation of the M_2^{6+} compounds $Pd_2(hpp)_4Cl_2$ [1386] and $Ir_2(hpp)_4Cl_2$ [1409]. However, since hpp forms numerous compounds of the type shown in Fig. 8-1b, the best was yet to come. But not until the next millennium; I shall return to this story in Chapter 9.

From 1986 to 1998, I was fortunate to have Xuejun Feng, a theoretical and computational chemist, in my group. Feng had spent his early years in China during that period called the cultural revolution, and only at the age of thirty had been able to go to Denmark where he did his doctoral research with Jens Peder Dahl at the Technical University of Denmark. During his years with me, he coauthored thirty three papers, and made valuable contributions to many more. No suitable position either in academe or in the chemical industry became available and in the end he took a position with a company that develops computer language and software for business applications. During the 1990s, his most outstanding contributions were extensive CASSCF calculations (1993) on the $Mo_2Cl_4(PR_3)_4$ type molecule [1200], that provided deep insight into the electronic structure as a function of internal rotation, and the first two DFT treatments (1998) of multiply bonded dimetal molecules [1334, 1364] which showed that DFT can provide very reliable structural predictions.

The Strange Case of a Mythical Compound. Apropos of supershort M–M bonds, I want to relate the instructive misadventure of the alleged Tc–Tc supershort bond that actually wasn't there [1266]. In 1989, the compound "$Cp*Tc_2O_3$ ($Cp* = C_5Me_5$)" was reported and a crystallographic study was said to show that it consisted of parallel chains of alternating Cp rings and O_3 triangles, with the planes of each perpendicular to the chain direction. Each intermediate position was said to be occupied by a Tc atom, as shown schematically below. While this structure seems very unlikely merely on qualitative grounds, its total absurdity was assured by an alleged Tc–Tc distance, through each O_3 triangle, of 1.867(4) Å. To cleanse this nonsense from the chemical literature, it seemed only necessary to reprepare the compound and see what was wrong with the structure determination. However, efforts by several experts in preparative inorganic chemistry failed to reveal a trace of the compound and there the matter languished for several years.

Finally, in 1994, I decided to make an indirect attack on the "Cp*Tc_2O_3" problem, based on my suspicion that this alleged compound was really Cp*TcO_3, for which there was a well-known rhenium analog. The crystal structure of Cp*ReO_3 had never been solved because of apparent twinning and/or disorder. Presumably the crystallographic work on Cp*TcO_3 had been botched by someone who had failed to see the signs of trouble. My idea was that since the rhenium compound was reproducible, we could solve its crystal structure and thus see how the structure of the Tc compound had gone so wrong.

With a beautiful crystalline sample of Cp*ReO_3 prepared by Tony Burrell (then at Los Alamos) and in collaboration with Vaclav Petricek, a Czech crystallographer I had gotten to know in France, and Lee Daniels, a former student who was then in charge of crystallography in my group, it was soon shown that the crystals were not only pseudomerohedral twins but also disordered to boot. With proper account taken of these complications, the piano stool structure of the Cp*ReO_3 molecule emerged clearly [1266]. Since the reported unit cell of the "Cp*Tc_2O_3" compound matched that of the Re compound very closely, we were on the verge of concluding that the reported Cp*Tc_2O_3 was Cp*TcO_3.

However, as in the best detective stories, and a few of O'Henry's short stories, there was a surprise ending yet to come. We noticed that in the *original* report of the "Cp*Tc_2O_3" structure, each Tc atom in the chain had been refined as a *full atom*, as required by the Cp*Tc_2O_3 formula. However, in our model of the structure of the Re compound, which was indubitably correct, each Re atom was disordered over two sites; each one had been satisfactorily refined as a *half atom*, to be consistent with the formula Cp*ReO_3.

There was but one inescapable conclusion. The crystallography reported for Cp*Tc_2O_3 was actually done on Cp*ReO_3. A rhenium atom has approximately twice the X-ray scattering power of a Tc atom and thus Re/2 and Tc are indistinguish-

Lee Daniels, who got his Ph.D. with me in 1984 and was in charge of X-ray work in my lab 1990–2002.

able by X-ray crystallography. The reason the synthesis of $Cp^*Tc_2O_3$ (or Cp^*TcO_3) could not be reproduced was then clear: no such compound had ever been made in the first place, and therefore no supershort Tc–Tc bond, with a length of 1.867 Å, ever existed.

This story illustrates the problem raised by the ready availability of X-ray data and black-box computer programs to people who don't think about what they are doing. Ironically, the incorrect outcomes are not due primarily to crystallographic incompetence as such, but to the inability of the chemists involved to recognize structural nonsense when they see it.

Polyhydrido Zirconium Clusters. During the 1980s, I had had four outstanding postdocs, all born in Poland. All of them had an extraordinary combination of ability for both synthetic and structural work. They were Stan Duraj, Piotr Kibala, Marek Matusz and Wieslaw Roth. Subsequently, Stan has had a very successful academic career at Cleveland State University, while the other three have all done important work in industrial laboratories.

The chemistry I want to describe here began with a serendipitous discovery by Piotr Kibala in 1988. He heated a solution of the green bioctahedral compound $Zr_2Cl_6(PMe_2Ph)_4$ and obtained a very small quantity (ca. 15 mg) of red crystals [996]. X-ray crystallography revealed a delightful structure consisting of an octahedron of Zr atoms, with each edge bridged by a Cl atom and a PMe_2Ph ligand at the external site of each Zr atom. Over the next few years, many repetitions of this procedure resulted often in no yield at all, but occasionally in similarly tiny amounts of the red compound.

A new graduate student who joined my group in January of 1990, William Wojtczak, was given the problem of finding a better way to synthesize octahedral Zr_6 clusters. He first found a way to make related compounds of the type $Zr_6X_{14}(PR_3)_4$ (with X = Cl or

Br and various phosphines) by first treating ZrX_4 with $(n\text{-}C_4H_9)_3SnH$ to produce a dark precipitate (still of unknown composition) which was then solubilized by addition of a phosphine. From these dark red solutions, we obtained compounds of the composition just stated, and were able to show that four such compounds had the central Zr_6Cl_{12} core with the six outer positions occupied by two X and four PR_3 ligands [1187]. It was later shown that if the dark red solution, containing excess phosphine, was treated with Na/Hg, Kibala's original compound could be obtained in 15–20% yield. [1248].

Since John Corbett had shown earlier that all the compounds containing $[Zr_6Cl_{12}]^{n+}$ clusters, which he made by the usual high-temperature sealed-tube reductions of $ZrCl_4$, had internal atoms such as Be, B, C or N, the absence of any such internal atoms in our compounds was remarkable. It bothered Corbett because he had argued that the $[Zr_6Cl_{12}]^{n+}$ units did not have enough electrons to be stable without those added by the internal atoms. However, since our compounds were made in an entirely different way (which French chemists referred to as "chimie douce") and were more reduced than Corbett's, I felt that our formulas were correct. I had not overlooked the idea that one or more hydrogen atoms might be in, or on, the Zr_6 octahedron. However, since there was no evidence for them, I did not propose that they were there, in spite of suggestions by Corbett and also Arndt Simon, that they were.

However, when a new postdoc, Linfeng Chen, got to work on the problem in 1994, the true character of these and related compounds as tetra- and penta-hydrido compounds was revealed and conclusively proven. Linfeng Chen (a Ph.D. from Rick Adams) was one of the most skillful preparative chemists I have ever had in my group; I wish I could have kept him forever, but I found him a job after two years. I will explicitly mention only two of his nine papers, one of which describes a neutron diffraction study carried out at Argonne by my friend Arthur Schultz [1299], and the other provides a well-referenced overview of most of his other work [1362].

What Linfeng Chen established was that the $[Zr_6X_{12}]$ cores all contain enough hydrogen atoms to give them a total of 14 core electrons. Thus, the anionic species are $[Zr_6X_{18}H_4]^{4-}$ and $[Zr_6X_{18}H_5]^{3-}$, and this can be confirmed by NMR spectroscopy on very pure specimens. The neutron diffraction work established where these hydrogen atoms are. As shown in Fig. 8-2 for the $[Zr_6Cl_{18}H_5]^{3-}$ ion, they lie just inside the faces of the Zr_6 octahedra, but the observed occupancy at each site is only fractional (either ½ or ⅝), and it also appears that these H atoms may by internally mobile. Whether they are exchangeable with external H atoms was never determined. Nowadays, about a decade later, there is considerable interest in metal clusters containing "lots of hydrogen" (see A. S. Weller et al., *Angew. Chem. Int. Ed.* **2005**, *44*, 6875) and it would be interesting to return to this chemistry. By the way, the original compound is presumably $Zr_6Cl_6H_2(PMe_2Ph)_6$, but there is no proof of this. I must acknowledge the early prescience of John Corbett and Arndt Simon; I should have listened to them.

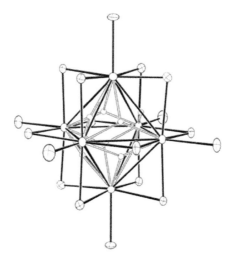

Figure 8-2. The $[Zr_6Cl_{18}H_5]^{3-}$ ion. The H atoms are seen just inside the triangular faces of the Zr_6 octahedron.

There would be hardly enough room in an entire chapter of this book to review all of the work done by Evgeny Dikarev and Marina Petrukhina. Evgeny was the young man who came from Moscow State University instead of Andrey Shevelkov in 1994, after my visit to Moscow, which I shall describe later. I know not what I might have missed when Andrey (or perhaps his wife) decided not to come, but he couldn't have been better than Evgeny. Moreover, two years later, Evgeny's wife Marina Petrukhina came too, thus giving me as good a husband and wife team as I can imagine. In addition to working separately and also at times together, both had exceptional ability to collaborate with others and make them far more productive than they probably would have been if left to themselves. Evgeny and Marina, separately and jointly, produced a total of 53 research publications.

Marina was trained as a physical chemist but her skill at preparative chemistry was immediately evident. Moreover, she had a gift for teaching and did an excellent job

Marina Petrukhina working in a glove box.

Evgeny Dikarev looking for crystals in a Schlenk tube.

one year in the undergraduate physical chemistry course. Evgeny's collaborations with W. Y. (Raymond) Wong, a postdoc from Hong Kong and Santiago (Santi) Herrero, a postdoc from Zaragoza, both led to extensive series of papers that contain superb chemistry. The cooperation of Evgeny with Raymond Wong proved to be exceptionally convenient for me because Raymond had a near-perfect command of English and thus their joint manuscripts arrived on my desk well organized and well written and virtually ready to submit. Evgeny and Marina are now at SUNY-Albany.

One of the particularly interesting lines of research pursued by Marina and Evgeny was solventless synthesis. This technique enables one to avoid interference or competition from solvent molecules when the objective is to combine a weak donor with a strong acceptor. It requires that both reactants are volatile at feasible operating temperatures. It proved to be an excellent way to carry out reactions of $Rh_2(O_2CCF_3)_4$ with CO [1430], alkynes [1437], S_8 [1473], I_2 [1445], DMSO to give different products than were obtainable in the liquid phase [1436], and a range of unsaturated and polycyclic aromatic hydrocarbons [1452,1485], such as the one shown in Fig. 8-3. In every case in the last study, the locus from which π-electron density was donated by the hydrocarbon to the dirhodium acceptor was just where a Hückel calculation said the highest π-bond order was.

With Carlos Murillo and Chun Lin in 1999.

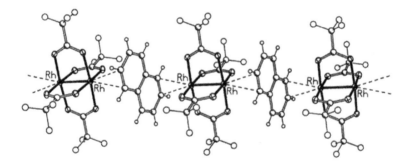

Figure 8-3. Infinite chains formed by solventless synthesis from $Rh_2(O_2CCF_3)_4$ and naphthalene. Note that the most electron rich bond of naphthalene is the donor, even though it is not the least sterically encumbered.

Beginning Supramolecular Chemistry. Toward the end of the nineties, an outstanding new postdoc, Chun Lin, arrived. Chun had received his Ph.D. working with my former student, Tong Ren. Chun had that wonderful combination of green fingers and a fertile imagination and he single-handedly initiated our studies of supramolecular assemblies that had dimetal units instead of single metal ions as vertices.

Having seen some of the beautiful supramolecular squares that had been created by Fujita and Stang, it had occurred to me that instead of using partially coordinated metal ions such as Pd^{2+} and Pt^{2+} and linking them by linear bidentate ligands such as piperidine or 4,4-bipyridine, as shown in Fig. 8-4, one might be able to use dimetal units already carrying two cisoid bridges and link them with dicarboxylates to form another class of supramolecular squares.

Figure 8-4. The type of molecular squares made by Stang and Fujita.

There were several immediately foreseeable advantages to the supramolecules I envisioned, the principal ones being that they would initially be neutral and hence soluble in many non- or semipolar solvents, and they would have redox chemistry so that we could then vary their charges at will.

It had also occurred to me that since dimetal units such as Mo_2^{4+} could be oxidized fairly easily electrochemically, it might be interesting to see how just two of them would communicate if they were suitably connected. This same idea had occurred to Malcolm Chisholm earlier and he had already made the first effort to do chemistry of this sort by linking two $(RCO_2)_3Mo_2$ units with a dicarboxylate bridge such as oxalate. However, it proved that assemblies in which both ligands and linkers are carboxylates present major difficulties because of the lability of all carboxylate ligands.

The approach we took when Chun arrived was to use dimetal units in which we had much less labile "spectator" ligands, and in the academic

With Tong Ren right after he passed the oral examination for admission as a Ph.D. candidate, March 1989.

year 1998–99 this idea opened up the whole area [1378, 1397]. The amount of work we have done in the area (much of it by Chun Lin himself) is very extensive and has been reviewed [1481, 1501]. I shall mention a few highlights in Chapter 9.

As a foretaste, Fig. 8-5 shows how pairs of dimolybdenum units may be linked by a dicarboxylate dianion to give a stable, well-behaved "dimer of dimers" in which a significant degree of electronic communication exists because the two Mo_2 units and the bridge are all coplanar. On the other hand we have shown that even when the distance between the two Mo_2 units becomes even shorter, if they are also made orthogonal to each other, communication is much reduced. Also shown in Fig. 8-5 is the first example of a molecular square made with dimetal rather than monometal corners.

Figure 8-5. Three of the earliest compounds that we made by connecting Mo_2(formamidinate)$_3$ units by (a) oxalate, (b) sulfate bridges, or (c) Mo_2(formamidinate)$_2$ units by oxalate bridges to form squares.

EMACs. Another major new line of work that began in my lab in the late 1990s was the study of EMACs, Extended Metal Atom Chains. Although the first such compound was made as early as 1968, it was only in 1991 that its correct structure, shown in Fig. 8-6, was reported. Even after that, it was only several years later when my interest, as well as that of Shie-Ming Peng in Taipei, led both of us to the deliberate extension of the work to provide many new compounds as well as structural and physical studies to elucidate their electronic structures.

Almost immediately remarkable results were obtained and I shall touch on a few highlights here. It was found [1320] that the cobalt analog to the nickel compound exists in two crystalline forms that may be obtained admixed from the same solution [1446]. They have almost identical structures, very similar to that shown in Fig. 8-6, except that in one the molecules have D_3 symmetry with equal Co to Co distances

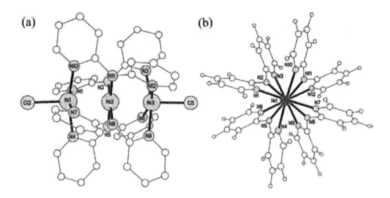

Figure 8-6. The structure of Ni$_3$(dpa)$_4$Cl$_2$: (a) sideview. (b) end view.

(both 2.318(1)) Å) while in the other, the molecules have C$_3$ symmetry with Co to Co distances of 2.285(1) Å and 2.459(1) Å. While it is not completely certain, it seems that only the symmetrical form persists in solution [1419]. Almost simultaneously we made the first trichromium compounds and found that while a few are symmetrical, most are not [1550]. The story of the chromium compounds is a very complex one. Indeed the whole field of EMACs remains very challenging, with abundant interesting phenomena to explain.

One of the attractive features of EMACs is that, in principle, they may be made very long, an effort so far pursued more by Shie-Ming Peng than by us. He has claimed a molecule with nine nickel atoms, although chemical characterization of it was not reported. Several molecules with five and seven metal atoms have been made both in my laboratory [1393] and in Taipei. The longer molecules are obtained by using surrounding ligands that are extensions of the dipyridylamide (which gives the trimetal compounds), as indicated by the following general formula:

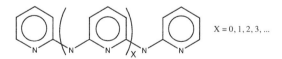

$$X = 0, 1, 2, 3, \ldots$$

Although work on EMACs began in 1997, it continued into the new millennium and is still proceeding in 2006. We have published more than 30 reports concerned with compounds of cobalt, chromium, nickel and copper. The main contributor to our work in recent years was John F. Berry, who is now in a tenure track position at the University of Wisconsin, after spending two years at the Max Planck Institute in Mülheim with Karl Wieghardt. The extension of EMAC chemistry to chromium compounds was first done in my laboratory [1343]. They have displayed a pronounced tendency to form chains with alternating short and long Cr to Cr distances, *i.e.*, $Cr\equiv Cr\cdots Cr\equiv Cr\cdots Cr$ and $Cr\equiv Cr\cdots Cr\equiv Cr\cdots Cr\equiv Cr\cdots Cr$ in the Cr_5^{10+} and Cr_7^{14+} compounds (see Fig. 8-7), while the Cr_3^{6+} compounds may be either of the $Cr\equiv Cr\cdots Cr$ or the evenly spaced kind [1550], depending on the attached axial ligands. Because the chains of metal atoms with alternating distances are often directionally disordered, very skillful work was required to get correct structures. Xiaoping Wang and John Berry became very expert at dealing with this problem.

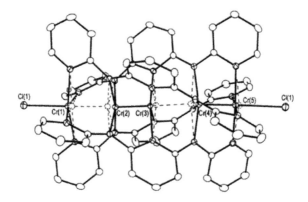

Figure 8-7. An example of the way in which chromium EMACs have alternating nonbonded and quadruply bonded distances in the metal chain. In this case the distances are from l to r: 2.60 Å, 1.87 Å, 2.60 Å, 1.95 Å [1418].

As the EMACs become longer, they more closely resemble metal wires surrounded by organic insulation. They are, in fact, what I have referred to [1529] as the *reductio ad ultimum* of an ordinary normal sized electric wire, as illustrated by Fig. 8-8. They also have functionalities at the ends so that it will be feasible to incorporate them into molecular scale circuits. The question that still needs to be answered is will they actually function as conductors?

The year 1999 ended with the submission of a publication [1422] based on work by Liz Hillard and Hong-Cai (Joe) Zhou, entitled "After 155 Years, A crystalline Chromium Carboxylate with a Supershort Cr–Cr Bond." Since 1844 when a French chemist, Eugene Peligot, reported the acetate, $Cr_2(O_2CCH_3)_4(H_2O)_2$, something on the order of a hundred $Cr_2(O_2CR)_4L_2$ compounds have been reported, but in every case either separate axial ligands (L) were present or the molecules formed infinite

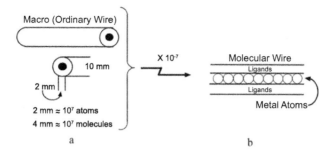

Figure 8-8. How a typical, common electric wire can be scaled down to a molecular scale wire.

chains in which the axial positions of each molecule were occupied by oxygen atoms belonging to its two neighbors. Therefore it had proved impossible to settle by X-ray crystallography the question of how short the Cr–Cr quadruple bond would be in a $Cr_2(O_2CR)_4$ compound if it were not lengthened by axial donors. My friend Manfred Fink, a physicist at UT-Austin, had performed an electron diffraction study of $Cr_2(O_2CCH_3)_4$ in the vapor phase above about 120 °C and extracted a Cr–Cr distance of 1.97 ± 0.02 Å, by very meticulous analysis of the radial distribution function. However, theoreticians, including my own esteemed colleague Michael Hall, were skeptical about the electron diffraction results, which admittedly are less precise in general than our usual X-ray-based structure analyses. Over many years, I had presented indirect arguments for a value below 2.00 Å and felt that Fink's work settled the matter in my favor, but it was clear that those who disagreed would never give in until confronted by a crystal structure, which they could not possibly refuse to accept. Finding a way to obtain such a structure had become an obsession with me. Over the years several strategies had been tried but failed.

Finally, I hit on the strategy of using R groups that would both confer good solubility in a hydrocarbon solvent and prevent the oxygen atoms on one molecule from approaching the axial positions of an adjacent one. This was different from trying to block the axial positions themselves — and it worked. We obtained the crystals we wanted and found a Cr–Cr distance of 1.9662(5) Å.

The decade of the 1990s thus closed on several research notes that were very satisfying for me. Part of the satisfaction was due not so much to what had already been done, but to what we could look forward to doing in the next millennium.

Other Activities During the 1990s

My Only Full Year Sabbatical. During the academic year 1989–90, I took the only sabbatical I have ever taken. I even managed despite my advanced age to obtain, for a second time, support from the Guggenheim Foundation. The year was spent mainly in Cambridge, England, where Dee and I had the pleasure of seeing a great deal of Jack and Freddie (Elfreda, if the truth be known) Lewis. We lived within five minutes walk of the chemistry department in a very small house owned by Malcolm Chisholm. The house was very comfortable, entirely satisfactory to our needs, and Dee said that it was so easy, and so much fun, to take care of that it made her feel again like a little girl "playing house." There was an area of small shops, only a short walk away, where we had the feeling of being in a small town. The railway station, where one could take a train to London that required only about one hour, was only ten minutes walk. It was an idyllic life.

At that time Jack Lewis was the Professor of Inorganic Chemistry, Ralph Raphael and Alan Battersby were Professors of Organic Chemistry, John Thomas was Professor of Physical Chemistry and David Buckingham was Professor of Theoretical Chemistry — all friends to be remembered with great affection and esteem.

Cambridge itself is a town we were already quite familiar with from many previous, shorter visits. Among these were two, that I have already mentioned, one in the Fall and one in the Spring of the academic year 1985–86, when I spent a total of six weeks as the A. R. Todd Professor. This is an annual appointment, tenable for three to six weeks, in honor of Alex Todd. It is primarily intended for organic chemists and I believe I may be the only non-organic chemist ever to have held it.

Dee and I both love Cambridge. It is a small city (I've walked it from end to end and side to side quite a few times) but it is the old central area where most of the colleges and shops are that defines its character. While some of the colleges appear superficially similar, each one has a distinct personality. I have had the privilege of visiting and dining in most of them, over the years. Of course, the one that is nearest to my heart is Robinson College, of which I have the special privilege, since 1992, of being an Honorary Fellow.

During that academic year, I spent a full twelve-month period out of the United States, with only short return visits, totaling no more than thirty-five days, for tax reasons. Steve Berry had found out a year or two earlier when he took a sabbatical in Germany, that if one is employed abroad for one year, the first seventy thousand dollars of salary is free of income tax, and he passed the good word to me. Restricting myself to such a small number of days back was not easy, especially since any day when even one hour was spent in the United States counted as a full day. One of my friends suggested that I could just "sneak in and out *via* Canada" but I'm not that kind of a risk-taker.

The requirement of staying out did become an onus when I learned in January of 1990 that I had been selected to received the Award in the Chemical Sciences from the NAS at the annual meeting in April, 1990. By then I had used up many of my 35 days of grace and already had tentative plans for the rest. However, there was no question of missing the meeting and I made a two-day visit to Washington. Also, sometime in the Winter I was notified that I was to share the King Faisal Prize in Science with two good friends, Ray Lemieux and Mostafa El Sayed. While I was not sure what exactly this entailed, I immediately started arranging to be in Riyadh at the appointed time in March. Dee, who had been spending some time at home, flew from Houston to Frankfurt where we met and were hosted for a day by Hans and Luise Bock. We then proceeded to Riyadh on Monday, March 5.

I had a very boozy dinner in the first class section of the Lufthansa flight and despite the sage advice of my nondrinking wife, I put a couple of mini bottles of scotch in my briefcase to help me deal with the dry spell ahead. I was assuming that as an honored guest of the royal family, my luggage would not be searched. I was wrong. I will never forget the look of horror (or mock horror) on the face of the customs agent as he held up the two cute little bottles and asked the rhetorical question: "What is this." Two well-armed, uniformed police immediately appeared and I began to think myself a real idiot — a view already suggested by Dee. However, two other men, in regular Arab garb and flashing identification

A picture of Ray Lemieux, Mostafa El-Sayed, one of the Saudi princes, and I when we received the King Faisal prize in 1990.

papers, also appeared on the other side of the barrier, and there were lengthy, vehement exchanges all around, while I stood there trying to seem as innocent and inconspicuous as possible. Finally, some consensus appeared to be reached. The two armed men stepped back while the customs man histrionically screwed the caps off the bottles and, holding one in each hand, arms extended, poured the contents into a large waste container. He then scolded me in Arabic and waved us through.

After this, all went rather smoothly. The personnel of the King Faisal Foundation had arranged some interesting sightseeing for all of the prize winners and their accompanying persons for a couple of days preceding the award ceremony. We stayed in a hotel that specialized in infidels, where food and service were high-standard, and the available fruit juices were of a variety and quality I have never seen elsewhere. Mostafa and I were the only Americans among all of the awardees and the American ambassador invited us for a short congratulatory visit at the embassy. He asked if there was anything we needed and Mostafa piped right up and said he would give anything for a cold beer. The ambassador said that even though according to diplomatic protocol the embassy and its grounds were U.S. territory, he kept no alcohol on the premises, although he had a satisfactory supply at home. So alas, Mostafa (and I) remained thirsty.

While we had some very interesting experiences in Riyadh, which for me included having one of the princes get a horse saddled up for me to take a ride on, it was with a sense of great relief that we felt our plane back to Frankfurt leaving the runway below us. The pervasive, menacing, rigidity of the cultural atmosphere gave me the creeps, to put it bluntly, and I shall never willingly go back into such a place again.

The First Contemporary Inorganic Chemistry Conference and the Initiation of the Cotton Medal. Toward the end of 1993, Carlos Murillo had the idea that a special symposium might be organized at Texas A&M in the Spring of 1995 to celebrate my 65th birthday. He enlisted the support of John Fackler and they chose the week in March, 12–15, 1995, when the university would be closed for Spring Break. This meant that hotels would have rooms and university facilities would be available. They planned a program of twenty-four speakers, all friends, former students or former postdocs of mine, who had achieved exceptional levels of recognition. I was certainly not a non-participating bystander. Carlos and John consulted me extensively about whom to invite. Carlos also had the idea of setting up an endowment for a gold metal, to be awarded for the first time in 1995, and then annually, for "excellence in chemical research." A great many of my friends and former students were asked to contribute to the endowment and over \$100,000 was raised. It was planned to make the first award at the meeting, which was called "Contemporary Inorganic Chemistry," held at the College Station Hilton and Conference Center. The speakers were Edwar Shamshoum, Jo Ann Canich and Scott Han, three former students who had done very well in major chemical companies; R. H. Holm, S. J. Lippard, R. D. Adams, G. G. Stanley, T. J. Marks, and W. G. Klemperer, former students in academia; F. Calderazzo, B. F. G. Johnson, R. Poli and R. A. Walton, former postdocs; F. G. A. Stone, W. A. Herrmann, M. H. Chisholm, R. J. Lagow, A. Müller, G. Jaouen, A. H. Cowley, R. Hoffmann, H. W. Roesky and J. Lewis, friends.

George Stanley made a very valuable contribution by taking a great many photographs of the participants and other attendees, and Alan Cowley distinguished himself by his graceful and witty remarks as master of ceremonies for the conference banquet.

In the meantime, Carlos had designed and commissioned the casting of the award Medal and formulated the rules for selection of recipients. My advice, based on my experience with several other ACS sectional medals was to have a local committee, supplemented by a few outside members, carry out the entire process, and that is how it was set up.

At the banquet where Jack Lewis and several others made opening comments, Alan called on Derek Barton, who first said that "since I know how much Al likes a good

1.	Unknown	15.	Mercedes Sanaú	29.	Günter Schmid
2.	Elizabeth Babaian-Kibala	16.	David Goodgame	30.	Brian Johnson
3.	Denise de O. Silva	17.	Rinaldo Poli	31.	Xuejun Feng
4.	Fausto Calderazzo	18.	Thomas Webb	32.	Richard A. Walton
5.	Irene Shim	19.	Susumo Kitigawa	33.	Victor W. Day
6.	Margaret Goodgame	20.	Stephen Koch	34.	B. Ray Stults
7.	Judith Eglin	21.	Linfeng Chen	35.	Jon McCleverty
8.	Avi Bino	22.	Guido Yagupsky	36.	Brian Kolthammer
9.	Timothy Felthouse	23.	Chris James	37.	Marek Matusz
10.	Larry Falvello	24.	Bruce Bursten	38.	Sheila Morehouse
11.	Gong Chen	25.	Piotr Kibala	39.	Lee Daniels
12.	Jian Lu	26.	Austin Reid	40.	Rudy Luck
13.	Stephen Lippard	27.	Giulio Deganello	41.	Michelle Millar
14.	Pascual Lahuerta	28.	Stan Duraj		

bottle, I shall present him with one." He then gave me an empty bottle of Chateau Barton, a fine Bordeaux wine. Then, turning serious, he said that he had been asked to announce that the selection committee, of which he was a member, after much deliberation had decided that the first F. A. Cotton Medal should go to F. A. Cotton. While, as most people who know me realize full well, I am not excessively modest, I was really taken by surprise, and somewhat embarrassed by this. I don't recall that my acceptance remarks were particularly articulate. Had I been aware of what was going to happen I would not have had so much to drink up to that time.

The F. A. Cotton Medal for Excellence in Chemical Research was next awarded to George Olah (who had been selected before his Nobel prize was announced). At the award symposium, Gabor Somorjai and Peter Stang also gave lectures, and I remember referring to this as a special meeting of the Hungarian Chemical Society. Since then, the medal has been awarded every year, and I take great pride in the distinction of the recipients, who are as follows:

George A. Olah	1996	Ada E. Yonath	2002
Pierre-Gilles de Gennes	1997	Gabor A. Somorjai	2003
JoAnne Stubbe	1998	Albert Eschenmoser	2004
Alexander Pines	1999	Richard H. Holm	2005
Tobin J. Marks	2000	Robin M. Hochstrasser	2006
Samuel J. Danishefsky	2001		

Gabor Somorjai, Peter Stang, me, and George Olah in 1996 when George received the Cotton Medal.

A Visit to Russia. My only visit to Russia resulted from a confluence of events, no one of which would have sufficed to bring it about. Beginning in the 1960s, there had been a good deal of shared interest between me and several Russian chemists concerning dimetal compounds of rhenium, technetium and rhodium, but practically no personal contact. In early 1992 I received a letter from a member of the Russian Academy of Sciences (to which I was later elected a Foreign Member in 1994), Vladimir Tretyakov, of the Lomonosov University in Moscow (also called Moscow State University). He said that they would like to offer me an honorary doctorate and also suggested that he had some younger colleagues who would like to talk with me about coming to my lab to do postdoctoral work.

Shortly after I received this invitation, I learned that the NAS would soon be sending a delegation to Moscow to make new arrangements with the Russian Academy for cooperative activities. It was necessary to replace the agreements that had previously existed with the Soviet Academy because the latter had ceased to exist following the dissolution of the USSR. When I mentioned my invitation to Frank Press (then president of the NAS) he offered me a place on our delegation, along with the NAS officers and several other NAS councilors. I then contacted Prof. Tretyakov and my visit to the university was scheduled to connect with the visit of the NAS delegation, in November 1992.

I did not travel directly to Russia with the other members of the NAS delegation, but went instead to spend some time in England before going on to Moscow. The rest of the delegation went first to St. Petersburg and I was to join them in Moscow. To get from London to Moscow, about a five-hour flight, I decided to go first class on British Air, feeling that would be the safest and most comfortable way to do it. It turned out otherwise.

After I checked in at the gate, I was told that there would be a delay because of mechanical problems. After about two hours, it was announced that our aircraft could not be made operational, but that another one was available at another gate, and about an hour later we were all on board. After about another half hour it was announced that it was no longer possible for the flight crew, who had brought this plane in from somewhere else, to fly it to Moscow, since they would then exceed the allowed time for continuous hours on duty. A new flight crew was promised within an hour and a half.

In the mean time, British Airways had generously authorized the cabin crew to regale us with booze and snacks.

When the news arrived that the new flight crew was now boarding, we were all relieved, but alas, too soon. We were told that we would all have to disembark because work rules now required the present *cabin* crew to go off duty. We then spent an hour or more in the lounge but finally, a new cabin crew, an eligible flight crew and a functioning aircraft were all available at the same place and the same time.

It was thus about six hours after the scheduled departure time that we finally took off. Now a new problem troubled me. There were many reports that taxis from Moscow airport to the city were very often driven by hooligans who would simply stop in some remote place to rob and beat up (or kill) the passenger. I supposed the person from the University who was going to pick me up at 8 pm was not going to wait until 2 am. I was not very relaxed as we soared through the night sky to Moscow.

When we entered the terminal at Moscow airport, I saw nothing to brace my spirits. It was a very dimly lit, forbidding looking place, with lots of unsavory-looking people hanging about. However, once I had cleared passport control, a slim, friendly young man, Andrey Shevelkov, greeted me in English and said he was there on behalf of Prof. Tretyakov to take me in to Moscow.

These were very difficult times for the Russian Academy and for Russian universities, because funding was down to a minimal subsistence level. The Academy could not afford to put us in a central hotel, so instead we stayed in a very decrepit one far from the center. The first morning I was alone at breakfast, but as I finished the rest of our delegation arrived. They had come by overnight train from St. Petersburg, but were not in good shape owing to the lack of heat on the train. Several had caught bad colds. The hotel coffee, even fortified by vodka, was not good, but we drank quite a bit of it and were then taken to the Academy building to start our meetings.

By the next day, new agreements had been made and a public signing ceremony was held in the grandiose Academy building, which had been built just before the collapse of the USSR for the Soviet Academy. The Russian Academy could not afford to use or maintain it and abandoned it soon after we met there. There was a lavish afternoon reception to celebrate the signing, but the food and refreshments were all paid for by the NAS because the Russian Academy could not afford it. I hope that by now their fortunes have improved.

The following day the members of the NAS delegation went their separate ways, many to visit other places, such as the research center at Novosibirsk, in Russia. I began my official visit to Moscow State University. The people there were extremely hospitable and I could not have asked for a more interesting tour of Moscow or more attentive guides. The D. Sc.(hc) was presented by the Rector of the university, a mathematician with excellent English and an impressive repertoire of humorous stories. There was a banquet at which the food was not very good but the excellence of the company and the vodka compensated.

One other occurrence that remains very clear was a lecture I presented with sentence by sentence translation into Russian by Andrey Shevelkov. I found this modus operandi very awkward as it seemed to shatter the natural flow of ideas. Only one more time, when I lectured in China a year later, the same procedure was followed, but I would never again agree to allow it. By the way, Shevelkov had been selected by Tretyakov to come to my laboratory the following year, but, as noted earlier, he did not and Evgeny Dikarev came instead.

Thanks especially to Boris Popovkin and several other members of the faculty, I did a lot of sightseeing. Central Moscow had an air of timeless solidity but not much joy. All of those I talked to were full of apprehension about the future because the economy was in crisis. I made several visits away from the center where I saw many square miles of enormous, gray, ugly blocks of flats. The interiors seemed even more depressing than the outside. Moscow in 1992 was a city that had nowhere to go but up. I hope it has.

A Visit to China. Owing to the fact that Professor Jiaxi Lu (deceased 2003) at Fujian Institute of Research on the Structure of Matter was working very actively in the area of molybdenum sulfide clusters, as we also were in the early nineties, we became acquainted by correspondence. I had also had an excellent former student of his, Maoyu Shang, as a postdoctoral fellow for several years. Shang made enormous contributions to our work, not only as a fine crystallographer but also as a gifted preparative chemist. He remained in this country as staff crystallographer at the University of Notre Dame, but is now retired. In August 1992 Prof. Lu made a tour of the United States, which he had not seen since he spent the war years as a postdoctoral fellow of Linus Pauling. As part of this trip he had visited me at A&M, given a lecture and visited us at the ranch. The following Winter Prof. Lu invited me to come to China and I did this in early September 1993. Professor Lu was a man of considerable influence then, not only because of his institute but also because he had recently completed a term as President of the Chinese Academy of Sciences (CAS). Owing to this, my tour was meticulously planned and organized.

I flew to Beijing *via* Honolulu and Tokyo and landed in Beijing on Thursday, September 2, where I was met by a young woman who acted as guide and translator and who had a Mercedes Benz sedan and driver at her disposal. I spent three days in Beijing, where I visited the Institute of Chemistry of the CAS, gave a lecture at the University and met several of the professors one day while the other two days were devoted to sightseeing. This included, of course, the Forbidden City, the Summer Palace, Tiananmen Square, the Ming Tomb, the Great Wall, and the Zoo, as well as various parts of the city. The trip to the wall was a lesson in what it means to be a VIP. Evidently the car that had been provided bore a license plate or other identification of VIP status because, as we got within a few miles of the wall, we were waved past a very long line of waiting cars and went straight to the reception center at the base of the wall. My guide and I spent several hours on the wall, had a bite to eat and then the waiting car whisked us back to Beijing. It is not hard to acquire a taste for that kind of treatment.

On the following Monday I flew from Beijing to Fuzhou where I spent several days at Jaixi Lu's institute. To catch my flight I had to be up at 4:30 am but since my body's sense of what time of day it was was so disoriented, this was of little moment. I was impressed by the fact that the flight left at 7:10 although it was scheduled to leave at 7:20! Here I lectured, spent many hours meeting with Professor Lu and his colleagues and students, including particularly Professors Jian-Quan Huang and Chun-Wai Liu. Professor Liu was one of those who had translated the second edition of my group theory book, and subsequently also the third edition. There were many interesting things to see in and around Fuzhou and my guide was a very amiable student, Jun Li, who subsequently came to the Untied States to do postdoctoral research with Bruce Bursten at Ohio State.

The high point of my visit to Fuzhou was the award of an honorary doctorate by the Fujian Institute. This was followed by a sumptuous dinner at a very elegant and exclusive dining club that Prof. Lu belonged to, in which an endless succession of courses were served, one at a time, by a small army of waiters and waitresses, all nattily dressed and wearing white gloves. I had attempted, throughout my stay to use chopsticks whenever I ate Chinese food. On the whole I managed, although not exactly expertly. It was at this dinner, I think, that I faced and met my greatest challenge—a hard-boiled quail egg. This small, slippery, ovoid object had to be handled just right to avoid the disaster of having it squirt out of the chopsticks to land who knows where. After this, nothing daunted me, although my technique still leaves much to be desired.

Following Fuzhou, Jun Li escorted me to Xiamen. Prof. Lu provided us with a car and a driver (not a Mercedes, this time). We traveled over one of the worst roads I have ever seen, with heavy construction of new factories going on over much of the way. Xiamen itself turned out to be a pleasant place and Jun Li stayed with me and we did quite a bit of sightseeing.

At Xiamen University I lectured and talked with a number of people in the chemistry department. I remember Professor Qianer Zhang who had been the mentor of one of my students, Jianriu Su, who showed me around the campus. As we were passing a building with all the windows open because of the warm weather, I heard a large chorus of voices speaking in unison and we went over to one of the windows to see what this was all about. It turned out to be a class in spoken English, being taught by a very earnest teacher, who would speak a phrase or sentence and have the class repeat it after him. The problem was that this teacher had a very severe accent, so that all these students were being drilled in poor pronunciation. This well meaning teacher was really doing them more harm than good. Clearly, what they needed were tapes made by someone who spoke correctly.

From Xiamen I then flew back to the north to spend several days in Nanjing where my principal host was Professor Xiao-Zeng You, who had also been the leader of the group of translators of my group theory book. Professor You had spent some time in the Untied States and his English was so good that we were able to converse in a very relaxed way. He most generously showed me many interesting things in and near Nanjing. I remember him warmly as a very gracious host.

For the final leg of my trip I was accompanied by one of the younger members of the faculty, Qing-Jin Meng, who was a frequent visitor to Brown University and thus very fluent. He also looked remarkably like Michael Chang, the tennis player. We went by train down to Shanghai, and I was hosted at Fudan University. I found Shanghai fascinating because of the splendid waterfront along the Yangtse River. It is lined by impressive buildings built by the British and recently a handsome embankment walkway had been created from which one could see the impressive traffic on the river as well as huge amounts of construction on the opposite bank. It had already been decided by the government that despite the soon-to-be accomplished repossession of Hong Kong, Shanghai would be the leading commercial and trade center of the PRC, and building activity was everywhere to be seen.

Finally, after two weeks in China, Prof. Huang took me to the Shanghai airport to catch a flight on Air China to Toyko, where I would make a good connection with an American carrier to the United States. Here I nearly had what would have been the only major glitch of the entire trip. The people in the Air China lounge spoke essentially no English and did not convey to me the fact that my flight was indefinitely delayed — which would, of course, have caused me to miss my connection in Japan. Fortunately, there was a polygot Dutchman in the lounge who understood Chinese and he explained to me the situation and also pointed out to me that at a gate very nearby was a United Airlines flight nearly ready to leave for Tokyo. I scampered over to that gate and found that they had a vacancy in First Class and so I was able to get to Tokyo comfortably and in time for my connection. Since, as is my unvarying habit, I was travelling with only one bag that I could carry, nothing had been checked, and I was able to make this change of flight without any problem about baggage.

I feel very fortunate to have had such a well-organized and fascinating trip to a country that was then clearly on the verge of explosive growth to become a major economic and political power in the world.

The NAS Council. The early 1990s saw the high noon of my activities in Washington, D.C. In addition to attending meetings of the NSB, I was elected to the Council of the National Academy of Sciences in 1991. The latter activity was one I found very much to my liking. The Council consists, *ex officio* of the five top officers, President, Vice President, Treasurer, Home Secretary, and Foreign Secretary, all of whose terms are more than three years. The remaining twelve counselors are elected for staggered three year terms. The quality level of the people I served with was the highest of any group I have ever belonged to. The President, for the first two years, was Frank Press, a man of enormous ability for such a job. His measured judgment, calm efficiency, and sincerity were rare assets.

The council faces a constantly changing agenda of important and often tricky issues, one of which, every twelve years, is seeing that a new President is nominated for a (usually) non-contested election. It does this by forming a nominating committee of other NAS members, but it then remains in close contact with their work, and has, in principle, the final decision. In my time we were pleased to approve the nomination of Bruce Alberts, who succeeded Frank Press in 1993.

One of the Academy President's most extensive duties is to be the CEO of the National Research Council (NRC), the operating arm of the NAS, NAE and Institute of Medicine, collectively called "The National Academies." Frank Press placed me on the NRC and also on the Executive Committee of the NAS. The purpose of the latter was to take urgent academy action in the event a meeting of the full council could not be arranged in time.

My experience on the NAS Council was one of the non-academic highlights of my professional life. What can be more enjoyable than working with a group of people who are all one's intellectual equals — or superiors — and all motivated to do the best possible things in furtherance of the scientific enterprise?

The Welch Prize. One of the most prestigious prizes in chemistry is the annual medal of the Robert A. Welch Foundation. In 1994 it was awarded jointly to Jack Halpern and me. An announcement was made here at a luncheon attended by Norm Hackerman

With Dee and my daughter Jennifer when the Welch Prize was announced, 1994.

With Jack Halpern on the evening in 1994 when we were jointly awarded the Robert A. Welch Prize.

and Norbert Dittrick and hosted by Ray Bowen, who was literally having his first lunch as President of the University. Ray stayed on the job for eight years, during which we became friends. His management style was a little aloof, but he and his Provost, Ron Douglas, did a good job, and A&M recovered from a rather lengthy period when the President's office, in my opinion, was not well run. It was also a period when Governor George W. Bush made some excellent appointments to the Board of Regents.

But to get back to the Welch Prize, it was awarded at the annual Welch Banquet in Houston in October. Jack and Helen, Dee and I were given top-of-the-line treatment and enjoyed the occasion very much. We had our (my, at least) only ride in a stretch limousine, where we realized that they are just as absurd when seen from the inside as when seen from the outside. One other unforgettable detail concerns the payment of U.S. income tax on the prize. Jack and I were equally astounded to find that the taxable value reported by the Welch foundation included the market price of the gold in the medal.

I might take this occasion to remark that I think it is an outrage for our government to apply a tax at the earned income level to unsolicited prizes such as the Welch prize, the Nobel prize, and a number of others. For one thing, no other nation that I know of is mean-spirited and curmudgeonly enough to do it; the others take pride in the once-in-a-lifetime prizes their citizens receive. I would also guesstimate that the amount of tax revenue raised this way is no more than one-ten thousandth of one percent of the federal budget.

The French Academy. A high point of my warm relationship with France occurred in 1995, when I was elected a Membre Associé of the Académie des Sciences, a branch of the Institut de France. The Institut is headquartered in a magnificent building on the

Leaving the podium under the great Coupole of the Institut de France after being inducted into the Académie de Sciences, 1995.

Rive Gauche at the south end of the Pont des Arts. The splendor of the setting, most particularly the meeting room under the great coupole is impossible to exaggerate. I will say something more about my visit to be inducted into the Academie along with my old friend Rex Richards and my (then) new friend Susan Solomon in Chapter 10. I first met Susan just in front of the hotel as I was returning from a walk somewhere. I wasn't sure who she was, and for some reason we greeted each in French and continued our conversation that way until Susan said to me, "We can talk in English, you know," which we then did.

The Priestley Medal. In 1997, I was genuinely surprised to be notified of my selection by the ACS board of directors to receive the Society's "highest honor," the Priestley Medal. The secret of my nomination had been well kept from me, and I had a distinct feeling that, after the debacle of 1982, I would never be recognized by the ACS except for research. I also told myself that that was OK with me — a fine example of what the shrinks call "being in denial." When I learned I had been selected, I was not only surprised but also extremely pleased. It was presented at the Spring national meeting in 1998.

The ACS is a huge, complex, and necessarily imperfect organization, and most, if not all, of its members have a mixed bag of feelings about it. My bottom line on the ACS has always been way into black figures. The health of the American chemical enterprise, in all its phases, would be inconceivable without the ACS. The health of the entire global chemical enterprise owes an enormous debt to the publishing activities of the ACS, that range from Chemical Abstracts through a long list of top quality journals to the wonderful data base called SciFinder. The ACS Awards program provides strong encouragement to chemists to be the best they can in a great variety of scholarly and professional areas.

I took seriously the challenge of presenting a non-trivial award address at the Spring 1998 national meeting. I believe I succeeded and have included the text as an Appendix in this book. I still believe, as I said in that address, that all history is cyclic and that in the future support for basic research will see a new day, even though, as I write this in 2006, the tide seems still to be running the other way. The major problem in American chemistry today is the increasing failure of the NSF to support basic research in chemistry.

Crete. A very memorable event in 1996 was a trip to Crete to receive an honorary D.Sc. As always when we go anywhere else in Europe, Dee and I stopped on the way to visit friends in England and France, and this time in Copenhagen as well and then reached Iraklion (Heraklion) *via* Athens, on October 3. We spent six days in Crete, where in addition to the locals there were some people we already knew: Mike Karabatsos and his wife Marianna, and Raphael Raptis and his wife Silvia. There is no doubt, although I never asked him point-blank, that Mike was responsible for the invitation. Mike, who was born in Greece, and I were in graduate school together and he had been a faculty member at Michigan State University since then, but he had also become an advisor on development of the relatively new University of Crete. Raphael had done his Ph.D. with John Fackler at A&M. I also met one of my future graduate students, Panagiotis (Panos) Angaridis (Ph.D. at A&M, 2002) on that trip.

Dancing with Raphael Raptis and Mike Karabatsos in Crête, 1996.

Dee and I thoroughly enjoyed sightseeing around Crete and the delightful Greek food. There are ruins all over the island and we spent a lot of time looking at those. One evening there was a departmental party at an open-air restaurant, with music and dancing.

My honorary degree diploma from the University of Crete makes an interesting addition to my collection of diplomas I cannot read at all. The others are from Lomonosov University, the Fujian Institute and Jaume I. On both of my diplomas from Israeli universities beside the Hebrew original, which I cannot read, they placed an English translation.

Other Awards. The John Scott Medal and Premium, which is presented annually by the City of Philadelphia, was established in the early 1800s by John Scott, an Edinburgh druggist, who called upon those "entrusted with the management of Dr. Franklin's legacy" to bestow it upon "ingenious men or women who make useful inventions." The first awards were made in 1834 for the inventions of a knitting machine and a door lock. More recently, however, people such as Marie Curie, Edwin Land and John Bardeen were chosen. At the time I was chosen in 1997, I had not yet heard of this award, but I soon came to understand that it is quite prestigious. The city of

A reception following the award of a D.Sc. in Paris. With me in the foreground are Gérard Jaouen and Jerzy Haber and in the rear between Gérard and me are Michel Che and Yves Jeannin.

Philadelphia administers it and provides a premium of $10,000. The presentation takes place in one of the buildings of the American Philosophical Society. My citation was for "seminal work in creating new paradigms for chemical bonding and synthesis in chemistry."

During the decade of the 1990s, several other good things happened which are worth mentioning but not at great length. I received a D.Sc.(hc) from the University of Rennes (July 1992), was elected an Honorary Fellow of Robinson College, Cambridge (1992), and elected a member of the American Philosophical Society (April 1992). The APS, which was founded by Benjamin Franklin in 1743 for the "promotion of useful knowledge" is certainly the oldest and probably most exclusive scholarly society in the United States. The biannual meetings in Philadelphia, in buildings close by Independence Hall, are always an intellectual feast. Also in 1992, I accepted a six-year reappointment to the National Science Board, but that may not have been one of my better ideas.

In 1994, I was elected a foreign member of both the Russian Academy of Science and the Royal Society of London. In April of that year I went to a meeting of the Swiss Chemical Society in Neuchatel to receive the Paracelsus Prize. I am not sure if this entitles me to call myself an alchemist as well as a chemist, but my visit to Neuchatel was certainly a great pleasure because of the extraordinary beauty of the location.

In this same decade my French connections were greatly deepened by election as foreign member (associé etranger) of the French Academy of Sciences in 1995 and by receiving an honorary D.Sc. from the University of Paris (Paris VI) in 1997. I was already somewhat familiar with these institutions, having lectured at both, but to be welcomed into the fold was an extraordinary pleasure in both cases. There was also architectural pleasure involved. The entire building that houses the Académie des Sciences, the Palais de l'Institut de France, stands on the south bank of the Seine, facing the east end of the Louvre, with the Pont des Arts connecting them. Everything inside the building is magnificent, but the "sale des séances publiques," surmounted by its vaulting coupole is undoubtedly the most spectacular meeting place for scholars in the world. The chambers in the Sorbonne in which the honorary doctorate was conferred are also magnificent. Anything that takes place in such surroundings seems important, whether it is or not.

The one who formally presented me for the honorary degree was Gérard Jaouen and I have several nice pictures showing him on that occasion, but I also present here another picture of him that I like very much, taken in his apartment outside Paris, exhibiting his technique at opening a bottle of champagne. Despite all the other things Gérard has done for me, I treasure him most as my gastronomic and oenological guru.

Gérard Jaouen showing me how to open a bottle of Champagne, 1996.

I should finally note that during the 1990s I was honored with D. Sc.(hc) degrees from the Universities of Zaragoza (1994) and Palermo (1997), and also with the Gold Medal of the American Institute of Chemists (1998). At Zaragoza it was Larry Falvello who proposed me, and the pleasure of the occasion was enhanced by its being shared with Gordon Stone. At Palermo, it was my former postdoctoral research associate, Professor Giulio Deganello who was responsible, and that occasion was enhanced by sharing it with Fred Basolo.

With Tobin Marks, Victor Day, and Steve LIppard in 1994. I include this picture for no other reason than that it shows me with three of my favorite former coworkers and best friends.

CHAPTER 9

THE NEW MILLENIUM

ONE RETIRES WHEN
ONE IS DEAD OR ILL.

—Eugene Ormandy (at age 74)

ANYTHING WORTH DOING
IS WORTH DOING TO EXCESS.

—Edwin H. Land

THE BEGINNING OF the third millenium* was the start of my eighth decade. It was a year in which, not too long ago, I might have been forced to retire. Instead, it was the year I moved into a newly completed ranch house, bought my fourth sports car (a BMW Z3), published the first of what are my latest 178 research papers, welcomed to my research group two students who by now are my 111th and 113th successful Ph.D. students, and received the Wolf Prize and the Lavoisier medal. While anyone who would say life begins at 70 is an imbecile, I am ready to testify that it sure as hell isn't over at 69.

As proof of this I cite a three week period that began a little over a month after my seventieth birthday. Early in 2000 I had received word that I would be the recipient of the Wolf Prize in Chemistry in May and also that the University Jaume I in Castellón, Spain wished to present me with an honorary Doctor of Science degree in June. Dee

---

* Not being an unmitigated dunce at arithmetic I am well aware that the year 2000 was the last year of the second millenium, not the first year of the third, but I am giving in to common usage.

A northern view of our second ranch house (taken in 2003 by Dennis Lichtenberger).

Hong-Cai Zhou and Beverly Moore presenting Dee with flowers at the announcement of the Wolf Prize.

and I immediately began planning a trip to Israel and Castellón (which is close to Valencia) with a few days in Paris sandwiched in between, and again as a last taste of high living before coming home to a hot Texas Summer.

The Wolf Prize not only provided a terrific start to the new millenium, it also provided Dee with her first visit to Israel, although she was well acquainted with most of my Israeli friends. The prize is administered by the government of Israel and they do it in grand style. We flew overnight to Paris and then on to the international airport east of Tel Aviv arriving on Thursday, May 18. The very long time required to clear security that is characteristic of this airport was not lessened because of our being official guests of the government, as I had hoped it might, but we were then met by Mr. Yaron Gruder, the Director-General of the Wolf Foundation, and were soon on the road up to Jerusalem. We stayed in the Laromme Hotel, a very nice one where I had stayed in 1993 when I was the Jeremy Musher lecturer.

On Friday, Dee and I were taken by a professional tour guide to see Masada and the environs of the Dead Sea, and we had dinner at the home of Raphy and Mira Levine, along with a few other guests including Dan Shechtman, the pioneer of quasicrystals. Among her galaxy of gifts, Mira is a superb cook. Dee and I went to bed in a deliciously relaxed state.

On Saturday the entire group of laureates and their guests were taken by bus on a tour of Jerusalem and also Bethlehem, and in the evening to a dinner at the Mt. Scopus campus. This gave me the chance to meet the other laureates. I already had a nodding acquaintance with Jean-Pierre Serre and Ray Davis, but meeting the others, especially Boulez and Muti was a great experience. Although Raoul Bott and I had both been members of the NAS for around 35 years, this was the first time we had actually talked. The entire group formed a very congenial party.

Sunday, May 21 was the big day. We had the morning and lunch time to ourselves. We were taken in mid-afternoon to the Knesset, where the ceremony took place from 6:00 to 7:30 pm. The great hall of the Knesset has three enormous and spectacular tapestries by Chagall on a long wall that is opposite to a glass wall that gives a splendid view of Jerusalem. There were short speeches by the President of Israel, the Minister of Education and several others. The prizes were then awarded by the President and several of us (myself, Khush, Bott, Davis, and Boulez, whose music of a highly abstract character had been played to open the ceremony) offered some remarks. Back at the Laromme Hotel there was a large reception. Peter Stang was with us all day on Sunday, and remained with us for the next few days and accompanied us by car up to Haifa on Wednesday. John Fackler was also on hand for the award ceremony.

The winners of the 2000 Wolf Prizes in the Knesset with the Chagall tapestries in the background. Left to right: J.-P. Serre (mathematics), myself, G.S. Khush (agriculture), R. Bott (mathematics), R. Muti (music), M. Koshiba (physics), P. Boulez (music), R. Davis, Jr. (physics).

On Monday, I gave a lecture at the Hebrew University, while Dee toured the old campus and saw the Dead Sea scrolls. We attended a reception at the U.S. embassy and were then driven to Beer Sheva. Dan Meyerstein organized a very nice day for us including a symposium in my honor at Ben Gurion University, local sightseeing for Dee, and a symposium dinner.

On Wednesday we were driven up to Haifa, along with Peter Stang. We spent a day and a half there, very nostalgic for me, especially because Zvi Dori (who had retired) joined us. We finished the week in a beach-front hotel in Tel Aviv and flew to Paris on Saturday.

Paris, as always, did not fail to provide a rich experience. We stayed at one of our favorite hotels, the Relais Christine, in the oldest part of the sixth arrondissement, and had meals at several favorite restaurants, including the Louis XIII which is quite nearby, and the magnificent Ledoyen just off the Champs Elysées. At the latter, which is my favorite restaurant in all the world, we were joined by Gérard and Anne Jaouen. We also had a wonderful dinner with Michel and Daniélle Che, chez eux, and spent many hours doing what we most love to do in Paris: just walking around feeling lucky to be alive in such a place. Then it was off to Spain.

We were met at Valencia airport by Pascual Lahuerta and Rosa Llusar. Pascual had long been the senior inorganic chemist in Valencia and Rosa was now a professor in Castellón. We had a wonderful visit in the region of Castellón and Valencia and enjoyed the elaborate doctoral ceremony in Castellón, none of which I could actually understand because it was all conducted in the regional language of Valenciano. We stopped in Paris for one more night on the way home, and arrived back at the ranch feeling that a total escape from reality now and then is a very good idea.

Marcetta Darensbourg and Bruce Bursten responding to George Stanley's request for smiles.

In March of 2000, Carlos Murillo again took the initiative in organizing an all-star symposium at A&M. Called Contemporary Inorganic Chemistry II, it featured twenty-four invited speakers (as well as poster sessions). Incorporated into these proceedings were the award of the sixth F. A. Cotton Medal, which went to Tobin Marks.

My research program since 1999 has been as productive as ever, although the size of the group has recently begun to decline as funding for truly basic research has become harder to get. I

Tobin Marks lecturing (2000).

have had no new funding from the NSF or any other government agency for several years, and the Welch Foundation terminated my annual grant (which never exceeded $62,500) in May of 2005. I am now supported entirely by the income from my Welch Chair endowment and funds

A pleasant moment shared with two very good friends, Gordon Stone and Marye Anne Fox, at the 2001 banquet of the Robert A. Welch Foundation.

Two of my best recent Ph.D.s Dino Villagrán (left) and Chun Y. Liu with Carlos Murillo in 2003.

provided by the university for the Laboratory for Molecular Structure and Bonding. NSF predoctoral fellowships to Elizabeth Hillard and John Berry, and a postdoctoral fellowship to James Donahue have helped to keep the show on the road. The talents of my coworkers are, of course, critical. This is especially true of two of the best students I have ever had—in half a century. With full appreciation of the excellence of the others, I have to mention the exceptional merits of Hong-Cai Zhou (Ph.D. 2000) and John F. Berry (Ph.D. 2004). It is students like these that make the professorial life worthwhile.

I must also mention three superb postdoctorals: Evgeny Dikarev, Marina Petrukhina and James Donahue, each of whom was prodigiously productive. All of those I have just mentioned as well as several others — Penglin Huang (02), Panagiotis Angaridis (02), Peng Lei (03), Chun Yuan Liu (05), Chad Wilkinson (05), Rongmin Yu (05), Dino Villagrán (05), and Sergey Ibragimov (06) — have also brought a wealth of talent and hard work to bear on the projects that I suggested to them. Since nearly everything done in my research program depends on having correct and accurate X-ray structures, not all of which are straightforward, the expertise of Xiaoping Wang (who succeeded Lee Daniels) made another critical contribution to our success.

The main subjects of our recent and current research have been various aspects of M–M bonding, because I think this is a phenomenon about which our ignorance still far exceeds our knowledge. As for the nature of the δ bond, I was pleased to be able to team up with Dan Nocera to write a comprehensive explanation of how it works based on a combination of his and our explorations of the manifold of four states that arise from the interaction of two electrons each initially occupying a $d\delta$ orbital [1450].

The principles of the 2-electron δ bond are exactly the same as those for any type of 2-electron bond; they were first explained by Charles Coulson and Inga Fischer in 1949 for H_2, when they computed the relative energies of the four lowest states as a function of internuclear distance. But of course unless molecular tweezers become available no direct experimental verification of their results will ever be possible. The beauty of the δ bond and the chemistry we have learned to do with Mo_2^{4+} compounds that contain it, is that we can vary the bond strength by varying the angle of internal rotation, designated χ, from 0° to 45°. As shown in Fig. 9-1, the experimental data verify the Coulson and Fischer model very nicely.

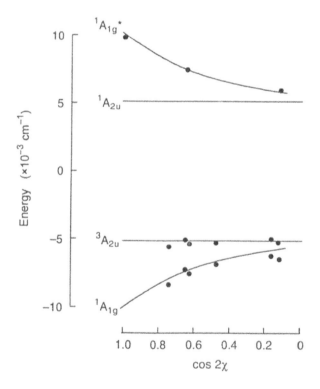

Figure 9-1. The Coulson/Fischer model of a 2-electron bond as fitted by experimental data for the δ bond in Mo_2^{4+} compounds. The energy of the A_{2u} state is the reference for the other energies. From Cotton and Nocera [1450].

Coming back now to the work initiated in 1997 by Chun Lin on supramolecules incorporating dimetal rather than nonmetal units, Chun Lin and others, especially Chun Yuan Liu, Jim Donahue, and Rongmin Yu, have produced a host of other

Jim Donahue.

compounds, employing mainly Mo_2, Rh_2, and Ru_2 units. The work took two main directions: (1) further study of how a pair of M_2 units may be electronically coupled to each other by various types and arrangements of bridging groups and (2) how elaborate supramolecular structures may be built up by linking M_2 units.

Jim Donahue, who got his Ph.D. at Harvard with Dick Holm, joined my group in 1999 and stayed nearly five years. He was an exceptionally diligent and clever synthetic chemist and now has a tenure track position at Tulane University. One major thrust of his work was to establish the relative abilities of tetradentate linkers to couple redox processes in two dimolybdenum units. He built on the key tactic embodied in Chun Lin's work, namely, that all ligands other than the linkers surrounding the dimolybdenum units (or any other dimetal units that might be chosen) should be kinetically inert. Thus, the dimolybdenum units and their associated spectator ligands were all of the type shown in Fig. 9-2. The linkers, thanks to Jim's skill and determination, were of many types. The majority were dicarboxylic acid dianions [1524, 1534], but he also made molecules with the tetrahedral dianions SO_4^{2-}, MoO_4^{2-}, and WO_4^{2-}, as linkers [1476]. His work is too voluminous to summarize further here, but one by-product I want to mention was our invention of trivial names for two diacids that didn't yet have any. So far as I know the conferring of names honoring Texas in Chemistry has been rare. I am proud to present our two contributions to what can be called *logonymics* in Fig. 9-3.

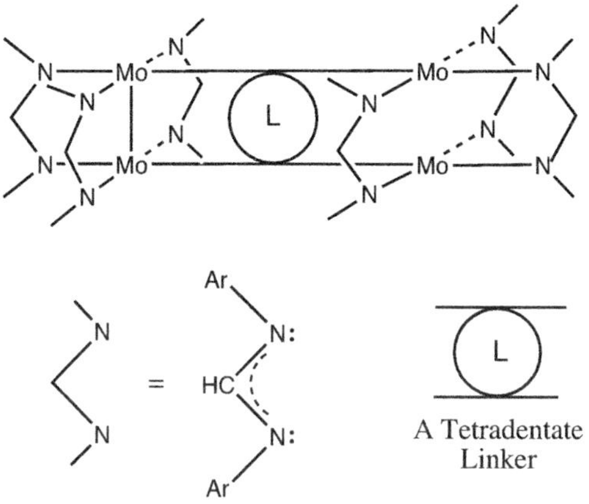

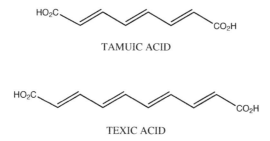

Figure 9-2. A schematic representation of a molecule with two M$_2$ units linked by a tetradentate linker. At lower left, the spectator ligand is defined as a diarylformamidinate anion.

HO$_2$C ⟍⟍⟍⟍⟍ CO$_2$H

TAMUIC ACID

HO$_2$C ⟍⟍⟍⟍⟍⟍ CO$_2$H

TEXIC ACID

Figure 9-3. Two examples of Texas-based logonymics.

In addition to the use of tetradentate linkers such as dicarboxylic acids to link M$_2$ units that are protected by three relatively non-labile bidentate ligands, as shown in Fig. 9-2, a vast array of compounds has been made by linking two or more M$_2$ units that are protected by only two cisoid spectator ligands. A few of the simpler compounds are shown schematically in Fig. 9-4. In the compounds of this type when the bridges (X) are single atoms, it has been found that communication between the M$_2$ units may occur as much or more by direct overlap of metal orbitals than *via* the bridging atoms [1586].

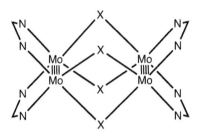

Figure 9-4. Linking of dimetal units by a set of four single-atom bridges.

Work that began in 1998 on the ability of the hpp ligand [1352], a special bicyclic guanidinate anion, to stabilize the most highly oxidized M_2^{n+} units yielded a spectacular result in 2002 when we discovered (in collaboration with Dennis Lichtenberger) that the $W_2(hpp)_4$ molecule, shown in Fig. 9-5 has an onset ionization potential of only 3.51 ± 0.05 eV [1512]. For all other closed shell molecules with known ionization potentials, the values are upward of 5 eV. In addition to making this discovery, we were able to provide a detailed explanation, by employing DFT calculations, for this extraordinary phenomenon. Even among open shell molecules or atoms, this low energy still easily holds the record, as witness $(\eta^5\text{-}C_5Me_5)_2Co$ (4.71 eV), and even beats the cesium atom (3.89 eV). Our first method of preparing $W_2(hpp)_4$ was laborious and relatively inefficient, but in January of 2006, we published a simple method [1565, 1574] (two steps from cheap and readily available starting materials) that gives yields of more than 90%.

Figure 9-5. The $W_2(hpp)_4$ molecule.

Beyond the uniquely interesting $W_2(hpp)_4$ molecule, the hpp ligand allowed us to prepare in 2003 the first examples of M_2^{n+} cores with $n = 7$, with M = Os [1516] and Re [1520]. The same M_2/hpp interactions that destabilize the LUMO in W_2^{4+}, Mo_2^{4+} and other species with δ and δ^* electrons will thus favor species that have lost such electrons. The high basicity of guanidinate anions, even without any specific orbital interactions, contributes to their exceptional ability to stabilize higher than usual positive charges on M_2^{n+} cores. An interesting example of this is provided by our isolation of the first (and still only) example of a paddlewheel compound with a Cr_2^{5+} core [1491, 1518]. Oddly, it was not by using the hpp ligand but another guanidinate anion that we accomplished this.

Just at the end of 1999, Liz Hillard and Joe Zhou had employed the 2,4,6-triisopropylbenzoate (TiPB) ion to provide a crystalline $Cr_2(O_2CR)_4$ compound in which there was no axial ligation, as I noted in Chapter 8. As a logical extension of this, I pointed out to Liz that although scores of $Cu_2(O_2CR)_4L_2$ compounds were known, just as in the chromium case, no $Cu_2(O_2CR)_4$ compound in which there were no separate axial ligands, nor any association of the molecules with each other, had ever been reported. I suggested that she try to make $Cu_2(TiPB)_4$ and see how the absence of any axial bonding would affect the Cu⋯Cu interaction. She succeeded in making the compound, but we were totally surprised by its structure, [1474] which is shown in Fig. 9-6. Instead of the classic "copper acetate" dinuclear structure, we found an equilateral triangle of square coordinated copper(II) ions. This provided a very interesting example of what is known as "spin frustration" and I joined forces with Naresh Dalal at Florida State University to characterize its magnetic properties very thoroughly [1523, 1526].

While the Wolf Prize and the accompanying travel in Israel have to be the non-research highlight of the new millenium, there have been other memorable occasions. Later in the year 2000 I enjoyed another visit to the charming city of Rennes to attend a meeting of the Société Francaise de Chimie, where I was presented with the Lavoisier Medal. There is no regular schedule for the presentation; it is the Société's highest award but they do it, apparently, only when the spirit moves them. Also in 2000 I gave the first of the annual Richard A. Walton lectures at Purdue University. These lectures honor not only Dick's personal accomplishments in research and teaching but also his effective service as department head.

I received the ACS Award in Organometallic Chemistry in 2001 and I was invited to publish my award address in *Inorganic Chemistry*, which I did in early 2002 [1492]. In this address, I focused on work done in my research group starting as early as 1966 and continuing right up to 1999. Major emphasis was placed on fluxionality and several kinds of bonding first reported from my laboratory. Hence the title: A Half-Century of Nonclassical Organometallic Chemistry — A Personal Perspective.

Other notable events in the new millenium were my election as a Foreign Honorary Member of the Chinese Academy of Sciences, a very enjoyable visit to Germany to be the Gauss Professor of the Göttingen Academy of Sciences, and honorary doctorates from North Carolina State University (2000), Ohio State University (2001), Drexel University (2002), and The Hebrew University of Jerusalem (2002). This last one afforded me another opportunity to see many of my friends there and it was a joy to find them all well and happy, considering the subliminal tension that is inescapable in Israel.

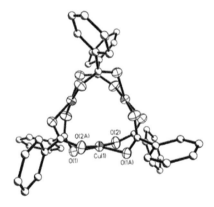

Figure 9-6. The $Cu_3(TiPB)_6$ molecule. The isopropyl groups and ring hydrogen atoms are omitted for clarity.

In April of 2002, I had the unpleasant experience of celebrating my 72nd birthday in the hospital being separated from my appendix. That this wretched organ, that never did me a bit of good anyway, should have suddenly decided to cause trouble after more than seven decades displeased me considerably, but it had to come out and out it came. The excision *per se* was actually no problem, but the aftermath was. It is a sad irony that if you are in a weakened condition, a hospital is not a good place to be. Hospitals are full of germs. I picked up a horrible bacterial infection which went on for over a week until I was finally given a big dose of some potent antibiotic. Still, I was by then in a rather wonky state as I set off for Israel to receive the honorary degree from the Hebrew University in Jerusalem, followed by the one from Drexel on the way home.

On the visit to Israel I did something I had long wanted to do — stay in the legendary King David Hotel. To add further to the pleasure, I had a room on the top floor with a view of the old city. In addition to a lovely

With Raphy Levine during my visit to Jerusalem in 2002 for the honorary degree from the Hebrew University.

evening with Raphy and Mira Levine there was a fabulous dinner party at Avi Bino's house where Zvi Dori, Mike and Ora Ardon, and a friend from my early MIT days that I hadn't seen for decades, Rolfe Herber, were present.

The honorary degree was presented as part of a graduation ceremony in an outdoor Greco-Roman theater which is part of the Mt. Scopus campus of the Hebrew University. Prime Minister Ariel Sharon was the speaker and there was simultaneous translation into English. The fact that Sharon would appear was not announced until a short time before the ceremonies were to start (presumably for security reasons) and this caused a delay, during which there was much speculation that he would use the occasion to make some important announcement. In fact, he did not. His speech was essentially standard graduation rhetoric. He did not stay for the rest of the ceremonies, but there was a break after his speech and he visited with all those on the stage and shook hands with everyone before leaving. While he was there, a squadron of sharpshooters were stationed conspicuously all around the perimeter of the theater. Without any doubt, it was the most memorably dramatic honorary degree presentation ever for me.

In closing this chapter I must cite one other publication which appeared in 2000. Although not from my laboratory, it gave me such exquisite pleasure that I can not fail to mention it. Just as I was indebted to my physicist friend Manfred Fink for providing the first proof in 1985 that the Cr–Cr bond in $Cr_2(O_2CCH_3)_4$ really is as short as I had always said it should be, I incurred a new debt to physicists, X.-B. Wang and L.-S. Wang, at Washington State University (see *J. Am. Chem. Soc.*, **2000**, *122*, 2096) for providing explicit proof, *via* photoelectron spectroscopy, that the $\sigma^2\pi^4\delta^2$ quadruple bond in $Re_2Cl_8^{2-}$ is just as I formulated it in 1964. Their result, which speaks for itself, is reproduced with permission in Fig. 9-7.

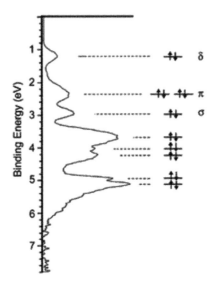

Figure 9-7. The photoelectron spectrum of $[Re_2Cl_8]^{2-}$ showing the assignment of the features to the molecular orbitals.

CHAPTER 10

MORE ABOUT PEOPLE

SAGE MIR MIT WEM DU
GEHST UND ICH WERDE
DIR SAGEN WER DU BIST

— Anonymous

WHILE I HAVE already mentioned many people, there are other observations I would still like to make about people I have known, and I present them here, blunt and artless in some cases, and in no particular order.

Scientists are people (though some non-scientists may demur) and people are interesting. Marie Curie was quoted by her biographer-daughter, Eve, as having said to an interviewer that he had to understand that "scientists are interested in things and not people." With all due respect to this remarkable person, I think she overstated her point. While science does try to produce "objective" descriptions of the world, science is done by people and to understand fully how science has evolved, one must be concerned with the people who did it.

I might begin the subject of people with remarks about my scientific ancestry. In some chemistry departments, one finds a so-called genealogy of their faculty. More than one of them traces me back to Wilkinson, Wilkinson back to Harry Julius Emeléus, Emeléus to some German chemist (it doesn't matter to whom, as will become apparent), that person further back and thus all of us to some remote intellectual ancestor such as Paracelsus. It isn't true. Wilkinson got his Ph.D. at Imperial College with a man called Briscoe, who didn't get a Ph.D. with anyone. Harry Emeléus was not Wilkinson's *Doctorvater*.

Meeting Famous People

That I am not a shy person has stood me in good stead by enabling me to meet, if only briefly, some very famous people. I have already related how I met Enrico Fermi in Rod Smythe's convertible. I also met and spoke with Niels Bohr. He came to give a lecture at MIT in about 1960 and after the lecture there was a reception for him in a lounge shared by the chemistry and physics departments at about the point where their realms met in Building 6. I saw Bohr sitting on one of the big leather sofas all alone so I went over and sat down beside him, and introduced myself. We chatted for quite some time, partly about what a wonderful place Copenhagen was, while he worked his way through about one pipeful of tobacco and one box of wooden matches.

My boldness seemed to encourage others and soon there was a steady relay of people coming to talk to him.

One more such experience occurred in Paris in the early 1990s. I and several of my friends from the ENSCP were having lunch in the cafeteria in the old Ecole Polytechnique building nearby and my friends pointed out an older man sitting alone at a table for two next to the wall, not far away. They said, "you know, that's Anatole Abragam." I said, "Really, why is he all alone?" The explanation was something along the lines that he was then retired and generally considered a rather forbidding figure by the younger people because he was so formidably intelligent. His first wife had recently died and it was believed that he came there often to get his main meal of the day. I said to my friends that I simply didn't believe that he enjoyed eating alone (few people do) so I picked up my tray, went over and asked if I might join him. I also said that, as I believed his English was so infinitely superior to my French, it would be a good idea if we spoke English. We then had a very pleasant conversation while we ate, mainly about people we both knew at Harvard and MIT. It was quite clear that he was then a lonely man and was by no means opposed to having company when he ate.

About five years later, when Rex Richards and I were both in Paris to be inducted into the Académie des Sciences, a massive strike of French transportation workers started the day we were to leave. I had known Rex for many years, beginning when I was a graduate student and had broken my leg skiing. I shall never forget his stopping by my lab to say hello and observing that my misfortune was "hard cheese." I had heard that expression on the lips of British comedian Terry-Thomas, but was amazed to discover that it was used by real people. Anyway, getting back to that day in Paris, I was wondering how in the world I was going to get to England when Rex said, "not to worry, I've got my car here, and I'll take you back to Oxford where you can get a train into London."

It was somewhat more complicated, however. Rex considered (not unreasonably) that there was a high probability that any car parked on the street in Paris would soon be dented. He had therefore made an arrangement with Anatole Abragam to park his car in the basement of the condominium where Anatole now lived with his second wife, while Anatole would leave his own car at the Collège de France, where he still had the right to park. Thus, Rex and his wife and I and our bags had to get from Place de l'Odeon to Anatole's apartment some 3-4 miles away. In view of Rex's bad back, walking was not possible, so we had to spend about an hour waiting for a taxi, during which we got some first-hand exposure to the great French sport of queue-jumping. I then met Anatole for the second time as we picked up the car. He was evidently a very happy man now with his new and very charming wife. The rest of our journey was problematic, too. Of the four boat services on the Calais-Dover run, three had already closed and space on the remaining one was short. I had to take an early boat in order to get to Cambridge in time for a dinner while Rex could not get a place for his car until two hours later.

I have seen Anatole Abragam several times since then at meetings of the Académie and always find that he has incisive observations concerning whatever we discuss. His autobiography, *De la physique avant toute chose* (Physics First), is a fascinating book which starts with his early childhood in Russia and explains his somewhat unconventional development as a physicist, with a serious interruption during WWII because, as a Jew, he and his wife had to go undercover in the south of France. I have heard that this fine book has now been translated into English, and I recommend it in either language.

Secretaries

In discussing important people in my life, I must mention secretaries. Many people have secretaries but not the kind I have been lucky enough to have. My first one at MIT was Nancy Dusenberry, who had been my wife's college roommate. Nancy had no formal secretarial training, but her personality, intelligence and enthusiasm made up for it. After a few years she married one of my first postdoctoral fellows, an Englishman named Anthony Blake, and they left for England *via* South America.

The next secretary, Anne Wood, was already the wife of an English postdoc, John S. Wood. She had been an executive secretary in England and for two years I had the most professional secretarial service imaginable — even at the uncompetitive wages paid by MIT. After a few years, John moved on to join the faculty at the University of Massachusetts.

Then, ca. 1962, came the first of the only three secretaries I had for over four decades. They were three of the best ones could hope for. Marilyn Milan was another trained pro, but also highly intelligent and very efficient. She could type like the wind. It was while Marilyn was with me that the Encyclopedia Brittanica offered me a contract to write a lengthy piece on the transition elements. Since I was badly in need of a quick few thousand bucks to buy another horse, I signed on. To accomplish this task efficiently, I sat back in my chair, feet on the desk, with Cotton and Wilkinson open on my lap, and dictated onto tape, simply editing and condensing what was on the pages to what I felt was appropriate for the encyclopedia. I was able to do this in a matter of 1-2 days, but preparing a typed manuscript was much more time consuming. Marilyn was so efficient that she had it done in a week or so, and a few weeks later I had my check. Soon after that I had acquired my excellent horse, which I named Brittanic Royalty, more commonly known around the barn as Roy. When I left MIT in January of 1972, Marilyn was assigned to John Waugh and some of the other physical chemists. I believe she later married and had a family. Having had such an excellent secretary who was also a long-termer, I was unwilling to accept anything less at A&M.

I have already described how I found my first A&M secretary, Irene Casimiro, and how she served admirably for nearly 20 years. One more thing that I want to say about her is that she was a biocybernetic marvel. She was the fastest typist I have ever seen, but the way she did it can only be explained by assuming that she was hardwired to be a typist. Irene could type from a handwritten or printed text without paying any attention to the content. She could carry on a conversation with someone in the office or on the phone without slowing down. It seems that the letters went straight from her eyes to her fingers. It was an uncanny thing to watch. However, she did not take to word processors, or anything else related to computers. As the 1990s began, she and I both realized that this was an untenable situation. She prepared to retire and I began a search for a new secretary.

With my next secretary, who worked on early drafts of this book, I found just what the times demanded. Ms. Beverly Moore was a master of everything related to word processing and computer graphics. She was with me for over a decade until on Labor Day of 2004 she suffered a very serious stroke. After she left I found that her idea of filing was to make piles of unrelated — and often unopened — documents. However, no real harm was done and the enthusiasm, dedication, and expertise she brought to her work more than offset whatever she lacked in routine office management. Beverly was a force of nature in her own eccentric way. I have always been fond of eccentrics. They are the spice of life.

Jack Lewis

I first met Jack in 1956 at Imperial College (IC). This was Geoff Wilkinson's first year back in England, where he had taken the chair in inorganic chemistry at IC, recently vacated by Harry J. Eméleus, who moved to Cambridge. Jack, then a fresh Ph.D. from Cliff Addison, had just been hired as a lecturer at IC. I immediately recognized Jack as a kindred spirit, not only in his approach to chemistry but also in general, and thus began one of the closest friendships of my life.

The venue of our friendship changed after one year, however. The same year Geoff was appointed to the chair at Imperial College, Ronald S. Nyholm arrived at University College (UC). Jack had more natural affinity with Nyholm's approach to chemistry and when Nyholm invited him to move to UC, he did that, thus establishing a close rapport with Nyholm that lasted until Nyholm's untimely death in a car crash just around the corner from Jack's house in Cambridge, in 1971.

My frequent visits to London to see Geoff thereafter included visits with Jack at UC, where Jack's career was advancing by leaps and bounds in the lively atmosphere created by Nyholm and his "Australian mafia." Chief among these were Brian Figgis, who went back to Australia, Tom Dunne, who later became department head at the

A picture with Jack and Freddie Lewis taken in the period of 1956–58. Apologies for the loss of Jack's dome, but it's too good a picture otherwise not to use it.

University of Michigan, Brice (Boz) Bosnich, recently retired from the University of Chicago, plus Ron (now Sir Ronald) Mason (a Welshman). Peter Pauling came, did a Ph.D. with Ron Nyholm (although he worked more with Mason), and stayed on for a while as a lecturer. This was as lively and talented a group as I have ever seen and my visits to UC were always a highlight. Incidentally, all those I have mentioned became lifelong friends, although I haven't seen Figgis for many years. I also got to know Ron Nyholm quite well and learned what outstanding abilities he had both for chemistry and for leadership.

Getting back to Jack Lewis, he soon showed that he was the most outstanding talent in the UK in the generation following Chatt, Nyholm and Wilkinson, and in 1961 he was appointed to the chair of inorganic chemistry in Manchester. Here Jack was able to establish his independence and thus in 1967, when he went back to UC he was no longer "under" Nyholm but in full charge of his own program. His reputation both as a scientist and as someone well endowed with many other abilities led to his being offered the opportunity to succeed Harry Eméleus at Cambridge in 1970. He became a Fellow of Sydney Sussex College, bought a house within walking distance of the new chemistry building, and he and Freddie, and their children, Ian and Penny, settled into a halcyon existence.

I might mention that in 1959 I had made an effort to attract Jack to MIT. He came for a semester as a visiting professor, but Ron Nyholm immediately reacted to this first danger signal that Jack might leave the UK and had him promoted to a higher rank at UC — the one with that curiously British title *Reader*.

Clearly, a man of Jack's virtues will never have a halcyon life because there are so many places where he is needed. In about 1974, a wealthy man named David Robinson approached Cambridge University with an offer to fund a new college. Jack was asked to take on the task of working with Mr. Robinson to plan and build the college and to become its first leader (Warden was the title Jack chose). Jack was selected because of his remarkable combination of international distinction in his field, common sense, and diplomacy.

Robinson College was a brilliant success and largely in recognition of his tireless work to achieve that, he was knighted in 1981. Throughout the whole process, Jack never slacked off on his chemistry and he produced several hundred research publications during his first decades in Cambridge.

With the college and his research both running smoothly, Jack was once again called on to do a major job, namely, to be the chairman of a Royal Commission on the environment. For his work on that Commission, he was awarded a peerage in 1989. He took the title Lord Lewis of Newnham, Newnham being the area of Cambridgeshire in which Robinson College is located.

In 1995 Jack reached the mandatory retirement ago for his professorial position, but was able to continue as Warden of Robinson until 2001. Since then, he has continued to be very active in affairs of

Jack Lewis at the time he became Emeritus Professor (1995).

the House of Lords, especially in matters concerning science, technology and the environment.

Jack has received both the ACS awards in inorganic chemistry (1971 and 1986), and he is a member of the U.S. National Academy of Sciences, The American Philosophical Society, and the American Academy of Arts and Sciences. He holds numerous honorary degrees, and in 2004 he received a Queen's Medal from the Royal Society.

I shall conclude this sketch by recounting an incident that very well illustrates Jack's remarkable capacity for keeping things in perspective. After Robinson College had been in full operation for a few years, an oil portrait of Mr. Robinson was hung at one end of the hall, and I remarked to Jack that there should be one of him at the other end. I made this comment several times, but he displayed no interest. I pointed out to him something that I had been told by a friend of mine who made his living painting portraits: if it is done after you're dead by a painter working only from photographs, the result is often very bad, to the extent that the surviving friends say something like, "It just doesn't capture the real him," and in some cases refuse to pay for it. "So," I said, "wouldn't that be a shame?" To which Jack retorted, "Well, I wouldn't really care, would I?" He did eventually give in and have a portrait done, and it now hangs in the hall.

Earl Muetterties

Earl Muetterties was another in my generation who deserves major credit for establishing the place of inorganic chemistry in the postwar period in the United States. His interests were wide-ranging and he had abilities as an organizer and administrator as well as for creative research.

I first met Earl, who was three years older than I, during my first and his last year of graduate study, 1951-52. He was a member of Eugene Rochow's group of graduate students, which was not, with the exceptions of Earl and Dietmar Seyferth, one that I found very impressive. While I found Earl likeable, I did not then take him seriously. He went to a job at DuPont, and I suppose I would have thought, had I thought about it at all, that that was the last I would hear of him.

I think that the challenge of succeeding in the high-powered environment of the DuPont Experimental Station transformed Earl into what he remained for the rest of his life: an extremely serious and highly focused researcher, and one eager to tackle important problems of all kinds. Succeed at DuPont he did, rising to become one of the Associate Directors of the Central Research Department at the Experimental Station at the age of only 38. He was a contemporary of another very good friend of mine, Howard Simmons. Howie, whom I later had the pleasure of seeing frequently at National Science Board meetings, stayed at DuPont until retirement, but managed to have a very wide-ranging intellectual life. One of our joint pleasures was to go book shopping during lunch hour when the NSF was still at 21st and G street in DC. Unfortunately, it seemed to have been one of the social mores at the Experimental Station during the Muetterties-Simmons days to smoke heavily, and Howie too paid the price for that (at age 68) although not so tragically early as Earl.

Over his DuPont years I kept in fairly close touch with Earl and for a while our research interests became very similar. In 1965 (*Inorg. Chem.*, **1965**, *4*, 769) he published a short but important note on what he called stereochemical nonrigidity, while at the same time, I was becoming more and more interested in what I (following Bill Doering) called fluxional molecules. This led to frequent exchanges of manuscripts.

We both continued to work in this general area for some years. I concentrated on organometallic compounds and metal carbonyls, while Earl homed in on a type of process called polytopal rearrangement, which is found among complexes with coordination numbers of 5, 7, 8, and higher.

I have mentioned that as early as 1958, Art Cope had attempted, with my enthusiastic support, to attract Earl from DuPont to MIT, but without success. Again, in about 1970, when Earl indicated to me that he was giving serious thought to leaving DuPont to take up an academic career, I tried to persuade him to come to MIT, even though I myself was considering a move to A&M. He made no decision until later, when my departure was a done deal, and he decided then (1973) to go to Cornell, largely, I think, because he and Roald Hoffmann had become close friends.

Soon after my own election to the National Academy of Sciences, Ted Cairns, Earl's boss at DuPont, and I nominated him for the NAS and in 1971 he was elected, at the early age of 44. It was extraordinary for a chemist in industry to be elected at all, let alone at such an early age. Clearly, Earl's accomplishments were recognized in academia, which facilitated his later transfer to that milieu. Howie Simmons was also elected to the NAS, in 1975.

One of my most vivid memories of Earl dates to a week we spent together in September 1981 at Villars, Switzerland. We had both been invited to give lectures in a summer school for Swiss chemistry students doing advanced work (the Troisiéme Cycle) prior to going on in research. On one afternoon, when we were both free to relax, we went for a long walk in the beautiful surroundings of Villars in the Suisse Romande, and we talked about a lot of things that naturally concerned 50-year-olds who had known each other for over two decades. Along relatively trivial lines, I remember that we established that each of us got the same consulting fees, which reassured us that neither of us was being underpaid. Earl had by then become very (or totally) deaf in one ear as a result of an infection from swimming, which was his preferred form of exercise. The main reason I remember this is because, as he explained it to me, he had simply been the victim of incompetent medical attention, and yet he showed no anger. It was a case of his doctor having fobbed him off, repeatedly, with the classic sort of "take some aspirin and call me later" advice. Earl, usually one who questioned whatever he was told, had unaccountably not done so in this matter until it was too late.

Most vivid in my memory, however, is my asking him, then as he had been for many years, a two-pack-a-day (or more) cigarette smoker, whether he didn't think this was risky. He said that, of course, he knew there was a risk, but the great pleasure he got from smoking was worth what he considered to be a relatively small risk. And so, he continued, I believe, until actually diagnosed with lung cancer.

Earl and I were complete opposites in many ways, apart from our both being consumed by our interest in inorganic chemistry. Earl was deeply religious (Roman Catholic), a liberal democrat, patient, and soft-spoken; he rarely criticized others openly, although there is no doubt he made his own judgements and was no pollyanna. He was also very loyal to — even protective of — his associates, which on a few occasions led to sharp disagreements between us. Not long before he died, he was very critical of the way I conducted my unsuccessful try for the ACS Presidency, and he told me so in no uncertain terms. A little later, when his illness was well advanced, I had my last conversation with him. It was one of the saddest occasions of my life, because at the age of only 56, he might well have had his finest work still ahead of him. I have always felt so terribly incapable of saying anything to someone in that position. All I can say now is that it was a privilege to have been his friend and a bitter disappointment that he departed so soon — so unnecessarily soon.

Geoffrey Wilkinson

GW was, beyond any comparison, that most self-contradictory person I have ever encountered. He was a mass of inconsistencies.

Geoff was an avowed atheist; he warmly recommended to me H. L. Mencken's *Treatise on the Gods*, as elaborate an apologia for atheism as I know, and once said to me that God was, to him, "three letters, g, o, and d." In spite of this and the fact that he showed no serious interest in music, one of my most vivid memories is of him working happily in his lab at Harvard and singing an aria from Bach's Magnificat at full volume. I might note that he also had an extensive repertoire of British labor songs, most of pronounced scatological and/or bawdy character. I learned all of them, and can still recall a few, with much amusement. Two of his (and my) favorites were entitled "Oh, they're shifting father's grave to build a sewer" and "My brother's a slum missionary." The latter goes on to say "he saves little girlies from sin. He'll save you a blonde for five shillings, (bleep, bleep) how the money rolls in."

Geoff would often remark that Harvard had "fired" him. This is not a fair comment, but Harvard did, in a sense, outsmart itself. Geoff's gift for chemistry, which was a very great one, was almost exclusively intuitive. He did not think quantitatively or mathematically. He knew very little math; he reasoned by analogy. It was this that, at least in part, cost him a promotion to tenure at Harvard. In spite of his brilliant and abundant accomplishments as an Assistant Professor, the senior faculty, who were, almost to a man, intellectual snobs, felt that chemistry would soon become so inherently quantitative, mathematical and physically oriented that a chemist of Geoff's stripe would eventually be left behind. While their reasoning may have been sound, their timing was pretty far off. For the next 35 years Geoff was able to maintain his research in the forefront of inorganic chemistry, although most of chemistry has become steadily more theoretical and physical.

Geoff could be, especially in his younger days, exceptionally generous. As for myself, I must acknowledge an extraordinary amount of warm hospitality by Geoff and Lise on my many trips to England. I have also mentioned his generosity in providing, out of his own research funds, a stipend which made it possible for Carl Ballhausen to come to Harvard for a postdoctoral year with Bill Moffitt. Geoff also had a knack for spotting unconventional or unappreciated talent, and offering opportunities to people who possessed it. John Osborn, for example, left Cambridge with a "lower second" (one step above a mere "pass degree") and therefore had scant prospect of doing graduate work in the laboratory of an outstanding chemist. Geoff took him. John played a key part in developing the "Wilkinson catalyst," and went on to have an excellent career. There was also Alan Davison, whose chances of going from Wales to do graduate work at a major place in England were not good. Geoff gave him his chance, and Alan went on to become a Professor at MIT and a Fellow of the Royal Society.

GW always had a tendency to be acerbic about other inorganic chemists, but in the early years, I saw this as nothing but a trivial and often amusing peccadillo. There were occasions when I shared his low opinion of someone else, although there were more times when I thought he was being ungenerous. As I have looked back recently, I have realized that I probably never heard GW praise another inorganic chemist, but he was quick to make deprecating or demeaning comments. A few examples, some going back to the 1960s, will illustrate my point.

Geoff relentlessly ridiculed Ronald Nyholm, in my presence. Nyholm was a superb lecturer or master of ceremonies and very active in writing reviews, reviews that were creative and tended to stimulate further work. Geoff resented those things in others at which he did not himself stand out, and christened Nyholm "the oracle of Gower Street" in reference to the location of University College where Nyholm had taken the inorganic chair at about the same time as Geoff went to Imperial College. Nyholm was also a skillful politician (which Geoff never was) and used his skill to accomplish many good things, including the merger of the Society of Chemical Industry and the Chemical Society into the present-day Royal Society of Chemistry. In due course, Nyholm was knighted and Geoff then started calling him Lord Brokendown, in reference to Nyholm's birthplace, a small mining town in Australia called Broken Hill. I never heard Geoff say a good word about Ron Nyholm.

Geoff never liked Luigi Venanzi, whose elegant dress and impeccable manners irritated the former working-class boy from a small town in Yorkshire, and he always called him "squeegy Luigi."

However, nothing quite matched his resentment of E. O. Fischer, whom he consistently referred to as "snake eyes." He didn't manage to cook up any insulting nicknames for Jack Lewis or Earl Muetterties, but said time and again that Jack "had no ideas" and that Earl made exaggerated and pretentious claims for his work. Gordon Stone was also accused of "having no ideas of his own" and GW christened him Tracker Stone, to imply that he was always on someone else's tracks.

Now a little of this sort of thing (of which I have given only a few examples) might have been harmless, human, and at times amusing, but the atmosphere in Geoff's office was thick with it. It could go further than mere words. Ron (now Sir Ronald) Mason was for a few years a lecturer at Imperial College, and served Geoff well by doing key structures for him. He also did important work of his own, especially after he moved to the chair in inorganic chemistry in Sheffield. Mason was clearly deserving of election to the Royal Society, but Geoff actively opposed it and was able to delay it until 1975. Once again, the reason for this that he gave me was "Mason has no ideas, he just does structures."

Geoff was obsessed with this "ideas" theme, in the sense that only he had good ideas while others (*e.g.*, Muetterties) had flawed ones or (*e.g.*, Lewis and Mason) had none at all. He very often observed when he saw a good paper by someone else that "I had that idea years ago, but never got around to working on it" or even "You know, that bugger picked that idea up from me."

Now, I'm not saying all this just to denigrate GW. The point is that, though I heard and observed all this for several decades, I was too naïve to realize that someday, it would be my turn in the barrel. And so, beginning in the late 1970s, I was surprised to see signs that soon became unmistakable that GW was not acting like a friend, though he continued to speak to me like one. I was told, for example, by a mutual friend, that Geoff had said something like "You know, Al got the whole idea for metal-metal multiple bonds from me." Now, if that weren't so hostile, it would be quite a laugh.

As already explained (Chapter 5) in 1959 GW published the preparation of "Molybdenum(II) Benzoate" for which he proposed several possible structures, all hopelessly wrong, with no metal-metal bonding in any of them. In 1964, he published the preparation of $[ReCl(OCOR)_2]_2$ compounds and stated "it is not necessary to invoke metal-metal bonding to account for the diamagnetism." As late as 1966, he reported some $Ru_2(O_2CR)_4Cl$ compounds, found them to be paramagnetic and concluded that "This suggests that the metal-metal distance in these systems is large enough to prevent direct orbital overlap." Thus even two years after I had clearly elucidated metal-metal multiple bonding in analogous compounds, GW still failed to recognize a metal-metal bond when he made one.

I do not think I was being oversensitive in being stung by his claim that I got my ideas about metal-metal bonding from him, a man who failed to recognize a metal-metal bond when he tripped over one — three times. Of course, he never tried to tell *me* that I had got the idea from him, only others. A very early indication that the mutuality was going out of our relationship occurred in 1974 when I was invited to be a Centenary Lecturer of the Royal Society of Chemistry. Someone told me that they were sorry I hadn't been invited earlier, but GW had always been backing someone else.

What really put the cap on the whole business was what went on in the 1980s with respect to the Royal Society. I did not actually learn the details until later, but during the entire time I was a nominee, GW always signed one or more other nominations, but *never once* signed mine.

Sometime around 1990 I simply stopped visiting, or even talking to Geoff, except about business regarding our books. I simply could not bring myself to feign friendship toward someone who I knew was systematically selling me down the river. GW, on the other hand, publicly complained about my giving him the cold shoulder, and I was asked, a number of times by people who had heard his complaint, "what was *my* problem." Except for a few *very* close friends, I have never before discussed the matter with anyone. In a sense, GW's hypocritically playing the victim was the worst part of all.

Needless to say, I have thought a lot about why he treated me as he did. On many occasions I had seen GW behave in a generous manner. I believe he was obsessed with being regarded as "the best." He could be generous to other people whom no one else was likely to rank as his equal or superior. To all who posed, in his mind, a threat to his status as "the best" he was hostile. As my reputation grew, I, like Fischer, Nyholm, and Muetterties before me, became a threat and he took whatever defensive action he could to diminish my prestige.

Geoff's animosity to other chemists whom he, apparently, feared would steal the spotlight, went beyond other inorganic chemists. When money was raised in 1998 to endow a Wilkinson Lecture (administered by the Royal Society of Chemistry) both Derek Barton and I were asked. I made a large contribution, and shortly afterward the subject of the award came up in a conversation with Derek. He told me that he also contributed, but not without some reluctance. It seems that at a going-away gathering for Derek at Imperial College, Geoff was asked to say a few words, and he did. Among them were some to the effect that not having Derek around any more would be fine with him. Perhaps it was simply a clumsy attempt at humor, but Derek didn't think so and was stung by it. I know that Geoff did not like the obvious contrast between his own intellectual abilities and Derek's, but I was a bit surprised to hear that he had lost control of himself to that extent.

I would like to remember Geoff as he was for about the first thirty five years that I knew him and just ignore the rest. But, of course, I can't completely do that.

Derek Barton

Derek was one of my friends whose intellectual superiority to myself I have no hesitation in admitting. But then, Derek was intellectually superior to nearly everyone. It was my good fortune that we became friends.

I first met him in London in 1956 when I spent a short period visiting Geoff Wilkinson at Imperial College. It was in June I believe, after Lise Wilkinson and the girls, Anne and Pernilla, had gone to Denmark for the Summer. One day Geoff informed me that Derek, whom I barely knew at that time, had invited us to his house

With Derek Barton at a reception honoring both of us for awards we had received in 1994.

for dinner and that Derek would drive me because Geoff had to walk home and get his own car. Thus, I had a ride in Derek's Jensen — an elegant and pricey two-seater convertible.

It was apparent to me at dinner that Derek's marriage was not a happy one. His wife seemed to be a cold fish indeed, and as soon as we finished eating, Derek, Geoff, and I retreated to the study for coffee and conversation. Conversation got off to a lively start because the most conspicuous thing on Derek's desk was a large picture of Bob Woodward. This immediately got a rise out of Geoff who made some irreverent crack about idolatry. Derek made a very serious reply, however, saying that there were times when he came into his study to work and he felt tempted to slack off for the evening. "Then," he said, "I look at that picture and say to myself, 'no doubt he is hard at it over there in Cambridge, so I'd better get busy'." Derek was quite candid with me years later here at A&M in saying that Bob Woodward was the only organic chemist he felt was his intellectual equal and he was always determined not to be outdone by him. I know it was a case of mutual respect.

An opportunity to get to know Derek better came a few years later when he spent about a month at MIT as the Arthur D. Little Visiting Professor. His office was just down the hall from mine and we talked quite often. His intellectually vigorous approach to chemistry — and everything else — had a formative effect on my own attitudes. There were some hard edges to Derek in those days, but I found him to be wonderful company.

Our contacts over the next few decades were infrequent and casual. On my regular visits to the Wilkinsons in London, he and his second wife, Christiane, were often dinner guests. Christiane and Lise hit it off well, although I don't think Derek and Geoff were especially compatible. When he was fifty nine Derek had taken early retirement from Imperial College and become codirector (along with Pierre Potier) of the Institut de Chimie des Substances Naturelles in Gif-sur-Yvette, near Paris. Not only did this make possible a nearly ten-year prolongation of his career, but it also gave him free rein to indulge his Francophilia (a taste I share) and his love of fine wine and food. But this, too, was due to end as he approached the age of sixty-nine.

It was Ian Scott who first learned that Derek was in no mood to retire and looked to the United States for new opportunities. He visited us here at A&M in the early Fall of 1984 and this convinced those in chemistry who might not have known it that he was still in fine fettle. I wrote in late October to urge him to consider moving here and simultaneously urged our administration, especially our President, Frank Vandiver, with whom I had a good relationship, to do what it took to get him. Ian Scott also worked very hard to make it happen, and in 1986 it did.

In his years at A&M Derek adhered to a schedule that would have broken many a younger man. He rose at, or before, four AM, read journals for a few hours, and then spent most of the rest of the day in the department, directing a very large research group. He was also an indefatigable globe trotter, who never complained of jetlag. He explained to me that "when you sleep only four hours out of twenty four, it doesn't matter a lot which four it is."

Derek was a remarkably resilient man. He was deeply affected by Christiane's death from cancer in the early 1990s, but this only increased his appetite for long distance travel, and his zest for life in general.

It was not too long after Christiane's death that Dee and I invited him for dinner. He said he would be delighted to come and that he would like to bring a friend. That is how we first met Judy Cobb, to whom he had just proposed marriage. The considerable difference in their ages was no barrier to success because of Derek's extraordinary vitality. His sudden death of a heart attack in March of 1998 was unexpected, mercifully quick, and deeply saddening to his legion of friends and admirers. His was one of the most successful scientific lives I know of.

Rick Adams

Richard Darwin Adams is one of those former students with whom I have maintained especially close relations ever since his student days. Rick started his Ph.D. with me in 1970 at MIT, moved with me to A&M and received his degree (from MIT) in 1973. I would have recommended him strongly to the best departments (and still would), but this was a time when the academic job market was very bleak. He first went to SUNY-Buffalo, but in 1975 he was offered a tenure-track Assistant professorship at Yale. Much to my chagrin (which is far too mild a word) he was not offered tenure, and in 1984 he moved to a tenured position at the University of South Carolina. He has been a strong member of that department and has received

Rick Adams, some time in the late 1990s.

numerous forms of recognition, including a senior Humboldt, a Chemical Pioneer award from the American Institute of Chemists (2000), the H. J. Albert Award of the International Precious Metals Institute (2005), the Southern Chemist Award (2000), the National ACS Award for Inorganic Chemistry (1999), and several other awards.

While Rick has done an impressive range of synthetic chemistry, his chief interest has always been in metal atom clusters, where he has studied the coordination and reactivity of organic ligands attached to metal carbonyl clusters. There is no doubt that he is the leading researcher in the world in this field. He has nearly 500 publications in refereed journals. He has also served for many years as the American regional editor of the *Journal of Organometallic Chemistry*.

Above and beyond all his scientific ability and accomplishments, I appreciate Rick as one of the most helpful and loyal friends any man could hope for.

Carlos Murillo

Shortly after I arrived at A&M, I had a letter from a Professor in Costa Rica, Sherman Thomas (a native African-Costa Rican), whom I had got to know several years earlier. He sent the letter to me at MIT and it was forwarded. He wrote to inform me that he had a very promising young student who would like to do his Ph.D. with me. I replied that I would be very interested so long as they realized that I was no longer at that world famous institution, MIT. Sherman said that this made no difference, so a few months later,

Carlos Murillo in the mid-1990s.

in September 1973, Carlos Murillo arrived. He already spoke decent English, having spent a year as an exchange student in high school in Illinois. Little did I imagine how lucky a day this was to be for me, and how long and valuable our friendship would be.

It was quickly evident that Carlos had the makings of an outstanding graduate student. I slowly realized that he had the ability to go far beyond that. In 1976, he finished his Ph.D. and went off to Princeton to spend a postdoctoral year with Malcolm Chisholm. During the four years he spent with me and then with Malcolm, he was the — or at least a — major contributor to sixteen important publications. He then went back to a position in the University of Costa Rica.

The conditions for research in that institution were poor at the time and he felt very frustrated. He spent May to August in 1981 on leave to do research at A&M, and he took a similar leave of several months in the Winter of 1983–84 and from August 1985 to January 1986. During all this time he had been doing a superb job of introducing undergraduates to research and sending many of them off for doctoral degrees in the United States and Europe. He is an exceptionally good teacher.

Finally in January of 1991, he came again to A&M and this time he stayed for good. He is now an Adjunct Professor, a Senior Lecturer, and the Executive Director of the Laboratory for Molecular Structure and Bonding. He is also my indispensable partner in research. He has not only great knowledge of chemistry but also that innate "feel" for it, which one is either born with or not. My research group would never have been as productive as it has been for the last decade and more without his inspired contributions, not to mention his managerial skills.

Carlos and I have now published more than 210 papers together and he is coauthor of both *Advanced Inorganic Chemistry*, 6th edition, and *Multiple Bonds Between Metal Atoms*, 3rd edition. We hope soon to start work on a midlevel (senior or first-year graduate) text on inorganic chemistry.

As I said earlier, on the day Carlos showed up to start graduate study at A&M, I had no idea what a lucky day that was for me. And blessings on Sherman Thomas, for sending him.

Larry Falvello

Lawrence Rocco Falvello spent about a decade (1981-91) with me and coauthored eighty-four research papers. During the time he was with me I lost all remaining capacity to supervise X-ray crystallographic work in a hands-on way. One generation of computer programs succeeded another, always becoming more complex, (albeit more powerful and sophisticated) and diffractometers also evolved rapidly. Of course, with Larry in charge I had no need to worry about details and technicalities any more, so I didn't try. Larry was a wonderful teacher and taught everyone else, *e.g.*, Rinaldo Poli, Marek Matusz, and many others. So, X-ray-wise, I became like an orchestra conductor

Larry Falvello and Milagros Tomás as they look today.

who couldn't actually play any of the instruments himself. When Larry left, I therefore needed another "chef de diffraction" and that was my former student Lee Daniels, who stayed for twelve years. In 2002 Xiaoping Wang, another former student, who had spent some time doing neutron diffraction with Art Schultz at Argonne came back and replaced Lee.

While Larry was here he met Milagros Tomás, who spent the academic year 1981–82 and the Summers of 1983 and 1984 doing research in my laboratory. They later married and both are now members of the professoriate in the University of Zaragoza.

Larry Falvello is a man of exceptional intellect, coupled with a very realistic attitude toward life. His understanding of diffraction physics is thorough and his broader knowledge of physical chemistry remarkable.

Achim Müller

In 1979, I was presented with an honorary doctorate by the University of Bielefeld, one of the relatively new German Universities. The way in which this came about was surprising to say the least. In 1978 I received, out of the blue, as the expression goes, a letter from a man I didn't know. His name is Achim Müller. Of course, today, every inorganic chemist knows who he is because of his remarkably creative accomplishments. I was surprised to hear that he wished to visit me, but I arranged for him to come and give us a seminar.

With Achim Müller at Hadrian's wall (1985).

In due course, his visit took place. We had dinner together after his seminar and I was impressed by his excellent English, delighted by his erudition and charmed by his energetic, guileless, and freewheeling personality. A lifelong and highly cherished friendship was cemented immediately on that first day we met. There is nobody like Achim. There used to be a regular feature in the *Readers Digest* (a magazine I read regularly as a child but gave up as an adult) called "The Most

Unforgettable Character I Have Met." For me Achim could be that man.

Achim told me during our first dinner that he was soon to be (or perhaps he already was) Dean of the Chemistry Faculty at Bielefeld, and that he intended to have me awarded an honorary doctorate. He did this, and in November of 1979, I had the pleasure of receiving it from his hands. To have someone you have never met, but who is himself an exceptionally talented person pick you out, sight unseen so to speak, and make a gesture like that is an incomparable pleasure.

Achim Müller looking satisfied with life in 1995.

On my first visit to Achim's home, I experienced the full force of his passions for Bach, Beethoven, Wagner, conservatism, and Cuban cigars. I will also never forget my first meeting with his wife May. It was immediately evident that she spoke English fluently, but what puzzled me was her accent, to which my somewhat stupid reaction was that it was unlike anything I had ever heard before in Germany. Finally, I decided that it would not be impolite to ask where she got it. Her reply, which made me feel like a prize numbskull, was: Well, I got it where I was born and raised, in Ireland. May, bless her heart, even looks Irish, but in the entire context of our meeting, I just didn't have wit enough to see the obvious. May had come to Germany as a cellist in a small orchestra and was wooed and won by Achim. She was also happier in the disciplined German culture than in the laissez-faire environment in which she grew up. The efficient and practical May is just the wife that Achim, the classic absent-minded professor, needs. I must also add that May, in her own way, is as charming a person as Achim.

Herbert Roesky

For me, Herbert is an icon in two respects. He is a peerless master of preparative inorganic and organometallic chemistry, and he is a model of loyal friendship.

I cannot think of any other chemist who has mastered the chemistry of so many

Herbert Roesky, lecturer par excellence.

elements, from all parts of the periodic table. In about 900 publications, he has made important and ingenious contributions to the chemistry of fluorine, silicon, aluminum, gallium, nitrogen, phosphorus, and all of the metallic elements with which these form compounds. Herbert's ideas for new compounds are always very fresh and original, and his synthetic methods elegant and efficient. He is also a master teacher of synthetic chemistry, and has written two unique monographs on the art and science of preparative chemistry. He is amazingly deft with his own hands and gives demonstration lectures that are nothing short of virtuosic.

Herbert has also been a very generous contributor to the human side of science as evidenced by his Presidency, beginning in 2001, of the Göttingen Academy of Sciences.

Wolfgang Herrmann

There is something inspiring about a person who gets off to a really fast start in life — and keeps going. Such a person is Wolfgang Herrmann. I first met Wolfgang in about 1980 when he was only in his 32nd year and already a professor at the University of Regensburg. By 1982, he had a chair at the University of Frankfurt (Main), and in 1985, he became E. O. Fischer's successor in the Technical University of Munich (TUM) at age 37.

In 1990, I began my series of six 1-month visits to the TUM as a von Humboldt Senior Fellow. Wolfgang had originally wanted to have me appointed for a year, but I convinced him that I simply could not spend that much time. Indeed, to this day, I am still one month short of six, and I hope he has forgiven me. My early visits were made under very pleasant circumstances. Wolfgang owned a small house in the farming village of Giggenhausen, about twenty minutes from Garching, and that is where

I lived, sometimes alone and sometimes with Dee. I spent many pleasant hours walking in the surrounding farm country and often drove into Munich. I was hard at work on the second edition of *Multiple Bonds Between Metal Atoms* and made good use of the excellent TUM chemistry library. I also

With Wolfgang Herrmann in 2000.

enjoyed my contacts with other members of the chemistry faculty, most particularly Hubert Schmidbaur, one of the very best chemists of his (and my) generation. My high regard for him and our friendship continues to the present.

My admiration for Wolfgang grew as I became aware of what a multitalented person he is. His creativity in chemistry extends to both fundamental problems and to industrial applications, and he has formidable administrative ability. The former boy wonder has become an important senior statesman. When I first began visiting the TUM in the early 1990s, he was already a Dean and since 1995 he has been President of the TUM. It depends on what happens on the national political scene, of course, but if the Christian Democratic party does well in the future, I would not be in the least surprised to see Wolfgang as a cabinet minister.

Joseph Chatt

Chatt was one of the major players in the post-WWII development of inorganic chemistry. I was fortunate to meet him early in my career. This was in 1954 when I was on my way home from Germany, at which time he was head of the inorganic chemistry department of an ICI basic research laboratory north of London. I can no longer remember how the contact was made, but he invited me to come visit his laboratory. I took a train from London to Welwyn Garden City and he had a chauffeured car pick me up at the railway station. His laboratory, which was located on a nearby estate

With Steve Lippard and Joseph Chatt at a celebration of the
sesquicentennial of the Royal Society of Chemistry (1991).

called "The Frythe," was modern and well equipped and he had several very able young coworkers, among whom were Luigi Venanzi, Sten Ahrland, and a Dr. Duncanson, whose given name I can no longer recall. They were working on some of the most interesting ideas then current in coordination chemistry.

Chatt was especially concerned with the kinetic trans effect and a host of other questions related to the behavior of platinum complexes. He was interested in how to stabilize metal to carbon bonds, and he devoted much attention to the nature of the then very puzzling metal-olefin complex called Zeise's salt. In the end, he did not solve the problem (Michael Dewar did), but his work focused attention on this important problem.

Joseph Chatt was a remarkably kind and patient man, universally respected for these qualities. Even when he disagreed strongly with someone, he was always the soul of politeness in expressing himself, although he would make his point of view perfectly clear.

One of Joseph's most long-lasting contributions to inorganic chemistry was a meeting at The Frythe in 1950, which he organized entirely on his own, to bring together some leading researchers in coordination chemistry. From this small beginning came the biennial International Conferences on Coordination Chemistry.

In 1962, Joseph became a professor at the University of Sussex and Director of the Unit of Nitrogen Fixation, supported at Sussex by the Ministry of Agriculture. Joseph was the only inorganic chemist prior to me to be awarded a Wolf Prize (1981). Chatt's career was one of considerable importance to inorganic chemistry, and to me.

Fausto Calderazzo

Fausto came to MIT in 1959–60, and I immediately recognized a man of exceptional talent and character. Fausto loves chemistry as much as anyone I know and in choosing his research topics, he stays off of bandwagons. His work is always original and interesting. He was one of three Italian postdoctorals that I had during my early years at MIT. The others were Franco Zingales (1959–60) and Adriano Sacco (1960–61), who are now retired.

Other Italians who followed at MIT in the period 1966–72 were Flavio Bonati and Alfredo Musco, both now deceased, Mauro Graziani (now at Trieste), and Giulio Deganello (now at Palermo). At A&M I have had Rinaldo Poli, 1985–87 (now at Toulouse), and Alfredo Burini, 1989–90 (now at Camarino).

With all due respect to those others I have just listed, Fausto is a very special friend. His finest qualities are independence of mind and the way he takes a principled approach rather than a self-interested one, to all questions. When he arrived at MIT in July 1959, he had already done some excellent work in the laboratory of Giulio Natta, whose team also included at that time such luminaries of the future as Raffaele Ercoli, Piero Pino, and Paolo Chini. He had, in fact, done kinetic studies on the reaction of dicyclopentadienylchromium with carbon monoxide. He was far more mature and experienced than any other postdoctoral coworker I had had previously, and with

Fausto Calderazzo and I in 2000.

his background he quickly became interested in a problem I suggested, namely, the carbonylation of metal carbonyl alkyls to give the corresponding metal carbonyl acyls, the so-called carbonyl insertion reaction. We published a thorough study of the reactions involving $CH_3Mn(CO)_5$ and $CH_3COMn(CO)_5$ [100], but the final answer to the mechanistic question was not found until some years later in Fausto's own laboratory. I think Fausto's solving of the carbonyl insertion problem is a modern landmark in fundamental organotransition metal chemistry, with considerable relevance to industrial homogeneous catalysis as well.

After returning to Milan and then several years at the Cyanamid European Research Institute in Geneva, he went to the Università di Pisa, where he has not only pursued his own research but also played a major role in setting up a curriculum in molecular sciences jointly with the Scuola Normale Superiore di Pisa, where the emphasis was more on physics and mathematics. Fausto takes the teaching of chemistry very seriously and has a very engaging lecturing style himself.

CHAPTER 11

A CONCLUDING MISCELLANY

Then pray that the road is long,
That the summer mornings are many,
That you will enter ports
seen for the first time
With such pleasure, with such joy.

— Constantine Cavafy

I THINK THAT the lines by Cavafy, whatever *he* intended them to mean, get to the heart of my life's journey. The road has been long, the Summer mornings have been many, and in my research, I have entered ports seen for the first time with pleasure and joy. I wouldn't change anything, even if I could.

Since one purpose of an autobiography is to present a self-portrait as well as an account of events, I have included here some recollections, observations, and notions that didn't seem to find a natural place in the previous chapters.

Let me begin with a quotation from one of my favorite authors of fiction:

> "What we want to avoid in this book [is] shoving in stuff just because it happened." *P. G. Wodehouse*

Wodehouse was quite right to recognize the temptation to "shove in stuff" because one does not have the self-discipline to just leave it out. That has been a problem in every book I have written. I have had suggestions galore from readers as to how my books could be improved, and 99% of them have been to put more things in. Anyone who has written a (good) book knows that what might be put in is never a problem. What to leave out is the problem.

In each of the preceding chapters, I allowed chronology to order the content — a logical but not very imaginative way to proceed. There are some things that don't come in naturally that way, and so in this last chapter, I am just "shoving them in" because I couldn't think of any more graceful way to include them. Having mentioned my previous books, I'll start with that.

Writing Books

Throughout my career, I have been a book writer. The most ambitious — and long-running — undertaking has been *Advanced Inorganic Chemistry*, now in its sixth edition. The idea for such a book goes back to 1952, during the second semester of my first year in graduate school. In the academic year 1951–52, Geoff Wilkinson taught inorganic chemistry for first-year graduate students at Harvard and I took the course.

Before the end of the Fall semester, I had chosen him as my Ph.D. mentor and begun doing a little research. He was not yet married, was relatively young, and had only three students, and we began developing a relationship that therefore was somewhat closer than would normally be the case.

Geoff had found it difficult, indeed impossible, to find a satisfactory textbook for his course. There were some very old-fashioned ones that he considered altogether unsuitable, and the massive, two-volume *Chemical Elements and Their Compounds* by N. V. Sidgwick, though a mine of information, was not appealing as a textbook. There really was only one option at the outset of the Fall semester and that was H. J. Eméleus and J. S. Anderson's *Modern Aspects of Inorganic Chemistry*, which was not really a textbook but a collection of reviews of interesting new areas of inorganic chemistry.

Early in 1952, a book by Therald Moeller of the University of Illinois entitled *Inorganic Chemistry, An Advanced Textbook* appeared. This book was good on general principles and the main group elements but only about 65 of its 910 pages dealt with the transition elements. It was during the Spring of 1952 that Geoff and I decided that as soon as we were both in a position to do so we would write a balanced book on inorganic chemistry aimed at the senior and first year graduate level in American universities.

The actual writing was long delayed since he was soon overwhelmed with research in metallocene chemistry and I was occupied with my Ph.D. research. It was, however, a serious commitment, and we both continued to plan on doing it. Real work did not begin until about 1959 when he was well settled down at Imperial College and I had got my feet on the ground at MIT.

The book was written, as were all later editions, with the two of us an ocean apart. I have often been told by other people that they find this amazing, particularly because, they say, there is no variation in style from one chapter to another. I think that is true, and on reflection, I too find it surprising. Of course, each of us edited what the other wrote, but this was not difficult or to any significant degree contentious. Somehow, despite the fact that in many ways we had different approaches to our own research, we had a spontaneous harmony in our approach to writing the book.

The first edition, which appeared in 1962, was well received, and understandably so. It was the first inorganic text, at any level, to present an introduction to ligand field theory, which by then had become very prominent in the research literature and had to be covered in a senior-level or graduate-level course in inorganic chemistry. It was a systematically organized book covering all of the elements from hydrogen to the actinides, with, in our judgement, a fair allotment of space to each one. The concepts of ligand field theory were not only expounded but also brought to bear on the actual magnetic and spectroscopic properties of each of the transition elements, as its chemistry was discussed. Other innovations were tables of the oxidation states and stereochemistries of each metallic element with examples of each and a strong emphasis on structures generally. These and other ways of treating the subject that were new at the time have continued to shape the coverage in all subsequent editions.

When I look back at that first edition today, with its 925 pages, including a good deal of bonding theory (notably ligand field theory) that has since been dropped because it is covered in lower level courses, and compare it with the current 6th edition that has 1325 pages, I am vividly reminded of how much we have learned about inorganic chemistry since 1962. Because of the rapid growth of the subject, references to the primary literature were first introduced in the second edition (1966) and this practice has continued to the present. The first edition gave only a few references to the secondary literature, *i.e.*, reviews and monographs. I have estimated that in editions two to six there are about 13,000 literature references, including about 4,100 in the sixth edition. With the third edition (1972) we adopted the policy of giving new references only for new material, and this policy has remained. Thus, over the years there is an extensive record of when and where new facts and new ideas were first reported.

The writing of the fourth (1980) and fifth (1988) editions became more and more laborious because of the ever accelerating rate of literature growth. The first edition was a true textbook, a truly teachable book, but to accommodate new descriptive material, more and more theoretical background had to be eliminated, until today there is very little. The book is now as much a reference book as it is a textbook. We felt we had to make that change in character in order to do two other things that we were committed to. One was to make each edition a book that would prepare the reader to comprehend the contemporaneous research literature and the other was to keep it within one volume, albeit a hefty one. By the time of the fifth edition it had become such a fat, heavy one that I jokingly proposed that for two dollars extra it would be available in a special binding with wheels and a pull strap, in the manner of airline luggage.

The sixth edition (1999) saw one more significant change. As we began to plan it, in about 1993 or 1994, we decided to include two more authors, to reduce the burden. I chose Carlos Murillo and Geoff chose Manfred Bochmann (both former students), and we divided the first draft responsibilities into four approximately equal portions. We also decided to ask Russell Grimes to revise and update the chapter on boron. This vast and idiosyncratic subject had always been a bête noire to both of us and we had passed it back and forth between us from one edition to the next like a hot potato. Russell was kind enough to say that he thought we had been doing a good job all along, but nonetheless, we were happy to hand it over to an expert.

Before leaving the subject of *Advanced Inorganic Chemistry*, I want to make an observation based on my experience with all the books I have written. The expression "to write a book" is deceptive. Taken at face value it describes a pleasant and satisfying activity. Planning, researching the material, and getting the words on paper are, on the whole, enjoyable. If that were all the phrase "write a book" meant, it would be joyous. Unfortunately, that part is only the beginning. The worst is yet to come, and it makes the life of a galley slave seem like beer and skittles. What follows the writing, *per se*, is the following list of horrors: Proof reading the typescript; preparing sketches of many of the illustrations or structural formulas; obtaining permission to reproduce illustrations from journals; proofreading galleys; trying to get the so-called artists the publishers use to produce drawings that at least come close to what you want; proofreading pages and fighting with the publisher about how figures and tables are positioned. When in a state of exhaustion, you are thanking heaven that the ordeal is over at last, the publisher reminds you that the index is still due — in two weeks.

One of the major problems is that no one — or almost no one — in the publisher's end of the operation has any understanding of what you have written and therefore any cockup that *can* occur usually does. The "artists" are robots who have no feeling at all for what they do. I can remember once, long ago, in the nth round of trying to get a drawing that looked right, I wrote on the $(n-1)^{th}$ version of the disaster that had been sent to me: "it should look like a tetrahedron." I found out that this was no help at all because no one in the "art" department had any idea what a tetrahedron was. Very often the copy editor was a young lady who had recently gotten her B.A. in English, and she dedicated herself to improving our prose and generally saving us from what she saw as our linguistic ineptitude. The one incident I shall never forget is that in checking proof I found that the Si–O–Si angle in siloxane was 140 °C. The "C" was our editor's well-meaning contribution! Luckily, I caught this, but I don't know how many other such helpful gestures may have gone unnoticed.

My second most ambitious, and also long-running, writing project is *Chemical Applications of Group Theory*, the first edition of which appeared in 1963, only a year after the first edition of *Advanced Inorganic Chemistry*. As I mentioned earlier (in Chapter 3) I began to learn a little about the use of group theory in chemistry as early as 1952 when the problem of formulating the bonding in ferrocene was first being addressed. There could have been no more persuasive object lesson in the power of symmetry analysis than that. A year or so later, having become still better acquainted with the subject, I was able to clarify the nature of the vibrational spectra of the (C_5H_5) $M(CO)_n$ molecules. Finally, the emergence of crystal (or ligand) field theory afforded me still another demonstration of the importance, indeed indispensability, of having a working knowledge of how to utilize molecular symmetry in analyzing problems in bonding and molecular dynamics. It was therefore very natural that soon after going to MIT I began to teach a course on the chemical applications of group theory.

As I had learned from my own experience, there was no book suitable for presenting this subject to chemists who were not inclined to be theorists. There were a few books written by and for theoretical physicists (for example one that a colleague of mine referred to as "that vile book by Weyl"), but these were entirely unsuitable. I therefore taught the course using only my own lecture notes. For the first two years, I attempted to "ease the pain" by starting with a discussion of the symmetries of real molecules and then trying to extract the basic principles indirectly from the examples. I found that this seemingly painless approach simply didn't work, and in about 1958 or 1959 inverted the whole process so as to begin with a purely mathematical approach. Thus, the plan was first to make clear what the basic concepts and properties of groups are, turn next to molecular symmetry groups, and only then to what group theory has to do with molecular behavior. After several years of experimenting with and tuning this approach I had a satisfactory set of lecture notes and began to polish them up for publication.

The book was published in 1963 and was well received. It sold well for a number of years, but by the late 1960s, it was clear to me that improvements were needed. Apart from many minor adjustments, two major additions were an explicit development of

projection operators and the introduction of the symmetry basis of the Woodward —
Hoffmann rules, which had come into being since the first edition was published. As
several people had pointed out to me, the omission of projection operators in the first
edition was a case of misplaced kindness; while they may seem somewhat forbidding,
once mastered they more than repay the trouble of learning how they work. Thus in
1971 the revised and augmented second edition appeared. While the first had been
accorded only one foreign translation (French), the second was translated into Polish,
Italian, Spanish, Japanese and Chinese. It was very satisfying to me to see this evidence
that the idea of a modicum of group theory as an essential element in a complete
chemical education had so thoroughly taken hold.

Finally, by the late 1980s, the role of crystallography had become so pervasive in
chemistry that I felt that some sort of introduction to spatial symmetry as well as point
symmetry should be part of the package. Therefore, the 3rd edition, which appeared in
1990, includes one more chapter on crystallographic symmetry.

I mentioned the foreign language translations of the second edition of *Chemical
Applications of Group Theory*. There were six for the first two editions, but so far, after
a decade, none for the 3rd edition. It is interesting that a similar pattern occurs for
Advanced Inorganic Chemistry over the first five editions. The years of publication and
the number of foreign language versions are as follows: 1962 — none, 1966 — seven,
1972 — four, 1980 — four, 1988 — none. I think what has happened is that from
the early 1960s the dominance of English as the *lingua franca* of science has reached
the point where all students of science who get beyond the elementary level are now
expected to (and do) acquire a working knowledge of English — at least technical
English. Thus it is no longer necessary, nor even desirable, for them to have more
advanced material in their own language. It seems to me that never before in the
history of modern science (some 300 years) has there been, as truly as there is today, a
single language of science.

Foreign language translations were ego builders but of little or no economic value.
Only the Japanese and German ones earned any significant amount of money. At
the bottom end of the spectrum in monetary terms were those published in India,
the USSR, and the PRC. The Indian editions were not actually translations but were
replications by an Indian publisher. For the rights to do this they offered $100 to be
divided among publisher and authors. When we complained, we were informed that

under *Indian* law, they had only to wait one year after making the offer and if we had not agreed they could publish as many copies as they pleased and pay nothing. We resolved that we preferred to be robbed outright rather than pretend we weren't being robbed, and did not sign.

The Chinese never even informed us, or the publishers, about what they were doing. I found out that they had pirated my group theory book one evening at a party in my honor at Larry Dahl's house in Wisconsin. I was introduced to a beaming Chinese visitor who asked for my signature on a Chinese translation of the book, and he proudly informed me that he was one of the translators. I made no secret of my displeasure and asked him if he understood that what had been done was simple robbery. Without any real embarrassment he informed me that this translation had been done during the reign of the Gang of Four, that the Gang of Four were very bad people, and so on. I did, at least, get him to give me a copy of the translation.

Some time later, I was sitting in my office at A&M and my colleague Abe Clearfield called to say that a visitor from the PRC was in his office and had a present for me. So I went down and was presented with a Chinese translation of the third edition of *Advanced Inorganic Chemistry*. I again informed this fellow, who was one of the seven translators, without mincing any words, that in the civilized world it was the custom to negotiate a contract for the right to publish a translation of someone else's book. I was expecting to hear the same malarkey about the Gang of Four, but no, there was an even more ingenious excuse for theft, roughly along the following lines. "Is very big book; need seven translators. Each one think other ones have permission."

The Soviets were less devious about translating the second edition. As I remember it, they just told Wiley they were going to do it, and did it. Geoff got very annoyed and wrote to the Soviet ambassador in London and said, among other things, that he considered it a supreme irony that those running what they were pleased to call a worker's paradise should see nothing wrong with stealing the fruits of the labor of workers like Cotton and Wilkinson. This led to our being told that if we were to visit the USSR we would be given some royalties in rubles. Geoff and Lise actually did go to Moscow for a short visit and discovered that the royalties were in non-convertible rubles — rubles that could not be taken out or converted even in the USSR to any other currency. I believe that they used their rubles to buy Lise a fur coat, which was the only thing they could find available for their non-convertible rubles that was of sufficient quality to be worth carrying home. The Bulgarians later followed their masters' example and published an unauthorized translation of the third edition.

In recent years, beginning in 1976, Geoff and I decided to write a book really aimed exclusively at the undergraduate student of inorganic chemistry (but not freshman chemistry!) and we tried to distill the essence of the subject into *Basic Inorganic Chemistry*. This is a very difficult task and I'm not sure we did the best possible job (in fact, I'm sure we didn't) but nevertheless the first edition was well received and was translated into eight foreign languages (including Malaysian, Indonesian, and Persian). With the collaboration of Paul L. Gaus of the College of Wooster, two more editions have appeared (1987, 1995).

To conclude, I want to return to a much earlier time. In the late 1950s, there was a great deal of interest in the improvement of the teaching of general (*i.e.*, high school and freshman) chemistry in the United States. One manifestation of this was the involvement of the National Science Foundation in the creation of a book called *Chemistry — An Experimental Science*. The genesis of the project was in a steering committee formed in 1959 with Glenn Seaborg as its chairman. Its purpose was to provide high schools with a new and contemporary approach to the teaching and learning of chemistry. The project was called *CHEM Study*. A large group of teachers from both high schools and universities was assembled to plan and write such a book. The leader of this writing team was George Pimentel and by the academic year 1960-61 the book was actually in print. It was revolutionary in its inductive strategy, emphasizing experimental facts while at the same time showing how principles and theories are deduced from facts. It was highly successful and deservedly so.

After about five years the NSF decided that having underwritten the pioneering effort and seen it to a successful conclusion, the time had come to spin it off into the private sector. They invited publishers to apply for the rights to utilize the material in the book for the purpose of creating new books with the same general approach. Each applicant had to submit the names of a writing (or rewriting) team and a plan for proposed modifications. The Houghton Mifflin Company in Boston invited me to work with a high school teacher from Beverly Hills High School, Lawrence D. Lynch, and entered the competition. As I recall, only two, or perhaps three applicants were to be chosen, and Houghton Mifflin was one of them. Thus, I started my initiation into the ways of the highly commercial world of school book publishing in about 1966. I soon learned that the key to success was to produce a book that the company's lobbyists could convince large school districts, especially whole states such as California and Texas, to adopt. In addition to being attractively designed and illustrated, which was the company's task, it was important that the writing style be considered pleasing by the customers and that was my job.

The plan of action was for me to have an office at the Houghton Mifflin company headquarters on Tremont Street where I would spend 2-3 mornings every week. They gave me a very nice office, on Tremont Street overlooking a historical graveyard and with a view of the gold dome of the statehouse atop Beacon Hill. They also hired a freelance editor to work with me, a middle-aged man named Bill Bixby, who was, in fact, a good companion and very helpful. Bill was a somewhat unbuttoned chap who had worked in many parts of the publishing industry. He and I hit it off beautifully and would often cap our morning's work with a two-martini lunch at a local Italian or seafood restaurant. This happy state of affairs lasted for about five months. The book that Pimentel had produced (for it was he who really wrote it) was not a very polished product (he had worked against a strict time deadline), but it was, like everything George did in his life, brimming with novel and creative ideas. I really enjoyed working it over. I came up with some very effective and interesting illustrative material, and the publisher spared no expense in providing extremely clear photographs, many in color, which was not yet common at the time. Each chapter opened with an image intended to arouse the student's interest and Bill Bixby and I did a lot of brainstorming to come up with these. My favorite is the one for the chapter on *Solids, Liquids and Solutions*. We reproduced Salvador Dali's *The Persistence of Memory* (the famous limp watches picture), which, as the caption points out "gains attention by contradicting the onlooker's sense of what is solid and what is not."

I was very proud of the book, which was titled *Chemistry — An Investigative Approach*, and still am, and it sold quite well. It provided much needed money to build a lovely Mees van der Rohe house on twelve acres in the elegant town of Sherborn, Massachusetts. I was also pleased to be associated with the same publisher as Winston Churchill, especially since Churchill, in his younger days, did exactly what I had done: when he needed money, he wrote a book.

Unfortunately, this beautiful story has a sad ending. After a few years, Houghton Mifflin allowed its good judgement to be overridden; I can only assume by greed.

They figured that they could increase the sales by appealing to a wider (*i.e.*, less elite) clientele and they decided to publish a second edition. I agreed, reluctantly, to work on it and another high school teacher was recruited. I asked to have Bill hired in again but they refused. I was never convinced with the reason they gave, I think the rather straightlaced organization was uncomfortable with a free spirit in the house. Anyway, with about as little help as before, this time from two high school teachers instead of one, I proceeded to do some revisions. I soon discovered that instead of Bixby, I had one (or more), in-house people "assisting" me. Everything I wrote was rewritten so that there was rarely a word of more than three syllables or a sentence that was more than a simple declarative one. While the illustrations were still quite good, the text as such was considerably "dumbed down." I protested, but to no avail. The net result was that the second edition was a flop.

My adventure in introductory level book writing included one other significant episode. As the writing of the first edition approached the end, I learned with surprise that the publisher expected one more thing from me: a set of "overhead visuals," *i.e.*, transparencies to provide additional illustrative material for the classroom teacher to use. By then I had already diverted too much time and energy from my duties at MIT and I said I would not do it. Houghton Mifflin then found someone else, but I insisted that I have final say on the acceptability of what this person did. The person they found was a man who taught chemistry and was in overall charge of science courses at the high school in a posh western suburb of Boston. One fine day his work was shown to me for my approval.

I have never seen such a pile of half-baked (and half something else) rubbish. At its best it was silly pap and at its worst flat out wrong. For example, there were drawings showing how orbital overlap led to the formation of s–s and p–p sigma bonds. Then, there was a drawing showing that two $p\pi$ orbitals on adjacent atoms failed to overlap and explaining that for this reason a π bond was not the result of overlap but of magnetic dipole interaction between the two adjacent p electrons. In the end, I had to do many of the overhead visuals myself to avoid utter disaster.

As I have mentioned earlier, my own experience with high school chemistry was at best underwhelming. This incident, however, has left me with lifelong cynicism, and the conviction that until there are nationwide standards for the quality of science *teachers*, money spent on trying to improve science curricula in high schools will be wasted.

Industrial Consulting

For much of my professional life I engaged in this activity and it helped my financial wellbeing a lot. I might note that my starting salary at MIT was $3,600 for nine months. Even in 1955 that did not support a very grand life style, although I was able to afford a Chevrolet two-door sedan and a modest bed-sitter (as the English call it) in the Back Bay section of Boston. Hence, a hundred dollars a day, ten days a year, was very welcome when I started in 1957 with the Union Carbide Research Institute (UCRI) in Tarrytown, New York. UCRI started that year with Verner Schomaker as its technical director, and thanks to a tip from my friend Peter Verdier, Verner made me his inorganic consultant. The major effort at UCRI was development of molecular sieve type aluminosilicates as catalysts. The fact that I never did research in that area in my own lab did not seem to disturb Verner. My other chief client at UCRI was a wonderfully inventive group leader called Jule Rabo.

Over the years my consulting with Carbide increased to include their main corporate research labs in West Virginia where the range of topics was very broad. For a time I was also a broad-ranging consultant for American Cyanamid in Stamford, Connecticut, and during the 1960s, for Polaroid Corporation. At Polaroid the sole topic was how to design transition metal complexes that could be used as pigments in their color film. In addition to all of the very welcome financial rewards, my consulting work brought me into contact with a great many ingenious chemists, some very astute research managers, and some interesting problems. I also learned at least as much from them as they learned from me, and I got a lot of wonderful free dinners, especially from Jule Rabo, a man of very refined taste, who became a valued friend. At the peak, I consulted about 25 days a year.

After I moved to Texas my activities as a consultant slacked off, and toward the end of the 1980s, I gave them up altogether. However, I kept in contact with industrial reality by gradually taking on more and more work as an expert witness for patent law firms. This was an activity I thoroughly enjoyed. It was ideally suited to my competitive nature and when I was on the stand particularly to my combative instincts. Several times when I signed a contract to work with a law firm for the first time I was asked if

I charged a higher rate when I testified and I cheerfully told them that my rate didn't change, but that if it had it would have gone down when I was on the stand because that was the part I liked best.

In some of the cases I worked on, hundreds of millions of dollars a year were at stake, and knowing that kicks the pulse rate up a bit. I also enjoyed the strategizing before the trials, however. It is a bizarre aspect of the American legal system that most patent cases are tried before a judge (or occasionally a jury) who is unlikely to know the formula for table salt. It is not only important to be right. You have to also be able to convince His Honor that you are right.

The most satisfying memory I have of being able to really earn the fees I charged has to do with a case involving an aluminosilicate catalyst. In this instance, the strategy was not preplanned but came to me as I was being cross-examined. As I have already admitted, it was my consulting and not my own research that taught me this area. The particular case I have in mind came to hinge on our opponent's argument that if the catalytically active dopant was present in an amount so low that there would be an average of less than one such atom per unit cell, the doping was insignificant and hence the material was not patentable. Scientifically, this is nonsense, but the question was how to make the judge recognize this? I explained it to him as follows. The form of alumina called corundum, an unremarkable white solid, sells for only about $20 for 100 grams, in chemically pure condition. However, a crystal of corundum that contains some chromium, at a level of only one atom per ten or more unit cells, is what is called a ruby, and sells for hundreds of dollars per fraction of a gram. Your Honor, valuable dopants come in very dilute forms. We won the case, and had quite a party at one of Wilmington's better restaurants that same evening.

Changing Times at Texas A&M University

When I joined the faculty of the Texas A&M University in February 1972, it was still relatively small: ca. 17,000 so-called full time equivalents. It has grown to be one of the giants (over 45,000), comparable to Ohio State, Michigan, Berkeley, and a few other state universities that are afflicted by gigantism. The increase in numbers has been accompanied by an enormous expansion of the physical plant.

At the same time, the university bureaucracy has grown out of all proportion to anything else. While the number of students has increased by a factor of less than three I would estimate that the bureaucracy is at least 15 times bigger. We have been relentlessly attacked by Parkinson's Law and the Peter Principle. An ever greater fraction of the state funds that come to the university are not going to support educational activities or research but to maintain growing legions of bureaucrats and armies of office workers "needed" by these bureaucrats.

When I came to A&M, there was a Dean of the College of Science, whose office was run by two secretaries. Today, the Dean of the College of Science has five associate Deans and the total number of secretarial personnel must be at least two dozen. There have been vastly increased expenditures to maintain all this increased office activity, more meetings, more memos, more rules, more unproductive fiddle-faddle of every kind.

The chemistry department has contributed fully to the general bloat of the bureaucracy. The growth of officialdom has considerably outstripped the growth in either the quantity or quality of the research enterprise or the scholarly output. Over the past 10–15 years the number of graduate students has declined a little but the size of the graduate office operation has roughly tripled. There were 54 tenure-track faculty in 1972, but over the last decade there have been only about 45. Since the building I am in was completed in 1972 there has been only one addition (1987) to the complex of chemistry buildings, and much of that one, although it is about as large as the 1972 building, consists of lecture rooms and undergraduate teaching laboratories.

The teaching of general chemistry has become an industry. The stack of commercially published material handed to each participant in the program is now over two feet high. There are books about how to use the books about how to use the books that are sold to the students.

A major concern of mine in departmental affairs has always been the strength of the inorganic division. The department has always had a divisional structure, which is to some extent necessary in such a large department, but this has meant that any one person's influence over faculty development does not usually extend much beyond one's own division. This tends toward the result that a strong division gets stronger while a weak division gets weaker. Certainly, the inorganic division has become stronger since I arrived. When I came it consisted of Minoru Tsutsui, Kurt Irgolic, Art Martell, Ralph Zingaro, and two assistant professors, neither of whom got tenure. In 1975, Mike Hall came as an Assistant Professor.

The first major move that Art and I made was to hire Abe Clearfield, in 1976. Abe and I had known each other from our days together at Temple University where in 1951 he took a Master's degree and I got my B.A. During the academic year 1974-75 Abe was a program officer rotator at NSF. On renewing our friendship, I realized that he would be a great addition to our faculty and Arthur agreed as soon as he had met him. Thus, after returning to his position at Ohio University in 1975-76, Abe came to A&M as a Full Professor. Abe was one of the pioneers of the now fashionable field of materials science.

Our next major additions were Don and Marcetta Darensbourg. In 1981 Don had spent part of his sabbatical in my lab and so he knew he liked it here. He and Marcetta were both at Tulane at the time and I think both felt a bit limited by the small scale of the department there. Thus, when Minoru died very unexpectedly of a heart attack, Arthur and I decided that adding both Don and Marcetta to our inorganic group would be a very good move and that was done in 1982.

In 1983, we had another great opportunity and were able to take advantage of it. The college of Science needed to find a new dean in 1983 and Arthur and I persuaded John Fackler to apply for the position. John had done his Ph.D. research with me at MIT from 1956 to 1960, and in 1982 he was Chairman of the chemistry department at Case Western Reserve University. My greatest problem in convincing John to be a candidate for Dean at A&M was his wife Naomi who had grown very fond of Cleveland. However, John applied, was chosen and started in 1983. He was also appointed Professor of Chemistry, but when he asked for laboratories and an office in the chemistry department, he was assured that these would be unnecessary, because as Dean he wouldn't have any time for research. When John said that if he couldn't have research labs he would not come at all, he got what he wanted. During his twelve years as Dean, John was one of the most active researchers in the department, and in 1987 was promoted to the rank of Distinguished Professor, a rank conferred only on those of exceptional research accomplishments.

At this point, we had an inorganic division that I regarded as, in a sense, my own family. But it was now a very mature and capable family, and thereafter collectively developed itself. In 1987 we added Tim Hughbanks, who became a Full Professor in 1997. In 1998 we hired one of Alan Cowley's former students, Francois Gabbai, who is now a Full Professor. In 2004 we hired an Assistant Professor, Raymond Schaak, who is doing extremely well.

Over this period of time, Kurt Irgolic went back home to Austria and he died there in a mountaineering accident in the Summer of 1999. Ralph Zingaro has become emeritus, but we still see him. As for Arthur Martell, his health began to decline in the late 1990s, and he died in 2003, three days short of his 87th birthday.

Animals

Without a long succession of dogs, cats, and horses, my life would have been like food without seasoning. The first horse Dee and I owned was a liver chestnut named Pal that I gave her as a wedding present, and the next one, which I have already mentioned, was my first field hunter Clarence. Since then, that is from the early 1960s on, we have had many more. Since many of my readers may have little or no interest in horses, I'll expatiate on only one more, Jogglebury, or JB as we always called him. He is the horse I'm on in the frontispiece. This big red chestnut gelding was bred in Ireland as a cross between a thoroughbred stallion and an Irish heavy horse mare. He had more jump in him than any other horse I ever had — not that he would always give it to me when I was on him. Several times he jumped over the lower part of the Dutch

Dee with Pal, a liver chestnut that I gave her when we were married (1959).

door out of his stall or over 4-5 foot paddock fences. Despite his sturdiness, he was also very fast. I took him to Texas with me and he had a long, pleasant retirement — still jumping out of his paddock when the spirit, or the scent of a mare in heat, moved him. He didn't seem to know he had been gelded. JB was the quintessential male animal, and except when I was mad at him, I just loved him.

Dee and I both participated very fully in the activities of the Norfolk Hunt Club and eventually became principal participants in the

Dee and I at a Hunt pace event.

maintenance of the pack of hounds and the conduct of the hunts. We both love dogs and having several dozen fox hounds to enjoy was wonderful. We also acquired some more horses and there was a period of one to two years just before we came to Texas when I was spending about twenty hours a week riding. I've never been any fitter in my life.

Nowadays, my pleasure with horses and other animals in largely vicarious. My daughter Jennifer runs the ranch, my wife keeps the books and looks after our dogs, and I play the role of benevolent but useless patriarch. Jennifer has been breeding and training horses for some years now, with excellent results. Watching an expert like Jennifer work with young horses is a life-enhancing experience. Occasionally, I have the privilege of pretending to be involved in the breeding program.

Innumerable cats have come and gone in the Cotton household, some as house pets, others kept outdoors. We even took

Getting to know some youngsters at the Norfolk kennels in about 1970.

two Siamese cats to Argentina in 1965. It is, however, dogs who have been our most constant companions. It was during our hunting days, as mentioned earlier, that we had mass contact with "dogs," which is a noun that one does *not* apply to fox hounds. To a fox hunter dogs are all those specimens of *canus domestica* that are not hunting hounds. Often, for emphasis the dismissive term curdog is used. Hunting hounds may be either doghounds or bitches — but they are never "dogs." Our taste in house dogs has run the gamut of sizes. Dee has always had a special fondness for big

With one of my daughter's colts in 1999.

dogs. We have had Rottweilers, bloodhounds, bassett hounds, miniature dachshunds, and all sorts in between, including mongrels. I love dogs mainly for their joie de vivre and for never letting adversity get in the way. How many times have I seen a dog with one bad leg run as fast as it could on three so as not to miss out on some fun. Based on experience, I recommend picking the runt of a litter. Runts are like jockeys: they're tough little guys, full of guts and heart, and they also tend to live a long time.

A natural extension of my love of animals has led me to collect bronze animals. In the period 1820 to 1880 there was a school of French sculptors who modeled animals, of almost *all* kinds — dogs, horses, cattle, sheep, buffalo, elephants, camels, birds, reptiles, and still others — in a very naturalistic way. These sculptors are known as *les animaliers.* I fell in love with their work before I could afford to buy any of it, but beginning with a small mare by Christophe Fratin in 1968, I have assembled a collection of bronze castings by such greats as Barye, Isador and Rosa Bonheur, Mêne, and Fremiet, to name a few. I have these bronzes all over the house so no matter where I sit, I can be looking at one. They give me great joy.

Some Recollections of Travel

Science more than any other modern intellectual activity knows no national boundaries. Therefore, being an academic scientist in the latter half of the twentieth

A chien basset by Jules Moigniez.

century offered practically unlimited opportunities for international travel and forming friendships in many other countries. Throughout the foregoing chapters I have described many trips.

My story as a traveling scientist began at the end of one era — that of no air travel — and carried me into the age of supersonic jet air travel. My first trip from Boston to the west coast and back (1952) was done by train and automobile and my first trip to Europe and back (1954) was done by steam ship. Between then and now I have flown in nearly every kind of commercial airliner from the unpressurized D-3 to the D-6, the Constellation, the Superconstellation, the 707, the 747, several airbuses, and the Concorde, to mention only the iconic ones.

Several of my earliest flights to Europe were made *via* an air service called MATS — Military Air Transport Service. This was operated by the Department of Defense, following WWII, to fly their own personnel, mainly staff officers. However, it was also made available to senior personnel in allied government operations such as government defense laboratories, and principal investigators of university research projects funded by agencies such as ARPA and the AEC. The flights to Europe were on Air Force Constellations which, because of their limited range, went first to Gander NF, where the tanks were filled to the brim, then to Ireland, where a landing was made in Shannon on nearly empty tanks, and finally on to England. Something like 20 hours on a bone-rattling, ear-pounding propeller airplane did not support the thesis that "getting there is half the fun." But it was cheap. Only those who took long trips during the prop plane era can really appreciate what a blessing the jet airplane is.

Travel by supersonic aircraft is — or was — the ne plus ultra. I was fortunate to make about a dozen transatlantic trips on the Concorde. Back in the long ago days when airlines were still regulated and ticket pricing was rational, there was a set of ratios between the classes of transatlantic service (coach: first class: Concorde) which was, approximately 2:4:5. Thus (or so I thought) if one were to pony up the cost of 1st class, it was silly not to go by Concorde. Every time I endure the long, monotonous flight back from Europe to the United States these days I regret that no one is enterprising enough to reestablish supersonic transatlantic service.

Foreign Students

One of the pleasures of doing research in academia is the opportunity to have foreign coworkers in the group. Over the years I have had quite a few foreign Ph.D. students, although the first one did not appear until 29th on my list and he, Peter Legzdins, did not come from any place more distant or exotic than Canada. Peter, who was later a postdoc with Wilkinson, has subsequently had a fine career at UBC in Vancouver. He has written a very useful monograph on metal nitrosyls. Within another year I had a second excellent student from Canada, Josef Takats, who went on to a career at the University of Alberta.

Four years or so later Alan Shaver came along. Alan showed me that the term "generation gap" described a real thing. I liked Alan from day one; I just couldn't get along with him. He was a hippie (although since he was totally clean and worked hard, maybe he wasn't a real hippie). To my enormous pleasure, Alan has gone on to have a fine career at McGill where he even became department head and then a dean. We have long since become good friends. Very recently I had one more Canadian student, David Maloney, who has the distinction, if that's the word for it, of being my 100th Ph.D.

Other foreign students in my MIT days were José Calderón, now a big shot (or so I'm told) in the Venezuelan National Petroleum Company, then Milorad Jeremic who went back to Belgrade and became a professor of the physical chemistry of macromolecules. Then came the high point from a purely personal point of view:

Carlos Murillo from Costa Rica. As is evident on many pages of this book, Carlos is a very special person to me.

Beginning in the early 1980s, I have had several students from Taiwan and India. I have also had one, Edwar Shamshoum, from Israel who was of all things a Christian Arab and as handsome as Omar Sharif. He was also a fine athlete. Ed is enjoying a very successful career in industrial research. One other foreign student from the 1980s that I must mention is a charming woman from Spain, Rosa Llusar. Rosa is unique in having received two Ph.D.s for work done with me. She did far more than was necessary for her A&M Dissertation, a fairly common situation among my students. Since she was planning to make a career in the Spanish university system, she used the results not included in her dissertation to write a very nice Tesis Doctoral at the University of Valencia. She subsequently did postdoctoral work with John Corbett and is now a Professor Titular at the University Jaume I in Castellón.

Beginning in the decade of the 1980s, the complexion of my research group has undergone a gradual evolution from American to Chinese. The decline in the number of Americans reflects a general trend found in all the sciences (and engineering) and across the nation. There were simply fewer American students going into science and engineering. However, for several reasons graduate schools needed to, wanted to, and did maintain the same numbers of students. This was done by making up the shortfall in Americans with an increased intake of foreigners. This increased need for foreign students coincided with increasing interest on the part of people in the People's Republic of China in coming to the United States for advanced study. At first there was the era of "visiting scholars" who were older people, mainly men, who had been repressed during the "great leap forward" and "cultural revolution" periods. Their pent up urge to catch up with the scientific progress in the West that had occurred while they had been repressed was suddenly released and they applied in large numbers to come to the West. Some went to iron curtain countries and to Western Europe, but the majority found a welcome in the Untied States and Canada.

This surge of older "visiting scholars" was but a forerunner to the far bigger surge of younger PRC scientists who applied for admission to Ph.D. programs. Chemistry professors in general were keen to welcome such students, who were typically very bright albeit somewhat poorly trained in experimental work owing to lack of modern instrumentation in the PRC at that time. I recall that in the early 1980s, when X-ray diffractometers were commonplace in the United States there were no more than two or three in the whole of China. I can remember several of my own PRC students who had scored 800 (maximum possible) on the quantitative GRE and who came in and aced the physical, organic and inorganic placement exams but flunked the one in analytical chemistry, because the instrumental analytical methods that were routine in the United States were not yet practiced or taught in the PRC. At A&M our procedure was to give a student who flunked a placement exam the choice of either taking a suitable makeup course or retaking the exam at the end of the semester. The PRC students usually opted to retake and almost without exception scored 100% on doing so.

By the late 1980s, most graduate programs in chemistry were flooded by applications from apparently highly qualified PRC students, to the point where they could have filled their entire annual quota for incoming students 3–5 times over from the PRC alone. Most departments decided to cap their intake at about 20% of the total.

My first student from the PRC was Bianxao Zhong who joined my group in 1987. Actually Zhong had come to the United States to study at the University of Kansas in 1988. He had also applied to A&M but had not been admitted so he went where he had been admitted and applied again to A&M. This time, with a proven record in the United States he was accepted. I believe that this pattern was not uncommon among those who know where they wanted to go but couldn't get there directly from China.

Zhong was a gift from heaven. I hadn't had such a smart student more than a few times before. He absorbed new knowledge like a sponge, was a gifted experimentalist, and never seemed to stop working. He was gentle, modest and personable — anybody's idea of the perfect coworker. He has been followed by sixteen others including Tong Ren, now at Purdue University, and Hongcai Zhou, now at Miami University in Ohio.

Altogether, I have had thirty seven foreign graduate students from seventeen different countries. Students from foreign countries, most of whom have remained (so far) in the United States have thus been just about one-third (31.8%) of my total.

Three Golden Rules

For many years there have been three golden rules in my research group, two of which are well-known biblical admonitions that I have adapted.

1 SIX DAYS SHALT THOU LABOR.
This means that we all (including me) work on Saturdays.

2 BY THEIR FRUITS SHALL YE KNOW THEM.
This means that *results* must be produced — not *excuses* for not having results.

3 IF IT ISN'T PUBLISHED, IT ISN'T FINISHED.
This means that writing a publishable report is an essential part of every research project.

EPILOGUE

I CLOSE BY returning to a point I addressed in the Prologue: choosing a title for this book was difficult. One American Scientist's Life in the Golden Age of Science could have been the title of this book — except that it is too long and pedantic. I have explained in detail what I mean by the golden age in my Priestley Medal Lecture [see *Chemical & Engineering News*, **1998**, *76(13)*, 39 and Appendix I]. It was the period between about 1950 and about 1990, when funding for truly basic research was more abundant than ever, before or since. Before the golden age began, funding for any kind of scientific research, pure or applied, was downright miserable, apart from wartime projects. Starting in about 1950 things got better. Beginning in about 1990 (or even earlier) basic research in industry (of the kind done in abundance at Bell Telephone Laboratories or the DuPont Experimental Station from, roughly, 1960 to 1980) began to die and is now dead or the next thing to it. More serious, however, has been a deliberate change in the philosophy at the NSF to require that proposed research be shown to serve some prestated societal need. The idea that research aimed only at "finding out how Nature works" deserves funding has come to be regarded as irresponsible if not outright subversive.

I think this change in philosophy has already had a corrupting effect on many young scientists. The less talented are terrified by the thought that if they can't convince funding agencies that someone will in the near future be able to make a buck out of what they propose to do they won't get funded. Many of the most talented are far more (if not entirely) driven by the urge to get patents, start companies, and get rich than by intellectual curiosity. Many of those who are both highly talented and idealistic don't go into the physical sciences at all any more. It is not an attractive situation. Despite my opening caveat about having been born at a bad time (the start of the Great Depression), I feel fortunate to have pursued my career when I did.

Indeed, I feel fortunate to have lived all phases of my life where and when I did. It has been a satisfying life, full of more good fortune than I had any right to expect, or to now take for granted. Elysian joys such as first sight of one's newborn children, the rapture of love, hearing Mahler's second symphony or the Goldberg variations, being carried over a fence on a powerful galloping horse, drinking Dom Perignon, or eating chocolate-covered candied orange rind are high points, but all the rest of my life has been rich and rewarding, too.

When Schiller wrote "wem der grosse Wurf gelungen eines Freundes Freund zu sein, wer ein holdes Weib errungen mische seinen Jubel ein" he was either enumerating two separate things or conflating them. In my life, they were conflated into one exceptional person, who has been the single best thing that ever happened to me. Her patience with my irascible personality over the past 47 years has been extraordinary and much appreciated.

What I have now written conveys, I believe, the important facts about my life — if indeed any of them are important — but does the narrative convey anything about me? How many autobiographies have I read, only to conclude with the realization that I still know very little about the human being who wrote it? I am certainly not driven by any unnatural urge to "bare my soul" but an autobiography should be more than a dry chronicle of events. In this sense, I may have failed — probably have — although I have tried.

It seems very fashionable nowadays to philosophize (even agonize) over the relationship of science and religion. I am as contemptuous of metaphysical nonsense (which includes religious ritual) as I am unmoved by metaphysical anguish. For me personally, religion is nonsense. Worse, however, is that in its organized form it is *pernicious* nonsense. The human misery religious fanaticism has caused, and continues to cause on a monstrous scale, is incalculable. Religious absolutism provides a cover for systematic, vicious inhumanity. It is also, of course, responsible for some of the most idiotic, though harmless, behavior, like high school football teams thanking God because they scored a touchdown.

While I have no quarrel with anyone else's right to believe in God, I find the proactive members of his fan club insufferable. I simply do not find it necessary to believe in the existence of any God or gods and I think the only strategy for dealing with real life is by a combination of logical thinking, tolerance for the weaknesses of other people, and a humane view of all other living creatures.

APPENDICES

Ph.D. Students

1. George, John W. — 1958
2. Leto, Joseph R. — 1958
3. Holm, Richard H. — 1959
4. Meyers, M. Douglas — 1959
5. Monchamp, Roch R. — 1959
6. Fackler, John P., Jr. — 1960
7. Horrocks, William DeW., Jr. — 1960
8. Faut, Owen D. — 1962
9. Soderberg, Roger H. — 1962
10. Bills, James L. — 1963
11. Haas, Terry E. — 1963
12. Schunn, Robert A. — 1963
13. Elder, Richard C. — 1964
14. Francis, Ronald — 1964
15. Bergman, John G. — 1965
16. Lippard, Stephen J. — 1965
17. Mague, Joel T. — 1965
18. Rauch, Francis C. — 1965
19. Wise, John J. — 1965
20. Robinson, William R. — 1966
21. Harris, Charles B. — 1966
22. Weaver, David L. — 1966
23. Eiss, Roger — 1967
24. Faller, John W. — 1967
25. Richardson, David C. — 1967
26. Zimmerman, Ralph A. — 1967
27. Edwards, W. Thomas, M.D. — 1968
28. Foxman, Bruce M. — 1968
29. Legzdins, Peter — 1968
30. Winquist, Bruce H. C — 1968
31. Bratton, W. Kevin. — 1969
32. LaPrade, Marie D. — 1969
33. Stokely, Peter F. — 1969
34. Takats, Josef — 1969
35. Arnone, Arthur — 1970
36. Marks, Tobin J. — 1970
37. Bier, C. James — 1971
38. Bullitt, Julian G. — 1971
39. Calderón, José L. — 1971
40. Pipal, J. Robert — 1971
41. Jeremic, Milorad — 1972
42. Lukehart, Charles M. — 1972
43. Norman, Joe G., Jr. — 1972
44. Shaver, Alan G. — 1972
45. Ucko, David A. — 1972
46. Adams, Richard D. — 1973
47. Klemperer, Walter G. — 1973
48. Mester, Zoltan C. — 1974
49. White, Alan J. — 1974
50. Troup, Jan M. — 1974
51. Tucker, Philip W. — 1975
52. Hunter, Douglas L. — 1975
53. Stanislowski, Anna G. — 1976
54. Shive, Larry W. — 1976
55. Murillo, Carlos A. — 1976
56. Legg, Margaret J. — 1977
57. Hanson, Brian E. — 1978
58. Bailey, Webb I., Jr. — 1979
59. Gage, Larry D. — 1979
60. Stanley, George G. — 1979
61. Hall, William T. — 1980
62. Najjar, Robert C. — 1982
63. Fang, Anne — 1982
64. Lay, Dennis G. — 1982
65. Wang, Wenning — 1983
66. Lewis, Gregg E. — 1983
67. Thompson, James L. — 1983
68. Daniels, Lee M. — 1984
69. Powell, Gregory L. — 1984
70. Han, Scott — 1984
71. Marler, David O. — 1984
72. Agaskar, Pradyot A. — 1985
73. Shamshoum, Edwar S. — 1985
74. Melton, Tammy J. — 1986
75. Lewis, Diane B. — 1986
76. Bancroft, Daniel P. — 1986

77. Diebold, Michael P. — 1987
78. Canich, Jo Ann M. — 1987
79. Llusar, Rosa M. — 1988
80. Abbott, Ronald G. — 1988
81. Zhong, Bianxao — 1989
82. Babaian-Kibala, Elizabeth — 1989
83. Torralba, Raymond — 1990
84. Chen, Jhy-Der — 1990
85. Ren, Tong — 1990
86. Mueller, Timothy — 1990
87. Son, Kyung-Ae — 1990
88. Czuchajowska-Wiesinger, Joanna — 1991
89. Wiesinger, Kenneth J. — 1991
90. James, Chris A. — 1992
91. Mandal, Sanjay K. — 1992
92. Miertschin, Charla S. — 1992
93. Sun, Zhong-Sheng — 1992
94. Kang, Seong-Joo — 1992
95. Hong, Bo — 1993
96. Wojtczak, William A. — 1993
97. Yao, Zhengui — 1993
98. Kim, Youngmee — 1993
99. Lu, Jian — 1994
100. Maloney, David J. — 1995
101. Su, Jianrui — 1996
102. Yokochi, Alex — 1997
103. Matonic, John H. — 1997
104. Wang, Xiaoping — 1998
105. Timmons, Daren — 1999
106. Zhou, Hongcai (Joe) — 2000
107. Huang, Penglin — 2002
108. Angaridis, Panagiotis — 2002
109. Lei, Peng — 2003
110. Hillard, Elizabeth A. — 2003
111. Berry, John F. — 2004
112. Liu, Chun Yuan — 2005
113. Wilkinson, Chad — 2005
114. Villagrán, Dino — 2005
115. Yu, Rongmin — 2005
116. Ibragimov, Sergey — 2006
117. Zhong Li — 2007
118. Qinliang Zhao — 2007
119. Mark Young — 2009

Postdoctorals

Bannister, Eric — 1958–59
Parish, Richard V. — 1958–59
Goodgame, David M. L. — 1959–61
Goodgame, Margaret — 1959–61
Zingales, Franco — 1959–60
Calderazzo, Fausto — 1959–60
Sacco, Adriano — 1960–61
Blake, Anthony B. — 1960–62
Chakravorty, Animesh — 1961–62
Bertrand, J. Aaron — 1961–62
Kraihanzel, Charles S. — 1961–62
Dunne, Thomas G. — 1961–63
Wood, John S. — 1962–64
Hazen, Edward E., Jr. — 1962–71
Johnson, Brian F. G. — 1963–64
Curtis, Neil F. — 1963–64
McCleverty, Jon A. — 1963–64
Szilard, Imre — 1963–65
Wing, Richard M. — 1963–65
Elder, Richard C. — 1964–65
Bennett, Michael J. — 1965–68
Oldham, Colin — 1965–66
Walton, Richard A. — 1965–66
Lippard, Stephen J. — 1965–66
Yagupsky, Guido — 1966–68
Bonati, Flavio — 1966–67
Hugel, René — 1966–67
Musco, Alfredo — 1966–68
Beauchamp, André L. — 1966–68
Brenčič, Jurij V. — 1967–69
Caulton, Kenneth G. — 1967–69
Reich, Charles R. — 1967–68
Richardson, David C. — 1967–69
Harris, Charles B. — 1966–67
Bermúdez de Brignole, Alicia — 1961–62
Graziani, Mauro — 1968–69
Edwards, W. Thomas — 1968–69

DeBoer, Barry G. — 1968–71
Jones, Hugh V. P. — 1968–71
LaPrade, Marie D. — 1969–70
Debeau, Martine L. — 1969–70
Rusholme, Geoffrey — 1967–71
Smith, James E. — 1970–71
Yonath, Ada — 1970–71
Day, Victor W. — 1970–72
Krucynski, Leonard — 1970–72
Bier, C. James — 1970–71
Pipal, J. Robert — 1970–71
Ciappenelli, Donald J. — 1970–71
Larsen, Sine — 1970–71
Deganello, Giulio — 1971–72
Hardcastle, Kenneth I. — 1971–72
Frenz, Bertram A. — 1971–74
Brice, Michael — 1972–75
Collins, Douglas M. — 1972–77
Lukehart, Charles M. — 1972–73
Webb, Thomas R. — 1972–75
la Cour, Troels — 1972–74
Carré, Francis — 1973–74
Lahuerta, Pascual — 1973–75, 1985
White, Alan J. — 1974–75
Swanson, Rosemarie — 1974–84
Stults, B. Ray — 1974–76
Jamerson, Jack D. — 1974–76
Kolb, John R. — 1974–76
Kalbacher, Barbara — 1974–76
Koch, Stephen A. — 1975–77
Millar, Michele — 1975–77
Extine, Michael W. — 1976–78
Rice, Catherine — 1976–78
Rice, Gary — 1976–78
Niswander, Ron H. — 1977–78
Sekutoski, Janine C. — 1977–78
McArdle, Patrick — 1977–78

Fanwick, Phillip E. — 1977–79
Jordanov, Jeanne — 1978–79
Bino, Avi — 1978–79
Kaim, Wolfgang — 1978–79
Felthouse, Timothy R. — 1978–80
Ilsley, William — 1978–80
Bursten, Bruce — 1978–80
Kolthammer, Brian — 1978–80
Baral, Sanjay — 1979–80
Larsen, Jan — 1980
Mott, Graham — 1980–82
Roth, Wieslaw J. — 1981–88
Tomás, Milagros — 1981–82
Falvello, Larry R. — 1981–91
Schwotzer, Willi — 1981–86
Reid, Austin H., Jr. — 1982–85
Chakravarty, Akhil R. — 1982–85
Kriechbaum, Gangolf — 1983–84
Duraj, Stan A. — 1983–84
Wang, Wenning — 1983–84
Lewis, Gregg E. — 1984–85
Meadows, James — 1983–84
Tocher, Derek — 1983–84
Shiu, Kom-Bei — 1984–85
Matusz, Marek — 1985–88
Dunbar, Kim R. — 1985–87
Kibala, Piotr A. — 1985–90
Sanaú, Mercedes — 1984–85
Poli, Rinaldo — 1985–87
Daniels, Lee M. — 1986–87
Verbruggen, Mark — 1986
Vidyasager, Kanamaluru — 1986–90
Barceló, Francisco — 1986–87
Feng, Xuejun — 1986–1998
Eagle, Cassandra — 1987–88
Luck, Rudy L. — 1987–90
Shang, Maoyu — 1987–93
Sandor, Robert — 1987–89
Price, Andrew — 1988–90
Gerards, Michael — 1988–90
Babaian-Kibala, Elizabeth — 1989–90
Burini, Alfredo — 1989–90
Ren, Tong — 1990–91
Chen, Jhy-Der — 1990–91

Labella, Luca — 1991
Eglin, Judith — 1990–92
Chen, Hong — 1990–93
James, Chris A. — 1992–93
Haefner, Steve — 1992–96
Chen, Gong — 1993
de O Silva, Denise — 1993–95
Dikarev, Evgeny — 1994–2002
Chen, Linfeng — 1994–96
Schmid, Günter — 1994–95
Kühn, Fritz E. — 1995–96
Jordon, Glenn T., IV — 1996–94
Wong, Raymond W. Y. — 1996–97
Petrukhina, Marina A. — 1996–2001
Pascual, Isabel — 1996–99
Herrero, Santiago — 1996–99
Matonic, John H. — 1997–98
Lin, Chun — 1997–2000
Wang, Li-Sheng — 1997–98
Barnard, Thomas S. — 1998
Wang, Xiaoping — 1998–2000
Gu, Jiande — 1998–2000
Reid, Steven M. — 1998–99
Stiriba, Salah-Eddine — 1998–99
Schooler, Paul — 1998–2000
Donahue, James P. — 1999–2004
Huang, Penglin — 2004–06
LeGall, Benoit — 2004–05
Liu, Chun Y. — 2005–06

Visitors

Morehouse, Sheila M. — 1963–64
Richardson, Jane S. — 1967–69
Pedersen, Erik — 1973–74
Dori, Zvi — 1977–78
Fitch, John W., III — 1978
Klein, Simonetta — 1980
Shim, Irene — 1980–81
Murillo, Carlos A. — 1981, 1984, 1986
Boeyens, Jan C. A. — 1984
Kitagawa, Susumu — 1986–87
Nawar, Nagwa — 1996
Guimet, Isabelle — 1998
Lu, Tongbu — 1998–99
Modec, Barbara — 1998–99
Yu, Shu-Yan — 2000
Roberts, Brian K. — 2003
Chao, Hui — 2004–2005

APPENDIX D

Priestly Lecture, 1998

Science Today — What Follows The Golden Age

Let me say first that this is not the sort of talk that a research scientist like me is used to giving. We scientists are normally very dependent upon slides or trasparencies to guide us through — and also we are used to talking about very technical matters to other scientists who are interested in these matters or students who have to at least pretend to be. Thus, I might be more comfortable with transparencies and material of a specialized nature. I am somewhat out of my element in giving a formal lecture, without visual aids, on a nontechnical subject, to a general audience. But let me try.

What Is a Golden Age?

My title refers to a golden age. What does that mean? A golden age is what someone defines it to be, and not everyone will necessarily agree. For example, we have all heard the years of the Athenian democracy called a golden age. No doubt the citizens of Athens would have agreed to that description, but the citizens were not even a majority of those who lived there. The rest were slaves, and one might not be surprised if they didn't share that view. So it first behooves me to say what I mean by the golden age of science and why I call it that.

I am referring to a period of time during which I have done most of my research. I started with it, but I have outlived it. I started my Ph.D. research in 1952, two years after the National Science Foundation came into being. The year before the Sputnik went up (1957, four decades ago), I was just getting my feet on the ground as an independent researcher, with the lofty rank of Instructor at MIT. From the point of view of academic research scientists such as I, these events shaped the next three decades. The NSF constituted a vehicle for the support of fundamental research and the Sputnik symbolized the motivation for support of basic research by the American taxpayer.

There was, of course, throughout this period — from the mid-1950s to the mid-1980s — also much more abundant support than ever before, by both government and industry, for research of an applied nature. But even modest support to academic scientists, to enable them to do the research they themselves considered important and interesting without regard to possible applications had never before existed in this country, with the exception of a very few special and quite limited cases of support by foundations, such as the Rockefeller Foundation.

The idea that university professors in large numbers could obtain very *generous* support for research on *fundamental* problems in pure science, based only on the say-so of other university professors (the peer review process) was a very new thing. I might note that somewhat the same idea became more or less a reality in the rest of the developed world as well, but in this talk I will restrict my attention to the events in the United States.

Before this golden age, U.S. academic scientists who wanted to do so-called "curiosity-driven" research had to do it on a shoestring. If they could interest some company in the possibility that a profitable application might result, they might get some money that way, but that did not leave them free to follow the science wherever it led. Now, beginning in the late 1950's and continuing unabated for about 30 years, the majority of professors in major universities were funded to do research of their own devising — research done, for the most part, simply to discover how Nature works — instead of to devise something practical, with practical meaning, ultimately, *marketable*.

This led to the greatest machine for the creation of new science (as well as new scientists) in the history of the world: The American Research University. Our system became the envy of the world and scientists from other nations (even those with pre-war track records far superior to our own — England, France, Germany, and Italy — were eager to come here. Many came as postdoctoral fellows, others as visiting professors, and many stayed to enrich our scientific enterprise permanently. Before the war, we went to Europe; now Europeans came to us.

Academic science as well as industrial science enjoyed enormous prestige with the general public and many of the best and brightest of our young people chose a career in science. Thus, in academia, professors had prestige, money, and an abundance of good students as well as support from postdoctoral co-workers. If that wasn't a golden age — for academic science in particular, but for science in general — I don't know what else you could have asked. It paid off in many ways. Let me mention a few.

For one, there is national pride — international bragging rights — prestige. Maybe an accountant wouldn't include that in arriving at the bottom line, but I believe that many Americans think national pride is worth something.

Evidence, of a sort, of our leadership — which was nearly absolute in every area of science for about three decades — is provided by the Nobel Prizes. These prizes are given in only a few fields and no doubt there are more deserving people who don't have them than do, but the indication they give of U.S. dominance in science is so dramatic that they are worth mentioning.

In physics, from 1901 to 1945, prizes for work done by scientists in the United States constituted 12% of the total — since 1946, 57%. Similarly in chemistry, from 1901 to 1945, prizes to Americans or those who did their work here, amounted to only 7%, but since then the figure is 50%.

More concretely, academic science has produced the research and high-tech workforce that is absolutely essential to our industrial research efforts. We have gotten not only new knowledge but research leadership in cutting-edge technology, in the chemical industry (which, let me remind you returns a huge trade surplus year in and year out, currently about $18B), in computing and information technology (also a trade balance asset), in aerospace technology (again a trade balance winner), in biotechnology (in which we lead the world, handily), in the pharmaceutical and medical fields, and in still other areas.

You can not be a winner in world technological competition with a workforce that just reads other nation's scientific publications and licenses their patents. It is essential that we produce the discoveries and the highly trained people ourselves. That, of course, does not guarantee commercial success, as we have seen in the case of consumer electronics. All the basic discoveries in solid state electronics were made in this country, but nearly all the consumer electronics (radios and television sets) being sold in the country today are made in Japan.

What led to this is complex, but it is important that we draw two lessons from it. One is while it can be a great advantage to be the scientific source of new technology, it is not enough. It is also necessary to deal effectively with the downstream problems of engineering, design, and manufacturing in an aggressive and competitive way. I think that lesson has been learned and needs no further comment.

However, there is a second point that worries me more, and that is that we not draw a false conclusion, namely, that the downstream function, engineering, design, and manufacturing are all we need and that doing pioneering, basic research is a waste of money. As I said earlier, just reading other nations' science and trying to commercialize it is not going to get us to, or keep us at, the top in technology. We need to make the early scientific breakthroughs ourselves and train our own people to the highest level in the process.

Is the Golden Age Over?

The answer is an unqualified "yes." What are the symptoms and what are the reasons? The symptoms are evident to anyone engaged in basic research and even to those in the more long-range type of applied research. The concept of curiosity-driven research is now openly attacked. The refrain is "why should the public pay unless they know what they are going to get out of it?" Scientists are generally inept — if not downright uninterested — in trying to explain to the public what they do and why they do it. They are even naïve enough to admit that they *enjoy* what they do! Thus we have the confrontation so aptly captured in the title of an Editorial in *Science* last May: *(The) scientifically illiterate vs. (the) politically clueless.* Blunt but accurate. Even those in industrial labs are finding it difficult to persuade management to fund long-range applied research. There seems to be a view that we don't need any more background knowledge — let alone basic knowledge. We don't need research, but rather development. Industrial support for what deserves to be called research (not development) is now no more that half of what it was a decade ago. The DOD and DOE are spending much less on basic research and scientists who seek funds for basic research are being driven to the NSF. The success rate for proposals at the NSF is declining steadily.

Because the NSF is the only agency whose explicit purpose is (or was) to support research in basic science, its ability to fulfill this role is critical. Yet even the NSF is beginning to succumb to popular and political pressure to be relevant, meaning to do nothing that cannot be justified by its purported promise of contributing directly to the short-term solution of an obvious problem in the lives of ordinary people.

Even ten years ago, I saw a small sign of weakening in the position of the NSF when the term peer review was replaced by merit review. Why? So far as I can recall because peer review was considered elitist. But elitist or not, peer review has a specific, prescriptive meaning. The merits of a proposal are to be judged by those who, like the proposer, are experts in the field and thus able to judge merit on the basis of intimate knowledge and experience. What does merit review really mean? Clearly, it means that merit is to be judged, but by whom and by what criteria? Perhaps by bureaucrats, perhaps by scientists of modest attainments with influence in Washington, or perhaps by politicians?

The sudden change in label did not, it must be admitted, cause a sudden change in the process. However, it provides cover and justification for slow evolution away from the ideal of well-informed experts judging proposals purely on scientific merit toward inexpert persons with extrascientific agendas making judgements that are intrinsically not scientifically informed or motivated.

The public today seems quite ambivalent about science. They thrill to exploits like the moon landing and the recent unmanned exploration of Mars, and it seems to be generally regarded as desirable to call all sorts of things science. We have in universities such as my own social scientists, political scientists, poultry scientists, and (I am *not* making this up!) tourism scientists!

But at the same time the public seems to shun science in many aspects of daily life — preferring totally unscientific, medieval and pre-medieval practices.

In spite of what we know about the planets and the universe altogether, there are evidently millions of people (including at least one president's wife and several senators and the late President Mitterand of France) who take astrology seriously. Let us be clear about what this means: These people actually believe that the planets (which simply follow Newton's simple inexorable and utterly impersonal laws of motion — with an occasional relativistic correction) adopt positions in the sky that determine what is going to happen to the personal lives of themselves and all of the billions of people on earth. They believe that, depending on which constellation a certain planet appears to be in, it is — or is not — the right time to propose marriage, ask for a raise or bet on a horse. Such outrageous ignorance must be very widespread because hundreds of newspapers and other periodicals offer a good living to the charlatans and/or crackpots who write regular columns on astrology. Millions of people still believe in crackpot medical ideas such as wearing copper bracelets, wearing little magnets, or immersing themselves in water that stinks like rotten eggs.

Not long ago, a large group of people were so totally possessed by irrational beliefs that they took their own lives. Despite the fact that they were intelligent enough to make a living programming computers, they actually believed that creatures more intelligent than themselves were circulating around the solar system in space ships waiting to welcome them into a glorious new and remote world. They believed this despite the fact that there is not one iota of scientific evidence to prove the existence of any form of life (let alone superintelligent life) now or ever, anywhere but here on earth.

There are of course less tragic — even amusing — examples of people who refuse to use scientific modes of thought. Take, for example, the weather forecaster who said (I heard him, in Boston some years ago) that we could expect the rest of the month to be quite rainy because the normal rainfall for this month was four inches and here we were at the 15th and there had been no rain yet. Or, take the case of the man who was concerned that there might be a bomb on board the aircraft on which he was going to fly. So he "reasoned" that the chance of there being two bombs aboard was so small that he could feel safe if he carried one on himself.

To take one more example that is not tragic, but not particularly funny either, there are, apparently many people who believe that a forked piece of wood held in someone's hand can tell you whether there is water under the ground. Of course, there are many cases where this activity "succeeds" because there is at least some water under the ground almost everywhere.

Science does not get a very good press as it used to, even in academe. Some of our outspoken academic brethren in the areas of art, literary criticism, sociology and philosophy now regularly and very vocally engage in what I believe is called "postmodernist criticism," also known as deconstructionism, structuralism or even social constructionism. These people assert that science is a hoax, that scientific descriptions of Nature are "social constructions," that science is no more objective than literary criticism or the writing of poetry, that it is just another "mythic system." In their enormous ignorance they make hare-brained statements like "Even Einstein showed that everything is relative." Actually, even a hare of normal intelligence would not say that.

Doubtless many of you have heard of a recent book by one John Horgan, a former writer for the *Scientific American*, called "The End of Science." Succinctly put, his thesis is that nothing important remains to be discovered. This is not the time or the place to discuss seriously and in detail the wrong-headedness of this. I cite it only to support my point that science is not getting a good press these days. Perhaps I will say one more thing before leaving Mr. Horgan to his melancholy fate as the Cassandra of the scientific age. Around the turn of the century, I believe, the then President of the United States received a very sincere letter from a man who counseled him to close down the patent office because it was clear that all worthwhile inventions had already been made.

I have raised but not yet answered the question of why the golden age of science ended. Two answers are commonly given, but I don't believe either one is really valid. One that we hear *ad nauseam* invokes the end of the cold war. What I think is absurd about this reason is that if basic science was needed to maintain our technical superiority in weapons and war material, why is science not equally necessary to maintain our technical strength in the international economy? The other justification commonly heard is that we need to cut everything in the budget in order to eliminate the deficit. I shall comment further on this at the end of my talk. I think that both of these alleged reasons for cutbacks in the support of scientific research are only rationalizations.

I think the true reason is that human history, in all its aspects, is innately cyclical. The bawdy 18th century was succeeded by the puritanical Victorian era which has, it its turn, been succeeded by the present era of overt amorality. Similarly, there are mood swings regarding the support of science. A prewar era of genteel poverty was succeeded by the well-funded golden age, and now we have entered an era in which the public is rejecting science, as I have already noted.

There is today much obvious evidence that curiosity-driven, i.e., truly fundamental, research is really being squeezed. There is a general public attitude that no research should be supported unless it can be shown in advance, in a way that is persuasive to people who are scientifically illiterate, that it will produce a practical payoff in the foreseeable future. This attitude is manifested in an act of congress, passed in 1993,

called GPRA, the Government Performance and Results Act. This act applies to the
NSF, where its effects may be a classic case of the law of unintended consequences. No
doubt this act was intended to force mission-oriented federal agencies to demonstrate
that they actually fulfill their missions. For an agency with a mission of producing a
material product, or at least a readily measurable outcome, this is fine. For the NSF,
whose mission is supposed to be the support of excellence in basic research, where a
year-by-year demonstration of effectiveness is obviously impossible, this is not fine. It
is an impossibility.

The inevitable result of conformity to GPRA must be to change, to some degree, the
way in which the NSF operates. That is a problem they are struggling with, and who
knows how it will come out. One thing that has already happened is that the formal
procedure for evaluating research proposals now has two criteria, not one. The one
that used to be the only one still exists, namely, that the proposed work is to be of the
highest interest, importance and quality scientifically. Now there is a second criterion
to be met, namely, that the proposed research have explicit and recognizable societal
value and that it be conducted in such a way as to contribute to societal goals. Exactly
what this is going to mean in practice I don't know. At best, it will be construed very
broadly and I can continue to work on the fundamental chemistry of the transition
metals, employing the best coworkers I can find. At worst, I had better start thinking
about something like how to prevent bridges from rusting since iron is a transition
metal.

What Lies Ahead

What lies ahead for American academic science in the immediate future? I think
what lies directly ahead for basic science is a period of difficult funding and relative
lack of respect. I think those of us in the sciences in academia will be more and
more pressed to do research that has, or at least claims to have, short-term, practical
applications rather than the fundamental work we ought to do. I also think that
the best and brightest young people will increasingly prefer the sure-fire, material
rewards of careers in investment banking, law, commerce, and management rather
than the intangible, intellectual rewards of unraveling Nature's secrets and creating
new paradigms of rational thought. Even those who do work for a Ph.D. will be
repelled by the difficulty they see their mentors experiencing in attempting to obtain
funding for basic research.

How long this period will last is quite uncertain. We are entering an era of human
history in which utterly new challenges to the survival of our race are emerging rapidly.
Only science will enable us to cope with many of them. In time I am sure that this
will be recognized and there may then be another Golden Age of Science. But in how

much time, who can say? For now we had better batten down the hatches and expect rough seas ahead for a while.

What Would I Like to See Happen Right Now?

Basic research should be supported more strongly than ever. I would suggest at twice the funding level we have today.

That statement may cause you to think: there's a crazy academic with no sense of economic or political reality. This country is out to balance its budget and this dreamer proposes an increased expenditure. He's off on cloud nine.

No, I'm not. I'll tell you why but first let me say that I want to exclude medical research from the discussion. Medical research can be very basic — and Harold Varmus, the head of NIH, is strongly committed to assuring that basic research is not short-changed in the NIH. Medical research has the built-in advantage that everyone can see why it is valuable. Even politicians have medical problems and are therefore supportive of medical research.

It is basic research in the other areas of science that doesn't sell so easily but clearly can make effective use of more support. And where should support for non-medical basic research come from? Mainly from the NSF. I do not mean to slight other supporters of a smaller scale. Indeed, I want to pay a special tribute to our own Robert A. Welch Foundation here in Texas. It is very heartening to see how they have staunchly continued to be very clear and explicit about the fact that their purpose is to support basic research in chemistry. Nonetheless, to paraphrase what they say at NPR "if NSF doesn't do it, who will?" on a national scale and in all the sciences.

I am advocating that the NSF budget for the support of basic research in the sciences on university campuses should be doubled. Is this fiscally realistic? I think so. Of the approximately \$3.5 B NSF budget, only about \$1.5 B goes for the type of research I am talking about. Well, \$1.5 B is around 0.1% of the entire national budget. Another \$1.5 B would be another 0.1%. Is that going to have any significant effect on balancing the budget? The budget balancing process will succeed only if the big items — those that soak up large percentages, not tenths of a percent — are brought into parity with tax revenues. Once the budget is balanced it can be kept that way only if we have the knowledge and highly trained people to assure our technical competitiveness and to deal efficiently with the problems of ecology, pollution and education that confront us. We can afford another 0.1% of the federal budget for those things. They are things we HAVE TO DO.

Recently, two senators, our own Phil Gramm and Joe Lieberman of Connecticut, have proposed doing that — and even more. I hope they can rally the political allies necessary to make it happen. In its 1999 budget proposal the Clinton administration has proposed only a 10% increase in the NSF budget with that mostly targeted to computer and engineering research and public science education. I would not regard that as satisfactory.

There is an immense amount still to be learned in the "pure" sciences of chemistry, physics, biology, geology, astronomy, as well as mathematics. I don't know whether it is possible ever to "know it all" but I am dead sure that we are a long way from knowing it all now. I am also dead sure that new insights into how Nature works will inevitably bear fruit in all areas of our lives, including the commercial ones.

This is not the time to "declare victory and go home." It is the time to get back into high gear and dig deeper. You can't find gold if you don't go prospecting.

APPENDIX E

Publications

1950–1955

1. Determination of Free Cyanide and Ammonia in Brass and Bronze Plating Baths, S. Heiman and F. A. Cotton, *Plating* **1950**, *Sept.*, 939-944.

2. The Heat of Formation of Ferrocene, F. A. Cotton and G. Wilkinson, *J. Amer. Chem. Soc.* **1952**, *74*, 5764-5765.

3. Bis-cyclopentadienyl Derivatives of Some Transition Elements, G. Wilkinson, P. L. Pauson, J. M. Birmingham and F. A. Cotton, *J. Amer. Chem. Soc.* **1953**, *75*, 1011-1012.

4. Bis-cyclopentadienyl Compounds of Rhodium(III) and Iridium(III), F. A. Cotton, R. O. Whipple and G. Wilkinson, *J. Amer. Chem. Soc.* **1953**, *75*, 3586-3587.

5. Bis-cyclopentadienyl Compounds of Nickel and Cobalt, G. Wilkinson, P. L. Pauson and F. A. Cotton, *J. Amer. Chem. Soc.* **1954**, *76*, 1970-1974.

6. Cyclopentadienyl Compounds of Manganese and Magnesium, G. Wilkinson and F. A. Cotton, *Chem. & Ind.* **1954**, 307-308.

7. Bis-cyclopentadienyl-Verbindungen von Chrom, Molybdän und Wolfram, F. A. Cotton and G. Wilkinson, *Z. Naturforschg.* **1954**, *9b*, 417-418.

8. On the Question of Octahedral Binding in Bis-cyclopentadienyl Compounds, F. A. Cotton and G. Wilkinson, *Z. Naturforschg.* **1954**, *9b*, 453-456.

9. Cyclopentadienyl-Carbon Monoxide and Related Compounds of Some Transition Metals, T. S. Piper, F. A. Cotton and G. Wilkinson, *J. Inorg. Nucl. Chem.* **1955**, *1*, 165-174.

10. Infrared Spectra and Structures of Cyclopentadienyl-Carbon Monoxide Compounds of V, Mn, Fe, Co, Mo, and W, F. A. Cotton, A. D. Liehr and G. Wilkinson, *J. Inorg. Nucl. Chem.* **1955**, *1*, 175-186.

11. The Thermodynamics of Chelate Formation. I. Experimental Determination of Enthalpies and Entropies in Diamine-Metal Ion Systems, F. A. Cotton and F. E. Harris, *J. Phys. Chem.* **1955**, *59*, 1203-1208.

12. Alkyls and Aryls of Transition Metals, F. A. Cotton, *Chem. Rev.* **1955**, *55*, 551-594.

1956–1959

13. Spectra and Structures of Metal-Carbon Monoxide Compounds. II. Manganese and Rhenium Decacarbonyls, F. A. Cotton, A. D. Liehr and G. Wilkinson, *J. Inorg. Nucl. Chem.* **1956**, *2*, 141-148.

14. On Manganese Cyclopentadienide and Some Chemical Reactions of Neutral *Bis*-cyclopentadienyl Metal Compounds, G. Wilkinson, F. A. Cotton and J. M. Birmingham, *J. Inorg. Nucl. Chem.* **1956**, *2*, 95-113.

15. Soft X-ray Absorption Edges of Metal Ions in Complexes. I. Theoretical Considerations, F. A. Cotton and C. J. Ballhausen, *J. Chem. Phys.* **1956**, *25*, 617-619.

16. Soft X-ray Absorption Edges of Metal Ions in Complexes. II. The Cu K Edge in Some Cupric Complexes, F. A. Cotton and H. P. Hanson, *J. Chem. Phys.* **1956**, *25*, 619-623.

17. Heats of Combustion and Formation of Metal Carbonyls. I. Chromium, Molybdenum and Tungsten Hexacarbonyls, F. A. Cotton, A. K. Fischer and G. Wilkinson, *J. Amer. Chem. Soc.* **1956**, *78*, 5168-5171.

18. The Thermodynamics of Chelate Formation. II. A Monte Carlo Study of the Distribution of Configurations in Short Chains, F. A. Cotton and F. E. Harris, *J. Phys. Chem.* **1956**, *60*, 1451-1454.

19. Bond Energies in Transition Metal Complexes, F. A. Cotton, *Acta Chem. Scand.* **1956**, *10*, 1520-1526.

20. Some Experimental Observations on Transition Metal Hydrocarbonyls, F. A. Cotton and G. Wilkinson, *Chem. & Ind.* **1956**, 1305-1306.

21. Infrared Spectrum of Iron Dodecacarbonyl, F. A. Cotton and G. Wilkinson, *J. Amer. Chem. Soc.* **1957**, *79*, 752.

22. Heats of Combustion and Formation of Metal Carbonyls. II. Nickel Carbonyl, A. K. Fischer, F. A. Cotton and G. Wilkinson, *J. Amer. Chem. Soc.* **1957**, *79*, 2044-2046.

23. K X-ray Absorption Edges of Cr, Mn, Fe, Co, Ni Ions in Complexes, F. A. Cotton and H. P. Hanson, *J. Chem. Phys.* **1957**, *26*, 1758-1759.

24. Chemical Shifts in $C_5H_5^-$ and $C_7H_7^+$ Ions: The Free Electron Model, J. R. Leto, F. A. Cotton and J. S. Waugh, *Nature* **1957**, *180*, 978-979.

25. The Structure and Bonding of Cyclopentadienylthallium and Bis-cyclopentadienylmagnesium, F. A. Cotton and L. T. Reynolds, *J. Amer. Chem. Soc.* **1958**, *80*, 269-273.

26. Soft X-ray Absorption Edges of Metal Ions in Complexes. III. Zinc(II) Complexes, F. A. Cotton and H. P. Hanson, *J. Chem. Phys.* **1958**, *28*, 83-87.

27. Some Two Center Overlap Integrals with AO's of Principal Quantum Number Four, L. Leifer, F. A. Cotton and J. R. Leto, *J. Chem. Phys.* **1958**, *28*, 364-365; *Erratum* **1958**, *28*, 1255.

28. Raman Spectra of Group VI Hexacarbonyls, A. Danti and F. A. Cotton, *J. Chem. Phys.* **1958**, *28*, 736.

29. Nuclear Resonance Spectrum of SF_4, F. A. Cotton, J. W. George and J. S. Waugh, *J. Chem. Phys.* **1958**, *28*, 994-995.

30. Properties of SF_4 and Its Adduct with BF_3, F. A. Cotton and J. W. George, *J. Inorg. Nucl. Chem.* **1958**, *7*, 397-403; *Erratum* **1960**, *12*, 386.

31. Proton Resonance Spectra and Structures of Mercury(II)-Olefin Addition Compounds, F. A. Cotton and J. R. Leto, *J. Amer. Chem. Soc.* **1958**, *80*, 4823-4826.

32. Simple Metal Orbital Treatment of the Binding of the Hydrogen Atom in Cobalt Carbonyl Hydride, F. A. Cotton, *J. Amer. Chem. Soc.* **1958**, *80*, 4425.

33. The Association of an Acetylene with Three Metal Atoms; A New Trinuclear Cobalt Complex, R. Markby, I. Wender, R. A. Friedel, F. A. Cotton and H. W. Sternberg, *J. Amer. Chem. Soc.* **1958**, *80*, 6529-6533.

34. Spectral Investigations of Metal Complexes of β-Diketones. I. Nuclear Magnetic Resonance and Ultraviolet Spectra of Acetylacetonates, R. H. Holm and F. A. Cotton, *J. Amer. Chem. Soc.* **1958**, *80*, 5658-5663.

35. Carbon-13 Nuclear Resonance Spectrum and Low-Frequency Infrared Spectrum of Iron Pentacarbonyl, F. A. Cotton, A. Danti, J. S. Waugh and R. W. Fessenden, *J. Chem. Phys.* **1958**, *29*, 1427-1428.

36. Chemical Demonstration of the Aromatic Nature of $(\pi\text{-}C_5H_5)Mn(CO)_3$, F. A. Cotton and J. R. Leto, *Chem. & Ind.* **1958**, 1368-1369.

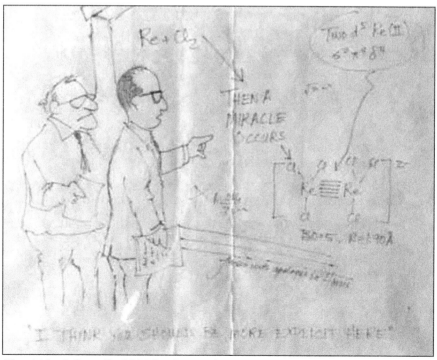

"Then A Miracle Occurs" by Bill (William) Jensen from the University of Cincinnati, with apologies to Sidney Harris.

37. Structure of the Reaction Product of Phenylacetylene with Iron Carbonyl, F. A. Cotton and J. R. Leto, *Chem. & Ind.* **1958**, 1592-1593.

38. Heats of Combustion and Formation of Bis-Benzenechromium, A. K. Fischer, F. A. Cotton and G. Wilkinson, *J. Phys. Chem.* **1959**, *63*, 154-155.

39. Heats of Combustion and Formation of Metal Carbonyls. III. Iron Pentacarbonyl; The Nature of the Bonding in Metal Carbonyls, F. A. Cotton, A. K. Fischer and G. Wilkinson, *J. Amer. Chem. Soc.* **1959**, *81*, 800-803.

40. The Raman and Infra-red Spectra and Structure of Di-(π-cyclopentadienyliron) Tetracarbonyl, F. A. Cotton, H. Stammreich and G. Wilkinson, *J. Inorg. Nucl. Chem.* **1959**, *9*, 3-7.

41. Acceptor Properties, Reorganization Energies, and π Bonding in the Boron and Aluminum Halides, F. A. Cotton and J. R. Leto, *J. Chem. Phys.* **1959**, *30*, 993-998.

42. Structure of the Reaction Product of Phenylacetylene with Iron Pentacarbonyl, J. R. Leto and F. A. Cotton, *J. Amer. Chem. Soc.* **1959**, *81*, 2970-2973.

43. The Preparation and Properties of a New Pentacyanomanganesenitric Oxide Anion, $[Mn(CN)_5NO]^{2-}$, and Some Observations on Other Pentacyanonitrosyl Complexes, F. A. Cotton, R. R. Monchamp, R. J. M. Henry and R. C. Young, *J. Inorg. Nucl. Chem.* **1959**, *10*, 28-38.

44. Infrared Spectra of Manganese Carbonyl Hydride and Deuteride, F. A. Cotton, J. L. Down and G. Wilkinson, *J. Chem. Soc.* **1959**, 833-837.

45. The Infra-red Spectrum and Structure of $K_4[Ni(CN)_3CO]_2$, W. P. Griffith, F. A. Cotton and G. Wilkinson, *J. Inorg. Nucl. Chem.* **1959**, *10*, 23-27.

46. New Tetrahedral Complex Cations with Phosphine Oxide Ligands, F. A. Cotton, E. Bannister, R. Barnes and R. H. Holm, *Proc. Chem. Soc.* **1959**, 158.

47. Magnetic Investigations of Spin-Free Cobaltous Complexes. I. Tetrahalo Cobalt(II) Ions, R. H. Holm and F. A. Cotton, *J. Chem. Phys.* **1959**, *31*, 788-792.

48. A Bridged Carbonyl Phosphine Complex of Nickel, L. S. Meriwether, E. C. Colthup, M. L. Fiene and F. A. Cotton, *J. Inorg. Nucl. Chem.* **1959**, *11*, 181-183.

49. High-Sensitivity Pressure Gauge Coupling Two Spoon Gauges, R. R. Monchamp, E. Bannister and F. A. Cotton, *Rev. Sci. Inst.* **1959**, *30*, 945-946.

50. Barrier to Rotation About the Metal-Cyclopentadienyl Bond, J. S. Waugh, J. H. Loehlin, F. A. Cotton and D. P. Shoemaker, *J. Chem. Phys.* **1959**, *31*, 1434-1435.

51. Rapid, Simple, and Inexpensive Preparation of $[^2H_3]$Methyl Iodide and $[^2H_6]$ Dimethyl Sulphoxide, F. A. Cotton, J. H. Fassnacht, W. D. Horrocks and N. A. Nelson, *J. Chem. Soc.* **1959**, 4138-4139.

52. The Reaction of Disulfur Decafluoride with Chlorine, J. W. George and F. A. Cotton, *Proc. Chem. Soc.* **1959**, 317.

1960

53. Bonding of Cyclo-octatetraene to Metal Atoms: Simple Theoretical Considerations, F. A. Cotton, *J. Chem. Soc.* **1960**, 400-406.

54. The Heat of Sublimation and the Metal-Metal Bond Energy in $Mn_2(CO)_{10}$, F. A. Cotton and R. R. Monchamp, *J. Chem. Soc.* **1960**, 533-535.

55. Magnetic Investigations of Spin-Free Cobaltous Complexes. II. Tetrahedral Complexes, R. H. Holm and F. A. Cotton, *J. Chem. Phys.* **1960**, *32*, 1168-1172.

56. β-Diketones and Their Metal Complexes. Part II. Acetylation of Fluoroacetone to 1-Fluoropentane-2,4-dione, J. P. Fackler, Jr. and F. A. Cotton, *J. Chem. Soc.* **1960**, 1435-1438.

57. Comparison of Calorimetric and Spectroscopic Entropies of Molybdenum Hexacarbonyl, R. R. Monchamp and F. A. Cotton, *J. Chem. Soc.* **1960**, 1438-1440.

58. The Stereochemistry of Five-Coordinate Compounds. Part I. Infrared Spectra of Some Iron(0) Compounds, F. A. Cotton and R. V. Parish, *J. Chem. Soc.* **1960**, 1440-1445.

59. Phosphine Oxide Complexes. Part I. Preparation and Properties of the Cation $[(Ph_3PO)_4Ni]^{2+}$, F. A. Cotton and E. Bannister, *J. Chem. Soc.* **1960**, 1873-1877.

60. Phosphine Oxide Complexes. Part II. Cations of the Type $[(Ph_3PO)_4M]$ with M = Mn(II), Fe(III), Co(II), Cu(II) and Zn(II), E. Bannister and F. A. Cotton, *J. Chem. Soc.* **1960**, 1878-1882.

61. High-Resolution Infrared Spectra and Structures of Cobalt Carbonyls, F. A. Cotton and R. R. Monchamp, *J. Chem. Soc.* **1960**, 1882-1885.

62. Triphenylphosphine Sulphide and Selenide as Ligands, E. Bannister and F. A. Cotton, *J. Chem. Soc.* **1960**, 1959-1960.

63. Magnetic Investigations of Spin-Free Cobaltous Complexes. III. On the Existence of Planar Complexes, F. A. Cotton and R. H. Holm, *J. Amer. Chem. Soc.* **1960**, *82*, 2979-2983.

64. Magnetic Evidence Indicative of Structures and Structural Changes in Complexes of Cobalt(II), F. A. Cotton and R. H. Holm, *J. Amer. Chem. Soc.* **1960**, *82*, 2983-2986.

65. Sulfoxides as Ligands. I. A Preliminary Survey of Methyl Sulfoxide Complexes, F. A. Cotton and R. Francis, *J. Amer. Chem. Soc.* **1960**, *82*, 2986-2991.

66. Phosphine Oxide Complexes. Part III. Bis(triphenylphosphine Oxide)dinitrato-complexes of Cobalt(II), Nickel(II), Copper(II), and Zinc(II), E. Bannister and F. A. Cotton, *J. Chem. Soc.* **1960**, 2276-2280.

67. The Effect of Complex-Formation by Phosphine Oxides on Their P-O Stretching Frequencies, F. A. Cotton, R. D. Barnes and E. Bannister, *J. Chem. Soc.* **1960**, 2199-2203.

68. Normal Co-ordinate Analysis and Force Constants for the Thionyl Halides, F. A. Cotton and W. D. Horrocks Jr., *Spectrochim. Acta* **1960**, *16*, 358-362.

69. A New Type of Tetrahedral Complex of Nickel(II), F. A. Cotton and
    D. M. L. Goodgame, *J. Amer. Chem. Soc.* **1960**, *82*, 2967.

70. The Preparation, Structures and Infrared Absorption of Salts of Cobalt(III)
    Hexafluoride Ion, M. D. Meyers and F. A. Cotton, *J. Amer. Chem. Soc.* **1960**, *82*,
    5027-5030.

71. Magnetic and Spectral Properties of the Spin-Free 3d[6] Systems Iron(II) and
    Cobalt(III) in Cobalt(III) Hexafluoride Ion: Probable Observation of Dynamic
    Jahn-Teller Effects, F. A. Cotton and M. D. Meyers, *J. Amer. Chem. Soc.* **1960**, *82*,
    5023-5026.

72. Dipole Moments and Structures of Some *Bis*-(trifluoroacetylacetonato) Complexes,
    R. H. Holm and F. A. Cotton, *J. Inorg. Nucl. Chem.* **1960**, *15*, 63-66.

73. Electronic Structure and Molecular Association of Some *Bis*-(β-Diketone)Nickel(II)
    Complexes, J. P. Fackler, Jr. and F. A. Cotton, *J. Amer. Chem. Soc.* **1960**, *82*, 5005.

74. Some Substituted Metal Carbonyls with Ligands Having Sulfur as the Donor Atom,
    F. A. Cotton and F. Zingales, *Chem. & Ind.* **1960**, 1219.

75. The Heat of Formation of Potassium Fluoroborate, J. L. Bills and F. A. Cotton,
    J. Phys. Chem. **1960**, *64*, 1477-1479.

76. Sulfoxides as Ligands. II. The Infrared Spectra of Some Dimethyl Sulfoxide
    Complexes, F. A. Cotton, R. Francis and W. D. Horrocks, Jr., *J. Phys. Chem.* **1960**,
    64, 1534-1536.

77. New Tetrahedral Complexes of Nickel(II), F. A. Cotton and D. M. L. Goodgame,
    J. Amer. Chem. Soc. **1960**, *82*, 5771-5774.

78. Tetrahedral Complexes of Nickel(II) Containing Triphenylarsine Oxide,
    D. M. L. Goodgame and F. A. Cotton, *J. Amer. Chem. Soc.* **1960**, *82*, 5774-5776.

79. π-Bonding in Tetrahedral Complexes, F. A. Cotton, *J. Chem. Soc.* **1960**, 5269-5270.

80. Tetrakis(triphenylphosphine)-silver(I) and -copper(I) Complexes, F. A. Cotton and
    D. M. L. Goodgame, *J. Chem. Soc.* **1960**, 5267-5269.

1961

81. Preparation, Spectra and Electronic Structures of Tetrahedral Nickel(II) Complexes
    Containing Triphenylphosphine and Halide Ions as Ligands, F. A. Cotton, O. D. Faut
    and D. M. L. Goodgame, *J. Amer. Chem. Soc.* **1961**, *83*, 344-351.

82. The Donor–Acceptor Properties of Isonitriles as Estimated by Infrared Study,
    F. A. Cotton and F. Zingales, *J. Amer. Chem. Soc.* **1961**, *83*, 351-355.

83. X-Ray Powder Data and Structures of Some Bis-(Acetylacetono)-Metal(II)
    Compounds and Their Dihydrates, R. H. Holm and F. A. Cotton, *J. Phys. Chem.*
    1961, *65*, 321-323.

84. Infrared and Raman Spectra and Normal Co-ordinate Analysis of Dimethyl Sulfoxide
    and Dimethyl Sulfoxide-d_6, W. D. Horrocks, Jr. and F. A. Cotton, *Spectrochim. Acta*
    1961, *65*, 134-147.

85. Interpretation of the Magnetic Resonance Spectrum of the Methylene Group in Certain Unsymmetrically Substituted Compounds, J. S. Waugh and F. A. Cotton, *J. Phys. Chem.* **1961**, *65*, 562-563.

86. Sulphoxides as Ligands — III. Dimethylsulphoxide Complexes of Co(II) and Ni(II) Containing Tetrahedral Tetrahalo Anions, F. A. Cotton and R. Francis, *J. Inorg. Nucl. Chem.* **1961**, *17*, 62-68.

87. Magnetic Investigations of Spin-Free Cobaltous Complexes. IV. Magnetic Properties and Spectrum of Cobalt(II) Orthosilicate, M. Goodgame and F. A. Cotton, *J. Phys. Chem.* **1961**, *65*, 791-792.

88. Magnetic Investigations of Spin-Free Cobaltous Complexes. V. Tetra-azido and Tetracyanato Cobaltate(II) Ions, F. A. Cotton and M. Goodgame, *J. Amer. Chem. Soc.* **1961**, *83*, 1777-1780.

89. Magnetic Investigations of Spin-Free Cobaltous Complexes. VI. Complexes Containing Phosphines and the Position of Phosphines in the Spectrochemical Series, F. A. Cotton, O. D. Faut, D. M. L. Goodgame and R. H. Holm, *J. Amer. Chem. Soc.* **1961**, *83*, 1780-1785.

90. Complexes of Tetrahydrothiophen Oxide, R. Francis and F. A. Cotton, *J. Chem. Soc.* **1961**, 2078-2081.

91. Phosphine Oxide Complexes. Part IV. Tetrahedral, Planar, and Binuclear Complexes of Copper (II) with Phosphine Oxides, and Some Arsine Oxide Analogues, D. M. L. Goodgame and F. A. Cotton, *J. Chem. Soc.* **1961**, 2298-2305.

92. Relative Bond Strengths in Trigonal Bipyramid Molecules, F. A. Cotton, *J. Chem. Phys.* **1961**, *35*, 228-231.

93. Molecular Association and Electronic Structures of Nickel(II) Chelates. I. Complexes of Pentane-2,4-dione and Some 1,5-Di-substituted Derivatives, F. A. Cotton and J. P. Fackler, Jr., *J. Amer. Chem. Soc.* **1961**, *83*, 2818-2825.

94. Phosphine Oxide Complexes. Part V. Tetrahedral Complexes of Manganese(II) Containing Triphenylphosphine Oxide, and Triphenylarsine Oxide as Ligands, D. M. L. Goodgame and F. A. Cotton, *J. Chem. Soc.* **1961**, 3735-3741.

95. Molecular Association and Electronic Structures of Nickel(II) Chelates. II. Bis-(3-Phenyl-2,4-pentanediono)nickel(II) and High Temperature Studies of Nickel Acetylacetonate, J. P. Fackler, Jr. and F. A. Cotton, *J. Amer. Chem. Soc.* **1961**, *83*, 3775-3778.

96. Magnetic Studies of High-Spin Cobaltous Compounds. VII. Some Thiocyanate Complexes, F. A. Cotton, D. M. L. Goodgame, M. Goodgame and A. Sacco, *J. Amer. Chem. Soc.* **1961**, *83*, 4157-4161.

97. Electronic Spectra of Some Tetrahedral Nickel(II) Complexes, D. M. L. Goodgame, M. Goodgame and F. A. Cotton, *J. Amer. Chem. Soc.* **1961**, *83*, 4161-4167.

98. The Electronic Structures of Tetrahedral Cobalt(II) Complexes, F. A. Cotton, D. M. L. Goodgame and M. Goodgame, *J. Amer. Chem. Soc.* **1961**, *83*, 4690-4699.

1962

99. Absorption Spectra and Electronic Structures of Some Tetrahedral Manganese(II) Complexes, F. A. Cotton, D. M. L. Goodgame and M. Goodgame, *J. Amer. Chem. Soc.* **1962**, *84*, 167-172.

100. Carbon Monoxide Insertion Reactions. I. The Carbonylation of Methyl Manganese Pentacarbonyl and Decarbonylation of Acetyl Manganese Pentacarbonyl, F. Calderazzo and F. A. Cotton, *Inorg. Chem.* **1962**, *1*, 30-36.

101. Substituted Metal Carbonyls with Ligands Having Sulfur as the Donor Atom, F. A. Cotton and F. Zingales, *Inorg. Chem.* **1962**, *1*, 145-147.

102. Absorption Intensities and Electronic Structures of Tetrahedral Cobalt(II) Complexes, F. A. Cotton and R. H. Soderberg, *J. Amer. Chem. Soc.* **1962**, *84*, 872-873.

103. Triphenylarsine and Triphenylarsine Oxide Complexes of Cobalt(II), D. M. L. Goodgame, M. Goodgame and F. A. Cotton, *Inorg. Chem.* **1962**, *1*, 239-245.

104. Preparation and Magnetic and Spectral Studies of Some Cobalt(II) Complexes of Benzimidazole, M. Goodgame and F. A. Cotton, *J. Amer. Chem. Soc.* **1962**, *84*, 1543-1548.

105. The Crystal and Molecular Structure of Di-(L-histidino)zinc(II) Dihydrate, R. H. Kretzinger, R. F. Bryan and F. A. Cotton, *Proc. Chem. Soc.* **1962**, 177.

106. The Preparation and Characterization of a Compound Containing Tetranitrato-cobaltate(II), F. A. Cotton and T. G. Dunne, *J. Amer. Chem. Soc.* **1962**, *84*, 2013.

107. Magnetic and Spectral Studies of the Structures of Some Methyl Isonitrile Complexes of Cobalt(II), A. Sacco and F. A. Cotton, *J. Amer. Chem. Soc.* **1962**, *84*, 2043-2047.

108. Selenocyanate Complexes of Cobalt(II), F. A. Cotton, D. M. L. Goodgame, M. Goodgame and T. E. Haas, *Inorg. Chem.* **1962**, *1*, 565-572.

109. Vibrational Spectra and Bonding in Metal Carbonyls. I. Infrared Spectra of Phosphine-Substituted Group VI Carbonyls in the CO Stretching Region, F. A. Cotton and C. S. Kraihanzel, *J. Amer. Chem. Soc.* **1962**, *84*, 4432-4438.

110. Preparation, Nuclear Resonance Spectra and Structure of Bis-(diethoxyphosphonyl)-methano Sodium, J. D. Baldeschwieler, F. A. Cotton, B. D. N. Rao and R. A. Schunn, *J. Amer. Chem. Soc.* **1962**, *84*, 4454-4456.

1963

111. Electron Spectra of β-Diketone Complexes. III. α-Substituted β-Diketone Complexes of Copper(II), J. P. Fackler, Jr., F. A. Cotton and D. W. Barnum, *Inorg. Chem.* **1963**, *2*, 97-101.

112. Electronic Spectra of β-Diketone Complexes. IV. γ-Substituted Acetylacetonates of Copper(II), J. P. Fackler, Jr. and F. A. Cotton, *Inorg. Chem.* **1963**, *2*, 102-106.

113. Reaction of Metal Carbonyls with β-Diketones to Produce β-Ketoenol Complexes, T. G. Dunne and F. A. Cotton, *Inorg. Chem.* **1963**, *2*, 263-266.

114. The Metal-Metal Bonded, Polynuclear Complex Anion in CsReCl₄, J. A. Bertrand, F. A. Cotton and W. A. Dollase, *J. Amer. Chem. Soc.* **1963**, *85*, 1349-1350.

115. The Crystal Structure and Molecular Structure of Dibenzenechromium, F. A. Cotton, W. A. Dollase and J. S. Wood, *J. Amer. Chem. Soc.* **1963**, *85*, 1543-1544.

116. Vibrational Spectra and Bonding in Metal Carbonyls. II. Infrared Spectra of Amine-Substituted Group VI Carbonyls in the CO Stretching Region, C. S. Kraihanzel and F. A. Cotton, *Inorg. Chem.* **1963**, *2*, 533-540.

117. The Crystal and Molecular Structure of Di-(L-Histidino)-Zinc (II) Dihydrate, R. H. Kretsinger, F. A. Cotton and R. F. Bryan, *Acta Cryst.* **1963**, *16*, 651-657.

118. Metal Salts and Complexes of Dialkoxyphosphonylacetylmethanide Ions, F. A. Cotton and R. A. Schunn, *J. Amer. Chem. Soc.* **1963**, *85*, 2394-2402.

119. The Crystal and Molecular Structure of Bis-(trimethylphosphine oxide)cobalt(II) Dinitrate, F. A. Cotton and R. H. Soderberg, *J. Amer. Chem. Soc.* **1963**, *85*, 2402-2406.

120. Calculation of Pseudo-Lattice Energies and the Estimation of Metal–Ligand Bond Energies in Some Tetrahedral Tetrachlorometallate(II) Complexes, A. B. Blake and F. A. Cotton, *Inorg. Chem.* **1963**, *2*, 906-911.

121. Some Phosphine Oxide and Arsine Oxide Complexes of Cobalt(II) Nitrate, F. A. Cotton, D. M. L. Goodgame and R. H. Soderberg, *Inorg. Chem.* **1963**, *2*, 1162-1165.

122. The Crystal Structure of Cesium Dodecachlorotrirhenate(III), a Compound with a New Type of Metal Atom Cluster, J. A. Bertrand, F. A. Cotton and W. A. Dollase, *Inorg. Chem.* **1963**, *2*, 1166-1171.

123. Stability Constants and Structure of Some Metal Complexes with Imidazole Derivatives, A. Chakravorty and F. A. Cotton, *J. Phys. Chem.* **1963**, *67*, 2878-2879.

1964

124. A Spectroscopic Study of the Polymeric Nature of Bis(acetylactonato)cobalt(II), F. A. Cotton and R. H. Soderberg, *Inorg. Chem.* **1964**, *3*, 1-5.

125. Relative Enthalpies of Formation of Some Tetrachlorometallate Ions, A. B. Blake and F. A. Cotton, *Inorg. Chem.* **1964**, *3*, 5-10.

126. A Molecular Orbital Treatment of the Bonding in Certain Metal Atom Clusters, F. A. Cotton and T. E. Haas, *Inorg. Chem.* **1964**, *3*, 10-17.

127. Molecular and Electronic Structures of Some Thiourea Complexes of Cobalt(II), F. A. Cotton, O. D. Faut and J. T. Mague, *Inorg. Chem.* **1964**, *3*, 17-21.

128. The Crystal and Molecular Structure of Bis(dipivaloylmethanido)zinc(II), F. A. Cotton and J. S. Wood, *Inorg. Chem.* **1964**, *3*, 245-251.

129. The Crystal and Molecular Structure of Trioxo(diethylenetriamine)molybdenum(VI), F. A. Cotton and R. C. Elder, *Inorg. Chem.* **1964**, *3*, 397-401.

130. The Heat of Formation of Germanium Dioxide, J. L. Bills and F. A. Cotton, *J. Phys. Chem.* **1964**, *68*, 802-806.

131. The Enthalpy of Formation of Tetraethylgermane and the Germanium-Carbon Bond Energy, J. L. Bills and F. A. Cotton, *J. Phys. Chem.* **1964**, *68*, 806-810.

132. Vibrational Spectra and Bonding in Metal Carbonyls. III. Force Constants and Assignments of CO Stretching Modes in Various Molecules; Evaluation of CO Bond Orders, F. A. Cotton, *Inorg. Chem.* **1964**, *3*, 702-711.

133. On the Question of the Hexachlororhenate(III) Ion, F. A. Cotton and B. F. G. Johnson, *Inorg. Chem.* **1964**, *3*, 780-782.

134. The Tetrameric Structure of Anhydrous, Crystalline Cobalt(II) Acetylacetonate, F. A. Cotton and R. C. Elder, *J. Amer. Chem. Soc.* **1964**, *86*, 2294.

135. A Test of the Mulliken–Wolfsberg–Helmholz LCAO–MO Treatment of Metal Complexes, F. A. Cotton and T. E. Haas, *Inorg. Chem.* **1964**, *3*, 1004-1007.

136. Cyclohexylaminebis(pentane-2,4-diono)cobalt(II): Existence of Monomer and Dimer, J. A. Bertrand, F. A. Cotton and W. J. Hart, *Inorg. Chem.* **1964**, *3*, 1007-1009.

137. Some Compounds Isostructural with Manganese Carbonyl, F. A. Cotton, T. G. Dunne, B. F. G. Johnson and J. S. Wood, *Proc. Chem. Soc.* **1964**, 175.

138. Eight-Coordinate Complexes of Cobalt(II). A Principle Influencing the Occurrence of High Coordination Numbers, F. A. Cotton and J. G. Bergman, *J. Amer. Chem. Soc.* **1964**, *86*, 2941-2942.

139. The Crystal and Molecular Structure of Tris(phenyldiethylphosphine)nonachloro-trirhenium(III), F. A. Cotton and J. T. Mague, *Inorg. Chem.* **1964**, *3*, 1094-1098.

140. The Crystal, Molecular and Electronic Structures of a Binuclear Oxomolybdenum(V) Xanthate Complex, A. B. Blake, F. A. Cotton and J. S. Wood, *J. Amer. Chem. Soc.* **1964**, *86*, 3024-3031.

141. The Occurrence of Re$_3$Cl$_9$ Clusters in Hexagonal Rhenium(III) Chloride, F. A. Cotton and J. T. Mague, *Proc. Chem. Soc.* **1964**, 233.

142. Metal Atom Clusters in Oxide Systems, F. A. Cotton, *Inorg. Chem.* **1964**, *3*, 1217-1220.

143. Mononuclear and Polynuclear Chemistry of Rhenium (III): Its Pronounced Homophilicity, F. A. Cotton, N. F. Curtis, C. B. Harris, B. F. G. Johnson, S. J. Lippard, J. T. Mague, W. R. Robinson and J. S. Wood, *Science* **1964**, *145*, 1305-1307.

144. Ligand Field Theory, F. A. Cotton, *J. Chem. Ed.* **1964**, *41*, 466-476.

145. Dimethyl- and Diethyldithiocarbamate Complexes of Some Metal Carbonyl Compounds, F. A. Cotton and J. A. McCleverty, *Inorg. Chem.* **1964**, *3*, 1398-1402.

146. The Existence of the Re$_3$Cl$_9$ Cluster in Anhydrous Rhenium(III) Chloride and Its Persistence in Solutions of Rhenium(III) Chloride, F. A. Cotton and J. T. Mague, *Inorg. Chem.* **1964**, *3*, 1402-1407.

147. The Occurrence of the Re_3Br_9 Group in Compounds Derived from Rhenium(III) Bromide, F. A. Cotton and S. J. Lippard, *J. Amer. Chem. Soc.* **1964**, *86*, 4497.

148. The Structure of the Deca(methylisonitrile)dicobalt(II) Cation; an Isostere of Dimanganese Decacarbonyl, F. A. Cotton, T. G. Dunne and J. S. Wood, *Inorg. Chem.* **1964**, *3*, 1495-1499.

149. The Identification and Characterization by X-Ray Diffraction of a New Binuclear Molybdenum(VI) Oxalate Complex, F. A. Cotton, S. M. Morehouse and J. S. Wood, *Inorg. Chem.* **1964**, *3*, 1603-1608.

150. Dichlorodinitrosylmolybdenum, Dichlorodinitrosyltungsten, and Some of Their Derivatives, F. A. Cotton and B. F. G. Johnson, *Inorg. Chem.* **1964**, *3*, 1609-1612.

151. Carbon Monoxide Insertion Reactions. II — The Reactions of Acetylmanganese Pentacarbonyl and Methylmanganese Pentacarbonyl with Triphenylphosphine, F. Calderazzo and F. A. Cotton, *Chemica Industria* **1964**, *46*, 1165-1169.

1965

152. The Preparation and Characterization of Compounds Containing both Hexabromorhenate(IV) and the Trirhenium Nonabromide Group, F. A. Cotton and S. J. Lippard, *Inorg. Chem.* **1965**, *4*, 59-65.

153. Some New Derivatives of the Octa-μ_3-chlorohexamolybdate(II), $[Mo_6C_8]^{4+}$, Ion, F. A. Cotton and N. F. Curtis, *Inorg. Chem.* **1965**, *4*, 241-244.

154. A Multiple Bond Between Technetium Atoms in an Octachloroditechnetate Ion, F. A. Cotton and W. K. Bratton, *J. Amer. Chem. Soc.* **1965**, *87*, 921.

155. The Crystal and Molecular Structure of *cis*-(Diethylenetriamine)molybdenum Tricarbonyl; the Dependence of Mo-C Bond Length on Bond Order, F. A. Cotton and R. M. Wing, *Inorg. Chem.* **1965**, *4*, 314-317.

156. The Crystal and Molecular Structure of Pentakis(methylisonitrile)cobalt(I) Perchlorate, F. A. Cotton, T. G. Dunne and J. S. Wood, *Inorg. Chem.* **1965**, *4*, 318-325.

157. Compounds Containing Dirhenium(III) Octahalide Anions, F. A. Cotton, N. F. Curtis, B. F. G. Johnson and W. R. Robinson, *Inorg. Chem.* **1965**, *4*, 326-330.

158. The Crystal and Molecular Structure of Dipotassium Octachlorodirhenate(III) Dihydrate, $K_2[Re_2Cl_8] \cdot 2H_2O$, F. A. Cotton and C. B. Harris, *Inorg. Chem.* **1965**, *4*, 330-333.

159. Metal-Metal Bonding in $[Re_2X_8]^{2-}$ Ions and Other Metal Atom Clusters, F. A. Cotton, *Inorg. Chem.* **1965**, *4*, 334-336.

160. Infrared Spectra (1000-200 cm^{-1}) of Some Transition-Metal Complexes with Tertiary Arsine Oxides, G. A. Rodley, D. M. L. Goodgame and F. A. Cotton, *J. Chem. Soc.* **1965**, 1499-1505.

161. Coordination Compounds of Thallium(III), F. A. Cotton, B. F. G. Johnson and R. M. Wing, *Inorg. Chem.* **1965**, *4*, 502-507.

162. Preparation and Spectra of Some Adducts of the Nonachlorotrirhenium and Nonabromotrirhenium Groups; Evidence of the Re$_3$Br$_9$ Group, F. A. Cotton, S. J. Lippard and J. T. Mague, *Inorg. Chem.* **1965**, *4*, 508-514.

163. Properties of Metal-to-Oxygen Multiple Bonds, Especially Molybdenum-to-Oxygen Bonds, F. A. Cotton and R. M. Wing, *Inorg. Chem.* **1965**, *4*, 867-873.

164. Response to Donohue's Note on Bond Angles in [MoO$_2$(C$_2$O$_4$)(H$_2$O)]$_2$O^{2-}, F. A. Cotton, S. M. Morehouse and J. S. Wood, *Inorg. Chem.* **1965**, *4*, 922.

165. Crystal Structure of Tetrameric Cobalt(II) Acetylacetonate, F. A. Cotton and R. C. Elder, *Inorg. Chem.* **1965**, *4*, 1145-1151.

166. Vibrational Spectra and Bonding in Metal Carbonyls. IV. CO and Re–Re Stretching Modes in the Decacarbonyls of Rhenium and Manganese, F. A. Cotton and R. M. Wing, *Inorg. Chem.* **1965**, *4*, 1328-1334.

167. New Rhenium(V) Oxyhalide Complexes, F. A. Cotton and S. J. Lippard, *Chem. Commun.* **1965**, 245-246.

168. An Authenticated Perchlorate Complex, F. A. Cotton and D. L. Weaver, *J. Amer. Chem. Soc.* **1965**, *87*, 4189.

169. The Molecular Structure of a Diamagnetic, Doubly Oxygen-Bridged, Binuclear Complex of Molybdenum(V) Containing a Metal–Metal Bond, F. A. Cotton and S. M. Morehouse, *Inorg. Chem.* **1965**, *4*, 1377-1381.

170. The Two-Electron Oxidation of Metal Atom Cluster Species of the Type [M$_6$X$_{12}$]$^{2+}$, R. E. McCarley, B. G. Hughes, F. A. Cotton and R. Zimmerman, *Inorg. Chem.* **1965**, *4*, 1491-1492.

171. Chemical and Structural Studies of Rhenium(V) Oxyhalide Complexes. I. Complexes from Rhenium(III) Bromide, F. A. Cotton and S. J. Lippard, *Inorg. Chem.* **1965**, *4*, 1621-1629.

172. Some Reactions of the Octahalodirhenium(III) Ions. I. Reactions with Phosphines, F. A. Cotton, N. F. Curtis and W. R. Robinson, *Inorg. Chem.* **1965**, *4*, 1696-1700.

173. Evidence for Multiple Bonding in Metal-to-Fluoroalkyl Bonds, F. A. Cotton and J. A. McCleverty, *J. Organometal. Chem.* **1965**, *4*, 490-491.

1966

174. Chemical and Structural Studies of Rhenium(V) Oxyhalide Complexes. II. M[ReX$_4$O] and M[ReX$_4$OL] Complexes from KReO$_4$, F. A. Cotton and S. J. Lippard, *Inorg. Chem.* **1966**, *5*, 9-16.

175. The Crystal and Molecular Structure of Tetragonal Ruthenium Dioxide, F. A. Cotton and J. T. Mague, *Inorg. Chem.* **1966**, *5*, 317-318.

176. Chemical and Structural Studies of Rhenium(V) Oxyhalide Complexes. III. The Crystal and Molecular Structure of Tetraphenylarsonium Oxotetrabromoacetonitrile-rhenate(V), F. A. Cotton and S. J. Lippard, *Inorg. Chem.* **1966**, *5*, 416-423.

177. The Crystal and Molecular Structure of Dimeric Bis(acetylacetonato)aquocobalt(II) and the Preparation of Some Other New Hydrates, F. A. Cotton and R. C. Elder, *Inorg. Chem.* **1966**, *5*, 423-429.

178. A Reply to the Note of Poilblanc and Bigorgne, Concerning Assignments of CO Fundamentals in $M(CO)_{6-n}L_n$ Molecules, F. A. Cotton, *J. Organometal. Chem.* **1966**, *5*, 293-294.

179. Cage Compounds Containing the Trirhenium(III) Cluster: $Re_3Br_3(AsO_4)_2(DMSO)_3$, F. A. Cotton and S. J. Lippard, *J. Amer. Chem. Soc.* **1966**, *88*, 1882-1885.

180. Metal Atom Cluster Compounds, F. A. Cotton, *McGraw-Hill Yearbook of Science and Technology*, McGraw Hill, N. Y. (1966).

181. A Solution to the Structural Dilemma of $Co_4(CO)_{12}$ Based on Valence Tautomerism and Steric Nonrigidity, F. A. Cotton, *Inorg. Chem.* **1966**, *5*, 1083-1085.

182. The Crystal and Molecular Structure of Bis(2,2,6,6-tetramethylheptane-3,5-dionato)-nickel(II), F. A. Cotton and J. J. Wise, *Inorg. Chem.* **1966**, *5*, 1200-1207.

183. The Crystal and Molecular Structure of Tetraphenylarsonium Tetranitratocobaltate(II); an Eight-Coordinate Cobalt(II) Complex, J. G. Bergman, Jr. and F. A. Cotton, *Inorg. Chem.* **1966**, *5*, 1208-1213.

184. The Electronic Structures of Bis(β-ketoenolato)copper(II) Complexes, F. A. Cotton and J. J. Wise, *J. Amer. Chem. Soc.* **1966**, *88*, 3451-3452.

185. The Preparation, Properties and Structure of Tetraphenylarsonium Tetrakis(trifluoroacetato)cobaltate(II), J. G. Bergman, Jr. and F. A. Cotton, *Inorg. Chem.* **1966**, *5*, 1420-1424.

186. Correlation of Nuclear Quadrupole Coupling Constants with Molecular Electronic Structure Using Molecular Orbital Theory, F. A. Cotton and C. B. Harris, *Proc. Nat. Acad. Sci.* **1966**, *56*, 12-15.

187. A Rhenium-to-Rhenium Triple Bond, M. J. Bennett, F. A. Cotton and R. A. Walton, *J. Amer. Chem. Soc.* **1966**, *88*, 3866-3867.

188. Structure and Bonding in Organo Derivatives of the Transition Elements, F. A. Cotton, *Proc. of The Robert A. Welch Foundation Conferences on Chemical Research IX, Organometallic Compounds*, Houston, Texas (1965).

189. Transition-Metal Compounds Containing Clusters of Metal Atoms, F. A. Cotton, *Quart Rev.* **1966**, *20*, 389-401.

190. Some Reactions of the Octahalodirhenate(III) Ions. II. Preparation and Properties of the Tetracarboxylato Compounds, F. A. Cotton, C. Oldham and W. R. Robinson, *Inorg. Chem.* **1966**, *5*, 1798-1802.

191. Reactions of Rhenium(III) Chloride with 2,2'-Bipyridyl, 1,10-Phenanthroline and Other Bidenate Donor Molecules, F. A. Cotton and R. A. Walton, *Inorg. Chem.* **1966**, *5*, 1802-1808.

192. Stereochemically Nonrigid Organometallic Compounds. I. π-Cyclopentadienyliron Dicarbonyl σ-Cyclopentadiene, M. J. Bennett, F. A. Cotton, A. Davison, J. W. Faller, S. J. Lippard and S. M. Morehouse, *J. Amer. Chem. Soc.* **1966**, *88*, 4371-4376.

193. Stereochemically Nonrigid Organometallic Compounds. II. 1,3,5,7-Tetramethylcyclooctatetraenemolybdenum Tricarbonyl, F. A. Cotton, J. W. Faller and A. Musco, *J. Amer. Chem. Soc.* **1966**, *88*, 4506-4507.

194. Stereochemically Nonrigid Organometallic Compounds. III. Concerning the Structure of Cyclooctatetraeneiron Tricarbonyl in Solution, F. A. Cotton, A. Davison and J. W. Faller, *J. Amer. Chem. Soc.* **1966**, *88*, 4507-4509.

195. Structures of Dimethyl Sulfoxide Complexes, M. J. Bennett, F. A. Cotton and D. L. Weaver, *Nature* **1966**, *212*, 286-287.

196. Crystalline Extracellular Nuclease of *Staphylococcus aureus*, F. A. Cotton, E. E. Hazen, Jr. and D. C. Richardson, *J. Biol. Chem.* **1966**, *241*, 4389-4390.

197. Molecular Structure and Bonding in Diethylenetriaminechromium Tricarbonyl: The Cr(0) and Approximate Mn(0) Radii, F. A. Cotton and D. C. Richardson, *Inorg. Chem.* **1966**, *5*, 1851-1854.

198. Nuclear Magnetic Resonance Studies of Eight-Coordinate β-Diketonate Complexes Related to Laser Chelates, F. A. Cotton, P. Legzdins and S. J. Lippard, *J. Chem. Phys.* **1966**, *45*, 3461-3462.

199. Semi-Empirical Calculation of the Electronic Structures of Complexes, F. A. Cotton, *Rev. Pure and Appl. Chem.* **1966**, *16*, 175-184.

200. Geometry of the $[Y(CF_3COCHCOCF_3)_4]^-$ Ion. A New Eight-Coordinate Stereoisomer in the Dodecahedral Class, S. J. Lippard, F. A. Cotton and P. Legzdins, *J. Amer. Chem. Soc.* **1966**, *88*, 5930-5931.

1967

201. Far-Infrared Spectra of Metal Atom Cluster Compounds. I. $Mo_6X_8^{4+}$ Derivatives, F. A. Cotton, R. M. Wing and R. A. Zimmerman, *Inorg. Chem.* **1967**, *6*, 11-15.

202. Tautomeric Changes in Metal Carbonyls. I. π-Cyclopentadienyliron Dicarbonyl Dimer and π-Cyclopentadienylruthenium Dicarbonyl Dimer, F. A. Cotton and G. Yagupsky, *Inorg. Chem.* **1967**, *6*, 15-20.

203. Stereochemically Nonrigid Organometallic Compounds. IV. Some Observations on π-Allylic Complexes, F. A. Cotton, J. W. Faller and A. Musco, *Inorg. Chem.* **1967**, *6*, 179-182.

204. A New Type of *bis*-(Acetylacetonato)metal(II) Polymer: The Trimer of *bis*-(Acetylacetonato)Zinc, M. J. Bennett, F. A. Cotton, R. Eiss and R. C. Elder, *Nature* **1967**, *213*, 174.

205. Some Reactions of the Octahalodirhenate(III) Ions. III. The Stability of the Rhenium–Rhenium Bond toward Oxygen and Sulfur Donors, F. A. Cotton, C. Oldham and R. A. Walton, *Inorg. Chem.* **1967**, *6*, 214-223.

206. The Stability and Reactivity of a New Form of Rhenium(IV) Chloride: Studies on Its Disproportionation in Solution, F. A. Cotton, W. R. Robinson and R. A. Walton, *Inorg. Chem.* **1967**, *6*, 223-228.

207. Transition Metal Complexes of the N-Cyanocarbimate Ion, $[S_2C_2N_2]^{2-}$, F. A. Cotton and J. A. McCleverty, *Inorg. Chem.* **1967**, *6*, 229-232.

208. Molecular Orbital Calculations for Complexes of Heavier Transition Elements. I. Study of Parameter Variations in the Case of Tetrachloroplatinate(II), F. A. Cotton and C. B. Harris, *Inorg. Chem.* **1967**, *6*, 369-376.

209. Molecular Orbital Calculations for Complexes of Heavier Transition Elements. II. Hexchloro Complexes of Tetravalent Rhenium, Osmium, Iridium and Platinum, F. A. Cotton and C. B. Harris, *Inorg. Chem.* **1967**, *6*, 376-379.

210. Structure and Bonding in Metal Carbonyls and Related Compounds, F. A. Cotton, *Helv. Chim. Acta, Fasc. Extraordinarius*, Alfred Werner Commemoration Volume, Verlag Helvetica Chimica Acta, Basel **1967**, 117-130.

211. Quadruple Bonds and Other Multiple Metal-to-Metal Bonds, F. A. Cotton, *Rev. Pure and Appl. Chem.* **1967**, *17*, 25-40.

212. Cycloheptatrienemolybdenum(0) Tricarbonyl (Preparation), F. A. Cotton, J. A. McCleverty and J. E. White, *Inorg. Synth.* **1967**, Vol. IX, McGraw-Hill, New York, 121-122.

213. Extended Hückel Calculations of the Molecular Orbitals in Bis(β-ketoenolate) Complexes of Copper(II) and Nickel(II), F. A. Cotton, C. B. Harris and J. J. Wise, *Inorg. Chem.* **1967**, *6*, 909-915.

214. The Electron Spin Resonance Spectrum of Bis(dipivaloylmethanido)copper(II), F. A. Cotton and J. J. Wise, *Inorg. Chem.* **1967**, *6*, 915-916.

215. Assignment of the Electronic Absorption Spectra of Bis(β-ketoenolate) Complexes of Copper(II) and Nickel(II), F. A. Cotton and J. J. Wise, *Inorg. Chem.* **1967**, *6*, 917-924.

216. Molecular Orbital Calculations for Complexes of Heavier Transition Elements. III. The Metal–Metal Bonding and Electronic Structure of $Re_2Cl_8^{2-}$, F. A. Cotton and C. B. Harris, *Inorg. Chem.* **1967**, *6*, 924-929.

217. Some Reactions of the Octahalodirhenate(III) Ions. IV. Reactions with Sodium Thiocyanate and the Preparation of Isothiocyanate Complexes of Rhenium(III) and Rhenium(IV), F. A. Cotton, W. R. Robinson, R. A. Walton and R. Whyman, *Inorg. Chem.* **1967**, *6*, 929-935.

218. Structures of Two Compounds Containing Strong Metal-to-Metal Bonds, M. J. Bennett, F. A. Cotton, B. M. Foxman and P. F. Stokely, *J. Amer. Chem. Soc.* **1967**, *89*, 2759-2760.

219. Some Reactions of the Octahalodirhenate(III) Ions. V. Polarographic Reduction, F. A. Cotton, W. R. Robinson and R. A. Walton, *Inorg. Chem.* **1967**, *6*, 1257-1258.

220. Triimerization of a *Bis*(β-ketophosphonyl) Metal Complex Involving Recyclizations, J. J. Bishop, F. A. Cotton, R. Eiss and R. P. Hugel, *Nature* **1967**, *214*, 111.

221. Tri(cyclopentadienylmanganese)tetranitrosyl. A Metal Cluster Compound with Doubly and Triply Bridging Nitrosyl Groups, R. C. Elder, F. A. Cotton and R. A. Schunn, *J. Amer. Chem. Soc.* **1967**, *89*, 3645-3646.

222. Some Reactions of the Octahalodirhenate(III) Ions. VI. Oxidation by Halogens to Give Nonahalodirhenate Species, F. Bonati and F. A. Cotton, *Inorg. Chem.* **1967**, *6*, 1353-1356.

223. Vibrational Spectra and Bonding in Metal Carbonyls. V. New Data for $XMn(CO)_5$ Molecules and Further Examination of Simplified Force Fields, F. A. Cotton, A. Musco and G. Yagupsky, *Inorg. Chem.* **1967**, *6*, 1357-1364.

224. Vibrational Spectra and Bonding in Metal Carbonyls. VI. Evidence for a π-Interaction Between Manganese Pentacarbonyl and the Perfluoromethyl Group, F. A. Cotton and R. M. Wing, *J. Organometal Chem.* **1967**, *9*, 511-517.

225. Compound Containing a Tetrahedral Cluster of Nickel Atoms, M. J. Bennett, F. A. Cotton and B. H. C. Winquist, *J. Amer. Chem. Soc.* **1967**, *89*, 5366-5372.

226. Stereochemically Nonrigid Organometallic Molecules. VII. Fluxional Behavior of Cyclooctatetraeneruthenium Tricarbonyl, W. K. Bratton, F. A. Cotton, A. Davison, A. Musco and J. W. Faller, *Proc. Nat. Acad. Sci.* **1967**, *58*, 1324-1328.

227. Octacarbonyldiiododiiron(I): Preparation and Properties, F. A. Cotton and B. F. G. Johnson, *Inorg. Chem.* **1967**, *6*, 2113-2115.

228. The Crystal and Molecular Structure of a Dimethyl Sulfoxide Complex of Iron(III) Chloride, M. J. Bennett, F. A. Cotton and D. L. Weaver, *Acta Cryst.* **1967**, *23*, 581-586.

229. Stereochemically Nonrigid Organometallic Molecules. VIII. Further Studies of σ-Cyclopentadienylmetal Compounds, F. A. Cotton, A. Musco and G. Yagupsky, *J. Amer. Chem. Soc.* **1967**, *89*, 6136-6139.230.

230. Stereochemically Nonrigid Organometallic Molecules. IX. Some Fluxional and Some Nonfluxional Compounds Derived from Cyclooctatetraene and Ruthenium Carbonyl, F. A. Cotton, A. Davison and A. Musco, *J. Amer. Chem. Soc.* **1967**, *89*, 6796-6797.

231. Stereochemically Nonrigid Organometallic Molecules. X. The Structure of Bis(cyclooctatetraene)triruthenium Tetracarbonyl, a Dissymmetric Metal Atom Cluster Compound Which Is Also a Fluxional Organometallic Molecule, M. J. Bennett, F. A. Cotton and P. Legzdins, *J. Amer. Chem. Soc.* **1967**, *89*, 6797-6798.

232. Two Independent Determinations of the Crystal and Molecular Structure of *trans*-Dichlorobis(dimethyl sulfoxide)palladium(II), M. J. Bennett, F. A. Cotton, D. L. Weaver, R. J. Williams and W. H. Watson, *Acta Cryst.* **1967**, *23*, 788-796.

1968

233. The Molecular Structure of a Trinuclear Cobalt(II) Complex of the Diethoxyphosphonylacetylmethane Anion, F. A. Cotton, R. Hügel and R. Eiss, *Inorg. Chem.* **1968**, *7*, 18-23.

234. The Structure of Hexa(acetylacetonato)aquotricobalt(II), F. A. Cotton and R. Eiss, *J. Amer. Chem. Soc.* **1968**, *90*, 38-46.

235. Stereochemically Nonrigid Organometallic Molecules. XI. The Molecular Structure, of (1,3,5,7-Tetramethylcyclooctatetraene)chromium Tricarbonyl, M. J. Bennett, F. A. Cotton and J. Takats, *J. Amer. Chem. Soc.* **1968**, *90*, 903-909.

236. Structural and Magnetic Study of Pentachloro-bis(1,5-dithiahexane)dirhenium, M. J. Bennett, F. A. Cotton and R. A. Walton, *Proc. Roy. Soc.* **1968**, *A303*, 175-192.

237. Stereochemically Nonrigid Organometallic Molecules. XII. The Temperature Dependence of the Proton Nuclear Magnetic Resonance Spectra of the 1,3,5,7-Tetramethylcyclooctatetraene Tricarbonyl Compounds of Chromium, Molybdenum and Tungsten, F. A. Cotton, J. W. Faller and A. Musco, *J. Amer. Chem. Soc.* **1968**, *90*, 1438-1444.

238. Stereochemically Nonrigid Organometallic Molecules. XIII. Preparation of 1,3,5,7-Tetramethylcyclooctatetraene Derivatives of the Iron Carbonyls, F. A. Cotton and A. Musco, *J. Amer. Chem. Soc.* **1968**, *90*, 1444-1448.

239. Stereochemically Nonrigid Organometallic Molecules. XIV. The Crystal and Molecular Structure of (1,3,5,7-Tetramethylcyclooctatetraene)diiron Pentacarbonyl, F. A. Cotton and M. D. LaPrade, *J. Amer. Chem. Soc.* **1968**, *90*, 2026-2031.

240. The Structure of (1,3,5-Trimethyl-7-methylene-1,3,5-cyclooctatriene)diiron Pentacarbonyl, an Anomalous Reaction Product of 1,3,5,7-Tetramethylcyclo-octatetraene with Polynuclear Iron Carbonyls, F. A. Cotton and J. Takats, *J. Amer. Chem. Soc.* **1968**, *90*, 2031-2036.

241. The Crystal and Molecular Structure of Nitropentamminocobalt(III) Bromide, F. A. Cotton and W. T. Edwards, *Acta Cryst.* **1968**, *B24*, 474-477.

242. The Crystal and Molecular Structure of Trirhenium Nonaiodide, M. J. Bennett, F. A. Cotton and B. M. Foxman, *Inorg. Chem.* **1968**, *7*, 1563-1569.

243. Some Reactions of the Octahalodirhenate(III) Ions. VII. Structural Characterization of Dichlorotetrabenzoatodirhenium(III), M. J. Bennett, W. K. Bratton, F. A. Cotton and W. R. Robinson, *Inorg. Chem.* **1968**, *7*, 1570-1575.

244. CO Force Constants and CO-CO Interaction Constants of Metal Carbonyls: A Reply to L. H. Jones, F. A. Cotton, *Inorg. Chem.* **1968**, *7*, 1683.

245. The Crystal and Molecular Structure of Trimeric Bis(acetylacetonato)zinc(II), M. J. Bennett, F. A. Cotton and R. Eiss, *Acta Cryst.* **1968**, *B24*, 904-913.

246. Fluxional Organometallic Molecules, F. A. Cotton, *Chem. Britain* **1968**, *4*, 345-347.

247. The Crystal Structure of Cesium Tetrakis(hexafluoroacetylacetonato)yttrate(III). A Novel Stereoisomer Having Dodecahedral Eight-Coordination, M. J. Bennett, F. A. Cotton, P. Legzdins and S. J. Lippard, *Inorg. Chem.* **1968**, *7*, 1770-1776.

248. An Example of the Monocapped Octahedral Form of Heptacoordination. The Crystal and Molecular Structure of Tris(1-phenyl-1,3-butanedionato)aquoyttrium(III), F. A. Cotton and P. Legzdins, *Inorg. Chem.* **1968**, *7*, 1777-1783.

249. The Structure of Oxopentachloropropionatobis(triphenylphosphine)dirhenium(IV), F. A. Cotton and B. M. Foxman, *Inorg. Chem.* **1968**, *7*, 1784-1792.

250. Stereochemically Nonrigid Organometallic Molecules. XV. The Structure of One of the Isomeric $C_8H_8Ru_2(CO)_6$ Molecules in the Crystalline Compound, F. A. Cotton and W. T. Edwards, *J. Amer. Chem. Soc.* **1968**, *90*, 5412-5417.

251. Stereochemically Nonrigid Organometallic Molecules. XVI. The Crystal and Molecular Structure of *p*-Methyl-π-benzyl-π-cyclopentadienyldicarbonylmolyb-denum, F. A. Cotton and M. D. LaPrade, *J. Amer. Chem. Soc.* **1968**, *90*, 5418-5422.

252. Some Reactions of the Octahalodirhenate(III) Ions. VIII. The Structure of a Phosphine-Substitution Product of Octachlorodirhenate(III), F. A. Cotton and B. M. Foxman, *Inorg. Chem.* **1968**, *7*, 2135-2140.

253. The Structure of Tetraphenylarsonium Bis(*N*-cyanodithiocarbimato)nickelate(II), F. A. Cotton and C. B. Harris, *Inorg. Chem.* **1968**, *7*, 2140-2144.

254. Fluxional Organometallic Molecules, F. A. Cotton, *Acc. Chem. Res.* **1968**, *1*, 257-265.

255. Proposed Nomenclature for Olefin-Metal and Other Organometallic Complexes, F. A. Cotton, *J. Amer. Chem. Soc.* **1968**, *90*, 6230-6232.

256. Stereochemically Nonrigid Organometallic Molecules. XVIII. A Compound Containing Three Nonequivalent Cyclopentadienyl Rings Which Are Rapidly Interconverting and Individually Rotating at Room Temperature, F. A. Cotton and P. Legzdins, *J. Amer. Chem. Soc.* **1968**, *90*, 6232-6233.

257. Stereochemically Nonrigid Organometallic Molecules. XVII. The Crystal and Molecular Structure of Bis(cyclooctatetraene)triruthenium Tetracarbonyl, M. J. Bennett, F. A. Cotton and P. Legzdins, *J. Amer. Chem. Soc.* **1968**, *90*, 6335-6340.

258. A Reinvestigation of the Crystal and Molecular Structure of Pentaphenylantimony, A. L. Beauchamp, M. J. Bennett and F. A. Cotton, *J. Amer. Chem. Soc.* **1968**, *90*, 6675-6680.

1969

259. Molecular Structure of Tetraphenylantimony Hydroxide, A. L. Beauchamp, M. J. Bennett and F. A. Cotton, *J. Amer. Chem. Soc.* **1969**, *91*, 297-301.

260. The Structure of Tetra-*n*-butyratodiruthenium Chloride, a Compound with a Strong Metal-Metal Bond, M. J. Bennett, K. G. Caulton and F. A. Cotton, *Inorg. Chem.* **1969**, *8*, 1-6.

261. The Octachlorodimolybdate(II) Ion. A New Species with a Quadruple Metal–Metal Bond, J. V. Brenčič and F. A. Cotton, *Inorg. Chem.* **1969**, *8*, 7-10.

262. The Crystal and Molecular Structure of (1,3,5-Cyclooctatriene)diiron Hexacarbonyl, F. A. Cotton and W. T. Edwards, *J. Amer. Chem. Soc.* **1969**, *91*, 843-847.

263. Stereochemically Nonrigid Organometallic Molecules. XIX. (1,2,3-*Trihapto*:4,5,6,7-*tetrahapto*cycloheptatrienyl)[(*pentahapto*cyclopentadienyl)dicarbonylmolybdenum] [tricarbonyliron]. Preparation, Structure and Temperature-Dependent Proton Magnetic Resonance Spectrum, F. A. Cotton and C. R. Reich, *J. Amer. Chem. Soc.* **1969**, *91*, 847-853.

264. The Structure of μ-Oxo-μ-chloro-di-μ-propionato-bis(chlorotriphenylphosphine-rhenium). A New Metal-Metal Bond, F. A. Cotton, R. Eiss and B. M. Foxman, *Inorg. Chem.* **1969**, *8*, 950-957.

265. Stereochemically Nonrigid Organometallic Molecules. XX. Proton Nuclear Magnetic Resonance Study of the Fluxional Behavior of Some Substituted (1,2,7-*Trihapto*benzyl)(*pentahapto*cyclopentadienyl)dicarbonyl Compounds of Molybdenum and Tungsten, F. A. Cotton and T. J. Marks, *J. Amer. Chem. Soc.* **1969**, *91*, 1339-1346.

266. The Preparation and Structural Characterization of Trirubidium Octachlorodi-molybdenum. A Binuclear Structure with Strong Metal–Metal Bonding, M. J. Bennett, J. V. Brenčič and F. A. Cotton, *Inorg. Chem.* **1969**, *8*, 1060-1065.

267. Stereochemically Nonrigid Organometallic Molecules. XXI. The Crystal and Molecular Structures of Tris(cyclopentadienyl)nitrosylmolybdenum, J. L. Calderon, F. A. Cotton and P. Legzdins, *J. Amer. Chem. Soc.* **1969**, *91*, 2528-2535.

268. Tetraiododibenzoatodirhenium(III), W. K. Bratton and F. A. Cotton, *Inorg. Chem.* **1969**, *8*, 1299-1304.

269. The Crystal and Molecular Structures of a Toluene-Containing Form of Di-μ-aquo-di-μ$_3$-hydroxo-deca(1,1,1-trifluoropentane-2,4-dionato)hexanickel(II), F. A. Cotton and B. H. C. Winquist, *Inorg. Chem.* **1969**, *8*, 1304-1312.

270. Strong Homonuclear Metal–Metal Bonds, F. A. Cotton, *Acc. Chem. Res.* **1969**, *2*, 240-247.

271. Stereochemically Nonrigid Organometallic Molecules. XXII. A Fluxional Indenyl-mercury Compound, F. A. Cotton and T. J. Marks, *J. Amer. Chem. Soc.* **1969**, *91*, 3178-3182.

272. Stereochemically Nonrigid Organometallic Molecules. XXV. The Low-Temperature NMR Spectrum of *cis*-(1,2,6-*trihapto*:-3,4,5-*trihapto*-1,3,5cyclooctatriene)hexacarbon-yldiiron, F. A. Cotton and T. J. Marks, *J. Organometal. Chem.* **1969**, *19*, 237-240.

273. Some Bonding Questions Prompted by Studies of the Fluxional Molecule Triscyclopentadienylnitrosylmolybdenum, F. A. Cotton, *Discussions of the Faraday Society* **1969**, *47*, 79-83.

274. A Rhodium(II)-to-Rhodium(II) Single Bond in Bis(triphenylphosphine)tetrakis(dimethylglyoximato)dirhodium, K. G. Caulton and F. A. Cotton, *J. Amer. Chem. Soc.* **1969**, *91*, 6517.

275. Stereochemically Nonrigid Organometallic Molecules. XXIII. The Crystal and Molecular Structures of (Cyclooctatetraene)tricarbonylruthenium, F. A. Cotton and R. Eiss, *J. Amer. Chem. Soc.* **1969**, *91*, 6593-6597.

276. Stereochemically Nonrigid Organometallic Molecules. XXIV. Preparation and Nuclear Magnetic Resonance Study of Some Fluxional Cyclooctatetraene Derivatives of Ruthenium Carbonyl, F. A. Cotton, A. Davison, T. J. Marks and A. Musco, *J. Amer. Chem. Soc.* **1969**, *91*, 6598-6603.

277. The Identification and Structural Characterization of $[enH_2]_2[Mo_2Cl_8]\cdot2H_2O$, a Compound Containing the Octachlorodimolybdate(II) Ion, J. V. Brenčič and F. A. Cotton, *Inorg. Chem.* **1969**, *8*, 2698-2703.

278. The Crystal and Molecular Structures of [*N,N*-Bis(2-diphenylphosphinoethyl)-ethylamine]tricarbonylchromium, F. A. Cotton and M. D. LaPrade, *J. Amer. Chem. Soc.* **1969**, *91*, 7000-7005.

279. The Extracellular Nuclease of *Staphylococcus Aureus*: Structures of the Native Enzyme and an Enzyme-Inhibitor Complex at 4 Å Resolution, A. Arnone, C. J. Bier, F. A. Cotton, E. E. Hazen, Jr., D. C. Richardson and J. S. Richardson, *Proc. Nat. Acad. Sci.* **1969**, *64*, 420-427.

280. New Evidence for the 1,2-Shift Pathway in Fluxional *Monohapto*cyclopentadienylmetal Compounds, F. A. Cotton and T. J. Marks, *J. Amer. Chem. Soc.* **1969**, *91*, 7523-7524.

281. An Infrared Study of the Structures of Cyclopentadienyl Compounds of Copper(I) and Mercury(II), F. A. Cotton and T. J. Marks, *J. Amer. Chem. Soc.* **1969**, *91*, 7281-7285.

282. Properties of Metal Ions: Physicochemical and Probe Properties, F. A. Cotton, *Proc. 4th Internatl. Congr. on Pharmacology*, Basel, Switzerland **1969**, 176-183.

1970

283. Structural Basis for the Acidity of Sulfonamides. Crystal Structures of Dibenzene-sulfonamide and Its Sodium Salt, F. A. Cotton and P. F. Stokely, *J. Amer. Chem. Soc.* **1970**, *92*, 294-302.

284. Stoichiometric and Structural Characterization of the Compound $(NH_4)_5Mo_2Cl_9\cdot H_2O$, J. V. Brenčič and F. A. Cotton, *Inorg. Chem.* **1970**, *9*, 346-351.

285. The Preparation of Some Compounds Containing Multiple Molybdenum–Molybdenum Bonds, J. V. Brenčič and F. A. Cotton, *Inorg. Chem.* **1970**, *9*, 351-353.

286. Hexakis(diethoxyphosphonylacetylmethano)trinickel, F. A. Cotton and B. H. C. Winquist, *Inorg. Chem.* **1970**, *9*, 688-690.

287. The Crystal and Molecular Structure of Bis(3,3',5,5'-tetramethyldipyrromethenato)-nickel(II), F. A. Cotton, B. G. DeBoer and J. R. Pipal, *Inorg. Chem.* **1970**, *9*, 783-788.

288. Characterization of the Trinegative Octachloroditechnetate Ion, W. K. Bratton and F. A. Cotton, *Inorg. Chem.* **1970**, *9*, 789-793.

289. Definitive Identification of the Structures of Dicyclopentadienyldiiron Tetracarbonyl in Solution, J. G. Bullitt, F. A. Cotton and T. J. Marks, *J. Amer. Chem. Soc.* **1970**, *92*, 2155-2156.

290. The Structure of Triphenylphosphine(*pentahapto*cyclopentadienyl)copper(I), F. A. Cotton and J. Takats, *J. Amer. Chem. Soc.* **1970**, *92*, 2353-2358.

291. Multiple Chromium(II)-Chromium(II) and Rhodium(II)-Rhodium(II) Bonds, F. A. Cotton, B. G. DeBoer, M. D. LaPrade, J. R. Pipal and D. A. Ucko, *J. Amer. Chem. Soc.* **1970**, *92*, 2926-2927.

292. Rapid Interchange of *Monohapto-* and *Pentahapto*cyclopentadienyl Rings in Tetracyclopentadienyltitanium, J. L. Calderon, F. A. Cotton, B. G. DeBoer and J. Takats, *J. Amer. Chem. Soc.* **1970**, *92*, 3801-3802.

293. A New Type of Fluxional Molecule. Bis-μ-dimethylgermyl-dicobalt Hexacarbonyl, R. D. Adams and F. A. Cotton, *J. Amer. Chem. Soc.* **1970**, *92*, 5003-5004.

294. Systematic Preparation and Characterization of *Pentahapto*cyclopentadienylcopper(I) Compounds, F. A. Cotton and T. J. Marks, *J. Amer. Chem. Soc.* **1970**, *92*, 5114-5117.

295. Some Reactions of the Octahalodirhenate(III) Ions. VIII. Definitive Structural Characterization of the Octabromodirhenate(III) Ion, F. A. Cotton, B. G. DeBoer and M. Jeremic, *Inorg. Chem.* **1970**, *9*, 2143-2146.

296. The Interpretation of a Spin-Tickling Experiment on (*Monohapto*cyclopentadienyl) (methyl)(dichloro)silane, F. A. Cotton and T. J. Marks, *Inorg. Chem.* **1970**, *9*, 2802-2804.

297. The Molecular Structure and Bonding in Hexamethyldialuminum, F. A. Cotton, *Inorg. Chem.* **1970**, *9*, 2804.

1971

298. Crystal and Molecular Structure of Chlorobis(dimethylglyoximato)triphenylphos-phinerhodium(III), F. A. Cotton and J. G. Norman, Jr., *J. Amer. Chem. Soc.* **1971**, *93*, 80-84.

299. High Resolution Structure of an Inhibitor Complex of the Extracellular Nuclease of *Staphylococcus aureus*. I. Experimental Procedures and Chain Tracing, A. Arnone, C. J. Bier, F. A. Cotton, V. W. Day, E. E. Hazen, Jr., D. C. Richardson, J. S. Richardson and A. Yonath, *J. Biol. Chem.* **1971**, *246*, 2302-2316.

300. Crystal and Molecular Structure of Bis(triphenylphosphine)tetrakis(dimethylgly-oximato)dirhodium. The Length of a Rhodium(II)-to-Rhodium(II) Single Bond, K. G. Caulton and F. A. Cotton, *J. Amer. Chem. Soc.* **1971**, *93*, 1914-1918.

301. Complexes of Cyclic 2-Oxacarbenes. I. A Spontaneous Cyclization to Form a Complex of 2-Oxacyclopentylidene, F. A. Cotton and C. M. Lukehart, *J. Amer. Chem. Soc.* **1971**, *93*, 2672-2676.

302. Staphylococcal Nuclease; X-Ray Structure, F. A. Cotton and E. E. Hazen, Jr., Chapter in *The Enzymes*, 3rd Ed., Vol. IV, P. D. Boyer, Ed., Academic Press, N. Y. **1971**, 153-175.

303. Stereochemically Nonrigid Organometallic Molecules. XXVII. The Fluxional Behavior of Tetra(cyclopentadienyl)titanium, J. L. Calderon, F. A. Cotton and J. Takats, *J. Amer. Chem. Soc.* **1971**, *93*, 3587-3591.

304. Stereochemically Nonrigid Organometallic Molecules. XXVIII. The Crystal and Molecular Structures of Tetra(cyclopentadienyl)titanium, J. L. Calderon, F. A. Cotton, B. G. DeBoer and J. Takats, *J. Amer. Chem. Soc.* **1971**, *93*, 3592-3597.

305. Stereochemically Nonrigid Organometallic Molecules. XXX. Observations on Some Cyclopentadienylmolybdenum Compounds, J. L. Calderon and F. A. Cotton, *J. Organometal. Chem.* **1971**, *30*, 377-380.

306. Tricarbonyl(cyclooctadiene) Complexes of Iron(0), Ruthenium(0) and Osmium(0), F. A. Cotton, A. J. Deeming, P. L. Josty, S. S. Ullah, A. J. P. Domingos, B. F. G. Johnson and J. Lewis, *J. Amer. Chem. Soc.* **1971**, *93*, 4624-4626.

307. Structural Characterization of a Novel Complex Derived from Tricarbonyl(cyclo-octadiene)ruthenium, F. A. Cotton, M. D. LaPrade, B. F. G. Johnson and J. Lewis, *J. Amer. Chem. Soc.* **1971**, *93*, 4626.

308. Some Observations on the Synthesis of Octamethylcyclotetraphosphazene and Hexamethylcyclotriphosphazene, F. A. Cotton and A. Shaver, *Inorg. Chem.* **1971**, *10*, 2362-2363.

309. Stereochemically Nonrigid Organometallic Molecules. XXIX. Cycloheptatrienediiron Hexacarbonyl, F. A. Cotton, B. G. DeBoer and T. J. Marks, *J. Amer. Chem. Soc.* **1971**, *93*, 5069-5075.

310. Vibrational Studies of Dinuclear Compounds Containing Quadruply Bonded Pairs of Molybdenum and Rhenium Atoms, W. K. Bratton, F. A. Cotton, M. DeBeau and R. A. Walton, *J. Coord. Chem.* **1971**, *1*, 121-131.

311. A New Preparation of Rhenium(IV) Chloride, (Mrs.) A. Brignole and F. A. Cotton, *Chem. Commun.* **1971**, 706.

312. Oxotriruthenium Cluster Complexes, F. A. Cotton, J. G. Norman, A. Spencer and G. Wilkinson, *Chem. Commun.* **1971**, 967-968.

313. The Crystal and Molecular Structures of Dichromium Tetra-Acetate Dihydrate and Dirhodium Tetra-Acetate Dihydrate, F. A. Cotton, B. G. DeBoer, M. D. LaPrade, J. R. Pipal and D. A. Ucko, *Acta Cryst.* **1971**, *B27*, 1664-1671.

314. A Crystallographic Study of the Metal-To-Metal Bond in Tetraallyldimolybdenum, F. A. Cotton and J. R. Pipal, *J. Amer. Chem. Soc.* **1971**, *93*, 5441-5445.

315. Preparation and Properties of Some Tungsten Carboxylates, F. A. Cotton and M. Jeremic, *Syn. Inorg. Metal-Org. Chem.* **1971**, *1*, 265-278.

316. The Structure of a Dinuclear Complex of Osmium(I) with a Metal-to-Metal Single Bond: Bis(μ-acetato)hexacarbonyldiosmium, J. G. Bullitt and F. A. Cotton, *Inorg. Chim. Acta* **1971**, *5*, 406-412.

317. Stereochemically Nonrigid Organometallic Molecules, XXXI. The Structures of (*Pentahapto*cyclopentadienyl)(*trihapto*cycloheptatrienyl)dicarbonylmolybdenum and its Tricarbonyliron Adduct, F. A. Cotton, D. G. DeBoer and M. D. LaPrade, *Proc. XXIIIrd International Congress of Pure and Applied Chemistry, 6*, Boston, Massachusetts **1971**, 1-30.

318. The Crystal and Molecular Structure of [1,2-Bis-(diphenylphosphino)ethane]-tetracarbonylchromium, M. J. Bennett, F. A. Cotton and M. D. LaPrade, *Acta Cryst.* **1971**, *B27*, 1899-1904.

319. Some Aspects of the Structure of Staphylococcal Nuclease, Part I. Crystallographic Studies, F. A. Cotton and E. E. Hazen, Jr., *Proc. Cold Spring Harbor Symposium, XXXVI, Structure and Function of Proteins at the Three Dimensional Level*, Cold Spring Harbor, NY **1971**, 243-255.

320. Molecular Structure of an Unusual Binuclear Manganese Complex with Highly Unsymmetrical Nitrosyl Bridges, J. L. Calderon, F. A. Cotton, B. G. DeBoer and (Miss) N. Martinez, *Chem. Commun.* **1971**, 1476-1477.

321. Application of a Dipole-Dipole Coupling Model to the Carbonyl Stretching Spectra of Some Dinuclear Metal Carbonyls, J. G. Bullitt and F. A. Cotton, *Inorg. Chim. Acta* **1971**, *5*, 637-642.

322. Molybdenum(II) Trifluoroacetate Dimer, F. A. Cotton and J. G. Norman, Jr., *J. Coord. Chem.* **1971**, *1*, 161-171.

1972

323. Highly Unsymmetrical Cyclopentadienylmetal Groups. The Structure of Bis(cyclo-pentadienyl)methylnitrosylmolybdenum, F. A. Cotton and G. A. Rusholme, *J. Amer. Chem. Soc.* **1972**, *94*, 402-406.

324. Cyclooctatriene–Bicyclooctadiene Equilibria. Effects of Additional Ring Fusions and Metal Binding, F. A. Cotton and G. Deganello, *J. Amer. Chem. Soc.* **1972**, *94*, 2142-2143.

325. Structural and Dynamic Properties of the *Pentahapto*cyclopentadienylmetal Dicarbonyl Dimers, J. G. Bullitt, F. A. Cotton and T. J. Marks, *Inorg. Chem.* **1972**, *11*, 671-676.

326. Complexes of Polypyrazolylborate Ligands. I. Conformational Isomerism in [Tetrakis(1-pyrazolyl)borate] (η^5-Cyclopentadienyl)dicarbonylmolybdenum, J. L. Calderon, F. A. Cotton and Alan Shaver, *J. Organometal. Chem.* **1972**, *37*, 127-136.

327. The Structure of Trimethylphenylammonium Nonachlorodirhodate(III) and a Survey of Metal–Metal Interactions in Confacial Biotahedra, F. A. Cotton and D. A. Ucko, *Inorg. Chim. Acta* **1972**, *6*, 161-172.

328. Complexes of Polypyrazolylborate Ligands. II. Conformational Isomerism in [Hydridotris(1-pyrazolyl)borate](η^5-cyclopentadienyl)dicarbonylmolybdenum and diethylbis(1-pyrazolyl)borate](η^5-cyclopentadienyl)dicarbonylmolybdenum, J. L. Calderon, F. A. Cotton and Alan Shaver, *J. Organometal. Chem.* **1972**, *38*, 105-156.

329. The Reaction of Diiron Nonacarbonyl with *cis*-bicyclo[6.2.0]deca-2,4,6-triene, F. A. Cotton and G. Deganello, *J. Organometal. Chem.* **1972**, *38*, 147-153.

330. Low-Valent Metal Isonitrile Complexes. I. The Structure of Di(μ-methylisonitrile) di-η^5-cyclopentadienyl)dinickel, R. D. Adams, F. A. Cotton and G. A. Rusholme, *J. Coord. Chem.* **1971**, *1*, 275-283.

331. The Structure of a Novel Complex Derived from Cyclooctadienetricarbonyl-ruthenium, F. A. Cotton and Marie D. LaPrade, *J. Organometal. Chem.* **1972**, *39*, 345-354.

332. Stereochemically Nonrigid Organometallic Molecules. XXXII. The Rearrangement of Trimethylgermyl-N-pyrazolides and N-imidazolides, F. A. Cotton and D. J. Ciappenelli, *Syn. Inorg. Metal-Org. Chem.* **1972**, *2*, 197-208.

333. Molybdenum(II) Trifluoroacetate Dimer. Bispyridine Adduct, F. A. Cotton and J. G. Norman, Jr., *J. Amer. Chem. Soc.* **1972**, *94*, 5697-5702.

334. Attainment of an Eighteen-Electron Configuration by Means of Hydrogen Bridging in a Purported Sixteen-Electron Pyrazolylborate Complex of Molybdenum; The Structure of (3,5-Dimethylpyrazolylborato)dicarbonyl-*trihapto*-cycloheptatrienyl)-molybdenum, F. A. Cotton, J. Calderon, M. Jeremic and A. Shaver, *J.C.S. Chem. Commun.* **1972**, 777-778.

335. Stereochemically Nonrigid Organometallic Molecules. XXXIII. Carbon-13 Nuclear Magnetic Resonance of (η^5-C$_5$H$_5$)(CO)$_2$(Fe(η^1-C$_5$H$_5$) from +52 to -88, D. J. Ciappenelli, F. A. Cotton and L. Kruczynski, *J. Organometal. Chem.* **1972**, *42*, 159-162.

336. New Evidence for Carbonyl- and Isonitrile-Bridged Transition States for Intramolecular Carbonyl Scrambling, R. D. Adams and F. A. Cotton, *J. Amer. Chem. Soc.* **1972**, *94*, 6193-6194.

337. Direct Evidence from Carbon-13 Nuclear Magnetic Resonance for Intramolecular Scrambling of Carbonyl Groups in a Metal Atom Cluster Carbonyl, Tetrarhodium Dodecacarbonyl, F. A. Cotton, L. Kruczynski, B. L. Shapiro and L. F. Johnson, *J. Amer. Chem. Soc.* **1972**, *94*, 6191-6193.

338. Complexes of Polypyrazolylborate Ligands. III. Structure of [Dihydrobis(3,5-dimethyl-1-pyrazolyl)borate](h^3-cycloheptatrienyl)dicarbonylmolybdenum in Solution, J. L. Calderon, F. A. Cotton and A. Shaver, *J. Organometal. Chem.* **1972**, *42*, 419-427.

339. Structural Characterization of a Basic Trinuclear Ruthenium Acetate, F. A. Cotton and J. G. Norman, Jr., *Inorg. Chim. Acta* **1972**, *6*, 411-419.

340. Low-Valent Metal Isonitrile Complexes. II. Preparation and Characterization of Some Di-(μ-alkylisonitrile)di(h^5-cyclopentadienyl)dinickel Compounds, R. D. Adams and F. A. Cotton, *Syn. Inorg. Metal-Org. Chem.* **1972**, *2*, 277-288.

341. Transition Metal Complexes Containing Carbenoid Ligands, F. A. Cotton and C. M. Lukehart, *Prog. Inorg. Chem.* **1972**, *16*, 487-613.

342. Complexes of Polypyrazolylborate Ligands. IV. Boron-Hydrogen-Molybdenum Bridge Bonding in the So-Called Sixteen Electron Complex [Dihydrobis-(3,5-dimethyl-1-pyrazolyl)borate](trihaptocycloheptatrienyl)dicarbonylmolybdenum in the Crystal, F. A. Cotton, M. Jeremic and A. Shaver, *Inorg. Chim. Acta* **1972**, *6*, 543-551.

1973

343. Effects of Additional Ring Fusions and Binding to Metal Atoms upon the Cyclooctatriene-Bicyclooctadiene Equilibrium, F. A. Cotton and G. Deganello, *J. Amer. Chem. Soc.* **1973**, *95*, 396-402.

344. The Molecular Aggregation of Anhydrous *bis*-(2,4-Pentanediono)-Nickel(II) In CCl$_4$ Solution, F. A. Cotton and C. M. Lukehart, *J. Coord. Chem.* **1973**, *2*, 243-244.

345. Structure of Dicyclopentadienylpentacarbonyldivanadium. A New Example of Grossly Unsymmetrical Carbonyl Bridging, F. A. Cotton, B. A. Frenz and L. Kruczynski, *J. Amer. Chem. Soc.* **1973**, *95*, 951-952.

346. An X-ray Structural Study of the Fluxional Molecule [Bis(Diphenylphosphino)-methane]tricarbonyliron, F. A. Cotton, K. I. Hardcastle and G. A. Rusholme, *J. Coord. Chem.* **1973**, *2*, 217-223.

347. The Chemistry of Rhenium(IV) Chloride. II. Structure of One of the Polymorphs (β) and Evidence for New Polymorph (γ), F. A. Cotton, B. G. DeBoer and Z. Mester, *J. Amer. Chem. Soc.* **1973**, *95*, 1159-1163.

348. Carbon-13 NMR Spectra of Chromium Pentacarbonyl Carbenoid Complexes, D. J. Ciappenelli, F. A. Cotton and L. Kruczynski, *J. Organometal. Chem.* **1973**, *50*, 171-174.

349. Fluxional Organometallic Molecules XXXIV. The Structure and Temperature-Dependent NMR Spectrum of *cis*-(1,2,6-*trihapto*-; 3,4,5-*trihapto*bicyclo[6.2.0]deca-1,3,5-triene)hexacarbonyldiiron(Fe-Fe), F. A. Cotton, B. A. Frenz, G. Deganello and A. Shaver, *J. Organometal. Chem.* **1973**, *50*, 227-240.

350. Biochemical Importance of the Binding of Phosphate by Arginyl Groups. Model Compounds Containing Methylguanidinium Ion, F. A. Cotton, E. E. Hazen, Jr., V. W. Day, S. Larsen, J. G. Norman, Jr., S. T. K. Wong and K. H. Johnson, *J. Amer. Chem. Soc.* **1973**, *95*, 2367-2369.

351. The Structure of a *Dihapto*(vinyl alcohol) Complex of Platinum(II), F. A. Cotton, J. N. Francis, B. A. Frenz and M. Tsutsui, *J. Amer. Chem. Soc.* **1973**, *95*, 2483-2486.

352. Compuestos Metal-ciclopentadienilo, F. A. Cotton, Anales de la Universidad Hispalense, Publicaciones de la Universidad de Sevilla **1973**, Serie: CIENCIAS — Num. *16*, 1-28.

353. Complexes of Cyclic 2-Oxacarbenes. II. Kinetics and Thermodynamics of Reactions Forming Complexes of Cyclic 2-Oxacarbenes, F. A. Cotton and C. M. Lukehart, *J. Amer. Chem. Soc.* **1973**, *95*, 3552-3564.

354. Structure of a Compound with a Molybdenum-to-Molybdenum Bond of Order Three and One-Half. F. A. Cotton, B. A. Frenz and T. R. Webb. *J. Amer. Chem. Soc.* **1973**, *95*, 4431-4432.

355. Further Exotic Products from the Reaction of Diiron Nonacarbonyl with Bicyclo[6.2.0]deca-2,4,6-triene. X-Ray Crystallography as a Practical Means of Cheap, Rapid, and Definitive Analysis, F. A. Cotton and J. M. Troup, *J. Amer. Chem. Soc.* **1973**, *95*, 3798-3799.

356. An Approach to Direct Evaluation of Pi Bonding in Metal Carbonyls, F. A. Cotton, W. T. Edwards, F. C. Rauch, M. A. Graham, R. N. Perutz and J. J. Turner, *J. Coord. Chem.* **1973**, *2*, 247-250.

357. Crystal and Molecular Structure of (Di-*tert*-butylstannylene)pyridinopentacarbonyl-chromium, M. D. Brice and F. A. Cotton, *J. Amer. Chem. Soc.* **1973**, *95*, 4529-4532.

358. Conformations of Fused Cycloalkanes in Organometallic Complexes. The Crystal and Molecular Structure of Tricyclo[6.3.0.0^{2,7}]undeca-3,5-dienetricarbonyliron, $(C_{11}H_{14})Fe(CO)_3$, F. A. Cotton, V. W. Day, B. A. Frenz, K. I. Hardcastle and J. M. Troup, *J. Amer. Chem. Soc.* **1973**, *95*, 4522-4528.

359. Structure of Methylguanidinium Dihydrogenorthophosphate. A Model Compound for Arginine-Phosphate Hydrogen Bonding, F. A. Cotton, V. W. Day, E. E. Hazen, Jr. and S. Larsen, *J. Amer. Chem. Soc.* **1973**, *95*, 4834-4840.

360. Structural and Dynamic Properties of Dicyclopentadienylhexacarbonyl-dimolybdenum in Various Solvents, R. D. Adams and F. A. Cotton, *Inorg. Chim. Acta* **1973**, *7*, 153-156.

361. Complexes of Polypyrazolylborate Ligands. V. Characterization of Acetyl(tri-1-pyrazolylborato)(dicarbonyl)iron, $(HBpz_3)(COCH_3)(CO)_2Fe$, in Solution and in the Crystal, F. A. Cotton, B. A. Frenz and A. Shaver, *Inorg. Chim. Acta* **1973**, *7*, 161-169.

362. X-Ray Structural and Raman Data, Including Remarkable Resonance Raman Effects, for Quadruple Molybdenum-to-Molybdenum Bonds, C. L. Angell, F. A. Cotton, B. A. Frenz and T. R. Webb, *J. Chem. Soc. Chem. Commun.* **1973**, 399-400.

363. Synthesis and Characterization of Octamethylcyclotetraphosphazenetricarbonylmolyb-denum, F. A. Cotton, G. A. Rusholme and A. Shaver. *J. Coord. Chem.* **1973**, *3*, 99-104.

364. On the Pathway of Bridge-Terminal Ligand Exchange in Some Binuclear Metal Carbonyls. Bis(*pentahapto*cyclopentadienyldicarbonyliron) and Its Di(methylisocyanide) Derivative and Bis(*pentahapto*cyclopentadienylcarbonylnitrosylmanganese), R. D. Adams and F. A. Cotton, *J. Amer. Chem. Soc.* **1973**, *95*, 6589-6594.

365. Intramolecular Ligand Scrambling *via* Bridged Transition States or Intermediates in Di(*pentahapto*cyclopentadienyl)(methylisocyanide)(pentacarbonyl)dimolybdenum. R. D. Adams, M. D. Brice and F. A. Cotton, *J. Amer. Chem. Soc.* **1973**, *95*, 6594-6602.

366. The Structure of Tricesium Octabromodimolybdate; An Example of a Space-Group Ambiguity, F. A. Cotton, B. A. Frenz and Z. C. Mester, *Acta Cryst.* **1973**, *B29*, 1515-1519.

367. Complexes of Polypyrazolylborate Ligands. VI. Some Complexes Containing Cycloheptatrienyl Groups and Their Tricarbonyliron Adducts. J. L. Calderon, F. A. Cotton and A. Shaver, *J. Organometal. Chem.* **1973**, *57*, 121-126.

368. Characterization of a Minor Product of the Reaction of Diiron Noncarbonyl with *cis*-Bicyclo[6.2.0]deca-2,4,6-triene: *trans*-di-μ-bis(2,3,4,5,6-*pentahapto*-tricyclo[6.2.0.0^{2,6}]deca-2,4-dien-6-yl)dicarbonyldiiron(Fe–Fe). F. A. Cotton, B. A. Frenz, J. M. Troup and G. Deganello, *J. Organometal. Chem.* **1973**, *59*, 317-327.

369. Métaux Carbonyles Stéréochimiquement non Rigides, F. A. Cotton, *Bull, Soc. Chim. France* **1973**, *9-10*, 2587-2592.

370. The Preparation and Structure of μ-Iodobis(η^5-cyclopentadienyldicarbonyliron) Fluoroborate, F. A. Cotton, B. A. Frenz and A. J. White, *J. Organometal. Chem.* **1973**, *60*, 147-152.

371. The Molecular Structure of a Product of the Reaction of Bicyclo[6.2.0]deca-2,4,6-triene with Diiron Nonacarbonyl, F. A. Cotton, B. A. Frenz and J. M. Troup, *J. Organometal. Chem.* **1973**, *61*, 337-346.

372. Structure and Dynamical Properties of $Rh_4(CO)_{12}$ in Solution: ^{13}C Nuclear Magnetic Resonance Study, J. Evans, B. F. G. Johnson, J. Lewis, J. R. Norton and F. A. Cotton, *J.C.S., Chem. Commun.* **1973**, 807.

373. Complexes of Polypyrazolylborate Ligands. VII. X-Ray Crystal Structure of a Complex Containing Both a Unidentate Pyrazole Ligand and a Bidentate Pyrazolylborate Ligand: Diethylbispyrazolylborato)(pyrazylato)(trihaptoallyl)-(dicarbonyl)molybdenum, F. A. Cotton, B. A. Frenz and A. G. Stanislowski, *Inorg. Chim. Acta.* **1973**, *7*, 503-508.

1974

374. Low-Valent Metal Isocyanide Complexes. III. Inversion at the Nitrogen Atoms of Bridging Isocyanide Ligands, R. D. Adams and F. A. Cotton, *Inorg. Chem.* **1974**, *13*, 249-253.

375. Low-Valent Metal Isocyanide Complexes. IV. Crystal and Molecular Structures of *cis-anti*-Bis(*pentahapto*cyclopentadienyl)dicarbonylbis(μ-methyl isocyanide)-diiron(Fe–Fe), (η^5-C$_5$H$_5$)$_2$Fe$_2$(CO)$_2$(μ-CNCH$_3$)$_2$, F. A. Cotton and B. A. Frenz, *Inorg. Chem.* **1974**, *13*, 253-256.

376. Low-Valent Metal Isocyanide Complexes. V. Structure and Dynamical Stereochemistry of Bis(*pentahapto*cyclopentadienyl)tricarbonyl(*tert*-butyl isocyanide)-diiron(Fe–Fe), (η^5-C$_5$H$_5$)$_2$Fe$_2$(CO)$_3$[CNC(CH$_3$)$_3$], R. D. Adams, F. A. Cotton and J. M. Troup, *Inorg. Chem.* **1974**, *13*, 257-262.

377. Unusual Structural and Magnetic Resonance Properties of Dicyclopentadienylhexa-carbonyldichromium, R. D. Adams, D. M. Collins and F. A. Cotton, *J. Amer. Chem. Soc.* **1974**, *96*, 749-754.

378. The Interaction of an Aliphatic Carbon-Hydrogen Bond with a Metal Atom. The Structure of (Diethyldi-1-pyrazolylborato)(*trihapto*-2-phenylallyl)(dicarbonyl)-molybdenum, F. A. Cotton, T. LaCour and A. G. Stanislowski, *J. Amer. Chem. Soc.* **1974**, *96*, 754-760.

379. A 2,2'-Bipyridine Derivative of Diiron Nonacarbonyl, $Fe_2(CO)_7$dipy. Some Observations on the Nature and Function of Semibridging Carbonyl Groups, F. A. Cotton and J. M. Troup, *J. Amer. Chem. Soc.* **1974**, *96*, 1233-1234.

380. A New Class of Binuclear Rhenium(II) Halide Species Containing a Strong Metal-Metal Bond: The Chemistry and Structure of Complexes of the Type $Re_2X_4(PR_3)_4$, F. A. Cotton, B. A. Frenz, J. R. Ebner and R. A. Walton, *J.C.S., Chem. Commun.* **1974**, 4-5.

381. Structural and Dynamical Properties of μ-Dimethylgermyl-μ-carbonyldicyclopenta-dienyldicarbonyldiiron, R. D. Adams, M. D. Brice and F. A. Cotton, *Inorg. Chem.* **1974,** *13*, 1080-1085.

382. Molecular Structures and Barriers to Internal Rotation in Bis(η^5-cyclopentadienyl)-hexacarbonylditungsten and Its Molybdenum Analog, R. D. Adams, D. M. Collins and F. A. Cotton, *Inorg. Chem.* **1974**, *13*, 1086-1090.

383. Accurate Determination of a Classic Structure in the Metal Carbonyl Field: Nonacarbonyldi-iron, F. A. Cotton and J. M. Troup, *J.C.S., Dalton* **1974**, 800-802.

384. Reactivity of Diiron Nonacarbonyl in Tetrahydrofuran. I. The Isolation and Characterization of Pyridinetetracarbonyliron and Pyrazinetetracarbonyliron, F. A. Cotton and J. M. Troup, *J. Amer. Chem. Soc.* **1974**, *96*, 3438-3443.

385. Interaction of an Aliphatic C-H Bond with a Metal Atom Can Compete Successfully with a Metal-Olefin Interaction, F. A. Cotton and V. W. Day, *J.C.S. Chem. Commun.* **1974**, 415-416.

386. The Preparation, Chemistry and Structure of the Lithium Salt of the Octamethyldimolybdate(II) Ion, F. A. Cotton, J. M. Troup, T. R. Webb, D. H. Williamson and G. Wilkinson, *J. Amer. Chem. Soc.* **1974**, *96*, 3824-3828.

387. Spectroscopic Studies of the Structural and Dynamical Properties of Bis(*pentahapto*-cyclopentadienyl)tricarbonyl(triphenylphosphito)diiron in Solution, F. A. Cotton, L. Kruczynski and A. J. White, *Inorg. Chem.* **1974**, *13*, 1402-1407.

388. Crystal and Molecular Structures of *cis*-Bis(*pentahapto*cyclopentadienyl)-tricarbonyl(triphenylphosphito)diiron, F. A. Cotton, B. A. Frenz and A. J. White, *Inorg. Chem.* **1974**, *13*, 1407-1411.

389. Further Refinement of the Molecular Structure of Triiron Dodecacarbonyl, F. A. Cotton and J. M. Troup, *J. Amer. Chem. Soc.* **1974**, *96*, 4155-4159.

390. The Preparation, Properties and Crystal Structure of Bis[(η^5-cyclopentadienyldi-carbonyliron)dimethylgermyl]oxide, [(η^5-C$_5$H$_5$)(CO)$_2$FeGe(CH$_3$)$_2$]$_2$O, R. D. Adams, F. A. Cotton and B. A. Frenz, *J. Organometal. Chem.* **1974**, *73*, 93-101.

391. X-Ray Crystal Structure of Hexakis(dimethylamino)dimolybdenum. A New Compound with a Metal-to-Metal Triple Bond, M. H. Chisholm, F. A. Cotton, B. A. Frenz and L. Shive, *J.C.S. Chem. Commun.* **1974**, 480-481.

392. Conformations of Cyclobutane, F. A. Cotton and B. A. Frenz, *Tetrahedron* **1974**, *30*, 1587-1594.

393. Preparation and Structure of the Fluxional Molecule (Tetraphenyldiphosphino-methane)heptacarbonyldiiron, F. A. Cotton and J. M. Troup, *J. Amer. Chem. Soc.* **1974**, *96*, 4422-4427.

394. Structure of Bis(methylguanidinium) Monohydrogen Orthophosphate. A Model for the Arginine-Phosphate Interactions at the Active Site of Staphylococcal Nuclease and Other Phosphohydrolytic Enzymes, F. A. Cotton, V. W. Day, E. E. Hazen, Jr., S. Larsen and S. T. K. Wong, *J. Amer. Chem. Soc.* **1974**, *96*, 4471-4478.

395. Mechanism of Fluxional Rearrangement in Cyclooctatetraenetricarbonyl-molybdenum. A Random Process *via* a Symmetrical Transition State, F. A. Cotton, D. L. Hunter and P. Lahuerta, *J. Amer. Chem. Soc.* **1974**, *96*, 4723-4724.

396. Structure of (2,4-Pentanedionato)(triphenylphosphine)ethylnickel(II) in the Crystalline State and in Solution, F. A. Cotton, B. A. Frenz and D. L. Hunter, *J. Amer. Chem. Soc.* **1974**, *96*, 4820-4825.

397. Structure and Bonding in Molecular Oxo-molybdenum Compounds, F. A. Cotton, *J. Less-Common Metals* **1974**, *36*, 13-22. Previously published in *Proceedings of the Climax First International Conference on the Chemistry and Uses of Molybdenum*, University of Reading, England, Sept. **1973**, P.C.H. Mitchell, Editor, page 6; Climax Molybdenum Company, Ltd., London.

398. The Crystal and Molecular Structure of a Complex Formed by Guanidine Hydrochloride and Glycylglycine, [C(NH$_2$)$_3$]Cl·C$_4$N$_2$O$_3$H$_8$, F. A. Cotton, T. LaCour, E. E. Hazen, Jr. and M. Legg, *Biochem. Biophys. Acta* **1974**, *359*, 7-12.

399. The Infrared Spectra of some Mono- and Binuclear Oxo-complexes of Molybdenum(IV), -(V) and -(VI); The Assignment of (LL)$_4$Mo$_2$O$_3$ Spectra, F. A. Cotton, D. L. Hunter, L. Ricard and R. Weiss, *J. Coord. Chem.* **1974**, *3*, 259-261.

400. A New Derivative of Triiron Dodecacarbonyl with Borderline Carbonyl Bridges, F. A. Cotton and J. M. Troup, *J. Amer. Chem. Soc.* **1974**, *96*, 5070-5073.

401. Rates or Rearrangement of Pyrazolylborate Complexes of Molybdenum which Contain Strong Aliphatic Hydrogen to Molybdenum Interactions. Estimates of the Strength of the Interaction, F. A. Cotton and A. G. Stanislowski, *J. Amer. Chem. Soc.* **1974**, *96*, 5074-5082.

402. Transition Elements and Their Compounds, F. A. Cotton, Encyclopaedia Britannica **1974**.

403. Conditions for the Occurrence of Bridging Carbonyl Groups, F. A. Cotton and D. L. Hunter, *Inorg. Chem.* **1974**, *13*, 2044.

404. A Novel New Metallocycle: Cyclohexa-1,4-(tetracarbonylchromium)-2,3,5,6-(dimethylarsenic), F. A. Cotton and T. R. Webb, *Inorg. Chim. Acta* **1974**, *10*, 127-131.

405. The Structure of a Tricarbonyliron Derivative of Barbaralone, F. A. Cotton and J. M. Troup, *J. Organometal. Chem.* **1974**, *76*, 81-88.

406. The Crystal and Molecular Structure of the Bis(tricarbonyliron) Derivative of 3,3'-Bis(bicyclo[4.2.0]octa-2,4-diene), F. A. Cotton and J. M. Troup, *J. Organometal. Chem.* **1974**, *77*, 83-39.

407. Conformations of Fused Cycloalkanes in Organometallic Complexes. II. The Structure of Bis(tricyclo3.0.0^{2,7}]undeca-3,5-diene)dicarbonyl molybdenum, $(C_{11}H_{14})_2Mo(CO)_2$, F. A. Cotton and B. A. Frenz, *Acta Cryst.* **1974**, *B30*, 1772-1776.

408. Conformations of Fused Cycloalkanes in Organometallic Compounds. III. The Crystal and Molecular Structures of Tricyclo[6.2.0.0^{2,7}]deca-3,5dienetricarbonyliron, $(C_{10}H_{12})Fe(CO)_3$, F. A. Cotton and J. M. Troup, *J. Organometal. Chem.* **1974**, *77*, 369-379.

409. X-Ray Structure of the Dimer of Bis(tetramethyldiphosphinoethane)ruthenium: Intermolecular Oxidative Addition of a C-H Bond to a Metal Atom, F. A. Cotton, B. A. Frenz and D. L. Hunter, *J.C.S., Chem. Commun.* **1974**, 755-756.

410. A Structural Explanation for the Infrared and Carbon-13 NMR Spectra of Triiron Dodecacarbonyl, F. A. Cotton and D. L. Hunter, *Inorg. Chim. Acta* **1974**, *11*, L9-L10.

411. Copper(I) and Copper(II) Complexes of Tetramethyldiphosphinedisulfide. I. Structural Characterization of the Dinuclear, Molecular Complex of Copper(I) Chloride, F. A. Cotton, B. A. Frenz, D. L. Hunter and Z. C. Mester, *Inorg. Chim. Acta* **1974**, *11*, 111-117.

412. Copper(I) and Copper(II) Complexes of Tetramethyldiphosphinedisulfide. II. Isolation and Characterization of a Polymeric Copper(II) Precursor, $(Me_4P_2S_2)CuCl_2$, to the Ultimate Copper(I) Product, $[(Me_4P_2S_2)CuCl]_2$, F. A. Cotton, B. A. Frenz, D. L. Hunter and Z. C. Mester, *Inorg. Chim. Acta* **1974**, *11*, 119-122.

413. Dimolybdenum Tetraacetate, F. A. Cotton, Z. C. Mester and T. R. Webb, *Acta Cryst.* **1974**, *30B*, 2768-2770.

414. Low-Valent Metal Isonitrile Complexes. VI. Syntheses and Mass Spectra of Methylisonitrile Substituted Derivatives of Tetracarbonyldicyclopentadienyldiiron, R. D. Adams and F. A. Cotton, *Syn. React. Inorg. Metal-Org. Chem.* **1974**, *4*, 477-489.

415. Carbon-13 Nuclear Magnetic Resonance Study of the Fluxional Behavior of Cyclooctatetraenetricarbonylchromium, -molybdenum and -tungsten and Tetramethylcyclooctatetraenetricarbonylchromium, F. A. Cotton, D. L. Hunter and P. Lahuerta, *J. Amer. Chem. Soc.* **1974**, *96*, 7926-7930.

416. The Structure and Photoelectron Spectra of Hexa(methylimido)tetraphosphorus. Evidence for Three-dimensional $p\pi$-$d\pi$ Delocalization, F. A. Cotton, J. M. Troup, F. Casabianca and J. G. Riess, *Inorg. Chim. Acta* **1974**, *11*, L33-L34.

1975

417. Reactivity of Diiron Nonacarbonyl in Tetrahydrofuran. III. A New Iron Carbonyl Complex of Acenaphthylene, $(C_{12}H_8)Fe(CO)_4$, F. A. Cotton and P. Lahuerta, *Inorg. Chem.* **1975**, *14*, 116-119.

418. Magnetic and Electrochemical Properties of Transition Metal Complexes with Multiple Metal-to-Metal Bonds. I. $[Tc_2Cl_8]^{n-}$ and $[Re_2Cl_8]^{n-}$ with $n = 2$ and 3, F. A. Cotton and E. Pedersen, *Inorg. Chem.* **1975**, *14*, 383-387.

419. Magnetic and Electrochemical Properties of Transition Metal Complexes with Multiple Metal-to-Metal Bonds. II. $[Ru_2(C_3H_7COO)_4]^{n+}$ with $n = 0$ and 1, F. A. Cotton and E. Pedersen, *Inorg. Chem.* **1975**, *14*, 388-391.

420. Magnetic and Electrochemical Properties of Transition Metal Complexes with Multiple Metal-to-Metal Bonds. III. Characterization of Tetrapotassium and Tripotassium Tetrasulfatodimolybdates, F. A. Cotton, B. A. Frenz, E. Pedersen and T. R. Webb, *Inorg. Chem.* **1975**, *14*, 391-398.

421. Magnetic and Electrochemical Properties of Transition Metal Complexes with Multiple Metal-to-Metal Bonds. IV. $[Mo_2Cl_8]^{n-}$ with $n = 3$ and 4 and $[Mo_2(C_3H_7COO)_4]^{n+}$ with $n = 0$ and 1, F. A. Cotton and E. Pedersen, *Inorg. Chem.* **1975**, *14*, 399-400.

422. Localized Scrambling of Carbonyl Ligands in Compounds with Metal-to-Metal Bonds, F. A. Cotton, D. L. Hunter and P. Lahuerta, *Inorg. Chem.* **1975**, *14*, 511-514.

423. Magnetic and Electrochemical Properties of Transition Metal Complexes with Multiple Metal-to-Metal Bonds. V. $[Re_2(C_6H_5COO)_4]^{n+}$ with $n = 1$ and 2 and $[Re_2Cl_4\{P(C_2H_5)_3\}_4]^{n+}$ with $n = 0$, 1 and 2, F. A. Cotton and E. Pedersen, *J. Amer. Chem. Soc.* **1975**, *97*, 303-308.

424. The Structure of Potassium Hydrogen Sulfate, F. A. Cotton, B. A. Frenz and D. L. Hunter, *Acta Cryst.* **1975**, *B31*, 302-304.

425. Synthesis and Structural Characterization of Sodium Tetra-μ-sulfato-dirhenate(III) Octahydrate, F. A. Cotton, B. A. Frenz and L. W. Shive, *Inorg. Chem.* **1975**, *14*, 649-652.

426. Effect of Tris(2,4-pentanedionato)chromium(III) on Carbon-13 Line Widths. Potential Dangers in Line Shape Analysis, F. A. Cotton, D. L. Hunter and A. J. White, *Inorg. Chem.* **1975**, *14*, 703-704.

427. A Fluxional Organometal Carbonyl Molecule with Several Scrambling Processes. Carbon-13 Nuclear Magnetic Resonance Study of (η^6-Cyclooctatriene)(hexacarbonyl)-diiron(Fe-Fe), F. A. Cotton, D. L. Hunter and P. Lahuerta, *J. Amer. Chem. Soc.* **1975**, *97*, 1046-1050.

428. Hexakis(dimethylamido)ditungsten. The First Structurally Characterized Molecule with an Unbridged Triple Bond Between Tungsten Atoms, F. A. Cotton, B. R. Stults, J. M. Troup, M. H. Chisholm and M. Extine, *J. Amer. Chem. Soc.* **1975**, *97*, 1242-1243.

429. A Remarkable Intramolecular Difference in Local Scrambling Rates for Two Mo(CO)₃ Groups and Absence of Metal-to-Metal Scrambling, F. A. Cotton, D. L. Hunter and P. Lahuerta, *J. Organometal. Chem.* **1975**, *87*, C42-C44.

430. A Bispyrazolylboratoallyldicarbonylmolybdenum Complex with a 16-Electron Configuration, F. A. Cotton, B. A. Frenz and C. A. Murillo, *J. Amer. Chem. Soc.* **1975**, *97*, 2118-2122.

431. Stereochemical Nonrigidity in Organometallic Compounds, F. A. Cotton, in "Dynamic Nuclear Magnetic Resonance Spectroscopy," **1975**, Chapter 10, pp. 377-440.

432. Stereochemical Nonrigidity in Metal Carbonyl Compounds, R. D. Adams and F. A. Cotton, in "Dynamic Nuclear Magnetic Resonance Spectroscopy," **1975**, Chapter 12, pp. 489-522.

433. Centenary Lecture. Quadruple Bonds and other Multiple Metal to Metal Bonds, F. A. Cotton, *Chem. Soc. Rev.* **1975**, *4*, 27-53.

434. Tumor-Inhibiting Properties of the Neutral *P-O*-Ethyl Ester of Adenosine 3':5'-Monophosphate in Correlation with Its Crystal and Molecular Structure, F. A. Cotton, R. G. Gillen, R. N. Gohil, E. E. Hazen, Jr., C. R. Kirchner, J. Nagyvary, J. P. Rouse, A. G. Stanislowski, J. D. Stevens and P. W. Tucker, *Proc. Nat. Acad. Sci. USA* **1975**, *72*, 1335-1339.

435. Fluxional Behavior of Some Dinuclear Iron and Cobalt Hexacarbonyl Compounds with Alkylsulfur and Dialkylphosphorus, -arsenic, -germanium and -tin Bridges, R. D. Adams, F. A. Cotton, W. R. Cullen, D. L. Hunter and L. Mihichuk, *Inorg. Chem.* **1975**, *14*, 1395-1399.

436. Conformations of Fused Cycloalkanes in Organometallic Complexes. IV. The Crystal and Molecular Structure of Tricyclo[6.4.0.0²,⁷]dodeca-3,5-dienetricarbonyl-iron, F. A. Cotton, V. W. Day and K. I. Hardcastle, *J. Organometal. Chem.* **1975**, *92*, 369-379.

437. The Structures of Two Reaction Products of Ditertiarybutylacetylene With Diiron Nonacarbonyl. A New Iron-Iron Double Bond, F. A. Cotton, J. D. Jamerson and B. R. Stults, *J. Organometal. Chem.* **1975**, *94*, C53-C55.

438. Quadruply Bonded Pairs of Metal Atoms Bridged by Amidines. I. Preparation and Structure of Tetrakis(*N*,*N*'-diphenylbenzamidinato)dimolybdenum(II), F. A. Cotton, T. Inglis, M. Kilner and T. R. Webb, *Inorg. Chem.* **1975**, *14*, 2023-2026.

439. Quadruply Bonded Pairs of Metal Atoms Bridged by Amidines. II. Preparation and Structures of Two Dirhenium Compounds, Including Shortest Re–Re Bond Yet Observed, F. A. Cotton and L. W. Shive, *Inorg. Chem.* **1975**, *14*, 2027-2031.

440. Structure of Tripotassium Octachloroditechnetate Hydrate, F. A. Cotton and L. W. Shive, *Inorg. Chem.* **1975**, *14*, 2032-2035.

441. Fluxionality in Organometallics and Metal Carbonyls, F. A. Cotton, *J. Organometal. Chem.* **1975**, *100*, 29-41.

442. The Molecular Structure of the Dimer of Bis(tetramethyldiphosphinoethane)-ruthenium that Results from Paired Oxidative Additions of C–H Bonds to Two Metal Atoms, F. A. Cotton, D. L. Hunter and B. A. Frenz, *Inorg. Chim. Acta* **1975**, *15*, 155-160.

443. A Carbon-13 Nuclear Magnetic Resonance Study of the Fluxional Character of the η^6-(Bicyclo[6.2.0]dodeca-2,4,6-triene)hexacarbonyldiiron(Fe–Fe) and a Triethylphosphine Derivative Thereof, and the Crystal Structure of the Latter, F. A. Cotton and D. L. Hunter, *J. Amer. Chem. Soc.* **1975**, *97*, 5739-5746.

444. Structure of Tetraphenylarsonium Tetrachloroferrate(III), F. A. Cotton and C. A. Murillo, *Inorg. Chem.* **1975**, *14*, 2467-2469.

445. A Carbon-13 NMR Study of the Rearrangement Pathway of the Fluxional Molecule Cyclopentadienylmercury Chloride and Some Related Compounds, F. A. Cotton, D. L. Hunter and J. D. Jamerson, *Inorg. Chim. Acta* **1975**, *15*, 245-247.

446. Reactivity of Diiron Nonacarbonyl in Tetrahydrofuran. IV. The Synthesis and Crystal and Molecular Structure of 1,3-Dithianeirontetracarbonyl, (*cyclo*-1,3-$C_4H_8S_2$)Fe(CO)$_4$, F. A. Cotton, J. R. Kolb and B. R. Stults, *Inorg. Chim. Acta* **1975**, *15*, 239-244.

447. A Refinement of the Crystal Structure of Tetraphenylmethane: Three Independent Redeterminations, A. Robbins, G. A. Jeffrey, J. P. Chesick, J. Donohue, F. A. Cotton, B. A. Frenz and C. A. Murillo, *Acta Cryst.* **1975**, *B31*, 2395-2399.

448. I: The Nucleotide Binding Site of Staphylococcal Nuclease and II: A New Approach to the Refinement of the Crystal Structures of Biological Macromolecules, D. M. Collins, F. A. Cotton, E. E. Hazen, Jr. and M. J. Legg, in "Structure and Conformation of Nucleic Acids and Protein-Nucleic Acid Interactions," pp. 317-331 **1975**.

449. Protein Crystal Structures: Quicker, Cheaper Approaches, D. M. Collins, F. A. Cotton, E. E. Hazen, Jr., E. F. Meyer, Jr., C. N. Morimoto, *Science* **1975**, *190*, 1047-1053.

450. Synthesis and Structure Proof of a New Ring System from the Reaction of Diironnonacarbonyl and Naphtho[*b*]cyclopropene, F. A. Cotton, J. M. Troup, W. E. Billups, L. P. Lin and C. V. Smith, *J. Organometal. Chem.* **1975**, *102*, 345-351.

1976

451. The Structure of Bis(diphenyldipyrazolylborato)nickel(II), F. A. Cotton and C. A. Murillo, *Inorg. Chim. Acta* **1976**, *17*, 121-124.

452. Metal Carbonyls: Some New Observations in an Old Field, F. A. Cotton, *Prog. Inorg. Chem.* **1976**, *21*, 2-28.

453. The Mechanism of Formation and Crystal Structure of *trans*-dichlorotetrakis(dimethylphosphine)ruthenium(II), RuCl$_2$[PH(CH$_3$)$_2$]$_4$, F. A. Cotton, B. A. Frenz and D. L. Hunter, *Inorg. Chim. Acta* **1976**, *16*, 203-207.

454. Structural and Carbon-13 Nuclear Resonance Studies of a Compound Containing the $Fe_2(CO)_6$ Entity. Hexacarbonyl[μ-(1,2,5-η:1,4,5-η)-3-oxo-1,2,4,5-tetraphenyl-1,4-pentadiene-1,5-diyl]diiron(*Fe–Fe*), F. A. Cotton, D. L. Hunter and J. M. Troup, *Inorg. Chem.* **1976**, *15*, 63-67.

455. Tetrakis(glycine)dimolybdenum Disulfate Tetrahydrate, F. A. Cotton and T. R. Webb, *Inorg. Chem.* **1976**, *15*, 68-71.

456. Preparation and Structure of a New Derivative of Tetrarhodium Dodecacarbonyl. Further Refinement of the Structure of Tetracobalt Dodecacarbonyl, F. H. Carre, F. A. Cotton and B. A. Frenz, *Inorg. Chem.* **1976**, *15*, 380-387.

457. Unusual Complexes formed by Reaction of Diiron Nonacarbonyl with 1-ene-3-yne Molecules, F. A. Cotton, J. D. Jamerson and B. R. Stults, *Inorg. Chim. Acta* **1976**, *17*, 235-242.

458. Further Evidence for the "Piano Stool" Transition State in η^6-Cyclooctatetraenemetal Fluxional Molecules, F. A. Cotton and J. R. Kolb, *J. Organometal. Chem.* **1976**, *107*, 113-119.

459. Tris(η^5-cyclopentadienyl)tricarbonyltricobalt. An Exceptionally Deformable Molecule, F. A. Cotton and J. D. Jamerson, *J. Amer. Chem. Soc.* **1976**, *98*, 1273-1274.

460. Oxidative Addition of Hydrohalic Acids to Dimolybdenum(II) Species. Reformulation of $Mo_2X_8^{3-}$ as $Mo_2X_8H^{3-}$, F. A. Cotton and B. J. Kalbacher, *Inorg. Chem.* **1976**, *15*, 522-524.

461. Structural and Dynamical Properties of Dicyclopentadienyltetracarbonyldiiron Derivatives with Linked Rings, F. A. Cotton, D. L. Hunter, P. Lahuerta and A. J. White, *Inorg. Chem.* **1976**, *15*, 557-564.

462. Carbon-13 Nuclear Magnetic Resonance Study of the Fluxional Behavior of Cyclooctatetraenetricarbonyliron and -ruthenium, F. A. Cotton and D. L. Hunter, *J. Amer. Chem. Soc.* **1976**, *98*, 1413-1417.

463. Calculation of the Ground State Electronic Structures and Electronic Spectra of Di- and Trisulfide Radical Anions by the Scattered Wave-SCF-Xα Method, F. A. Cotton, J. B. Harmon and R. M. Hedges, *J. Amer. Chem. Soc.* **1976**, *98*, 1417-1424.

464. Metal-Metal Multiple Bonds in Organometallic Compounds. I. (Di-*tert*-butylacetylene)hexacarbonyldiiron and -dicobalt, F. A. Cotton, J. D. Jamerson and B. R. Stults, *J. Amer. Chem. Soc.* **1976**, *98*, 1774-1779.

465. Investigations of Quadruple Bonds by Polarized Crystal Spectra. I. The Interpretation of the Spectrum of Tetra(*n*-butylammonium) Octachlorodirhenate. The Disordered Crystal Structure, F. A. Cotton, B. A. Frenz, B. R. Stults and T. R. Webb, *J. Amer. Chem. Soc.* **1976**, *98*, 2768-2773.

466. Investigations of Quadruple Bonds by Polarized Crystal Spectra. 2. Quadruply Bonded Tetra-μ-glycine-dimolybdenum(II) Sulfate Tetrahydrate, F. A. Cotton, D. S. Martin, T. R. Webb and T. J. Peters, *Inorg. Chem.* **1976**, *15*, 1199-1201.

467. Structure and Bonding of a Compound Containing Two Quadruply Bonded Rhenium(II) Atoms. Solution of an Unusual Form of Crystallographic Disorder, F. A. Cotton, B. A. Frenz, J. R. Ebner and R. A. Walton, *Inorg. Chem.* **1976**, *15*, 1630-1633.

468. The Molybdenum-Molybdenum Triple Bond. 1. Hexakis(dimethylamido)dimolybdenum and Some Homologues: Preparation, Structure, and Properties, M. H. Chisholm, F. A. Cotton, B. A. Frenz, W. W. Reichert, L. W. Shive and B. R. Stults, *J. Amer. Chem. Soc.* **1976**, *98*, 4469-4476.

469. The Tungsten-Tungsten Triple Bond. 1. Preparation, Properties, and Structural Characterization of Hexakis(dimethylamido)ditungsten(III) and Some Homologues, M. H. Chisholm, F. A. Cotton, M. Extine and B. R. Stults, *J. Amer. Chem. Soc.* **1976**, *98*, 4477-4485.

470. The Tungsten-Tungsten Triple Bond. 2. Preparation, Structure, and Dynamical Behavior of Dichlorotetrakis(diethylamido)ditungsten, M. H. Chisholm, F. A. Cotton, M. Extine, M. Millar and B. R. Stults, *J. Amer. Chem. Soc.* **1976**, *98*, 4486-4491.

471. Electronic Spectra of Crystal of Dimolybdenum Tetraformate and the Tetrasulfa-todimolybdenum(II) Ion, F. A. Cotton, D. S. Martin, P. E. Fanwick, T. J. Peters, T. R. Webb, *J. Amer. Chem. Soc.* **1976**, *98*, 4681-4682.

472. Insertion Reactions of Carbon Dioxide with Triply Bonded Ditungsten Dialkylamido Compounds. Structures and Dynamic Properties of the Resulting Carbamato Complexes, M. H. Chisholm, M. Extine, F. A. Cotton, B. R. Stults, *J. Amer. Chem. Soc.* **1976**, *98*, 4683-4684.

473. Bonding Implications of the Resistance of Dimolybdenum(II) to Attach Axial Ligands. Structures of Two Polyprazolylborate Complexes, D. M. Collins, F. A. Cotton and C. A. Murillo, *Inorg. Chem.* **1976**, *15*, 1861-1866.

474. The Scrambling of Carbonyl Groups in Guaiazulenehexacarbonyldimolybdenum and Two Isomeric Triethylphosphine Substitution Products, F. A. Cotton, P. Lahuerta and B. R. Stults, *Inorg. Chem.* **1976**, *15*, 1866-1871.

475. Preparation and Structure of a Multiply Bonded Re_2^{5+} Species Bridged by Two Diphosphinomethane Ligands, F. A. Cotton, L. W. Shive and B. R. Stults, *Inorg. Chem.* **1976**, *15*, 2239-2244.

476. A New Structural Type for $M_3(CO)_{12}$ Molecules and Their Derivatives. The Molecular Structure of $Ru_3(CO)_{10}$(1,2-diazine), F. A. Cotton and J. D. Jamerson, *J. Amer. Chem. Soc.* **1976**, *98*, 5396-5397.

477. The Tungsten–Tungsten Triple Bond. 3. Dimethyltetrakis(diethylamido)ditungsten. Structure and Dynamical Solution Behavior, M. H. Chisholm, F. A. Cotton, M. Extine, M. Millar and B. R. Stults, *Inorg. Chem.* **1976**, *15*, 2244-2252.

478. The Tungsten–Tungsten Triple Bond. 4. Structural Characterization of Hexakis(tri-methylsilylmethyl)ditungsten and Preparation of Bis-μ-(trimethylsilylmethylidyne)-tetrakis(trimethylsilylmethyl)ditungsten, M. H. Chisholm, F. A. Cotton, M. Extine and B. R. Stults, *Inorg. Chem.* **1976**, *15*, 2252-2257.

479. The Preparation and Crystal Structure of Dimolybdenum Tetraformate; Photoelectron Spectra of this and Several Other Dimolybdenum Tetracarboxylates, F. A. Cotton, J. G. Norman, Jr., B. R. Stults and T. R. Webb, *J. Coord. Chem.* **1976**, *5*, 217-223.

480. Carbonyl Scrambling in Azulenehexacarbonyldimolybdenum and Its Tungsten Analogue with Guaiazulene. Structure of Guaiazulenehexacarbonylditungsten, F. A. Cotton and B. E. Hanson, *Inorg. Chem.* **1976**, *15*, 2806-2809.

481. Structure of Tetrakis(benzoato)dimolybdenum(II), D. M. Collins, F. A. Cotton and C. A. Murillo, *Inorg. Chem.* **1976**, *15*, 2950-2951.

482. Synthesis and Structure of Dilithium Octamethyldirhenate(III), F. A. Cotton, L. D. Gage, K. Mertis, L. W. Shive and G. Wilkinson, *J. Amer. Chem. Soc.* **1976**, *98*, 6922-6926.

483. Unexpectedly Complex Structure and Dynamic Behavior of Bis(η^5-cyclopentadienyl)-dicarbonyl Molybdenum-μ-hex-3-yne (Mo–Mo), W. I. Bailey, Jr., F. A. Cotton, J. D. Jamerson and J. R. Kolb, *J. Organometal. Chem.* **1976**, *121*, C23-C26.

1977

484. Localized and Internuclear Carbonyl Scrambling in Azulenepentacarbonyl-diiron(Fe–Fe) and Its Ruthenium Analogue, F. A. Cotton, B. E. Hanson, J. R. Kolb and P. Lahuerta, *Inorg. Chem.* **1977**, *16*, 89-92.

485. The Tungsten–Tungsten Triple Bond. 5. Chlorine Atom Substitution Reactions Involving Dichlorotetrakis(diethylamido)ditungsten. Preparation, Properties, Structures, and Dynamical Solution Behavior of Bis(trimethylsilylmethyl)-, Dibromo-, and Diiodotetrakis(diethylamido)ditungsten, M. H. Chisholm, F. A. Cotton, M. W. Extine, M. Millar and B. R. Stults, *Inorg. Chem.* **1977**, *16*, 320-328.

486. Crystallographic Proof of the Stability of a Quadruple Bond Between Tungsten Atoms, D. M. Collins, F. A. Cotton, S. Koch, M. Millar and C. A. Murillo, *J. Amer. Chem. Soc.* **1977**, *99*, 1259-1261.

487. μ-Allene-bis(cyclopentadienyl)tetracarbonyldimolybdenum; a Bridging Allene Ligand, M. H. Chisholm, L. A. Rankel, W. I. Bailey, Jr., F. A. Cotton and C. A. Murillo, *J. Amer. Chem. Soc.* **1977**, *99*, 1261-1262.

488. Hexakis(neopentoxy)dimolybdenum. Preparation, Characterization and Reactions with Lewis Bases and Carbon Dioxide, M. H. Chisholm, W. W. Reichert, F. A. Cotton and C. A. Murillo, *J. Amer. Chem. Soc.* **1977**, *99*, 1652-1654.

489. The Tungsten–Tungsten Triple Bond. 6. Hexakis(*N,N*-dimethylcarbamato)-ditungsten and Dimethyltetrakis(*N,N*-diethylcarbamato)ditungsten. Structures and Dynamical Solution Behavior, M. H. Chisholm, F. A. Cotton, M. W. Extine and B. R. Stults, *Inorg. Chem.* **1977**, *16*, 603-611.

490. The Role of Arginine Residues at Enzyme Active Sites. The Interaction Between Guanidinium Ions and *p*-Nitrophenyl Phosphate and its Effect on the Rate of Hydrolysis of the Ester, F. A. Cotton, T. La Cour, E. E. Hazen, Jr. and M. J. Legg, *Biochim. Biophys. Acta* **1977**, *481*, 1-5.

491. Preparation of Several Polypyrazolylborato Compounds. Structures of Two Phenyltrispyrazolylborato Complexes, F. A. Cotton, C. A. Murillo and B. R. Stults, *Inorg. Chim. Acta* **1977**, *22*, 75-80.

492. Further Studies of the Electronic Spectra of $Re_2Cl_8^{2-}$ and $Re_2Br_8^{2-}$. Assignment of the Weak Bands in the 600-350-nm Region. Estimation of the Dissociation Energies of Metal-Metal Quadruple Bonds, W. C. Trogler, C. D. Cowman, H. B. Gray and F. A. Cotton, *J. Amer. Chem. Soc.* **1977**, *99*, 2993-2996.

493. A New Mode of Carbonyl Scrambling. Structure and Dynamics of (1,2-Diazine)-heptacarbonyldiiron(Fe-Fe), F. A. Cotton, B. E. Hanson, J. D. Jamerson and B. R. Stults, *J. Amer. Chem. Soc.* **1977**, *99*, 3293-3297.

494. The Carbonyl Scrambling Processes in the Isomeric Pentacarbonylguaiazulenediiron and Homologous Ruthenium Molecules; a Novel Mechanism for the Internuclear Processes, F. A. Cotton, B. E. Hanson, J. R. Kolb, P. Lahuerta, G. G. Stanley, B. R. Stults and A. J. White, *J. Amer. Chem. Soc.* **1977**, *99*, 3673-3683.

495. The Molybdenum-Molybdenum Triple Bond. 2. Hexakis(alkoxy)dimolybdenum Compounds: Preparation, Properties, and Structural Characterization of $Mo_2(CCH_2CMe_3)_6$, M. H. Chisholm, F. A. Cotton, C. A. Murillo and W. W. Reichert, *Inorg. Chem.* **1977**, *16*, 1801-1808.

496. Crystal and Molecular Structure of Anhydrous Tetraacetatodichromium, F. A. Cotton, C. E. Rice and G. W. Rice, *J. Amer. Chem. Soc.* **1977**, *99*, 4704-4707.

497. The Crystal and Molecular Structure of Tetrameric Bis (2,4-Pentanedionato) Iron(II): An Unusual Iron-Carbon Interaction, F. A. Cotton and G. W. Rice, *Nouv. J. Chim.* **1977**, *1*, 301-305.

498. Preparation and Characterization of Compounds Containing the Octamethyl-ditungsten(II) Anion and Partially Chlorinated Analogues, F. A. Cotton, S. Koch, K. Mertis, M. Millar and G. Wilkinson, *J. Am. Chem. Soc.* **1977**, *99*, 4989-4992.

499. Molybdenum–Molybdenum Bonds, F. A. Cotton, *J. Less-Common Metals* **1977**, *54*, 3-12.

500. Local and Internuclear Carbonyl Scrambling in Pentacarbonyl-7*H*-indenediiron, $(C_9H_8)Fe_2(CO)_5$, and Tetracarbonyl-7H-indene(triethylphosphine)diiron, $(C_9H_8)Fe_2(CO)_4(PEt_3)$, F. A. Cotton and B. E. Hanson, *Inorg. Chem.* **1977**, *16*, 1861-1865.

501. Structure of the High-Temperature Form of Osmium(IV) Chloride, F. A. Cotton and C. E. Rice, *Inorg. Chem.* **1977**, *16*, 1865-1867.

502. Cesium Octachlorodirhenate(III) Hydrate. The Correct Structure and Its
     Significance with Respect to the Nature of Metal-Metal Multiple Bonding,
     F. A. Cotton and W. T. Hall, *Inorg. Chem.* **1977**, *16*, 1867-1871.

503. Crystal Absorption Spectra for Potassium Octachlorodimolybdate(II) Dihydrate,
     P. E. Fanwick, D. S. Martin, F. A. Cotton and T. R. Webb, *Inorg. Chem.* **1977**, *16*,
     2103-2106.

504. The Molecular Structure and Dynamic NMR Spectrum of bis[η^5-cyclopentadienyl)-
     dicarbonylmolybdenum]-μ-ethyne(Mo–Mo), W. I. Bailey, Jr., D. M. Collins and
     F. A. Cotton, *J. Organometal. Chem.* **1977**, *135*, C53-C56.

505. Spectroscopic Study of the $Tc_2Cl_8^{3-}$ Ion, F. A. Cotton, P. E. Fanwick, L. D. Gage,
     B. Kalbacher and D. S. Martin, *J. Am. Chem. Soc.* **1977**, *99*, 5642-5645.

506. Detailed Electronic Description of Triple Bonds Between Transition Metal Atoms and
     Verification by Photoelectron Spectroscopy, F. A. Cotton, G. G. Stanley,
     B. J. Kalbacher, J. C. Green, E. Seddon and M. H. Chisholm, *Proc. Natl. Acad. Sci.
     USA* **1977**, *74*, 3109-3113.

507. Self-Consistent Field Xα Scattered-Wave Treatment of the Electronic Structures of
     Octachlorodimetalate Anions of Technetium and Tungsten, F. A. Cotton and
     B. J. Kalbacher, *Inorg. Chem.* **1977**, *16*, 2386-2396.

508. The Tungsten-to-Tungsten Triple Bond. 7. Replacement of Dimethylamido by Chloro
     Groups in Hexakis(dimethylamido)ditungsten and -dimolybdenum Compounds.
     Preparation, Properties, and Structure of Dichlorotetrakis(dimethylamido)ditungsten
     and -dimolybdenum, M. Akiyama, M. H. Chisholm, F. A. Cotton, M. W. Extine and
     C. A. Murillo, *Inorg. Chem.* **1977**, *16*, 2407-2411.

509. The Crystal and Molecular Structures of Bis(2,4-pentanedionato)chromium,
     F. A. Cotton, C. E. Rice and G. W. Rice, *Inorg. Chim. Acta* **1977**, *24*, 231-234.

510. Structure and Stereodynamic Behavior of (1,2-Diazine)decacarbonyl
     Triangulotriruthenium. Evidence for Hidden Processes in Fluxional Molecules,
     F. A. Cotton, B. E. Hanson and J. D. Jamerson, *J. Amer. Chem. Soc.* **1977**, *99*,
     6588-6594.

511. Existence of Direct Metal-to-Metal Bonds in Dichromium Tetracarboxylates,
     F. A. Cotton and G. G. Stanley, *Inorg. Chem.* **1977**, *16*, 2688-2671.

512. A Quadruple Bond Between Tungsten Atoms in an Air-Stable Compound,
     F. A. Cotton and S. A. Koch, *J. Am. Chem. Soc.* **1977**, *99*, 7371-7372.

513. Exceedingly Short Metal-to-Metal Multiple Bonds, F. A. Cotton, S. Koch and
     M. Millar, *J. Am. Chem. Soc.* **1977**, *99*, 7372-7374.

514. Preparation and Fluxional Behavior of (1,2-Diazine)decacarbonyl-*triangulo*-trios-
     mium, F. A. Cotton and B. E. Hanson, *Inorg. Chem.* **1977**, *16*, 2820-2822.

515. The Preparation and Molecular Structure of Tetrapivalatodichloroditechnetate (III)
     and its Dirhenium Congener: A Tc-Tc Quadruple Bond, F. A. Cotton and
     L. D. Gage, *Nouv. J. Chim.* **1977**, *1*, 441-442.

516. The Probable Existence of a Triple Bond Between Two Vanadium Atoms, F. A. Cotton and M. Millar, *J. Am. Chem. Soc.* **1977**, *99*, 7886-7891.

517. Preparation of Decacarbonyl[bis(diphenylphosphino)methane]triruthenium and the Elucidation of Its Structure by Dynamic Carbon-13 NMR Spectroscopy, F. A. Cotton and B. E. Hanson, *Inorg. Chem.* **1977**, *16*, 3369-3371.

518. Tetrakis(2,4,6-trimethoxyphenyl)dichromium. A Homologous New Compound with an Exceedingly Short Bond, F. A. Cotton and M. Millar, *Inorg. Chim. Acta* **1977**, *25*, L105-L106.

519. Observations on the Stereodynamic Behavior of Mono- and Polynuclear Iron Carbonyl Compounds Including a Survey of $M(CO)_3$ Scrambling Processes, F. A. Cotton and B. E. Hanson, *Israel J. Chem.* **1977**, *15*, 165-173.

520. The Chemical and Structural Characteristics of the *Closo*-phosphorimides and their Derivatives, J. G. Riess, M. Postel, F. Jeanneaux, F. A. Cotton, B. R. Stults and C. E. Rice, 1st International Congress on Phosphorus Compounds **1977**, 51-82, October 17-21.

1978

521. Insensitivity of the Molybdenum-to-Molybdenum Quadruple Bond in the Dimolybdenum Tetracarboxylates to Axial Coordination and Changes in Inductive Effects, F. A. Cotton, M. W. Extine and L. D. Gage, *Inorg. Chem.* **1978**, *17*, 172-176.

522. Sensitivity of the Chromium-Chromium Quadruple Bond in Dichromium Tetracarboxylates to Axial Coordination and Changes in Inductive Effects, F. A. Cotton, M. W. Extine and G. W. Rice, *Inorg. Chem.* **1978**, *17*, 176-188.

523. Preparation and Structure of a Quadruply Bonded Dimolybdenum Compound Containing the 7-Azaindolyl Ligand, F. A. Cotton, D. G. Lay and M. Millar, *Inorg. Chem.* **1978**, *17*, 186-188.

524. The Molybdenum–Molybdenum Triple Bond. 3. A Triple Bond Between Two Four-Coordinated Molybdenum(III) Atoms. Structural Characterization of the Bis(dimethylamine) Adduct of Dimolybdenum Hexatrimethylsiloxide, M. H. Chisholm, F. A. Cotton, M. W. Extine and W. W. Reichert, *J. Am. Chem. Soc.* **1978**, *100*, 153-157.

525. Reactions of Metal-to-Metal Multiple Bonds. 1. µ-Allenebis(cyclopentadienyl)-tetracarbonyldimolybdenum and -ditungsten Compounds. Preparation Properties, and Structural Characterization, W. I. Bailey, Jr., M. H. Chisholm, F. A. Cotton, C. A. Murillo and L. A. Rankel, *J. Am. Chem. Soc.* **1978**, *100*, 802-807.

526. Reactions of Metal-to-Metal Multiple Bonds. 2. Reactions of Bis(cyclopentadienyl)-tetracarbonyldimolybdenum with Small Unsaturated Molecules. Structural Characterization of µ-Dimethylaminocyanamide-bis(cyclopentadienyl)tetracarbonyl-dimolybdenum, M. H. Chisholm, F. A. Cotton, M. W. Extine and L. A. Rankel, *J. Am. Chem. Soc.* **1978**, *100*, 807-811.

527. Crystal and Molecular Structure of Tris[tetra-μ-formato-diaquodichromium(II)] Decahydrate: A Case of an Unusually Good False Minimum in a Structure Solution, F. A. Cotton and G. W. Rice, *Inorg. Chem.* **1978**, *17*, 688-692.

528. Reactions and Reaction Products of Bromine and Iodine with Tetrakis(ethylxanthato)-dimolybdenum (Mo–Mo), F. A. Cotton, M. W. Extine and R. H. Niswander, *Inorg. Chem.* **1978**, *17*, 692-696.

529. Detailed Structure of Bis(μ-trimethylsilylmethylidyne)tetrakis(trimethylsilylmethyl)-ditungsten(W-W), M. H. Chisholm, F. A. Cotton, M. W. Extine and C. A. Murillo, *Inorg. Chem.* **1978**, *17*, 696-698.

530. The Molybdenum–Molybdenum Triple Bond. 4. Insertion Reactions of Hexa-kis(alkoxy)dimolybdenum Compounds with Carbon Dioxide and Single Crystal X-Ray Structural Characterization of $Mo_2(O_2COBu\text{-}t)_2(OBu\text{-}t)_4$, M. H. Chisholm, F. A. Cotton, M. W. Extine and W. W. Reichert, *J. Amer. Chem. Soc.* **1978**, *100*, 1727-1734.

531. An Unusual, Trinuclear, Mixed-Valence Reaction Product of (Dicyclopentadienyl-chromium with Trifluoroacetic Acid: Bis(cyclopentadienyl)hexakis(trifluoroacetato)-trichromium, F. A. Cotton and G. W. Rice, *Inorg. Chim. Acta* **1978**, *27*, 75-78.

532. Reactions of Triply Bonded Dimetal Compounds. Reversible Addition of Carbon Monoxide to a Hexakis(alkoxy)dimolybdenum Compound. A Molecule with a Carbonyl-Bridged Metal-Metal Double Bond, M. H. Chisholm, R. L. Kelly, F. A. Cotton and M. W. Extine, *J. Am. Chem. Soc.* **1978**, *100*, 2256-2257.

533. Molecular and Electronic Structure of Tetrakis(dimethylamido)molybdenum(IV), M. H. Chisholm, F. A. Cotton and M. W. Extine, *Inorg. Chem.* **1978**, *17*, 1329-1332.

534. $La_8Ru_4O_{21}$: A Mixed Valence Ternary Ruthenium Oxide of a New Hexagonal Structure Type, F. A. Cotton and C. E. Rice, *J. Solid State Chem.* **1978**, *24*, 359-365.

535. The Crystal Structure of $La_3Ru_3O_{11}$: A New Cubic $KSbO_3$ Derivative Oxide with No Metal-Metal Bonding, F. A. Cotton and C. E. Rice, *J. Solid State Chem.* **1978**, *25*, 137-142.

536. Reactions of Metal-to-Metal Multiple Bonds. 3. Addition of Nitric Oxide to Hexakis(alkoxy)dimolybdenum Compounds. Preparation and Properties of Bis(nitrosyl)hexakis(alkoxy)dimolybdenum Compounds and Structural Characterization of the Isopropoxy Derivative, M. H. Chisholm, F. A. Cotton, M. W. Extine and R. L. Kelley, *J. Am. Chem. Soc.* **1978**, *100*, 3354-3358.

537. Molecular and Electronic Structure of Tetraallyldirhenium, F. A. Cotton and M. W. Extine, *J. Am. Chem. Soc.* **1978**, *100*, 3788-3792.

538. The Electronic Structures of the Tetra(μ-Sulfato)Dimolybdenum Ions, T. F. Block, R. F. Fenske, D. L. Lichtenberger, F. A. Cotton, *J. Coord. Chem.* **1978**, *8*, 109-112.

539. Bis(dimethylamido)tris(*N,N*-dimethylcarbamato)tantalum(V). Structure and Dynamical Solution Behavior of a Compound Containing Seven-Coordinate Tantalum, M. H. Chisholm, F. A. Cotton and M. W. Extine, *Inorg. Chem.* **1978**, *17*, 2000-2003.

540. Further Studies of Cr-Cr Bond Lengths in Quadruply Bonded Dinuclear Compounds, F. A. Cotton and G. W. Rice, *Inorg. Chem.* **1978**, *17*, 2004-2009.

541. Structural and Dynamic Study of Bis(tetraethyl pyrophosphite) pentacarbonyl-diiron(Fe–Fe) and (Tetraethylpyrophosphite)heptacarbonyldiiron(Fe–Fe), F. A. Cotton, R. J. Haines, B. E. Hanson and J. C. Sekutowski, *Inorg. Chem.* **1978**, *17*, 2010-2014.

542. Exceedingly Short Metal-Metal Bond in a Bis(*o*-alkoxyphenyl) dicarboxylatodi-chromium Compound, F. A. Cotton and M. Millar, *Inorg. Chem.* **1978**, *17*, 2014-2017.

543. X-ray Crystallographic Structural Studies of the Quadruply Bonded Octamethyl-ditungsten Anion and a Related Species Containing a Mixture of Chloro and Methyl Ligands, D. M. Collins, F. A. Cotton, S. A. Koch, M. Millar and C. A. Murillo, *Inorg. Chem.* **1978**, *17*, 2017-2020.

544. Rational Preparation and Structural Study of a Dichromium *o*-Oxyphenyl Compound: The Shortest Metal-to-Metal Bond Yet Observed, F. A. Cotton and S. Koch, *Inorg. Chem.* **1978**, *17*, 2021-2024.

545. Discovering and Understanding Multiple Metal-to-Metal Bonds, F. A. Cotton, *Acc. Chem. Res.* **1978**, *11*, 225-232.

546. A Triad of Homologous, Air-Stable Compounds Containing Short, Quadruple Bonds Between Metal Atoms of Group 6, F. A. Cotton, P. E. Fanwick, R. H. Niswander and J. C. Sekutowski, *J. Am. Chem. Soc.* **1978**, *100*, 4725-4732.

547. A Trinuclear Molybdenum(IV) Cluster Compound Having an Unusual Structure and Unusual Stability, A. Bino, F. A. Cotton, Z. Dori, *J. Am. Chem. Soc.* **1978**, *100*, 5252-5253.

548. Tetrakis(2-methoxy-5-methylphenyl)dichromium, F. A. Cotton, S. A. Koch and M. Millar, *Inorg. Chem.* **1978**, *17*, 2084-2086.

549. Dichromium and Dimolybdenum Compounds of 2,6-Dimethoxyphenyl and 2,4,6-Trimethoxyphenyl Ligands, F. A. Cotton, S. A. Koch and M. Millar, *Inorg. Chem.* **1978**, *17*, 2087-2093.

550. Preparation and Structures of the Homologous Quadruply Bonded Tris(cycloocta-tetraene)dimolybdenum and -ditungsten, F. A. Cotton, S. A. Koch, A. J. Schultz and J. M. Williams, *Inorg. Chem.* **1978**, *17*, 2093-2098.

551. Tetrachlorobis[1,2-bis(diphenylphosphino)ethane]dirhenium. A Rhenium-Rhenium Triple Bond in a Staggered Conformation, F. A. Cotton, G. G. Stanley and R. A. Walton, *Inorg. Chem.* **1978**, *17*, 2099-2102.

552. A Thiol Complex of Technetium Pertinent to Radiopharmaceutical Use of ^{99m}Tc, J. E. Smith, E. F. Byrne, F. A. Cotton and J. C. Sekutowski, *J. Am. Chem. Soc.* **1978**, *100*, 5571-5572.

553. The Molybdenum-to-Molybdenum Triple Bond. 5. Preparation and Structure of Dimethyltetrakis(dimethylamido)dimolybdenum, M. H. Chisholm, F. A. Cotton, M. W. Extine and C. A. Murillo, *Inorg. Chem.* **1978**, *17*, 2338-2340.

554. Reactions of Metal-to-Metal Multiple Bonds. 4. μ-Acetylene-bis(cyclopentadienyl)-tetracarbonyldimolybdenum Compounds. Preparations, Properties, Structural Characterizations, and Dynamical Solution Behavior, W. I. Bailey, Jr., M. H. Chisholm, F. A. Cotton and L. A. Rankel, *J. Am. Chem. Soc.* **1978**, *100*, 5764-5773.

555. Structure and Bonding in Octaisopropoxydimolybdenum(IV), M. H. Chisholm, F. A. Cotton, M. W. Extine and W. W. Reichert, *Inorg. Chem.* **1978**, *17*, 2944-2946.

556. Preparation and Characterization of Di-μ-sulfido Binuclear Compounds of W(IV) and W(V). Unambiguous Examples of Formal Single and Double Bonds Between Tungsten Atoms, A. Bino, F. A. Cotton, Z. Dori and J. C. Sekutowski, *Inorg. Chem.* **1978**, *17*, 2946-2950.

557. Tungsten Pentachloride, F. A. Cotton and C. E. Rice, *Acta Cryst.* **1978**, *B34*, 2833-2834.

558. Chemistry of Compounds Containing Metal-to-Metal Triple Bonds Between Molybdenum and Tungsten, M. H. Chisholm and F. A. Cotton, *Acc. Chem. Res.* **1978**, *11*, 356-362.

559. Ground State Electronic Structures of Some Metal Atom Cluster Compounds, F. A. Cotton and G. G. Stanley, *Chem. Phys. Lett.* **1978**, *58*, 450-453.

560. Di(η^5-cyclopentadienyl)pentacarbonyldivanadium. A Prototypal Example of Semibridging Carbonyl Groups, F. A. Cotton, L. Kruczynski and B. A. Frenz, *J. Organometal. Chem.* **1978**, *160*, 93-100.

561. Structure, Bonding, and Chemistry of *closo*-Tetraphosphorus Hexakis(methylimide), $P_4(NCH_3)_6$, and Its Derivatives. 1. The Structures of the Tetra-*P*-thio and Tetra-*P*-oxo Derivatives, F. Casabianca, F. A. Cotton, J. G. Riess, C. E. Rice and B. R. Stults, *Inorg. Chem.* **1978**, *17*, 3232-3236.

562. A Well-Charactarized Compound Containing a Heteronuclear (Molybdenum–Tungsten) Quadruple Bond, F. A. Cotton and B. E. Hanson, *Inorg. Chem.* **1978**, *17*, 3237-3240.

563. Two Bis(diphenylphosphinomethane) (DPPM) Complexes of Quadruply Bonded Dimolybdenum(II): $Mo_2(DPPM)_2X_4$, X = Cl, NCS, E. H. Abbott, K. S. Bose, F. A. Cotton, W. T. Hall and J. C. Sekutowski, *Inorg. Chem.* **1978**, *17*, 3240-3245.

564. A New Class of Trinuclear Tungsten(IV) Cluster Compounds with W–W Single Bonds, A. Bino, F. A. Cotton, Z. Dori, S. Koch, H. Küppers and M. Millar, *Inorg. Chem.* **1978**, *17*, 3245-3253.

565. Two New Molybdenum Complexes and Their Structures, F. A. Cotton, P. E. Fanwick and J. W. Fitch, III, *Inorg. Chem.* **1978**, *17*, 3254-3257.

566. Two Quadruply-Bonded Dimolybdenum Tetra(dithiocarboxylato) Compounds, F. A. Cotton, P. E. Fanwick, R. H. Niswander and J. C. Sekutowski, *Acta Chem. Scand.* **1978**, *A32*, 663-671.

567. Thermochemistry of Some Metal-to-Metal Triple Bonds, J. A. Connor, G. Pilcher, H. A. Skinner, M. H. Chisholm, F. A. Cotton, *J. Am. Chem. Soc.* **1978**, *100*, 7738-7739.

568. A New Example of a Dichromium Compound with a Supershort CrCr Quadruple Bond, F. A. Cotton, B. E. Hanson and G. W. Rice, *Angew. Chem.* **1978**, *90*, 1015-1016; *International Edition in English*, *17*, 953.

569. Structure, Bonding, and Chemistry of *closo*-Tetraphosphorus Hexakis(methylimide), $P_4(NCH_3)_6$, and Its Derivatives. 2. Polymorphs of the Monothio Derivative, F. A. Cotton, J. G. Riess, C. E. Rice and B. R. Stults, *Inorg. Chem.* **1978**, *17*, 3521-3525.

570. A Binuclear Tantalum Compound Containing Bridging 1,2-Dimethyl-1,2-diimidoethene Formed by Dimerization of Acetonitrile, F. A. Cotton and W. T. Hall, *Inorg. Chem.* **1978**, *17*, 3525-3528.

571. Reactions of Transition-Metal-Nitrogen σ Bonds. 5. Carbonation of Tetrakis(diethylamido)chromium(IV) to Yield Binuclear Chromium(III) and -(II) Carbamato Complexes, M. H. Chisholm, F. A. Cotton, M. W. Extine and D. C. Rideout, *Inorg. Chem.* **1978**, *17*, 3536-3540.

572. Homologous Chromium, Molybdenum, and Tungsten Compounds with Very Short Quadruple Bonds, F. A. Cotton, R. H. Niswander and J. C. Sekutowski, *Inorg. Chem.* **1978**, *17*, 3541-3545.

573. Staphylococcal Nuclease Reviewed: A Prototypic Study in Contemporary Enzymology. I. Isolation; Physical and Enzymatic Properties, P. W. Tucker, E. E. Hazen, Jr. and F. A. Cotton, *Mol. & Cell. Biochem.* **1978**, *22*, 67-77.

1979

574. Staphylococcal Nuclease Reviewed: A Prototypic Study in Contemporary Enzymology. II. Solution Studies of the Nucleotide Binding Site and the Effects of Nucleotide Binding, P. W. Tucker, E. E. Hazen, Jr. and F. A. Cotton, *Mol. & Cell. Biochem.* **1979**, *23*, 3-16.

575. Staphylococcal Nuclease Reviewed: A Prototypic Study in Contemporary Enzymology. III. Correlation of the Three-Dimensional Structure with the Mechanisms of Enzymatic Action, P. W. Tucker, E. E. Hazen, Jr. and F. A. Cotton, *Mol. & Cell. Biochem.* **1979**, *23*, 67-86.

576. Staphylococcal Nuclease Reviewed: A Prototypic Study in Contemporary Enzymology. IV. The Nuclease as a Model for Protein Folding, P. W. Tucker, E. E. Hazen, Jr. and F. A. Cotton, *Mol. & Cell. Biochem.* **1979**, *23*, 131-141.

577. Reactions of Metal-to-Metal Multiple Bonds. 5. Addition of Nitric Oxide to Hexa-*tert*-butoxyditungsten. Preparation, Properties and Structural Characterization of Tri-*tert*-butoxy(nitrosyl)(pyridine)tungsten, M. H. Chisholm, F. A. Cotton, M. W. Extine and R. L. Kelly, *Inorg. Chem.* **1979**, *18*, 116-119.

578. Dicyclopentadienyldi-*tert*-butoxydichromium. Preparation, Properties, Structure, and Reactions with Small Unsaturated Molecules, M. H. Chisholm, F. A. Cotton, M. W. Extine and D. C. Rideout, *Inorg. Chem.* **1979**, *18*, 120-125.

579. Crystal and Molecular Structure of Bis(pentamethylcyclopentadienyl)dicarbonyl-dicobalt, W. I. Bailey, Jr., D. M. Collins, F. A. Cotton, J. C. Baldwin and W. C. Kaska, *J. Organometal. Chem.* **1979**, *165*, 373-381.

580. A Partly Staggered Quadruple Bond in Tetrabromobis(1-diphenylphosphino-2-diphenylarsinoethane)dimolybdenum(II), F. A. Cotton, P. E. Fanwick, J. W. Fitch, H. D. Glicksman and R. A. Walton, *J. Am. Chem. Soc.* **1979**, *101*, 1752-1757.

581. Variable Electron Population of Discrete Metal Atom Clusters of the M_3X_{13} Type, A. Bino, F. A. Cotton and Z. Dori, *Inorg. Chim. Acta* **1979**, *33*, L133-L134.

582. Some New Types of Quadruply Bonded Dirhenium Compounds Containing Bridging Carboxylato Groups, F. A. Cotton, L. D. Gage and C. E. Rice, *Inorg. Chem.* **1979**, *18*, 1138-1142.

583. 1,3-Diphenyltriazine Complexes of Dichromium(II), Dimolybdenum(II), and Chromium(III), F. A. Cotton, G. W. Rice and J. C. Sekutowski, *Inorg. Chem.* **1979**, *18*, 1143-1149.

584. A Quadruply Bonded Dimolybdenum Compound with Bridging Pyrimidinethiol Ligands, F. A. Cotton, R. H. Niswander and J. C. Sekutowski, *Inorg. Chem.* **1979**, *18*, 1149-1151.

585. A Set of Homologous Quadruply Bonded Molecules of Dichromium(II), Dimolybdenum(II), and Ditungsten(II) with the Ligand 2,4-Dimethyl-6-hydroxypyrimidine, F. A. Cotton, R. H. Niswander and J. C. Sekutowski, *Inorg. Chem.* **1979**, *18*, 1152-1159.

586. A New Compound Containing the Tetrasulfatodimolybdenum Anion with a Bond Order of 3.5, A. Bino and F. A. Cotton, *Inorg. Chem.* **1979**, *18*, 1159-1161.

587. Tetradecaisopropoxydihydridotetratungsten(IV). Oxidative Addition of PrO-H across a Tungsten-to-Tungsten Triple Bond, M. Akiyama, D. Little, M. H. Chisholm, D. A. Haitko, F. A. Cotton and M. W. Extine, *J. Am. Chem. Soc.* **1979**, *101*, 2504-2506.

588. The "Supershort" Chromium-to-Chromium Quadruple Bond: Its Occurrence in a Tetracarboxamidatodichromium(II) Compound, A. Bino, F. A. Cotton, W. Kaim, *J. Am. Chem. Soc.* **1979**, *101*, 2506-2507.

589. Mixed Amino Acid-Thiocyanato Complexes of Dimolybdenum(II), A. Bino and F. A. Cotton, *Inorg. Chem.* **1979**, *18*, 1381-1386.

590. Oxidative Addition to a Quadruple Bond. The Complete X-ray Structure of $[Mo_2Cl_8H]^{3-}$, A. Bino and F. A. Cotton, *Angew. Chem.* **1979**, *91*, 356-357, *International Edition in English* **1979**, *18*, 332-333.

591. Some Reactions of the Octahalodirhenate(III) Ions. 10. Further Study of the Tetrakis(pivalato)dirhenium Dihalides, D. M. Collins, F. A. Cotton and L. D. Gage, *Inorg. Chem.* **1979**, *18*, 1712-1715.

592. Preparation and Structure of Dichlorotetrakis(*o*-oxypyridine)dirhenium(III), F. A. Cotton and L. D. Gage, Inorg. Chem. **1979**, 18, 1716-1718.

593. Polymorphs of Tetrakis(glycine)tetrachlorodimolybdenum(II), A. Bino, F. A. Cotton and P. E. Fanwick, *Inorg. Chem.* **1979**, *18*, 1719-1722.

594. An EDTA Complex Containing Two Triangular Trimolybdenum Clusters with Mo-Mo Single Bonds. A. Bino, F. A. Cotton and Z. Dori, *J. Am. Chem. Soc.* **1979**, *101*, 3842-3847.

595. The Unexpected Formation of a Mixed Bis(η^5-pentamethylcyclopentadienyl)tetracarbonyl(μ^3-ethylidene)triangulotricobalt Compound and its Structural Characterization, W. I. Bailey, Jr., F. A. Cotton and J. D. Jamerson, *J. Organometal. Chem.* **1979**, *173*, 317-324.

596. An Easily Prepared, Air-Stable Compound with a Triple Metal-to-Metal (MoMo) Bond, A. Bino and F. A. Cotton, *Angew. Chem.* **1979**, *91*, 496-497, *Angew. Chem. Int. Ed. Engl.* **1979**, *18*, 462-463.

597. A Complex Reaction Product of Dimolybdenum Tetraacetate with Aqueous Hydrochloric Acid. Structural Characterization of the Hydrido-Bridged $[Mo_2Cl_8H]^{3-}$ Ion, the $[Mo(O)Cl_4(H_2O)]^-$ Ion, and an $H_5O_2^+$ Ion with an Exceptionally Short Hydrogen Bond, A. Bino and F. A. Cotton, *J. Am. Chem. Soc.* **1979**, *101*, 4150-4154.

598. Staphylococcal Nuclease: Proposed Mechanism of Action Based on Structure of Enzyme-thymidine 3',5'-Bisphosphate-calcium Ion Complex at 1.5-Å Resolution, F. A. Cotton, E. E. Hazen, Jr. and M. J. Legg, *Proc. Natl. Acad. Sci. USA* **1979**, *76*, 2551-2555.

599. The Reaction of the Octachlorodirhenate(III) Ion with Methyl Isocyanide giving the Pentachloro(methylisocyano)rhenium(IV) Ion, F. A. Cotton, P. E. Fanwick and P. A. McArdle, *Inorg. Chim. Acta* **1979**, *35*, 289-292.

600. The Tungsten-Tungsten Triple Bond. 8. Dinuclear Alkoxides of Tungsten(III) and Structural Characterization of Hexaisopropoxybis(pyridine)ditungsten, the First Compound with Four-Coordinated Tungsten Atoms United by a Triple Bond, M. Akiyama, M. H. Chisholm, F. A. Cotton and M. W. Extine, D. A. Haitko, D. Little and P. E. Fanwick, *Inorg. Chem.* **1979**, *18*, 2266-2270.

601. Structural Studies of Some Multiply Bonded Diruthenium Tetracarboxylate Compounds, A. Bino, F. A. Cotton and T. R. Felthouse, *Inorg. Chem.* **1979**, *18*, 2599-2604.

602. Reactions of Tantalum(III) with Alkynes and Nitriles, F. A. Cotton and W. T. Hall, *J. Am. Chem. Soc.* **1979**, *101*, 5094-5095.

603. Low Temperature Electronic Spectra of Solid Tetrakis(dimethyl-dimethylenephosphonium)dimolybdenum and -dichromium, F. A. Cotton and P. E. Fanwick, *J. Am. Chem. Soc.* **1979**, *101*, 5252-5255.

604. Preparation of the Oxotetrabromo- and Oxotetraiodomolybdenum(V) Ions from Dimolybdenum Tetraacetate. Structures by X-ray Crystallography. A. Bino and F. A. Cotton, *Inorg. Chem.* **1979**, *18*, 2710-2713.

605. Tetrakis(dimethylphosphoniumdimethylido)dichromium and -dimolybdenum. 1. Crystal and Molecular Structures, F. A. Cotton, B. E. Hanson, W. H. Ilsley and G. W. Rice, *Inorg. Chem.* **1979**, *18*, 2713-2717.

606. [*N*-(2-Pyridyl)acetamido-*N*,*N*']dimolybdenum: The (New) Shortest Known Molybdenum–Molybdenum Quadruple Bond, F. A. Cotton, W. H. Ilsley and W. Kaim, *Inorg. Chem.* **1979**, *18*, 2717-2719.

607. Reformulation from X-ray Crystallography of a Dinuclear Thiocyanate Complex of Rhenium. The First Observation of Solely N-Bonded, Bridging Thiocyanate, F. A. Cotton, A. Davison, W. H. Ilsley and H. S. Trop, *Inorg. Chem.* **1979**, *18*, 2719-2723.

608. Strong Metal-to-Metal Quadruple Bonds in a Series of Five Isostructural Compounds as Indicated by Photoelectron Spectroscopy, B. E. Bursten, F. A. Cotton, A. H. Cowley, B. E. Hanson, M. Lattman and G. G. Stanley, *J. Am. Chem. Soc.* **1979**, *101*, 6244-6249.

609. Dimolybdenum Tetraacetate and the Octachlorodimolybdate(II) Anion as Reagents for Preparing Mononuclear Molybdenum(II) Complexes. Geometry of the Hepta(methyl-isocyano)molybdenum(II) Cation, P. Brant, F. A. Cotton, J. C. Sekutowski, T. E. Wood and R. A. Walton, *J. Am. Chem. Soc.* **1979**, *101*, 6588-6593.

610. Chelate Rings in a Quadruply Bonded Dimetal Compound: Preparation and Structure of Diacetatodi(4-phenylimino-2-pentanonato)dimolybdenum, F. A. Cotton, W. H. Ilsley and W. Kaim, *Inorg. Chim. Acta* **1979**, *37*, 267-272.

611. Preparation and Structural Characterization of Salts of Oxotetrachlorotechnetium(V), F. A. Cotton, A. Davison, V. W. Day, L. D. Gage and H. S. Trop, *Inorg. Chem.* **1979**, *18*, 3024-3029.

612. Tetraacetanilidodichromium and -dimolybdenum: Another Supershort Chromium–Chromium Bond with an Unexpected Structural Difference, A. Bino, F. A. Cotton and W. Kaim, *Inorg. Chem.* **1979**, *18*, 3030-3034.

613. Preparation of Tetraammonium Octakis(isothiocyanato)dimolybdenum(II) and Structural Characterization of Two Crystalline Hydrates, A. Bino, F. A. Cotton and P. E. Fanwick, *Inorg. Chem.* **1979**, *18*, 3558-3562.

614. The Tetrakis(hydrogen phosphato)dimolybdenum Ion $[Mo_2(HPO_4)_4]^{2-}$. Compounds with a Metal-Metal Triple Bond Which Are Easily Prepared and Permanently Stable in Air, A. Bino and F. A. Cotton, *Inorg. Chem.* **1979**, *18*, 3562-3565.

615. A Tetrakis(amidinato)dichromium Complex with a "Supershort" Chromium–Chromium Quadruple Bond, A. Bino, F. A. Cotton and W. Kaim, *Inorg. Chem.* **1979**, *18*, 3566-3568.

616. A New Mixed-Ligand Complex of the Quadruply Bonded Ditungsten Unit. Bis(*N*,*N*'-diphenylacetamidinato)bis(2,4-dimethyl-6-hydroxyprimidinato) ditungsten Bis(tetrahydrofuranate), F. A. Cotton, W. H. Ilsley and W. Kaim, *Inorg. Chem.* **1979**, *18*, 3569-3571.

617. Reactions of Metal-to-Metal Multiple Bonds. 6. Reversible Carbonylation of Hexakis(*tert*-butoxy)dimolybdenum (M≡M). A Carbonyl-Bridged Metal-to-Metal Double Bond, $Mo_2(OBu-t)_6(CO)(M=M)$, M. H. Chisholm, F. A. Cotton, M. W. Extine and R. L. Kelly, *J. Am. Chem. Soc.* **1979**, *101*, 7645-7650.

618. Tetracarbanilanilidodichromium(II): A Clue to the Understanding of the CrCr Quadruple Bond, F. A. Cotton, W. H. Ilsley and W. Kaim, *Angew. Chem. 91*, 937; *Angew. Chem. Int. Ed. Engl.* **1979**, *18*, 874-875.

619. Molybdenum-to-Molybdenum Bonds: A Spectrum of Multiplicities from Less Than One to Greater Than Four, A. Bino and F. A. Cotton, *Proc. of the Climax Third Int. Conf. on the Chem. and Uses of Molybdenum*, (H. F. Barry and P. C. H. Mitchell-Eds.). Climax Molybdenum Company, Ann Arbor, Michigan **1979**, 1-8.

1980

620. A Linear, Trinuclear, Mixed-Valence Chloro Complex of Ruthenium, $[Ru_3Cl_{12}]^{4-}$, A. Bino and F. A. Cotton, *J. Am. Chem. Soc.* **1980**, *102*, 608-611.

621. Electronic Structures and Photoelectron Spectra of the Metal Atom Cluster Species Re_3Cl_9, Re_3Br_9, and $[Re_3Cl_{12}]^{3-}$, B. E. Bursten, F. A. Cotton, J. C. Green, E. A. Seddon and G. G. Stanley, *J. Am. Chem. Soc.* **1980**, *102*, 955-968.

622. Structural Characterization of Four Quadruply Bonded Ditungsten Compounds and a New Trinuclear Tungsten Cluster, F. A. Cotton, T. R. Felthouse and D. G. Lay, *J. Am. Chem. Soc.* **1980**, *102*, 1431-1433.

623. Structural Characterization of Two Tetrakis(μ-carbonato) Complexes of Dirhodium(II), F. A. Cotton and T. R. Felthouse, *Inorg. Chem.* **1980**, *19*, 320-323.

624. Structural Studies of Three Tetrakis(carboxylato)dirhodium(II) Adducts in Which Carboxylate Groups and Axial Ligands are Varied, F. A. Cotton and T. R. Felthouse, *Inorg. Chem.* **1980**, *19*, 323-328.

625. Pyridine and Pyrazine Adducts of Tetrakis(acetato)dichromium, F. A. Cotton and T. R. Felthouse, *Inorg. Chem.* **1980**, *19*, 328-331.

626. Spectroscopic and Quantum Theoretical Studies of Species with Metal-to-Metal Bonds, F. A. Cotton, *J. Mol. Struct.* **1980**, *59*, 97-108.

627. Novel Solvents for the Catalytic Process for Producing Polyhydric Alcohols, L. Kaplan and F. A. Cotton, United States Patent 4,188,335, February 12 **1980**. Assignee: Union Carbide Corporation.

628. Tetra(2-oxopyridinato)chloroditechnitate. A New Compound with a Tc-Tc Bond of Order 3.5 and a Remarkable Vibronic Absorption Spectrum, F. A. Cotton, P. E. Fanwick and L. D. Gage, *J. Am. Chem. Soc.* **1980**, *102*, 1570-1577.

629. Tetrakis[bis(difluorophosphino)methylamine]dichlorodimolybdenum. A New Type of Compound with a Metal-Metal Triple Bond, F. A. Cotton, W. H. Ilsley and W. Kaim, *J. Am. Chem. Soc.* **1980**, *102*, 1918-1923.

630. A Compound with an Osmium to Osmium Triple Bond, F. A. Cotton and J. L. Thompson, *Inorg. Chim. Acta* **1980**, *44*, L247-L248.

631. Preparation and Structural Characterization of Polypeptide Complexes of Dimolybdenum(II). 1. A Tetrakis(glycylglycine) Complex, A. Bino and F. A. Cotton, *J. Am. Chem. Soc.* **1980**, *102*, 3014-3017.

632. The Structure of Tetrachlorotetrakis(diethylsulfide)dimolybdenum(II), F. A. Cotton and P. E. Fanwick, *Acta Cryst.* **1980**, *B36*, 457-459.

633. Crystal Structure and Polarized, Low-Temperature Electonic Absorption Spectrum of Tetrakis(*L*-leucine)dimolybdenum(II) Dichloride Bis(*p*-toluenesulfonate) Dihydrate, A. Bino, F. A. Cotton and P. E. Fanwick, *Inorg. Chem.* **1980**, *19*, 1215-1221.

634. Sensitivity of the Cr-Cr Quadruple Bond to Axial Interactions in Dichromium(II) Compounds, F. A. Cotton, W. H. Ilsley and W. Kaim, *J. Am. Chem. Soc.* **1980**, *102*, 3464-3474.

635. Lewis Basicity of a Chlorine Atom Bound to an Aliphatic Carbon Atom as Sensed by a Quadruple Chromium-Chromium Bond. Structures of $M_2[(2,6\text{-xylyl})NC(CH_3)O]_4$ ·$2CH_2Cl_2$, M = Cr, Mo, F. A. Cotton, W. H. Ilsley and W. Kaim, *J. Am. Chem. Soc.* **1980**, *102*, 3475-3479.

636. Bonding in Metal Atom Cluster Compounds from the *d*-Orbital Overlap Model to SCF-Xα-SW Calculations, B. E. Bursten, F. A. Cotton and G. G. Stanley, *Israel J. Chem.* **1980**, *19*, 132-142.

637. A Mixed-Ligand Complex of Quadruply Bonded Ditungsten(II): Bis(2,4-dimethyl-6-oxopyrimidine)bis(1,3-diphenyltriazino)ditungsten, F. A. Cotton, W. H. Ilsley and W. Kaim, *Inorg. Chem.* **1980**, *19*, 1450-1452.

638. Homologous Chromium, Molybdenum, and Tungsten Derivatives of 6-Chloro-2-hydroxypyridine. Inductive Effects on the Metal–Metal Bond Length, F. A. Cotton, W. H. Ilsley and W. Kaim, *Inorg. Chem.* **1980**, *19*, 1453-1457.

639. Fluxional Molecules: Reversible Thermal Intramolecular Rearrangements of Metal Carbonyls, F. A. Cotton and B. E. Hanson, in "Rearrangements in Ground and Excited States." Vol. 2, Essay 12, Academic Press, Inc., pp. 379-421 **1980**.

640. Molecular Orbital and Spectroscopic Studies of Triple Bonds Between Transition-Metal Atoms. 1. The d^3-d^3 Mo_2L_6 Compounds (L = OR, NR_2, CH_2R), B. E. Bursten, F. A. Cotton, J. C. Green, E. A. Seddon and G. G. Stanley, *J. Am. Chem. Soc.* **1980**, *102*, 4579-4588.

641. Ambidentate Character of Dimethyl Sulfoxide in Adducts of Tetrakis(propionato)- and Tetrakis(trifluoroacetato)dirhodium(II), F. A. Cotton and T. R. Felthouse, *Inorg. Chem.* **1980**, *19*, 2347-2351.

642. Reactions of Niobium(III) and Tantalum(III) Compounds with Acetylenes. 1. Preparation and Structure of Pyridinium Tetrachloro(pyridine)(tolane)tantalate, $[pyH][TaCl_4(py)(PhC{\equiv}CPh)]$, F. A. Cotton and W. T. Hall, *Inorg. Chem.* **1980**, *19*, 2352-2354.

643. Reactions of Niobium(III) and Tantalum(III) Compounds with Acetylenes. 2. Preparation and Structure of $Ta_2Cl_4(\mu\text{-}Cl)_2(\mu\text{-}Me_3CC\equiv CCMe_3)(THF)_2$. The Shortest Known Ta-Ta bond, F. A. Cotton and W. T. Hall, *Inorg. Chem.* **1980**, *19*, 2354-2356.

644. *o*-Phenylenebis(dimethylarsine)decacarbonyltriiron. Preparation and Structure of a Compound with Two Semibridging Carbonyl Ligands, A. Bino, F. A. Cotton, P. Lahuerta, P. Puebla and R. Usón, *Inorg. Chem.* **1980**, *19*, 2357-2359.

645. Some Reactions of the Octahalodirhenate(III) Ions. 11. Two New Quadruply Bonded Dirhenium Amidinato-Bridged Complexes, F. A. Cotton, W. H. Ilsley and W. Kaim, *Inorg. Chem.* **1980**, *19*, 2360-2364.

646. Coordination Compounds with Metal to Metal Bonds: The Constructive Interaction of Theory and Experiment, F. A. Cotton, *Pure Appl. Chem.* **1980**, *52*, 2331-2337.

647. Dimolybdenum: Nature of the Sextuple Bond, B. E. Bursten, F. A. Cotton and M. B. Hall, *J. Am. Chem. Soc.* **1980**, *102*, 6348-6349.

648. A Triple Bond Between Osmium Atoms. Preparation and Structure of Dichloro-tetrakis(2-hydroxypyridinato)diosmium(III), F. A. Cotton and J. L. Thompson, *J. Am. Chem. Soc.* **1980**, *102*, 6437-6441.

649. Dependence of Stability, Bond Strength and Electronic Structure of Dimetal Units upon Atomic Number, Oxidation Number and Chemical Environment, B. E. Bursten and F. A. Cotton, *Faraday Symposium* **1980**, *14*, 180-193.

650. Hexaisopropoxydinitrosyldichromium and Its Reactions with Nitrogen Donor Ligands, D. C. Bradley, C. W. Newing, M. H. Chisholm, R. L. Kelly, D. A. Haitko, D. Little, F. A. Cotton and P. E. Fanwick, *Inorg. Chem.* **1980**, *19*, 3010-3014.

651. Two New Amidato-Bridged Dimolybdenum(II) Compounds: Tetrakis[*N*-(2,6-dimethylphenyl)formamido]dimolybdenum Bis(tetrahydrofuranate) and Tetrakis(*N*-phenylpivalamido)dimolybdenum, F. A. Cotton, W. H. Ilsley and W. Kaim, *Inorg. Chem.* **1980**, *19*, 3586-3589.

652. Trinuclear Clusters of the Early Transition Elements, A. Müller, R. Jostes and F. A. Cotton, *Angew. Chem.* **1980**, *92*, 921; *Angew. Chem, Int. Ed. Eng.* **1980**, *19*, 875-882.

1981

653. The Crystal Structure of Dichlorobis(triphenylphosphine)(norbornadiene)ruthenium, D. E. Bergbreiter, B. E. Bursten, M. S. Bursten and F. A. Cotton, *J. Organometal. Chem.* **1981**, *205*, 407-415.

654. A New Aqueous Chemistry of Organometallic, Trinuclear Cluster Compounds of Molybdenum, A. Bino, F. A. Cotton and Z. Dori, *J. Am. Chem. Soc.* **1981**, *103*, 243-244.

655. Chemical and X-Ray Structural Studies on the (Acetato)- and (Trifluoroacetato) penta-carbonylmetalates of Chromium and Molybdenum, F. A. Cotton, D. J. Darensbourg and B. W. S. Kolthammer, *J. Am. Chem. Soc.* **1981**, *103*, 398-405.

656. Tetrakis(2-amino-4-methylbenzothiazolato)dimolybdenum Tetrahydrofuranate, Tris(2-amino-4-methylbenzothiazolato)(acetato)dimolybdenum Bis(tetrahydro-furanate), and Tetrakis(2-amino-4-chlorobenzothiazolato)dimolybdenum Bis(tetra-hydrofuranate). Quadruply Bonded Compounds with Bridging Ligands Derived from a Five-Membered Ring System, F. A. Cotton and W. H. Ilsley, *Inorg. Chem.* **1981**, *20*, 572-578.

657. π Acidity of Tris(2-cyanoethyl)phosphine. X-ray Structural Studies of M(CO)$_5$P(CH$_2$CH$_2$CN)$_3$ (M = Cr, Mo) and Mo(CO)$_5$P(C$_6$H$_5$)$_3$, F. A. Cotton, D. J. Darensbourg and W. H. Ilsley, *Inorg. Chem.* **1981**, *20*, 578-583.

658. Seven Dinuclear Rhodium(II) Complexes with *o*-Oxypyridine Anions as Ligands, F. A. Cotton and T. R. Felthouse, *Inorg. Chem.* **1981**, *20*, 584-600.

659. Molecular and Chain Structures of Four Tetrakis(μ-propionato)dirhodium(II) Complexes with Axial Nitrogen-Donor Ligands, F. A. Cotton and T. R. Felthouse, *Inorg. Chem.* **1981**, *20*, 600-608.

660. A Dodecahedral, Eight-Coordinate Chelate Complex, Tetrakis(2-mercaptopyrimidin-ato)tungsten(IV), F. A. Cotton and W. H. Ilsley, *Inorg. Chem.* **1981**, *20*, 614-616.

661. (Acetato)tris(*N,N'*-diphenylacetamidinato)dimolybdenum and Bis(acetato)bis[*N,N'*-bis(2,6-xylyl)acetamidinato]dimolybdenum Tetrakis(tetrahydrofuranate). Two New Mixed-Ligand Quadruply Bonded Dimolybdenum Compounds, F. A. Cotton, W. H. Ilsley and W. Kaim, *Inorg. Chem.* **1981**, *20*, 930-934.

662. Reactions of Phosphines with Tetrakis(trifluoroacetato)dimolybdenum(II), F. A. Cotton and D. G. Lay, *Inorg. Chem.* **1981**, *20*, 935-940.

663. The Tetradecaisopropoxydihydridotetratungsten Story, M. Akiyama, M. H. Chisholm, F. A. Cotton, M. W. Extine, D. A. Haitko, J. Leonelli and D. Little, *J. Am. Chem. Soc.* **1981**, *103*, 779-784.

664. Reactions of Niobium(III) and Tantalum(III) Compounds with Acetylenes. 3. Preparation and Structure of [TaCl$_2$(SC$_4$H$_8$)(Me$_3$CC≡CMe)]$_2$(μ-Cl)$_2$, F. A. Cotton and W. T. Hall, *Inorg. Chem.* **1981**, *20*, 1285-1287.

665. X-Ray Molecular Structures of Mn(CO)$_5$(O$_2$CCF$_3$) and Mn(CO)$_3$(C$_5$H$_5$N)$_2$(O$_2$CCF$_3$), F. A. Cotton, D. J. Darensbourg and B. W. S. Kolthammer, *Inorg. Chem.* **1981**, *20*, 1287-1291.

666. Dichromium(II) Compounds Containing 2-Phenylbenzoic Acid (biphCO$_2$H): [Cr$_2$(O$_2$Cbiph)$_4$]$_2$ and Cr$_2$(O$_2$Cbiph)$_4$(THF)$_2$, F. A. Cotton and J. L. Thompson, *Inorg. Chem.* **1981**, *20*, 1292-1296.

667. Reactions of Niobium(III) and Tantalum(III) Compounds with Acetylenes. 4. Polymerization of Internal Acetylenes, F. A. Cotton, W. T. Hall, K. J. Cann and F. J. Karol, *Macromolecules* **1981**, *14*, 233-236.

668. The Structure of 3,4,5,6-Tetrahaptocyclodeca-1,3,5,7-Tetraene-tricarbonyliron, a Pi Complex of a Ten-Membered, Monocyclic Polyolefin, F. A. Cotton and J. M. Troup. *J. Organometal. Chem.* **1981**, *212*, 411-418.

669. Metal-Metal Multiple Bonds and Metal Clusters, F. A. Cotton "Reactivity of Metal-Metal Bonds," M. H. Chisholm, Ed., ACS Symposium Series, No. 155 **1981**, 1-16.

670. The μ-Oxo-decachloroditantalum(V) Ion, F. A. Cotton and R. C. Najjar, *Inorg. Chem.* **1981**, *20*, 1866-1869.

671. Further Studies of the Unusual Nature of Tris(β-cyanoethyl)phosphine: Structures of the Phosphine and the Phosphine Oxide, F. A. Cotton, D. J. Darensbourg, M. F. Fredrich, W. H. Ilsley and J. M. Troup, *Inorg. Chem.* **1981**, *20*, 1869-1872.

672. A Trinuclear Tungsten(IV) Cluster Compound with a Capping Chlorine Atom and Three Bridging Oxygen Atoms, F. A. Cotton, T. R. Felthouse and D. G. Lay, *Inorg. Chem.* **1981**, *20*, 2219-2223.

673. Molecular and Electronic Structures of Two Quadruply Bonded Ditungsten Compounds and a Dimolybdenum Homologue, F. A. Cotton, M. W. Extine, T. R. Felthouse, B. W. S. Kolthammer and D. G. Lay, *J. Am. Chem. Soc.* **1981**, *103*, 4040-4045.

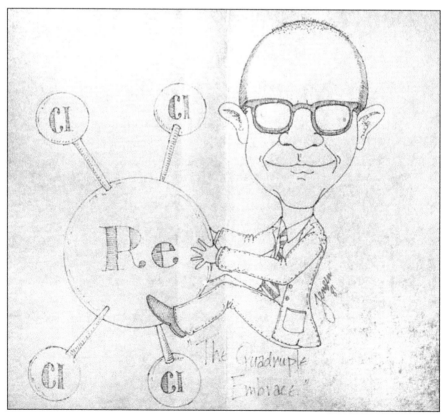

"The Quadruple Embrace" by Bill (William) Jensen from the University of Cincinnati, 1981.

674. Tetraamidodichromium(II) Compounds and Their Dihalomethane Adducts. Structures of $Cr_2[(2,6\text{-xylyl})NC(CH_3)O]_4\cdot 1.5C_6H_5CH_3$ and $M_2[(2,6\text{-xylyl})\text{-}NC(CH_3)O]_4\cdot 2CH_2BR_2$, M = Cr, Mo, S. Baral, F. A. Cotton and W. H. Ilsley, *Inorg. Chem.* **1981**, *20*, 2696-2703.

675. Crystal and Molecular Structure of Tetrakis(trifluoroacetato)bis(dimethylsulfone)-dirhodium(II). A Compound with Axially Coordinated Sulfone Ligands, F. A. Cotton and T. R. Felthouse, *Inorg. Chem.* **1981**, *20*, 2703-2708.

676. Binuclear Chloro Complexes of Tantalum(III) with Tantalum–Tantalum Double Bonds, F. A. Cotton and R. C. Najjar, *Inorg. Chem.* **1981**, *20*, 2716-2719.

677. Reactions of $M_2Cl_4(PR_3)_4$ (M = Mo and W) with Carbon Monoxide, F. A. Cotton, D. J. Darensbourg and B. W. S. Kolthammer, *J. Organometal. Chem.* **1981**, *217*, C14-C16.

678. Crystal and Molecular Structures of the Bis(triphenylphosphine) and Bis(triphenyl-phosphite) Adducts of Tetrakis(trifluoroacetato)dirhodium(II), F. A. Cotton, T. R. Felthouse and S. Klein, *Inorg. Chem.* **1981**, *20*, 3037-3042.

679. Electronic Structure of Phosphine Adducts of Tetrakis(carboxylato)dirhodium(II). Pronounced Influence of Axial Ligands, B. E. Bursten and F. A. Cotton, *Inorg. Chem.* **1981**, *20*, 3042-3048.

680. Tetrachlorotetraethoxodi-μ-ethoxo-ditungsten(W–W). Structure, Bonding, and Improved Preparation, F. A. Cotton, D. DeMarco, B. W. S. Kolthammer and R. A. Walton, *Inorg. Chem.* **1981**, *20*, 3048-3051.

681. Structural Characterization of the Octachloroditechnetate(III) Ion in Its Tetra-*n*-butylammonium Salt, F. A. Cotton, L. Daniels, A. Davison and C. Orvig, *Inorg. Chem.* **1981**, *20*, 3051-3055.

682. Experimental and Theoretical Evidence for Double Bonds Between Metal Atoms. Dinuclear Alkoxo-Bridged Ditungsten(IV,IV) Complex, $W_2Cl_4(OR)_4(ROH)_2$, L. B. Anderson, F. A. Cotton, D. DeMarco, A. Fang, W. H. Ilsley, B. W. S. Kolthammer and R. A. Walton, *J. Am. Chem. Soc.* **1981**, *103*, 5078-5086.

683. Eighty Years of Coordination Chemistry, F. A. Cotton, in *Transition Metal Chemistry*; A. Müller and E. Diemann, Eds., Verlag Chemie, Weinheim **1981**, 1-8.

684. Recent Chemistry of Compounds with Multiple Bonds Between Metal Atoms: The Metals of Transition Group V, F. A. Cotton, in *Transition Metal Chemistry*, A. Müller and E. Diemann, Eds., Verlag Chemie, Weinheim **1981**, 51-58.

685. Trinuclear Mo–Mo-Bonded Cluster Cation with One Oxygen Atom Cap and One Ethylidyne Cap, A. Bino, F. A. Cotton, Z. Dori and B. W. S. Kolthammer, *J. Am. Chem. Soc.* **1981**, *103*, 5779-5784.

686. Structures of the Bridged and Chelated Isomers of Tetrachlorobis [1,2-bis(diphenyl-phosphino)ethane]ditungsten(II), F. A. Cotton and T. R. Felthouse, *Inorg. Chem.* **1981**, *20*, 3880-3886.

687. A Molybdenum–Molybdenum Quadruple Bond Bridged by Four Carboxyl Groups but Entirely Isolated from Axial Coordination, F. A. Cotton and J. L. Thompson, *Inorg. Chem.* **1981**, *20*, 3887-3890.

688. A Quadruply Bonded Dimolybdenum Compound Containing Only Chelating Ligands: Bis(diethyl-2-pyrazolylhydroxoborato)bis(diethyldi-2-pyrazolylborato)-dimolybdenum, F. A. Cotton, B. W. S. Kolthammer and G. N. Mott, *Inorg. Chem.* **1981**, *20*, 3890-3895.

689. Preparation and Structural Characterization of the Fully Chelated Dimolybdenum(II) Complex Tetrakis[2-((dimethylamino)methyl)phenyl]dimolybdenum(II), F. A. Cotton and G. N. Mott, *Inorg. Chem.* **1981**, *20*, 3896-3899.

690. Trimolybdenum Cluster Compounds with Two Capping Ethylidyne Groups, M. Ardon, A. Bino, F. A. Cotton, Z. Dori, M. Kaftory, B. W. S. Kolthammer, M. Kapon and G. Reisner, *Inorg. Chem.* **1981**, *20*, 4083-4090.

691. Molecular and Electronic Structures of Three Pyridine- and Piperidine-Substituted Chromium Carbonyl Compounds: $Cr(CO)_5(C_5H_5N)$, $Cr(CO)_5(C_5H_{10}NH)$, and *cis*-$Cr(CO)_4(C_5H_{10}NH)[P(OMe)_3]$, F. A. Cotton, D. J. Darensbourg, A. Fang, B. W. S. Kolthammer, D. Reed and J. L. Thompson, *Inorg. Chem.* **1981**, *20*, 4090-4096.

692. Length of a Tungsten–Phosphine Bond Free of Excessive Steric Interactions: Crystal Structure of $W(CO)_5P(CH_3)_3$, F. A. Cotton, D. J. Darensbourg and B. W. S. Kolthammer, *Inorg. Chem.* **1981**, *20*, 4440-4442.

693. The Structure of Tetra-μ-acetato-bis(acetonitrile)dirhodium(II) (Rh–Rh), F. A. Cotton and J. L. Thompson, *Acta Cryst.* **1981**, *B37*, 2235-2236.

1982

694. A Second-Order Jahn-Teller Effect in a Tetranuclear Metal Atom Compound. F. A. Cotton and A. Fang, *J. Am. Chem. Soc.* **1982**, *104*, 113-119.

695. 2-[(Dimethylamino)methyl]phenyl Compounds of Chromium(III) and Dichrom-ium(II) by Reaction of 2-Lithio-N,N-dimethylbenzylamine with Dichromium(II) Tetraacetate, F. A. Cotton and G. N. Mott, *Organometallics* **1982**, *1*, 38-43.

696. [*o*-Phenylenebis(dimethylarsine)]heptacarbonyldiiron. Preparation, Crystal Structure, and Fluxionality in Solution, W. I. Bailey, Jr., A. Bino, F. A. Cotton, B. W. S. Kolthammer, P. Lahuerta, P. Puebla and R. Usón, *Inorg. Chem.* **1982**, *21*, 289-294.

697. Steric Contributions to the Solid-State Structure of Bis(phosphine) Derivatives of Molybdenum Carbonyl. X-ray Structural Studies of *cis*-$Mo(CO)_4[PPh_{3-n}Me_n]_2$ ($n = 0$, 1, 2), F. A. Cotton, D. J. Darensbourg, S. Klein and B. W. S. Kolthammer, *Inorg. Chem.* **1982**, *21*, 294-299.

698. Structure of a Novel Tetrahedral Boron Complex, Bis(acetato)(acetylacetonato)-boron(III), $B(O_2CMe)_2(acac)$, F. A. Cotton and W. H. Ilsley, *Inorg. Chem.* **1982**, *21*, 300-302.

699. Survey of the Bonding in Several Structural Types of Trinuclear Molybdenum and Tungsten Cluster Compounds, B. E. Bursten, F. A. Cotton, M. B. Hall and R. C. Najjar, *Inorg. Chem.* **1982**, *21*, 302-307.

700. Comparative Studies of Mo-Mo and W-W Quadruple Bonds by SCF-Xα-SW Calculations and Photoelectron Spectroscopy, F. A. Cotton, J. L. Hubbard, D. L. Lichtenberger and I. Shim, *J. Am. Chem. Soc.* **1982**, *104*, 679-686.

701. Preparation and Structural Characterization of the Isostructural Dichromium(II) and Dimolybdenum(II) Complexes, $M_2(O_2CCH_3)_2[o\text{-}(NMe_2)C_6H_4CH_2]_2$, Containing Two Four Atom Bridging Ligands and Surprisingly Short M–M Quadruple Bonds, F. A. Cotton and G. N. Mott, *Organometallics* **1982**, *1*, 302-305.

702. Crystal and Molecular Structure of Tetrakis(trifluoroacetato)bis- (dimethyl-$d_6$sul-foxide)dirhodium(II): A Structural Deuterium Isotope Effect in an Oxygen-Bonded Sulfoxide Adduct, F. A. Cotton and T. R. Felthouse, *Inorg. Chem.* **1982**, *21*, 431-435.

703. Structure of Yttrium Octachloroditechnetate Nonahydrate, F. A. Cotton, A. Davison, V. W. Day, M. F. Fredrich, C. Orvig and R. Swanson, *Inorg. Chem.* **1982**, *21*, 1211-1214.

704. Axially Coordinated Tetrakis[N-(2,6-dimethylphenyl)acetamido]dimolybdenum(II) Compounds. Structures of $Mo_2[(xylyl)NC(CH_3)O]_4 \cdot 2THF$ and $Mo_2[(xylyl)\text{-}NC(CH_3)O]_4 \cdot L$ (L = NC_5H_5, $NC_5H_4CH_3$), S. Baral, F. A. Cotton, W. H. Ilsley and W. Kaim, *Inorg. Chem.* **1982**, *21*, 1644-1650.

705. Structural Characterizations of *cis*-$Mo(CO)_4(PPhMe_2)(NHC_5H_{10})$ and *cis*-$Mo(CO)_4(PPhMe_2)(PPh_3)$ and Their Solution Reactivities toward Carbon Monoxide, F. A. Cotton, D. J. Darensbourg, S. Klein and B. W. S. Kolthammer, *Inorg. Chem.* **1982**, *21*, 1651-1655.

706. Solid-State and Solution Structures of $[PNP][W(CO)_5O_2CCH_3]$ and $[PNP]\text{-}[W(CO)_4(PEt_3)O_2CCH_3]$ and the CO-Labilizing Ability of the Acetato Ligand in These Anionic Derivatives, F. A. Cotton, D. J. Darensbourg, B. W. S. Kolthammer and R. Kudaroski, *Inorg. Chem.* **1982**, *21*, 1656-1662.

707. An Accurately Characterized Diplatinum(III) Bond: The Structure of $Na_2[Pt_2(HPO_4)_4(H_2O)_2]$, F. A. Cotton, L. R. Falvello and S. Han, *Inorg. Chem.* **1982**, *21*, 1709-1710.

708. Triangular Trinuclear Molybdenum(IV) Cluster Compounds with Two Capping Oxygen Atoms, M. Ardon, A. Bino, F. A. Cotton, Z. Dori, M. Kaftory and G. Reisner, *Inorg. Chem.* **1982**, *21*, 1912-1917.

709. Bonds Between Metal Atoms. A New Mode of Transition Metal Chemistry, F. A. Cotton and M. H. Chisholm, *Chem. & Eng. News* **1982**, *60*, 40-54.

710. Preparation and Characterization of bis(2-thiopyrimidinato)bis(pyridine)dioxodi(μ-oxo)dimolybdenum(V), F. A. Cotton and W. H. Ilsley, *Inorg. Chim. Acta.* **1982**, *59*, 213-217.

711. X-Ray Structural Studies of *cis*-Mo(CO)$_4$(PR$_3$)$_2$ (R = Me, Et, *n*-Bu) Derivatives and Their Relationship to Solution Isomerization Processes in These Octahedral Species, F. A. Cotton, D. J. Darensbourg, S. Klein and B. W. S. Kolthammer, *Inorg. Chem.* **1982**, *21*, 2661-2666.

712. Structures of Two Dirhodium(II) Compounds Containing Hydrogen-Bonded Nitroxyl Groups: Tetrakis(trifluoroacetato)bis(2,2,6,6-tetramethyl-4-hydroxypiperi-dinyl-l-oxy)dirhodium(II) and Tetrakis(trifluoroacetato)diaquadirhodium(II) Di-*tert*-butyl Nitroxide Solvate, F. A. Cotton and T. R. Felthouse, *Inorg. Chem.* **1982**, *21*, 2667-2675.

713. New Trinuclear, Oxo-Centered, Basic Carboxylate Compounds of Transition Metals. 1. Trichromium(II,III,III) Compounds, F. A. Cotton and W. Wang, *Inorg. Chem.* **1982**, *21*, 2675-2678.

714. Dinuclear, Metal–Metal-Bonded Platinum(III) Compounds. 1. Preparation and Structure of K$_2$[Pt$_2$(SO$_4$)$_4$(OSMe$_2$)$_2$]·4H$_2$O, F. A. Cotton, L. R. Falvello and S. Han, *Inorg. Chem.* **1982**, *21*, 2889-2891.

715. Structure, Bonding, and Chemistry of *closo*-Tetraphosphorus Hexakis(methylimide), P$_4$(NCH$_3$)$_6$, and Its Derivatives. 3. Structures of the Dithio and Trithio Derivatives, F. A. Cotton, J. G. Riess, C. E. Rice and B. R. Stults, *Inorg. Chem.* **1982**, *21*, 3123-3126.

716. New Trinuclear, Oxo-Centered, Basic Carboxylate Compounds of Transition Metals. 2. Synthesis and X-ray Structure of V$_3$(O)$_3$(THF)(C$_6$H$_5$CO$_2$)$_6$, a Compound with a Deviant Structure, F. A. Cotton, G. E. Lewis and G. N. Mott, *Inorg. Chem.* **1982**, *21*, 3127-3130.

717. Structures and Infrared Spectra of Tricyclopentadienyltricobalt Tricarbonyl and Its Methylcyclopentadienyl Analogue, W. I. Bailey, Jr., F. A. Cotton, J. D. Jamerson and B. W. S. Kolthammer, *Inorg. Chem.* **1982**, *21*, 3131-3135.

718. New Trinuclear, Oxo-Centered, Basic Carboxylate Compounds of Transition Metals. 3. Syntheses and X-ray Studies of the Trivanadium(III,III,III) and Trivana-dium(II,III,III) Compounds [V$_3$(μ_3-O)(CH$_3$CO$_2$)$_6$(CH$_3$COOH)$_2$(THF)]$^+$ VCl$_4$(CH$_3$COOH)$_2$]$^-$ and V$_3$(μ_3-O)(CF$_3$CO$_2$)$_6$(THF)$_3$ with "Classical" Triangular Structures, F. A. Cotton, G. E. Lewis and G. N. Mott, *Inorg. Chem.* **1982**, *21*, 3316-3321.

719. Structural Study of Two Crystal Forms of [Mo$_3$O(CCH$_3$)(O$_2$CCH$_3$)$_6$(H$_2$O)$_3$](CF$_3$SO$_3$)-(CF$_3$SO$_3$H)$_2$·5H$_2$O, A. Bino, F. A. Cotton, Z. Dori, L. R. Falvello and G. M. Reisner, *Inorg. Chem.* **1982**, *21*, 3750-3755.

720. Structural and Bonding Studies of the Hexachlorobis(μ-chloro)(μ-hydrido)dimolyb-denum(III) Ion, [Mo$_2$Cl$_8$H]$^{3-}$, A. Bino, B. E. Bursten, F. A. Cotton and A. Fang, *Inorg. Chem.* **1982**, *21*, 3755-3759.

721. A New Type of Trinuclear Cluster Compound: The Incomplete Bicapped (or Hemicapped) Structure in [W$_3$O(O$_2$CCH$_3$)$_6$(H$_2$O)$_3$]ZnBr$_4$·8H$_2$O, M. Ardon, F. A. Cotton, Z. Dori, A. Fang, M. Kapon, G. M. Reisner and M. Shaia, *J. Am. Chem. Soc.* **1982**, *104*, 5394-5398.

722. Susceptibility of the Tungsten–Tungsten Quadruple Bond to Oxidative Addition. Structure of an HCl Adduct, F. A. Cotton and G. N. Mott, *J. Am. Chem. Soc.* **1982**, *104*, 5978-5982.

723. Hexachloro(dimethylsulfide)bis(hexamethyltriaminophosphine)ditantalum (Ta=Ta), [(Me$_2$N)$_3$P]Cl$_2$Ta(μ-Cl)$_2$(μ-Me$_2$S)TaCl$_2$[(Me$_2$N)$_3$P]: Preparation and Structure, F. A. Cotton, L. R. Falvello and R. C. Najjar, *Inorg. Chim. Acta* **1982**, *63*, 107-111.

724. Designed Synthesis of an Alkylidyne-Capped Trinuclear (Mo, Mo, W) Metal Cluster, F. A. Cotton and W. Schwotzer, *Angew. Chem.* **1982**, *94*, 652; *Int. Ed. Engl.* **1982**, *21*, 629-630.

725. Preparation and Structure of Ditungsten Tetrabenzoate Bis(tetrahydrofuranate), F. A. Cotton and W. Wang, *Inorg. Chem.* **1982**, *21*, 3859-3860.

726. Preparation and Characterization of a Compound Containing the Octachloroditungstate (W–W) Ion, [W$_2$Cl$_8$]$^{4-}$, F. A. Cotton and G. N. Mott, R. R. Schrock and L. G. Sturgeoff, *J. Am. Chem. Soc.* **1982**, *104*, 6781-6782.

727. Bonding in the Diruthenium Molecule by ab Initio Calculations, F. A. Cotton and I. Shim, *J. Am. Chem. Soc.* **1982**, *104*, 7025-7029.

728. Ditungsten(IV) Alkoxides as Reagents for Carbon–Carbon Bond Formation via the Reductive Coupling of Ketones, F. A. Cotton, D. DeMarco, L. R. Falvello and R. A. Walton, *J. Am. Chem. Soc.* **1982**, *104*, 7375-7376.

729. Two Reactions of Bis(diphenylphosphino)acetylene with Ta$_2$Cl$_6$(SMe$_2$)$_3$. Dinuclear Products with Coupled Acetylenes and $\overline{P-Ta-C}$ or $\overline{P-Ta\equiv C}$ Rings, F. A. Cotton, L. R. Falvello and R. C. Najjar, *Organometallics* **1982**, *1*, 1640-1644.

1983

730. Structure, Bonding, and Chemistry of *closo*-Tetraphosphorus Hexakis(methylimide), P$_4$(NCH$_3$)$_6$, and Its Derivatives. 4. Spectroscopic Studies and General Conclusions, F. A. Cotton, J. G. Riess and B. R. Stults, *Inorg. Chem.* **1983**, *22*, 133-136.

731. A Bis(diphosphine)hexachloroditantalum Complex with Ta-Ta Double Bond: Hexachlorobis[bis(dimethylphosphino)ethane]ditantalum(III), F. A. Cotton, L. R. Falvello and R. C. Najjar, *Inorg. Chem.* **1983**, *22*, 375-377.

732. Dinuclear and Polynuclear Oxovanadium(IV) Compounds. 1. Synthesis and Structural Study of V$_2$O$_2$Cl$_4$(μ-Hmhp)$_3$, a Novel Complex Containing Three Neutral Bridging Ligands, F. A. Cotton, G. E. Lewis and G. N. Mott, *Inorg. Chem.* **1983**, *22*, 378-382.

733. Preparation and Structure of Crystalline Tetrakis(D-mandelato)dimolybdenum(II) Bis(tetrahydrofuranate), F. A. Cotton, L. R. Falvello and C. A. Murillo, *Inorg. Chem.* **1983**, *22*, 382-387.

734. Reactions of Mo$_2$(O-*i*-Pr)$_6$ with *trans*-WBr(CO)$_4$(CPh) and Pyridine. Structural Characterizations of WBr(CO)$_2$(py)$_2$(CPh) and Mo$_2$(O-*i*-Pr)$_6$(py)$_2$(CO), F. A. Cotton and W. Schwotzer, *Inorg. Chem.* **1983**, *22*, 387-390.

735. Synthesis and Structure of Tetrakis(2,4,6-trimethoxyphenyl)divanadium(II) Bis(tetrahydrofuranate), F. A. Cotton, G. E. Lewis and G. N. Mott, *Inorg. Chem.* **1983**, *22*, 560-561.

736. Preparation and Structures of Two Compounds Containing the Tetra(μ-salicylato)-dimolybdenum(II) Unit, F. A. Cotton and G. N. Mott, *Inorg. Chim. Acta* **1983**, *70*, 159-166.

737. High- and Low-Valence Metal Cluster Compounds: A Comparison, F. A. Cotton, in Inorganic Chemistry: Toward the 21st Century, M. H. Chisholm, Ed., ACS Symposium Series, No. 211 **1983**, pp. 209-219.

738. Comparison of the Eight-Coordinate Structures of d^1 TaCl$_4$(dmpe)$_2$ and d^0 TaCl$_4$(dmpe)$_2^+$, F. A. Cotton, L. R. Falvello and R. C. Najjar, *Inorg. Chem.* **1983**, *22*, 770-774.

739. An Unusual Ditantalum (Ta=Ta) Compound with a Bridging Oxo Ligand, F. A. Cotton and W. J. Roth, *Inorg. Chem.* **1983**, *22*, 868-870.

740. A New Cyclotetramolybdenum Diyne, Mo$_4$Cl$_8$[P(OCH$_3$)$_3$]$_4$, F. A. Cotton and G. L. Powell, *Inorg. Chem.* **1983**, *22*, 871-873.

741. Some Reactions of Tetrakis(μ-acetato)dichromium(II) with Potentially Chelating Ligands: Two Unexpected Products, F. A. Cotton and G. N. Mott, *Inorg. Chem.* **1983**, *22*, 1136-1139.

742. Isomeric Bis[(phenylazo)acetaldoximato]platinum(II) Compounds, D. Bandyopadhyay, P. Bandyopadhyay, A. Chakravorty, F. A. Cotton, L. R. Falvello, *Inorg. Chem.* **1983**, *22*, 1315-1321.

743. Reaction of Halide Ions with Tetrakis(μ-trifluoroacetato)dimolybdenum(II): Crystal Structures of (*N-n-*Bu$_4$)$_2$[Mo$_2$(O$_2$CCF$_3$)$_4$X$_2$] (X = Br, I) and (*N-n-*Bu$_4$)$_2$[Mo$_2$(O$_2$CCF$_3$)$_2$Br$_4$], F. A. Cotton and P. E. Fanwick, *Inorg. Chem.* **1983**, *22*, 1327-1332.

744. Molecular Orbital and Spectroscopic Studies of Triple Bonds Between Transition-Metal Atoms. 2. The d^5-d^5 Re$_2$Cl$_4$(PR$_3$)$_4$ Compounds, B. E. Bursten, F. A. Cotton, P. E. Fanwick, G. G. Stanley and R. A. Walton, *J. Am. Chem. Soc.* **1983**, *105*, 2606-2611.

745. Nyholm Lecture: Synergic Interplay of Experiment and Theory In Studying Metal-Metal Bonds of Various Orders, F. A. Cotton, *Chem. Soc. Rev.* **1983**, *12*, 35-51.

746. Preparation and Structure of the Bridged Form of Mo$_2$Cl$_4$[(CH$_3$)$_2$PCH$_2$-CH$_2$P(CH$_3$)$_2$]$_2$. Dependence of the Mo$_2$ Quadruple Bond Length on Torsional Angle, F. A. Cotton and G. L. Powell, *Inorg. Chem.* **1983**, *22*, 1507-1510.

747. A Molecular Orbital Calculation of the [Re$_2$Cl$_8$]$^{2-}$ Ion by the Relativistic SCF-Xα-SW Method. Redetermination and Reassignment of the Electronic Absorption Spectrum, B. E. Bursten, F. A. Cotton, P. E. Fanwick and G. G. Stanley, *J. Am. Chem. Soc.* **1983**, *105*, 3082-3087.

748. Alcohol Exchange and Oxidation Reactions of Doubly Bonded Ditungsten(IV) Complexes of the Type $W_2Cl_4(OR)_4(ROH)_2$: The Structural Identification of Novel Hydrogen-Bonded Bridging Systems, F. A. Cotton, L. R. Falvello, M. F. Fredrich, D. DeMarco and R. A. Walton, *J. Am. Chem. Soc.* **1983**, *105*, 3088-3097.

749. Reactions of *tert*-Butyl Isocyanide with a Binuclear Niobium(III) Compound, F. A. Cotton and W. J. Roth, *J. Am. Chem. Soc.* **1983**, *105*, 3734-3735.

750. Ortho-Metallation at a Multiply Bonded Dirhenium Center: The First Such Example Occurring at a Multiple Bond of the M_2L_8 Type, T. J. Barder, S. M. Tetrick, R. A. Walton, F. A. Cotton and G. L. Powell, *J. Am. Chem. Soc.* **1983**, 105, 4090-4091.

751. Structural Characterization of a Doubly-Bonded Diniobium Compound, Bis-(1,2-bisdiphenylphosphinoethane)hexachlorodiniobium(III), F. A. Cotton and W. J. Roth, *Inorg. Chim. Acta* **1983**, *71*, 175-178.

752. Structural Characterization of the 3,4-dimethylpyridinium Salt of the $[Pt_2(HPO_4)_4(3,4\text{-}Me_2py)_2]^{2-}$ Ion, F. A. Cotton, S. Han, H. L. Conder and R. A. Walton, *Inorg. Chim. Acta* **1983**, *72*, 191-193.

753. Dinuclear and Polynuclear Oxovanadium(IV) Compounds. 2. A Complicated Sodium Oxovanadium(IV) Trifluoroacetate Compound, $Na_4(VO)_2(CF_3CO_2)_8(THF)_6(H_2O)_2$, F. A. Cotton, G. E. Lewis and G. N. Mott, *Inorg. Chem.* **1983**, *22*, 1825-1827.

754. Dinuclear, Metal-Metal-Bonded Platinum(III) Compounds. 2. Synthesis and Properties of Complexes Containing the $[Pt_2(HPO_4)_4(B)_2]^{2-}$ Anions (B = a Heterocyclic Tertiary Amine) or the $[Pt_2(H_4PO_4)(HPO_4)_3(py)_2]^-$ Ion, H. L. Conder, F. A. Cotton, L. R. Falvello, S. Han and R. A. Walton, *Inorg. Chem.* **1983**, *22*, 1887-1891.

755. *cis*-Dihalobis(arylazooxime)ruthenium(II): Synthesis, Structure, and Reactions, A. R. Chakravarty, A. Chakravorty, F. A. Cotton, L. R. Falvello, B. K. Ghosh and M. Tomás, *Inorg. Chem.* **1983**, *22*, 1892-1896.

756. Ground-State Electronic Structures and Other Electronic Properties of the Octahedral and Oligooctahedral Ruthenium Complexes Hexachlororuthenium(III), Nonachlorodiruthenium(III), and Dodecachlorotriruthenium(II,III), B. E. Bursten, F. A. Cotton and A. Fang, *Inorg. Chem.* **1983**, *22*, 2127-2133.

757. Synthesis and Structural Characterization of $[Re_2Cl_4(PMe_2Ph)_4]^{n+}$ (*n* = 0, 1, or 2): A Series of Complexes Possessing Metal-Metal Bond Orders of 4, 3.5, and 3 and the Dependence of Bond Length upon Bond Order, F. A. Cotton, K. R. Dunbar, L. R. Falvello, M. Tomás and R. A. Walton, *J. Am. Chem. Soc.* **1983**, *105*, 4950-4954.

758. Novel Products in the Carbonylation of the Triply Bonded Ditungsten Hexaisopropoxide Dipyridine Adduct. 1. Synthesis and Structure of $[(i\text{-}PrO)_3W(\mu\text{-}CO)(\mu\text{-}O\text{-}i\text{-}Pr)W(O\text{-}i\text{-}Pr)_2py]_2$, F. A. Cotton and W. Schwotzer, *J. Am. Chem. Soc.* **1983**, *105*, 4955-4957.

759. Novel Products in the Carbonylation of the Triply Bonded Ditungsten Hexaisopropoxide Dipyridine Adduct. 2. Preparation and Structure of $W_2(O\text{-}i\text{-}Pr)_6(CO)_4$, a Molecule Containing Two Atoms of the Same Metallic Element with Oxidation Numbers Differing by Six Units, F. A. Cotton and W. Schwotzer, *J. Am. Chem. Soc.* **1983**, *105*, 5639-5641.

760. A Triply Bonded $Re_2X_4L_4$ Compound with Cis Stereochemistry at Each Metal Atom, N. F. Cole, F. A. Cotton, G. L. Powell and T. J. Smith, *Inorg. Chem.* **1983**, *22*, 2618-2621.

761. Structural Characterization of the Nonachloroditungsten(III,IV) Ion, F. A. Cotton, L. R. Falvello, G. N. Mott, R. R. Schrock and L. G. Sturgeoff, *Inorg. Chem.* **1983**, *22*, 2621-2623.

762. New Preparative Route to Bi-oxo-capped Trinuclear Molybdenum(IV) Clusters by Reduction of Molybdate(VI). Structure of $[Mo_3O_2(O_2CCH_3)_6(H_2O)(OH)_2]\cdot16H_2O$, A. Birnbaum, F. A. Cotton, Z. Dori, D. O. Marler, G. M. Reisner, W. Schwotzer and M. Shaia, *Inorg. Chem.* **1983**, *22*, 2723-2726.

763. Reactions of Hexa-*tert*-butoxyditungsten(W≡W) with Diphenylacetylene. Syntheses and Structures of $W_2(OCMe_3)_4(\mu\text{-}C_6H_5C)_2$ and $W_2(OCMe_3)_4C_6H_5CCC_6H_5)_2$, F. A. Cotton, W. Schwotzer and E. S. Shamshoum, *Organometallics* **1983**, *2*, 1167-1171.

764. A New Type of Triangular Tritungsten Cluster Compound from Reaction of 3-Hexyne with Hexa-*tert*-butoxyditungsten, F. A. Cotton, W. Schwotzer and E. S. Shamshoum, *Organometallics* **1983**, *2*, 1340-1343.

765. Multiple Metal-Metal Bonds, F. A. Cotton, *J. Chem. Educ.* **1983**, *60*, 713-720.

766. Treatment of the Octachlorodirhenate Ion with High-Pressure Carbon Monoxide; Preparation and Structure of $[cis\text{-}Re(CH_3CN)_4(CO)_2]_2[ReCl_6]$, F. A. Cotton, L. M. Daniels and C. D. Schmulbach, *Inorg. Chim. Acta* **1983**, *75*, 163-167.

767. Two Novel Examples of Hydroxylation of Aromatic Rings in Coordination Chemistry, P. Bandyopadhyay, D. Bandyopadhyay, A. Chakravorty, F. A. Cotton, L. R. Falvello and S. Han, *J. Am. Chem. Soc.* **1983**, *105*, 6327-6329.

768. Structural Studies of the Vanadium(II) and Vanadium(III) Chloride Tetrahydrofuran Solvates, F. A. Cotton, S. A. Duraj, M. W. Extine, G. E. Lewis, W. J. Roth, C. D. Schmulbach and W. Schwotzer, *J. Chem. Soc., Chem. Commun.* **1983**, *23*, 1377-1378.

769. Comproportionation Synthesis of Bi-μ_3-oxo-capped Trimolybdenum(IV) Clusters: Structural Characterization of $[Mo_3O_2(O_2CCH_3)_6(H_2O)_3](CF_3SO_3)_2$, F. A. Cotton, Z. Dori, D. O. Marler and W. Schwotzer, *Inorg. Chem.* **1983**, *22*, 3104-3106.

770. Preparation and Structure of Bis(bis(diphenylphosphino)methane) hexachlorodiniobium(III), $Nb_2Cl_6(dppm)_2$, F. A. Cotton and W. J. Roth, *Inorg. Chem.* **1983**, *22*, 3654-3656.

771. Preparation, Structures, and Spectra of Tetrakis(6-fluoro-2-oxypyridine)-dichromium, -dimolybdenum, and -ditungsten: A Series of Polar Quadruple Bonds, F. A. Cotton, L. R. Falvello, S. Han and W. Wang, *Inorg. Chem.* **1983**, *22*, 4106-4112.

772. The Structure of Trinuclear Palladium(II) Acetate as Crystallized from Benzene, F. A. Cotton and S. Han, *Revue de Chimie Minerale* **1983**, *20*, 496-503.

773. The Structure of Rhenium Pentacarbonyl Chloride, Re(CO)$_5$Cl, F. A. Cotton and L. M. Daniels, *Acta Cryst.* **1983**, *C39*, 1495-1496.

1984

774. The Structure of Tetrakis(acetato)bis(4-pyridinecarbonitrile)dirhodium(II) Acetonitrile Solvate, [Rh$_2$(O$_2$CCH$_3$)$_4$(C$_6$H$_4$N$_2$)$_2$]·CH$_3$CN, F. A. Cotton and T. R. Felthouse, *Acta Cryst.* **1984**, *C40*, 42-45.

775. A Neutron Diffraction Crystallographic Study of the Tetramethylammonium Salt of the Hexachlorobis(μ-chloro)(μ-hydrido)dimolybdenum(III) Ion, [Mo$_2$Cl$_8$H]$^{3-}$, F. A. Cotton, P. C. W. Leung, W. J. Roth, A. J. Schultz and J. M. Williams, *J. Am. Chem. Soc.* **1984**, *106*, 117-120.

776. A Linear μ-Oxo-Diosmium(IV) Molecule in Which Two Bis(diphenylphosphino)-methane Ligands Bridge an Entire M-O-M Unit: Os$_2$(μ-O)(μ-Ph$_2$PCH$_2$PPh$_2$)$_2$Cl$_6$, A. R. Chakravarty, F. A. Cotton and W. Schwotzer, *Inorg. Chem.* **1984**, *23*, 99-103.

777. Synthesis and Structural Characterization of the Dirhenium Polyhydride Complex [Re$_2$(μ-H)$_3$H$_2$(PPh$_3$)$_4$(CN-*t*-Bu)$_2$]PF$_6$, J. D. Allison, F. A. Cotton, G. L. Powell and R. A. Walton, *Inorg. Chem.* **1984**, *23*, 159-164.

778. Preparation and Structural Characterization of Three Tetracarboxylato Dirhodium (Rh–Rh) Compounds with Bulky Ligands, F. A. Cotton and J. L. Thompson, *Inorg. Chim. Acta* **1984**, *81*, 193-203.

779. Molybdenum Carbonyl Complexes of Diphenylphosphinous Acid and Diphenylphos-phinous Acid Derivatives. Crystal and Molecular Structures of [Et$_4$N][Mo(CO)$_4$-(PPh$_2$O)$_2$H]·0.37CH$_2$Cl$_2$, Mo(CO)$_4$(PPh$_2$OH)(PPh$_2$OCH$_2$CH$_2$NMe$_2$)·H$_2$O and Mo(CO)$_4$(PPh$_2$SPPh$_2$), F. A. Cotton, L. R. Falvello, M. Tomás, G. M. Gray and C. S. Kraihanzel, *Inorg. Chim. Acta* **1984**, *82*, 129-139.

780. The Reaction of Tetra(μ-trifluoroacetato)dimolybdenum with Trimethylchlorosilane and Propionitrile: Products with Novel Stereochemistries, P. Agaskar and F. A. Cotton, *Inorg. Chim. Acta* **1984**, *83*, 33-39.

781. Theoretical and Experimental Studies of the Electronic Structure of the Mo$_3$(μ$_3$-O)-(μ$_3$-OR)(μ-OR)$_3$(OR)$_6$ Type of Triangular Metal Atom Cluster Compound, M. H. Chisholm, F. A. Cotton, A. Fang and E. M. Kober, *Inorg. Chem.* **1984**, *23*, 749-754.

782. A Metal-Metal-Bonded Dinuclear Phosphine Complex of Niobium(IV) Chloride, (NbCl$_2$(PMe$_2$Ph)$_2$)$_2$(μ-Cl)$_4$, F. A. Cotton and W. J. Roth, *Inorg. Chem.* **1984**, *23*, 945-947.

783. Reactivity of 2-(Diphenylphosphino)pyridine toward Complexes Containing the Quadruply Bonded Re$_2$$^{6+}$ Core: Ortho Metalation and Redox Chemistry, T. J. Barder, F. A. Cotton, G. L. Powell, S. M. Tetrick and R. A. Walton, *J. Am. Chem. Soc.* **1984**, *106*, 1323-1332.

784. Configuration Chirality of Metal-Metal Multiple Bonds: Preparation and Circular Dichroism Spectrum of Tetrachlorobis[(S,S)-2,3-(diphenylphosphino)butane]dimolybdenum (Mo$_2$Cl$_4$(S,S-dppb)$_2$), P. A. Agaskar, F. A. Cotton, I. F. Fraser and R. D. Peacock, *J. Am. Chem. Soc.* **1984**, *106*, 1851-1853.

785. New Preparations and Molecular Structures of *cis*-MCl$_2$(Ph$_2$PCH$_2$PPh$_2$)$_2$, M = Ru, Os and *trans*-RuCl$_2$(Ph$_2$PCH$_2$PPh$_2$)$_2$, A. R. Chakravarty, F. A. Cotton and W. Schwotzer, *Inorg. Chim. Acta* **1984**, *84*, 179-185.

786. Reactions of Niobium(III) and Tantalum(III) Compounds with Acetylenes. 5. Preparation and Structure of [NbCl$_2$(SC$_4$H$_8$)(PhCCPh)]$_2$)(μ-Cl)$_2$ and its Relationship to Other Alkyne Complexes of Niobium(III) and Tantalum(III), F. A. Cotton and W. J. Roth, *Inorg. Chim. Acta* **1984**, *85*, 17-21.

787. Uranium-to-Uranium Bonds: Do They Exist? F. A. Cotton, D. O. Marler and W. Schwotzer, *Inorg. Chim. Acta* **1984**, *85*, L31-L32.

788. Mixed-Metal Atom Cluster Compound Containing Silver(I) and Platinum(II), R. Usón, J. Forniés, M. Tomás, F. A. Cotton and L. R. Falvello, *J. Am. Chem. Soc.* **1984**, *106*, 2482-2483.

789. *ortho*-Metallation at Dimetallic Centres: Syntheses and X-Ray Characterizations of Os$_2$(O$_2$CMe)$_2$[Ph$_2$P(C$_6$H$_4$)]$_2$Cl$_2$ and Rh$_2$(O$_2$CMe)$_2$[Ph$_2$P(C$_6$H$_4$)]$_2$(2MeCO$_2$H), A. R. Chakravarty, F. A. Cotton and D. A. Tocher, *J. Chem. Soc., Chem. Commun.* **1984**, 501-502.

790. Multiply Bonded Dimetal Complexes Containing Bis(diphenylphosphino)methane Bridges: Complexes Possessing Rhenium-Rhenium Double Bonds and a Tungsten-Tungsten Single Bond, T. J. Barder, F. A. Cotton, D. Lewis, W. Schwotzer, S. M. Tetrick and R. A. Walton, *J. Am. Chem. Soc.* **1984**, *106*, 2882-2891.

791. Preparative, Structural, and Spectroscopic Studies of Tetrakis(carboxylato)ditungsten(II) Compounds with W-W Quadruple Bonds, F. A. Cotton and W. Wang, *Inorg. Chem.* **1984**, *23*, 1604-1610.

792. Formation of Isomeric Dimolybdenum(II) Compounds Using the Potentially Ambidentate Ligands [R$_2$PC(S)NR']$^-$ (R/R' = Ph/Ph, Ph/Me) and [Me$_2$NC(S)NPh]$^-$ H. P. M. M. Ambrosius, F. A. Cotton, L. R. Falvello, H. T. J. M. Hintzen, T. J. Melton, W. Schwotzer, M. Tomás and J. G. M. van der Linden, *Inorg. Chem.* **1984**, *23*, 1611-1616.

793. Isolation and Characterization of [Zn$_2$(O$_2$CCH$_3$)$_3$][Mo$_3$O$_2$(O$_2$CCH$_3$)$_6$(OCH$_3$)$_3$]·2H$_2$O, an Anionic Trimolybdenum(IV) Cluster Compound Containing an Unusual Cation, A. Birnbaum, F. A. Cotton, Z. Dori and M. Kapon, *Inorg. Chem.* **1984**, *23*, 1617-1619.

794. Syntheses, Molecular Structures and Electrochemical Behavior of Two Isomeric Ru(PhNpy)$_2$(PPh$_3$)$_2$ Complexes, A. R. Chakravarty, F. A. Cotton and E. S. Shamshoum, *Inorg. Chim. Acta* **1984**, *86*, 5-11.

795. Oxidative Addition of Nitrosobenzene Fragments Across the Triply Bonded Ditungsten Hexa-*tert*-butoxide Molecule. The Preparation and Structure of [W(OCMe$_3$)$_2$(NPh)]$_2$(μ-O)(μ-OCMe$_3$)$_2$, F. A. Cotton and E. S. Shamshoum, *J. Am. Chem. Soc.* **1984**, *106*, 3222-3225.

796. Preparation of Thiolate-Bridged Dimolybdenum Complexes from Mo-Mo Quadruple Bonds by Both Conventional and Unconventional Reactions, F. A. Cotton and G. L. Powell, *J. Am. Chem. Soc.* **1984**, *106*, 3371-3372.

797. Ruthenium(IV) in Centrosymmetric RuX$_2$N$_2$O$_2$ Coordination: Synthesis, Structure, and Redox Properties of Dihalobis(triazene 1-oxidato)ruthenium Species, S. Bhattacharya, A. Chakravorty, F. A. Cotton, R. Mukherjee and W. Schwotzer, *Inorg. Chem.* **1984**, *23*, 1709-1713.

798. Structures of *trans*-Bis[(phenylazo)acetaldoximato]platinum(II) and -palladium(II): A Case of Nonplanar Tetracoordination in a Bis Complex of Palladium(II), D. Bandyopadhyay, P. Bandyopadhyay, A. Chakravorty, F. A. Cotton and L. R. Falvello, *Inorg. Chem.* **1984**, *23*, 1785-1787.

799. Synthesis and Structural Characterization of Bis(μ-Sulfato) Hexakis (Pyridine)dimo-lybdenum(II) Dipyridinate, Mo$_2$(SO$_4$)$_2$(py)$_6$·2py, A Partially-substituted Member of the M$_2$(SO$_4$)$_4$$^{n\pm}$ Class of Bridged Dimetal Systems, F. A. Cotton and A. H. Reid, Jr., *Nouv. J. de Chimie* **1984**, *8*, 203-206.

800. Response of the Chromium-Chromium Bond Length in Cr$_2$(O$_2$CR)$_4$L$_2$ Compounds to Changes in R and L, F. A. Cotton and W. Wang, *Nouv. J. de Chimie* **1984**, *8*, 331-340.

801. The Structure of U$_3$(O)(OCMe$_3$)$_{10}$, an Unusual Trinuclear U(IV) Oxo-Alkoxide, F. A. Cotton, D. O. Marler and W. Schwotzer, *Inorg. Chim. Acta* **1984**, *95*, 207-209.

802. New Bi-Oxo-Capped Triangular Trinuclear Cluster Compounds of Niobium, F. A. Cotton, S. A. Duraj and W. J. Roth, *J. Am. Chem. Soc.* **1984**, *106*, 3527-3531.

803. Structural and Mass Spectral Studies on Some Multiply Bonded Diosmium Tetra-carboxylate Compounds, F. A. Cotton, A. R. Chakravarty, D. A. Tocher and T. A. Stephenson, *Inorg. Chim. Acta* **1984**, *87*, 115-119.

804. Dinuclear Metal-Metal-Bonded Platinum(III) Compounds. 4. Structural Studies of Several Compounds with Sulfate or Hydrogenphosphate Ions, D. P. Bancroft, F. A. Cotton, L. R. Falvello, S. Han and W. Schwotzer, *Inorg. Chim. Acta* **1984**, *87*, 147-153.

805. Synthesis and Molecular Structure of Dichlorotetrakis(benzamidato)diosmium(III), A. R. Chakravarty, F. A. Cotton and D. A. Tocher, *Inorg. Chim. Acta* **1984**, *89*, L15-L16.

806. Structure and Bonding in Dirhodium(II) Tetrasalicylate, D. P. Bancroft, F. A. Cotton and S. Han, *Inorg. Chem.* **1984**, *23*, 2408-2411.

807. Carbon-Carbon Bond Formation in the Reductive Coupling of Ketones by the Doubly Bonded Ditungsten(IV) Alkoxides $W_2Cl_4(\mu\text{-OR})_2(OR)_2(ROH)_2$, L. B. Anderson, F. A. Cotton, D. DeMarco, L. R. Falvello, S. M. Tetrick and R. A. Walton, *J. Am. Chem. Soc.* **1984**, *106*, 4743-4749.

808. A New Double Bond Metathesis Reaction: Conversion of an Nb=Nb and an N=N Bond in to Two Nb=N Bonds, F. A. Cotton, S. A. Duraj and W. J. Roth, *J. Am. Chem. Soc.* **1984**, *106*, 4749-4751.

809. The Structure of Tetrakis(acetonitrile)tetrachlorouranium(IV), $[UCl_4(C_2H_3N)_4]$, F. A. Cotton, D. O. Marler and W. Schwotzer, *Acta Cryst.* **1984**, *C40*, 1186-1188.

810. Precise Structural Characterizations of the Hexaaquovanadium(III) and Diaquo-hydrogen Ions. X-Ray and Neutron Diffraction Studies of $[V(H_2O)_6][H_5O_2](CF_3SO_3)_4$, F. A. Cotton, C. K. Fair, G. E. Lewis, G. N. Mott, F. K. Ross, A. J. Schultz and J. M. Williams, *J. Am. Chem. Soc.* **1984**, *106*, 5319-5323.

811. New Bromo Complexes of Osmium(IV) and Osmium(III): $[Os_2Br_{10}]^{2-}$ and $OsBr_3(PPh_3)_2(CH_3CN)$, F. A. Cotton, S. A. Duraj, C. C. Hinckley, M. Matusz and W. J. Roth, *Inorg. Chem.* **1984**, *23*, 3080-3083.

812. Displacive Transfer of a Phenyl Group from Triphenylphosphine to a Metal Atom: Synthesis and Molecular Structure of $Ru_2Ph_2(PhCONH)_2[Ph_2POC(Ph)N]_2$, A. R. Chakravarty, F. A. Cotton and D. A. Tocher, *J. Am. Chem. Soc.* **1984**, *106*, 6409-6413.

813. Preparation of the Crystalline β (Bridged) Isomer of $Mo_2Cl_4(dppe)_2$ and Determination of Its Crystal Structure, P. A. Agaskar and F. A. Cotton, *Inorg. Chem.* **1984**, *23*, 3383-3387.

814. Further Studies of the Phosphine Complexes of Niobium(IV) Chloride, F. A. Cotton, S. A. Duraj and W. J. Roth, *Inorg. Chem.* **1984**, *23*, 3592-3596.

815. New Routes to the Preparation of the Aquomolybdenum(IV) Ion by Comproportionation Reactions, F. A. Cotton, D. O. Marler and W. Schwotzer, *Inorg. Chem.* **1984**, *23*, 3671-3673.

816. Reactions of Dinuclear Niobium(III) and Tantalum(III) Compounds with Alkyl Isocyanides to Give Dinuclear Products with Dimerized Isocyanides, F. A. Cotton, S. A. Duraj and W. J. Roth, *J. Am. Chem. Soc.* **1984**, *106*, 6987-6993.

817. Structural Characterization of a Crystalline Monomeric Trialkoxytungsten Alkyne: Tri-*tert*-butoxybenzylidynetungsten, $(Me_3CO)_3WCPh$, F. A. Cotton, W. Schwotzer and E. S. Shamshoum, *Organometallics* **1984**, *3*, 1770-1771.

818. Synthesis and Molecular Structure of $Ru_2(C_5NH_4NH)_6(PMe_2Ph)_2$, a Ru-Ru Bonded Molecule with 2-Pyridinylamide Ligands in Three Coordination Modes and Other Novel Features, A. R. Chakravarty, F. A. Cotton and D. A. Tocher, *Inorg. Chem.* **1984**, *23*, 4030-4033.

819. New Preparative Methods, Structure, and Nitration of Bi-oxo-capped Trimolybdenum(IV) and Tritungsten(IV) Cluster Cations, F. A. Cotton, Z. Dori, D. O. Marler and W. Schwotzer, *Inorg. Chem.* **1984**, *23*, 4033-4038.

820. Comparative Study of Structures, Including Jahn-Teller Effects, in the Saccharinate Complexes, [M(C$_7$H$_4$NO$_3$S)$_2$(H$_2$O)$_4$]·2H$_2$O, of Chromium and Zinc, as Well as Other Divalent Metal Ions, F. A. Cotton, G. E. Lewis, C. A. Murillo, W. Schwotzer and G. Valle, *Inorg. Chem.* **1984**, *23*, 4038-4041.

821. An Octanuclear Basic Benzoate Containing Four Vanadium(III) and Four Zinc(II) Atoms: [VZnO(O$_2$CC$_6$H$_5$)$_3$(THF)]$_4$·2THF, F. A. Cotton, S. A. Duraj and W. J. Roth, *Inorg. Chem.* **1984**, *23*, 4042-4045.

822. Preparation and Structural Characterization of Phosphine Adducts of Tantalum(IV) Chloride, F. A. Cotton, S. A. Duraj and W. J. Roth, *Inorg. Chem.* **1984**, *23*, 4046-4050.

823. Preparation and Characterization of *trans*-Diiodotetra-*tert*-butoxy-tungsten (VI): The First Structurally Well-Characterized W^{VI}-I Bonds, F. A. Cotton, W. Schwotzer and E. S. Shamshoum, *Inorg. Chem.* **1984**, *23*, 4111-4113.

824. A Novel Dinuclear Vanadium(II) Compound with Bridging Chlorine Atoms, Bridging Diphosphinomethane Ligands Bidentate Tetrahydroborate Ligands, F. A. Cotton, S. A. Duraj and W. J. Roth, *Inorg. Chem.* **1984**, *23*, 4113-4115.

825. Dinuclear Uranium Alkoxides. Preparation and Structures of KU$_2$(OCMe$_3$)$_9$, U$_2$(OCMe$_3$)$_9$ U$_2$(OCHMe$_2$)$_{10}$, Containing [U(IV),U(IV)],[U(IV),U(V)] [U(V),U(V)], Respectively, F. A. Cotton, D. O. Marler and W. Schwotzer, *Inorg. Chem.* **1984**, *23*, 4211-4215.

826. 2-Anilinopyridine Complexes of Dimolybdenum(II) and Ditungsten(II), A. R. Chakravarty, F. A. Cotton and E. S. Shamshoum, *Inorg. Chem.* **1984**, *23*, 4216-4221.

827. Correlation of Mo-Mo Quadruple Bond Length with Angle of Internal Rotation, Using Data for 10 Compounds, F. L. Campbell, III, F. A. Cotton and G. L. Powell, *Inorg. Chem.* **1984**, *23*, 4222-4226.

828. Insertion Reactions of Hexaalkoxydimolybdenum and -ditungsten Compounds with Organic Isocyanates. Syntheses and Structures of W$_2$(OCMe$_3$)$_4$[N(C$_6$H$_5$)C(O)OCMe$_3$]$_2$ and Mo$_2$(O-*i*-Pr)$_4$[N(C$_6$H$_5$)C(O)O-*i*-Pr], M. H. Chisholm, F. A. Cotton, K. Folting, J. G. Huffman, A. L. Ratermann and E. S. Shamshoum, *Inorg. Chem.* **1984**, *23*, 4423-4427.

829. Molecular and Electronic Structures of Two 16-Electron Complexes of Tungsten(II): WBr$_2$(CO)$_2$(C$_7$H$_8$) (C$_7$H$_8$ = Norbornadiene) and WBr$_2$(CO)$_2$(PPh$_3$)$_2$, F. A. Cotton and J. H. Meadows, *Inorg. Chem.* **1984**, *23*, 4688-4693.

830. Synthesis and Characterization of an Unusual Asymmetric Diosmium Complex, Os$_2$Cl$_3$(PhNpy)$_3$, A. R. Chakravarty, F. A. Cotton and D. A. Tocher, *Inorg. Chem.* **1984**, *23*, 4693-4697.

831. An Ortho-Metalation Reaction of Os$_2$(O$_2$CCH$_3$)$_4$Cl$_2$: Syntheses and Crystal Structure Characterizations of Os$_2$(O$_2$CR)$_2$[(C$_6$H$_5$)$_2$P(C$_6$H$_4$)]$_2$Cl$_2$ (R = CH$_3$ (l), C$_2$H$_5$ (2)), A. R. Chakravarty, F. A. Cotton and D. A. Tocher, *Inorg. Chem.* **1984**, *23*, 4697-4700.

832. Synthesis and Structure of New Trinuclear Cluster Compounds $[M_3(\mu_3\text{-O})_2\text{-}$ $(O_2CC_3H_7)_6(H_2O)_3]^{2+}$ (M = Mo, W). Comparison of Mo and W Bond Radii as a Function of M-M Bond Order, F. A. Cotton, Z. Dori, D. O. Marler and W. Schwotzer, *Inorg. Chem.* **1984**, *23*, 4738-4742.

833. Preparation, Structure, and Properties of the Polar Dirhodium(II) Tetrakis(6-fluoro-2-oxypyridinate) Molecule, F. A. Cotton, S. Han and W. Wang, *Inorg. Chem.* **1984**, *23*, 4762-4765.

1985

834. Syntheses, Molecular Structures, and Properties of Two Polar Diruthenium(II,III) Complexes of 2-Hydroxypyridine and 2-Anilinopyridine, A. R. Chakravarty, F. A. Cotton and D. A. Tocher, *Inorg. Chem.* **1985**, *24*, 172-177.

835. $\delta\to\delta^*$ Transition Energies as a Function of δ-Bond Strength: An Extrapolative Assessment of the Ground-State Electron Correlation Energy, F. L. Campbell, III, F. A. Cotton and G. L. Powell, *Inorg. Chem.* **1985**, *24*, 177-181.

836. Structural and Electrochemical Characterization of the Novel Ortho-Metalated Dirhodium(II) Compounds $Rh_2(O_2CCH_3)_2[(C_6H_5)_2P(C_6H_4)]_2\cdot2L$, A. R. Chakravarty, F. A. Cotton, D. A. Tocher and J. H. Tocher, *Organometallics*, **1985**, *4*, 8-13.

837. Structural Characterization and Infrared Studies of Tungsten Bromo Carbonyl Compounds, F. A. Cotton, L. R. Falvello and J. H. Meadows, *Inorg. Chem.* **1985**, *24*, 514-517.

838. A Novel d^{10}-d^3-d^{10} Trinuclear Bimetallic Linear Complex of Zinc and Vanadium, F. A. Cotton, S. A. Duraj, W. J. Roth and C. D. Schmulbach, *Inorg. Chem.* **1985**, *24*, 525-527.

839. Two Compounds Containing the Tris(μ-chloro)hexakis(tetrahydrofuran)divanadium(II) Cation. Preparation, Structures Spectroscopic Characterization, F. A. Cotton, S. A. Duraj and W. J. Roth, *Inorg. Chem.* **1985**, *24*, 913-917.

840. Bonding in the Diniobium Molecule by All-Electron ab Initio Calculations, F. A. Cotton and I. Shim, *J. Phys. Chem.* **1985**, *89*, 952-956.

841. Addition of Di-*p*-tolylcarbodiimide to Ditungsten Hexa-*tert*-butoxide. Preparation and Structure of $[W(OCMe_3)_3]_2(\mu\text{-}C_7H_7NCNC_7H_7$, F. A. Cotton, W. Schwotzer and E. S. Shamshoum, *Organometallics* **1985**, *3*, 461-465.

842. Unusual Reactivity Associated with the Triply Bonded Re_2^{4+} Core: Facile Reaction of $Re_2Cl_4(Ph_2PCH_2PPh_2)_2$ with Nitriles to Afford the Cations, $[Re_2Cl_3(Ph_2PCH_2\text{-}PPh_2)_2(NCR)_2]^+$, T. J. Barder, F. A. Cotton, L. R. Falvello and R. A. Walton, *Inorg. Chem.* **1985**, *24*, 1258-1263.

843. New Isomeric Form of the "$M_2(OC_5H_3NCl)_4$" Core: A Polar Arrangement of the Four 6-Chloro-2-hydroxypyridinato (chp) Ligands in a Chlorodiruthenium(II,III) Complex, $Ru_2Cl(chp)_4$, A. R. Chakravarty, F. A. Cotton and D. A. Tocher, *Inorg. Chem.* **1985**, *24*, 1263-1267.

844. Syntheses and Molecular Structures of $Os_2X_2(PhCONH)_4$ (X = Cl, Br) Compounds, A. R. Chakravarty, F. A. Cotton and D. A. Tocher, *Inorg. Chem.* **1985**, *24*, 1334-1338.

845. Oxalato Complexes of the Trinuclear Aquo Ion of Molybdenum(IV), $[Mo_3O_4]^{4+}$, E. Benory, A. Bino, D. Gibson, F. A. Cotton and Z. Dori, *Inorg. Chim. Acta* **1985**, *99*, 137-142.

846. M_3X_{13} Type of Equilateral-Triangular Metal Atom Cluster Compound with Several Unprecedented Features: $[Mo_3(\mu_3\text{-}CH_3C)(\mu_2\text{-}Br)_3(\eta^2,\mu_2\text{-}CH_3CO_2)_3(H_2O)_3]ClO_4\cdot 4H_2O$, A. Birnbaum, F. A. Cotton, Z. Dori, M. Kapon, D. Marler, G. M. Reisner and W. Schwotzer, *J. Am. Chem. Soc.* **1985**, *107*, 2405-2410.

847. New Compounds with Platinum to Silver Bonds Unsupported by Covalent Bridges, R. Usón, J. Forniés, M. Tomás, J. M. Casas, F. A. Cotton and L. R. Favello, *J. Am. Chem. Soc.* **1985**, *107*, 2556-2557.

848. Reactions of $Re_2Cl_4(dppm)_2$ with Carbon Monoxide That Proceed with Retention of the Metal-Metal Bond: Synthesis of $Re_2Cl_4(dppm)_2(CO)_n$ (*n* = 1, 2) and the Structural Characterization of $Cl_2Re(\mu\text{-}Cl)(\mu\text{-}CO)(\mu\text{-}dppm)_2ReCl(CO)$, F. A. Cotton, L. M. Daniels, K. R. Dunbar, L. R. Falvello, S. M. Tetrick and R. A. Walton, *J. Am. Chem. Soc.* **1985**, *107*, 3524-3530.

849. Multiple Bonds Between Vanadium Atoms, F. A. Cotton, M. P. Diebold and I. Shim, *Inorg. Chem.* **1985**, *24*, 1510-1516.

850. Molecular Mechanics of Low Bond Order Interactions in Tetrakis(carboxylato)dimetal Systems, J. C. A. Boeyens, F. A. Cotton and S. Han, *Inorg. Chem.* **1985**, *24*, 1750-1753.

851. Preparación del Diclorotetrapiridinavanadio(II). Una Ruta Sintética Mejorada, F. A. Cotton and C. A. Murillo, *Ing. Cienc. Quim.* **1985**, *9*, 5-6.

852. Coordination of the 2-Anilinopyridine Ligand via Ortho Metalation: Synthesis and Crystal and Molecular Structure of $RhCl_2py_2[(C_6H_4)NHpy]$, A. R. Chakravarty, F. A. Cotton and D. Tocher, *Organometallics* **1985**, *4*, 863-866.

853. Preparation and Structure of $[U_2(C_6Me_6)_2Cl_7]^+$, the First Uranium(IV) Complex with a Neutral Arene in η^6-Coordination, F. A. Cotton and W. Schwotzer, *Organometallics* **1985**, *4*, 942-943.

854. Two Compounds Containing a Divanadium Tetrabenzoate Frame and Cyclopenta-dienyl or Pentamethylcyclopentadienyl Ligands, F. A. Cotton, S. A. Duraj and W. J. Roth, *Organometallics* **1985**, *4*, 1174-1177.

855. Solid-State Structure of Ferrocenecarboxylic Acid, $[Fe(C_5H_4CO_2H)(C_5H_5)]$, F. A. Cotton and A. H. Reid, Jr., *Acta Cryst.* **1985**, *C41*, 686-688.

856. Structure of Dichlorotris(triphenylphosphine)osmium(II), $[OsCl_2\{P(C_6H_5)_3\}_3]$, A. R. Chakravarty, F. A. Cotton and D. A. Tocher, *Acta Cryst.* **1985**, *C41*, 698-699.

857. Structure of Di-μ-chloro-tetrachlorobis(tetrahydroduran)[μ-(dimethyl sulfide)] diniobium(Nb–Nb), $[Nb_2Cl_6(C_4H_8O)_2(C_2H_6S)]$, and Di-$\mu$-chloro-tetrachlorobis-(tetrahydrofuran)[μ-(dimethyl sulfide)]ditantalum(Ta–Ta), $[Ta_2Cl_6(C_4H_8O)_2(C_2H_6S)]$, F. A. Cotton, S. A. Duraj and W. J. Roth, *Acta Cryst.* **1985**, *C41*, 878-881.

858. Structure of Tetrakis[dimethyl(phenyl)phosphine]mercury(II) Decachloro-μ-oxo-ditantalate(V), [Hg{P(CH$_3$)$_2$(C$_6$H$_5$)}$_4$][Ta$_2$Cl$_{10}$O], F. A. Cotton, S. A. Duraj and W. J. Roth, *Acta Cryst.* **1985**, *C41*, 881-883.

859. A Facile and Economical Synthesis of Ru$_3$(CO)$_{12}$. First Observation of the Removal of Carboxylate Bridges from a Multiply-Bonded Dimetal Center by Carbon Monoxide, F. A. Cotton, A. H. Reid, Jr. and D. A. Tocher, *Synth. React. Inorg. Met.-Org. Chem.* **1985**, *15(5)*, 637-640.

860. Syntheses, Structures and Redox Properties of Tetrakis(μ-benzamidato)dirhodium(II) Complexes, A. R. Chakravarty, F. A. Cotton, D. A. Tocher and J. H. Tocher, *Inorg. Chim. Acta* **1985**, *101*, 185-196.

861. A New Reaction Converting the Mo$_3$(μ$_3$-S)(μ-S$_2$)$_3$ Core to the Mo$_3$(μ$_3$-S)(μ-S)$_3$ Core: Structure of the {Mo$_3$(μ$_3$-S)(μ-S)$_3$-[N(CH$_2$CO$_2$)$_3$]$_3$(H)$_2$}$^{3-}$ Ion, F. A. Cotton, R. Llusar, D. O. Marler, W. Schwotzer, Z. Dori, *Inorg. Chim. Acta* **1985**, *102*, L25-L27.

862. Structural Characterization of the Triply Bonded Dirhenium(II) Complexes Re$_2$Cl$_4$(μ-Ph$_2$PCH$_2$PPh$_2$)$_2$ and α-Re$_2$Cl$_4$(Me$_2$P(CH$_2$)$_2$PMe$_2$)$_2$, T. J. Barder, F. A. Cotton, K. R. Dunbar, G. L. Powell, W. Schwotzer and R. A. Walton, *Inorg. Chem.* **1985**, *24*, 2550-2554.

863. Mechanistic Insight into the Reaction of Phenyl Isocyanate with Ditungsten Hexa-*tert*-butoxide: Isolation and Characterization of Early Intermediates, F. A. Cotton and E. S. Shamshoum, *J. Am. Chem. Soc.* **1985**, *107*, 4662-4667.

864. Synthesis and Structure of a Binuclear Ruthenium 4-Chlorobenzamidato Complex, A. R. Chakravarty, F. A. Cotton and D. A. Tocher, *Polyhedron* **1985**, *4*, 1097-1102.

865. Further Studies of Phosphine Adducts of Niobium(IV) and Tantalum(IV) Chlorides: New Seven- and Eight-Coordinate Compounds with Trimethylphosphine: NbCl$_4$(PMe$_3$)$_3$ and Ta$_2$Cl$_8$(PMe$_3$)$_4$, F. A. Cotton, M. P. Diebold and W. J. Roth, *Polyhedron* **1985**, *4*, 1103-1108.

866. Electrochemical Studies on Three Polar Diruthenium(II,III) Complexes, A. R. Chakravarty, F. A. Cotton, D. A. Tocher and J. H. Tocher, *Polyhedron* **1985**, *4*, 1475-1478.

867. A Binuclear Tantalum(III) Complex with a Bridging Carboxylato Ligand, F. A. Cotton, M. P. Diebold, S. A. Duraj and W. J. Roth, *Polyhedron* **1985**, *4*, 1479-1484.

868. Preparation, Molecular Structure and Electronic Structure of the Rhombic, Six-Coordinate Niobium(IV) Complex NbCl$_2$(Bu^{t}C(O)CHC(O)Bu^t)$_2$, F. A. Cotton, M. P. Diebold and W. J. Roth, *Polyhedron* **1985**, *4*, 1485-1491.

869. The Structure of (Dichloro) (Di-μ-acetato)(bis-triphenyl phosphine) Dimolybdenum(II), F. A. Cotton and G. L. Powell, *Polyhedron* **1985**, *4*, 1669-1671.

870. Preparation, Structures and Reactivity of Two New Addition Compounds of Carbodi-imides with Ditungsten Hexa-*t*-Butoxide, F. A. Cotton and E. S. Shamshoum, *Polyhedron* **1985**, *4*, 1727-1734.

871. A Dichromium(II) Compound with an Elongated Metal-Metal Bond: Cr$_2$(oxindolate)$_4$-(oxindole)$_2$, F. A. Cotton and W. Wang, *Polyhedron* **1985**, *4*, 1735-1739.

872. Reaction Products of 1,2-Bis(diphenylphosphino)ethane with the Octabromodimolyb-date((II) Ion, P. A. Agaskar, F. A. Cotton, D. R. Derringer, G. L. Powell, D. R. Root and T. J. Smith, *Inorg. Chem.* **1985**, *24*, 2786-2791.

873. Synthesis, Structure, and Properties of Chlorobis(acetato)bis(6-methyl-2-hydroxypyr-idinato)diruthenium(II,III), the First Ru_2^{5+} Compound with a Mixed Set of Bridging Ligands, A. R. Chakravarty, F. A. Cotton and D. A. Tocher, *Inorg. Chem.* **1985**, *24*, 2857-2861.

874. Metal-Metal Multiple Bonds in Dinuclear Clusters, F. A. Cotton and R. A. Walton, *Structure and Bonding* **1985**, *62*, 1-49.

875. Further Study of Trinuclear Palladium(II) Acetate: Crystals from Dichloromethane, F. A. Cotton and S. Han, *Revue de Chimie Minerale* **1985**, *22*, 277-284.

876. The Crystal and Molecular Structure of β-Mo_2Br_4(*bis*(diphenylphosphino)ethane)$_2$, P. A. Agaskar and F. A. Cotton, *Revue de Chimie Minerale* **1985**, *22*, 302-311.

877. Binuclear Alkoxide Complexes of Niobium and Tantalum in Lower Oxidation States, F. A. Cotton, M. P. Diebold and W. J. Roth, *Inorg. Chem.* **1985**, *24*, 3509-3510.

878. The Aryl Group Transfer Process from a Triarylphosphine to a Chlorotetrakis(amidato)-diruthenium(II,III) Species: Syntheses and Molecular Structures of $Ru_2(R)_2(R'CONH)_2$-$[R_2POC(R')N]_2$ ($R = C_6H_5$, $R' = 3,5$-$(OCH_3)_2C_6H_3$; $R = p$-$C_6H_4CH_3$, $R' = C_6H_5$, A. R. Chakravarty and F. A. Cotton, *Inorg. Chem.* **1985**, *24*, 3584-3589.

879. Further Studies of the Reactions of Ditungsten Hexa-t-Butoxide with Acetylenes. Isolation and Characterization of $WO(OCMe_3)_4$(THF), $[W_3(OCMe_3)_5(\mu$-$O)(\mu$-$CC_3H_7O]_2$ and $W(CPh)(OCMe_3)_3$, F. A. Cotton, W. Schwotzer and E. S. Shamshoum, *J. Organometal. Chem.* **1985**, *296*, 55-68.

880. Mononuclear and Binuclear Cationic Complexes of Vanadium(II), F. A. Cotton, S. A. Duraj, L. E. Manzer and W. J. Roth, *J. Am. Chem. Soc.* **1985**, *107*, 3850-3855.

881. A Novel Disproportionation Reaction That Occurs at a Multiply Bonded Dimetal Center. The Isolation of Dirhenium(IV,II) Alkoxide Complexes of the Type $(RO)_2Cl_2ReReCl_2(PPh_3)_2$ ($R = Me$, Et, or *n*-Pr), A. R. Chakravarty, F. A. Cotton, A. R. Cutler, S. M. Tetrick and R. A. Walton, *J. Am. Chem. Soc.* **1985**, *107*, 4795-4796.

882. Mixed-ligand Complexes of Diruthenium(II,III): Syntheses and Molecular Structures of Chloro-bis-(acetato)-bis-(2-anilino-pyridinato)(2-anilino-pyridine)diruthenium·methylene Chloride and Chloro(acetato)tris(6-chloro-2-oxypyridinato)diruthenium·methylene Chloride, A. R. Chakravarty and F. A. Cotton, *Inorg. Chim. Acta* **1985**, *105* 19-29.

883. Replacement of Acetate Ions in Dimolybdenum Tetraacetate by Acetonitrile Molecules: Crystal Structures of Two Compounds Containing the *cis*-$[Mo_2(O_2CCH_3)_2(CH_3CN)_4]^{2+}$ Cation, F. A. Cotton, A. H. Reid, Jr. and W. Schwotzer, *Inorg. Chem.* **1985**, *24*, 3965-3968.

884. Reactions of the Dicarbonyl Complex $Re_2Cl_4(dppm)_2(CO)_2$ with Nitriles and Isocyanides. Synthesis of $[Re_2Cl_3(dppm)_2(CO)_2L]^{n+}$ (n = 0, 1; L = RCN, RNC) and the Structural Characterization of $[Re_2Cl_3(dppm)_2(CO)_2(NCC_2H_5)]PF_6$, F. A. Cotton, K. R. Dunbar, L. R. Falvello and R. A. Walton, *Inorg. Chem.* **1985**, *24*, 4180-4187.

885. The $Mo_3S_4^{4+}$ Aquo Ion, F. A. Cotton, Z. Dori, R. Llusar and W. Schwotzer, *J. Am. Chem. Soc.* **1985**, *107*, 6734-6735.

886. The Cuboidal $Mo_4S_4^{6+}$ Aquo Ion and Its Derivatives, F. A. Cotton, M. P. Diebold, Z. Dori, R. Llusar and W. Schwotzer, *J. Am. Chem. Soc.* **1985**, *107*, 6735-6736.

887. Edge-Sharing Bioctahedral Dimolybdenum(III) Molecules with μ-RS Groups. Direct Experimental Evidence for Spin-State Equilibria, F. A. Cotton, M. P. Diebold, C. J. O'Connor and G. L. Powell, *J. Am. Chem. Soc.* **1985**, *107*, 7438-7445.

888. The First Alkylidyne-Capped Tritungsten(IV) Cluster Compounds: Preparation, Structure, and Properties of $[W_3O(CCH_3)(O_2CCH_3)_6(H_2O)_3]Br_2\cdot 2H_2O$, F. A. Cotton, Z. Dori, M. Kapon, D. O. Marler, G. M. Reisner, W. Schwotzer and M. Shaia, *Inorg. Chem.* **1985**, *24*, 4381-4384.

889. Steric and Electronic Factors Influencing the Structures of Bridged (β-Type) $M_2Cl_4(LL)_2$ (M = Mo, Re) Compounds: A Refined Correlation of Bond Length with Torsion Angle, F. L. Campbell, III, F. A. Cotton and G. L. Powell, *Inorg. Chem.* **1985**, *24*, 4384-4389.

890. Vanadium(II) and Niobium(III) Edge-Sharing Bioctahedral Complexes that Contain Bis(dimethylphosphino)methane Bridges, F. A. Cotton, S. A. Duraj, L. R. Falvello and W. J. Roth, *Inorg. Chem.* **1985**, *24*, 4389-4393.

891. Synthesis, X-Ray Structure, and Chemical Reactivity of the Tetranuclear Cluster $(NBu_4)_2[Pt_2Ag_2Cl_4(C_6F_5)_4]$. X-Ray Structure of the Binuclear Compound $(NBu)_4$-$[PtAgCl_2(C_6F_5)_2PPh_3]$, R. Usón, J. Forniés, B. Menjon, F. A. Cotton, L. R. Falvello and M. Tomás, *Inorg. Chem.* **1985**, *24*, 4651-4656.

892. Preparation and Structures of Novel Di- and Trinuclear Clusters of Iridium(II) without Carbonyl Ligands, F. A. Cotton, P. Lahuerta, M. Sanaú and W. Schwotzer, *J. Am. Chem. Soc.* **1985**, *107*, 8284.

893. Preparation and Properties of $NbBr_4(PMe_2Ph)_3$ and $NbBr_3(PMe_2Ph)_3$, F. A. Cotton, M. P. Diebold and W. J. Roth, *Inorg. Chim. Acta* **1985**, *105*, 41-49.

894. Structure of a Diruthenium(II,III) Complex with Benzamidato Bridging Ligands, A. R. Chakravarty and F. A. Cotton, *Polyhedron* **1985**, *4*, 1957-1958.

895. Structural Characterization of $Mn_2(CO)_6(μ-Ph_2PCH_2PPh_2)_2$, F. A. Cotton and K.-B. Shiu, *Gazz. Chim. Ital.* **1985**, *115*, 705-709.

1986

896. A Series of Edge-Sharing Bioctahedral, M-M Bonded Molecules: Non-monotonic Bond Length Variation and Its Interpretation, A. R. Chakravarty, F. A. Cotton, M. P. Diebold, D. B. Lewis and W. J. Roth, *J. Am. Chem. Soc.* **1986**, *108*, 971-976.

897. Syntheses, Structures and Solid-State ^{13}C NMR of Two η^6-Arene Uranium(IV) Complexes, [U(C$_6$Me$_6$)Cl$_2$(μ-Cl)$_3$UCl$_2$(C$_6$Me$_6$)]AlCl$_4$ and U(C$_6$Me$_6$)Cl$_2$(μ-Cl)$_3$UCl$_2$(μ-Cl)$_3$UCl$_2$(C$_6$Me$_6$), G. C. Campbell, F. A. Cotton, J. F. Haw and W. Schwotzer, *Organometallics* **1986**, *5*, 274-279.

898. Solid-State Structure of α-Mo$_2$Cl$_4$(dppe)$_2$ and Its Transformation of β-Mo$_2$Cl$_4$(dppe)$_2$. Evidence for the Internal Flip Mechanism, P. A. Agaskar and F. A. Cotton, *Inorg. Chem.* **1986**, *25*, 15-19.

899. A New Mixed-Ligand Diruthenium(II) Compound. Synthesis and Characterization of [Ru$_2$Cl(Me$_2$PCH$_2$PMe$_2$)$_2$(PhNpy)$_2$][BPh$_4$], A. R. Chakravarty, F. A. Cotton and L. R. Falvello, *Inorg. Chem.* **1986**, *25*, 214-219.

900. Chiroptical Properties of Compounds Containing Metal-Metal Bonds. Syntheses, Structures, and the Measurement and Interpretation of Electronic and Circular Dichroism Spectra of Rh$_2$[(S)-mandelate]$_4$(EtOH)$_2$ and Rh$_2$[(R)-α-methoxy-α-phenylacetate]$_4$(THF)$_2$, P. A. Agaskar, F. A. Cotton, L. R. Falvello and S. Han, *J. Am. Chem. Soc.* **1986**, *108*, 1214-1223.

901. A Dinuclear, Metal-Metal Bonded, Carboxylato-Bridged Niobium(III) Complex, F. A. Cotton, M. P. Diebold, M. Matusz and W. J. Roth, *Inorg. Chim. Acta* **1986**, *112*, 147-152.

902. Oxidative Addition of Diphenyl Disulfide Across a Ta=Ta Bond. Preparation and Characterization of [TaCl$_3$(Me$_2$S)]$_2$(μ-SPh)$_2$, G. C. Campbell, J. A. M. Canich, F. A. Cotton, S. A. Duraj and J. F. Haw, *Inorg. Chem.* **1986**, *25*, 287-290.

903. A New Diruthenium(II,III) Compound, Ru$_2$(C≡CPh)(PhNpy)$_4$·2CH$_2$Cl$_2$, with an Axial η^1-Acetylide Ligand, A. R. Chakravarty and F. A. Cotton, *Inorg. Chim. Acta* **1986**, *113*, 19-26.

904. Comparative Structural Studies of the First Row Early Transition Metal(III) Chloride Tetrahydrofuran Solvates, F. A. Cotton, S. A. Duraj, G. L. Powell and W. J. Roth, *Inorg. Chim. Acta* **1986**, *113*, 81-85.

905. The Crystal Structure of Dirhodium Tetrabutyrate, F. A. Cotton and K.-B. Shiu, *Revue de Chim. Minerale* **1986**, *23*, 14-19.

906. The Reaction of Carbon Dioxide with Ditungsten Hexa-*tert*-Butoxide, R. G. Abbott, F. A. Cotton and E. S. Shamshoum, *Gazz. Chim. Ital.* **1986**, *116*, 91-95.

907. Highlights of Recent Research on Compounds with Mo-Mo Bonds, F. A. Cotton, *Polyhedron* **1986**, *5*, 3-14.

908. The Preparation of Ta$_2$Cl$_6$(PhN)$_2$(Me$_2$S)$_2$ by Reaction of Ta$_2$Cl$_6$(Me$_2$S)$_3$ with PhNNPh: Crystal Structure of the Product, J. A. M. Canich, F. A. Cotton, S. A. Duraj and W. J. Roth, *Polyhedron* **1986**, *5*, 895-898.

909. Diamagnetic Anisotropy of the Mo≡Mo Bond in Mo$_2$(O$_2$CR)$_4$ Type Complexes, P. A. Agaskar and F. A. Cotton, *Polyhedron* **1986**, *5*, 899-900.

910. Synthesis and Molecular Structure of (NBu$_4$)$_x$[Pt(C$_6$Cl$_5$)$_2$(μ-Cl)$_2$Ag]$_x$: A Novel Chain Polymeric Compound, R. Usón, J. Forniés, M. Tomás, J. M. Casas, F. A. Cotton and L. R. Falvello, *Polyhedron* **1986**, *5*, 901-902.

911. Preparation and Structural Characterization of $Os_2Cl_4(Ph_2Ppy)_2(O_2CCH_3)$: A Mixed Ligand Compound with an Os_2^{5+} Core and a Bond Order of 2.5, F. A. Cotton, K. R. Dunbar and M. Matusz, *Polyhedron* **1986**, *3*, 903-905.

912. A Mixed Oxo-Thio-Bridged Tritungsten Triangular Cluster Complex, Z. Dori, F. A. Cotton, R. Llusar and W. Schwotzer, *Polyhedron* **1986**, *5*, 907-909.

913. Synthesis and Structure of the Ortho-Metalated Dirhodium(II) Compound $Rh_2(O_2CCH_3)_3[(C_6H_4)P(C_6H_5)(C_6F_4Br)]\cdot P(C_6H_5)_2(C_6F_4Br)$, F. Barceló, F. A. Cotton, P. Lahuerta, R. Llusar, M. Sanaú, W. Schwotzer and M. A. Ubeda, *Organometallics* **1986**, *5*, 808-811.

914. Tetramethyldiplatinum(III) (Pt-Pt) Complexes with 2-Hydroxypyridinato Bridging Ligands, D. P. Bancroft, F. A. Cotton, L. R. Falvello and W. Schwotzer, *Inorg. Chem.* **1986**, *25*, 763-770.

915. Preparation and X-Ray Structures of Compounds Containing the Tetrakis(methyl-hydroxypyridinato)- and Tetrakis(chlorohydroxypyridinato)dipalladium(II) Molecules, D. P. Bancroft, F. A. Cotton, L. R. Falvello and W. Schwotzer, *Inorg. Chem.* **1986**, *25*, 1015-1021.

916. Reactions of Small Molecules with $Re_2Cl_4(PEt_3)_4$. 1. Structural and Spectroscopic Studies of Several Products Resulting from the Reaction with Carbon Monoxide, S. Bucknor, F. A. Cotton, L. R. Falvello, A. H. Reid, Jr. and C. D. Schmulbach, *Inorg. Chem.* **1986**, *25*, 1021-1027.

917. Interview with F. A. Cotton, *J. Chem. Educ.* **1986**, *63*, 259-262.

918. Synthesis, Spectroscopy, and X-ray Structure of $Os_2Cl(chp)_4$: An Unusual Os_2^{5+} Complex with a Polar Arrangement of 6-Chloro-2-hydroxypyridinato Ligands and One Axial Chloride, F. A. Cotton, K. R. Dunbar and M. Matusz, *Inorg. Chem.* **1986**, *25*, 1585-1589.

919. Preparation and Structural Characterization of $Os_2Cl_4(chp)_2(L)$ (chp = 6-Chloro-2-hydroxypyridinato; L = H_2O, Pyridine): A New Class of $M_2X_4(LL)_2$ Complexes Possessing an Eclipsed Conformation Where LL is a Substituted Hydroxypyridinato Ligand, F. A. Cotton, K. R. Dunbar and M. Matusz, *Inorg. Chem.* **1986**, *25*, 1589-1594.

920. Proposed Reformulation of the Recently Reported $TaBr_3(PMe_2Ph)_2$, F. A. Cotton, M. P. Diebold and W. J. Roth, *Inorg. Chem.* **1986**, *25*, 1728.

921. Syntheses, Structures, and Circular Dichroism Spectra of β-$Mo_2X_4(S,S\text{-}dppb)_2$ (X = Cl, Br; S,S-dppb = (2S,3S)-Bis(diphenylphosphino)butane), P. A. Agaskar, F. A. Cotton, I. F. Fraser, L. Manojlovic-Muir, K. W. Muir and R. D. Peacock, *Inorg. Chem.* **1986**, *25*, 2511-2519.

922. Synthesis and Characterization of a Confacial Bioctahedral Tantalum(II) Dimer with a Formal Triple Metal-Metal Bond, F. A. Cotton, M. P. Diebold and W. J. Roth, *J. Am. Chem. Soc.* **1986**, *108*, 3538-3539.

923. $Sm(\eta^6\text{-}C_6Me_6)(\eta^2\text{-}AlCl_4)_3$: The First Structure of a Rare Earth Complex with a Neutral π-Ligand, F. A. Cotton and W. Schwotzer, *J. Am. Chem. Soc.* **1986**, *108*, 4657.

924. Mixed Carbonyl-Isocyanide and Carbonyl-Nitrile Complexes Derived from the Reactions of the Multiply Bonded Dirhenium(II) Complexes $Re_2X_4(dppm)_2(CO)$ (X = Cl or Br; dppm = $Ph_2PCH_2PPh_2$). The Structural Characterization of $Cl_2Re(\mu\text{-}Cl)(\mu\text{-}dppm)_2\text{-}$ $ReCl(CO)$ and $Cl_2Re(\mu\text{-}Cl)(\mu\text{-}CO)(\mu\text{-}dppm)_2ReCl(CNxylyl)$, F. A. Cotton, K. R. Dunbar, A. C. Price, W. Schwotzer and R. A. Walton, *J. Am. Chem. Soc.* **1986**, *108*, 4843-4850.

925. The Multiply Bonded Octahalodiosmate(III) Anions. 2. Structure and Bonding, P. A. Agaskar, F. A. Cotton, K. R. Dunbar, L. R. Falvello, S. M. Tetrick and R. A. Walton, *J. Am. Chem. Soc.* **1986**, *108*, 4850-4855.

926. Decarbonylation of Molybdenum(II) Carbonyl Complexes: A New Route to Quadruply Bonded Molybdenum(II) Dimers, F. A. Cotton and R. Poli, *J. Am. Chem. Soc.* **1986**, *108*, 5628-5629.

927. $[U_3(\mu^3\text{-}Cl)_2(\mu^2\text{-}Cl)_3(\mu^1,\eta^2\text{-}AlCl_4)_3(\eta^6\text{-}C_6Me_6)_3][AlCl_4]$ — A Trinuclear Arene Complex of U(III), F. A. Cotton, W. Schwotzer and C. Q. Simpson II, *Angew. Chem. Int. Ed. Engl.* **1986**, *25*, 637-639; *Angew. Chem.* **1986**, *98*, 652.

928. New Chemistry of Oxo Trinuclear, Metal-Metal Bonded Niobium Compounds, F. A. Cotton, M. P. Diebold, R. Llusar and W. J. Roth, *J. Chem. Soc., Chem. Commun.* **1986**, 1276-1278.

929. Synthesis and Characterization of Four Vanadium(II) Compounds, Including Vanadium(II) Sulfate Hexahydrate and Vanadium(II) Saccharinates, F. A. Cotton, L. R. Falvello, R. Llusar, E. Libby, C. A. Murillo and W. Schwotzer, *Inorg. Chem.* **1986**, *25*, 3423-3428.

930. Four Compounds Containing Oxo-Centered Trivanadium Cores Surrounded by Six μ_2,η^2-Carboxylato Groups, F. A. Cotton, M. W. Extine, L. R. Falvello, D. B. Lewis, G. E. Lewis, C. A. Murillo, W. Schwotzer, M. Tomás and J. M. Troup, *Inorg. Chem.* **1986**, *25*, 3505-3512.

931. Preparation of Iridium(III) Complexes Containing the Phosphine $P(o\text{-}BrC_6F_4)(Ph)_2$. A New Example of η^2-Coordination and Halocarbon Binding in an (*o*-Haloaryl)-phosphine, F. A. Cotton, P. Lahuerta, M. Sanaú, W. Schwotzer, and I. Solana, *Inorg. Chem.* **1986**, *25*, 3526-3528.

932. Preparation and Properties of the Tribromide of *trans*-Dibromotetrakis(acetonitrile)-vanadium(III), $[VBr_2(CH_3CN)_4]Br_3$. A Symmetric Br_3^- Ion, F. A. Cotton, G. E. Lewis and W. Schwotzer, *Inorg. Chem.* **1986**, *25*, 3528-3529.

933. Derivatization of the Cuboidal Mo_4S_4-Aquo Ion: Preparation and Structure of $Mo_4S_4(HB(pz)_3)_4(pz)$ (pz = Pyrazolyl, $C_3N_2H_3^-$), F. A. Cotton, Z. Dori, R. Llusar and W. Schwotzer, *Inorg. Chem.* **1986**, *25*, 3529-3532.

934. Reactivity of the Dirhenium(III) Carboxylate Complexes $Re_2(O_2CCH_3)_2X_4(H_2O)_2$ (X = Cl or Br) toward Monodentate Phosphines. A Novel Disproportionation Reaction Leading to Dirhenium(IV,II) Alkoxide Complexes of the Type $(RO)_2X_2ReReX_2(PPh_3)_2$ (X = Cl or Br; R = Me, Et, or Pr), A. R. Chakravarty, F. A. Cotton, A. R. Cutler and R. A. Walton, *Inorg. Chem.* **1986**, *25*, 3619-3624.

935. Thermal Reaction of $Mo_2I_4(CO)_8$ with PMe_2Ph, PEt_2Ph and Pyridine. Formation of Metal-Metal Quadruple Bonds *vs.* Disproportionation. X-Ray Crystal Structures of $MoI_3(PMe_2Ph)_2(POMe_2Ph)$ and $[PHEt_2Ph][MoI_4(PEt_2Ph)_2]$, F. A. Cotton and R. Poli, *Inorg. Chem.* **1986**, *25*, 3624-3629.

936. Reactions of the Dirhenium(II) Complexes $Re_2X_4(dppm)_2$ (X = Cl or Br; dppm = $Ph_2PCH_2PPh_2$) with Isocyanides. 3. Dinuclear Species Containing Two or Three Isocyanide Ligands, L. B. Anderson, T. J. Barder, F. A. Cotton, K. R. Dunbar, L. R. Falvello and R. A. Walton, *Inorg. Chem.* **1986**, *25*, 3629-3636.

937. Synthesis, Characterization, and Molecular Structure of a Novel Triply Bonded Dirhenium(II) Complex that Contains Three Intramolecular Phosphine Bridging Ligands: $Re_2Cl_4(\mu-Me_2PCH_2PMe_2)_3$, L. B. Anderson, F. A. Cotton, L. R. Falvello, W. S. Harwood, D. Lewis and R. A. Walton, *Inorg. Chem.* **1986**, *25*, 3637-3641.

938. Quadruply Bonded $Mo_2I_4(dppe)_2$ (dppe = Bis(diphenylphosphino)ethane): Twisted and Eclipsed Rotational Conformations and Their Significance, F. A. Cotton, K. R. Dunbar and M. Matusz, *Inorg. Chem.* **1986**, *25*, 3641-3649.

939. Oxidative Fragmentation of the Cuboid Mo_4S_4 Cluster Core: Synthesis and Structures of $[Mo_3(\mu_3-S)(\mu-S)_3([9]aneN_3)_3]^{4+}$ and $\{[MoO([9]aneN_3)]_2(\mu-S)_2\}^{2+}$, F. A. Cotton, Z. Dori, R. Llusar and W. Schwotzer, *Inorg. Chem.* **1986**, *25*, 3654-3658.

940. Synthesis and Molecular Structure of $Mo_2I_4(dppm)_2 \cdot 2C_7H_8$ (dppm = Bis(diphenyl-phosphino)methane), F. A. Cotton, K. R. Dunbar and R. Poli, *Inorg. Chem.* **1986**, *25*, 3700-3703.

941. Products of the Reaction Between $Mo_2I_4(CO)_8$ and Bis(dimethylphosphino)methane. X-Ray Crystal Structure of $MoI_2(CO)(dmpm)_2 \cdot C_7H_8$, F. A. Cotton and R. Poli, *Inorg. Chem.* **1986**, *25*, 3703-3705.

942. Air Oxidation of $Ir_2(Cl)_2(COD)_2$ Revisited. The Structures of $[Ir(\mu^2-Cl)(COD)]_2$ (Ruby Form) and its Oxidation Product, $Ir_2Cl_2(COD)_2(\mu^2-OH)_2(\mu^2-O)$, F. A. Cotton, P. Lahuerta, M. Sanaú and W. Schwotzer, *Inorg. Chim. Acta* **1986**, *120*, 153-157.

943. Iridium(I) Dimers with Bridging *N,N'*-di-*p*-tolylformamidine. X-ray Molecular Structure of $Ir_2(\mu-p-CH_3C_6H_4NCHN-p-C_6H_4CH_3)(\mu-NH-p-C_6H_4CH_3)(C_8H_{14})_2$, F. A. Cotton and R. Poli, *Inorg. Chim. Acta* **1986**, *122*, 243-248.

944. Synthesis and Molecular Structure of the Polar, Binuclear Complex, Monochlorotetra-(6-fluoro-2-oxopyridine)diruthenium(II,III), A. R. Chakravarty, F. A. Cotton and W. Schwotzer, *Polyhedron* **1986**, *5*, 1821-1827.

945. Synthesis, Properties, and Structural Characterization of the Bis(dimethylphosphino)-methane Complex $Mo_2(\mu-dmpm)_2Cl_4$, F. A. Cotton, L. R. Falvello, W. S. Harwood, G. L. Powell and R. A. Walton, *Inorg. Chem.* **1986**, *25*, 3949-3953.

946. Synthesis, Structure, and Reactivity of $(NBu_4)_x[Pt(C_6Cl_5)_2(\mu-Cl)_2Ag]_x$, a Novel Chain Polymeric Pt-Ag Compound. Structure of $(NBu_4)[PtAgCl_2(C_6Cl_5)_2(PPh_3)]$ and $[Pt(C_6Cl_5)_2(\mu-Cl)_2Ag_2(PPh_2Me)]_2$, R. Usón, J. Forniés, M. Tomás, J. M. Casas, F. A. Cotton and L. R. Falvello, *Inorg. Chem.* **1986**, *25*, 4519-4525.

947. Single Crystal X-Ray Study of Solid Solutions of a Jahn-Teller Compound in an Undistorted Host: The Cr, Zn Saccharinate Hexahydrate System, F. A. Cotton, L. R. Falvello, C. A. Murillo and G. Valle, *Z. Anorg. Allg. Chem.* **1986**, *540/541*, 67-79. (Dedicated to P. Hagenmuller).

1987

948. Metal Metal Bonds in Edge Sharing Bioctahedra, F. A. Cotton, in *Understanding Molecular Properties*, J. Avery, *et al.*, Eds., D. Riedel Publishing Co. **1987**, pp. 17-26.

949. Preparation and Structure of $Nb_3Cl_8(CNCMe_3)_5$, F. A. Cotton and W. J. Roth, *Inorg. Chim. Acta* **1987**, *126*, 161-166.

950. Synthesis, Molecular Structure and Spectroscopic Characterization of $Mo(CO)_2I_2(\eta^2$-dppm)$(\eta^1$-dppm), F. A. Cotton and M. Matusz, *Polyhedron* **1987**, *6*, 261-267.

951. Mixed-Ligand Systems Containing Quadruple Bonds. Capture and Structural Characterization of an Intermediate in the Ligand Exchange Process Leading to New Carboxylates of the Dimolybdenum(4^+) Unit. Synthesis and X-Ray Crystallographic and Electrochemical Studies of $Mo_2[(\eta^5\text{-}C_5H_4CO_2)Fe(\eta^5\text{-}C_5H_5)]_2(O_2CCH_3)_2(C_5H_5N)_2$ and $[Mo_2[(\eta^5\text{-}C_5H_4CO_2)Fe(\eta^5\text{-}C_5H_5)]_4(ax\text{-}CH_3CN)(ax\text{-}DMSO)](DMSO)_2$, F. A. Cotton, L. R. Falvello, A. H. Reid, Jr. and J. H. Tocher, *J. Organometal. Chem.* **1987**, *319*, 87-97.

952. N,N'-Di-*p*-tolylformamidinato-Bridged Mixed-Valence Iridium(I)–Iridium(III) Dimers Containing a Metal-Metal Dative Bond. Synthesis, Molecular Structure, and Physicochemical Properties, F. A. Cotton and R. Poli, *Inorg. Chem.* **1987**, *26*, 590-595.

953. New Mode of Bonding of Bis((diphenylphosphino)methyl)phenylphosphine, dpmp: Synthesis and Structure of a Sterically Crowded, Triply Bonded Dirhenium Complex, $[Re_2Cl_3(dpmp)_2]X$ (X = Cl, PF_6), F. A. Cotton and M. Matusz, *Inorg. Chem.* **1987**, *26*, 984-988.

954. Conversion of an Electron-Rich Triple Bond to a Double Bond by Oxidative Addition of Diphenyl Diselenide to $Re_2Cl_4(\mu\text{-}dppm)_2$. Preparation and Characterization of $Re_2Cl_4(\mu\text{-}SePh)_2(\mu\text{-}dppm)_2$ (dppm = Bis(diphenylphosphino)methane), F. A. Cotton and K. R. Dunbar, *Inorg. Chem.* **1987**, *26*, 1305-1309.

955. Isolation and Structure of $Os_2Cl_4[(C_6H_5)_2P(C_6H_4)]_2$. An M_2L_8 Compound with an Unprecedented Geometry and a Short Os-Os Bond, F. A. Cotton and K. R. Dunbar, *J. Am. Chem. Soc.* **1987**, *109*, 2199-2200.

956. Discrete Trinuclear Complexes of Niobium Related to the Local Structure in Nb_3Cl_8, F. A. Cotton, M. P. Diebold and W. J. Roth, *J. Am. Chem. Soc.* **1987**, *109*, 2833-2834.

957. Reactions of M-M Bonded Dinuclear Zirconium(III) Complexes with Vicinal Dihaloalkanes and Olefins: A New Type of Metal Olefin Complex, F. A. Cotton and P. A. Kibala, *Polyhedron* **1987**, *6*, 645-646.

958. The Structures of Metal-Metal Bonded Edge-Sharing Bioctahedral Complexes, F. A. Cotton, *Polyhedron* **1987**, *6*, 667-677.

959. Further Studies on the Ortho-Metallation Reaction of Dirhodium Tetraacetate: Syntheses and Structures of $Rh_2(O_2CCH_3)_2[P(C_6H_4)(C_6H_5)(C_6F_4Br)]_2(H_2O)$ and $Rh_2(O_2CCH_3)_2[P(C_6H_4)(C_6H_5)(C_6F_4Br)]_2$, F. Barcelo, F. A. Cotton, P. Lahuerta, M. Sanaú, W. Schwotzer and M. A. Ubeda, *Organometallics* **1987**, *6*, 1105-1110.

960. Heterobinuclear PtAg Compounds with Platinum-Silver Bonds Unsupported by Covalent Bridges. Molecular Structure of $(C_6F_5)_3(SC_4H_8)PtAgPPh_3$, F. A. Cotton, L. R. Falvello, R. Usón, J. Forniés, M. Tomás, J. M. Casas and I. Ara, *Inorg. Chem.* **1987**, *26*, 1366-1370.

961. Isolation and Structure of the Novel Dirhodium(II) Compound $Rh_2(dmpm)_2$-$[(C_6H_5)_2P(C_6H_4)]_2Cl_2$ with Bridging Bis(dimethylphosphino)methane and Ortho-Metalated Triphenylphosphine Ligands, F. A. Cotton and K. R. Dunbar, *J. Am. Chem. Soc.* **1987**, *109*, 3142-3143.

962. Reactions of Iodine with Triphenylphosphine and Triphenylarsine, F. A. Cotton and P. A. Kibala, *J. Am. Chem. Soc.* **1987**, *109*, 3308-3312.

963. Preparation and Structural Characterization of Os(II) and Ru(II) Complexes, $MCl_2(vdpp)_2$ (M = Os or Ru, vdpp = 1,1-Bis(diphenylphosphino)ethene, F. A. Cotton, M. P. Diebold and M. Matusz, *Polyhedron* **1987**, *6*, 1131-1134.

964. Reactions Between $W_2I_4(CO)_8$ and Phosphines Under Forcing Conditions: Attempted Decarbonylation to Tungsten(II) Dimers with Quadruple Metal-Metal Bond. X-Ray Molecular Structure of $WI_2(CO)(dppm)_2$, F. A. Cotton, L. R. Falvello and R. Poli, *Polyhedron* **1987**, *6*, 1135-1142.

965. Synthesis, Molecular Structure and Physicochemical Properties of $M_2(form)_4$ (M = Ni, Pd; form = *N,N'*-Di-*p*-tolylformamidinato). Attempts to Generate a Palladium(III) Dimer, F. A. Cotton, M. Matusz and R. Poli, *Inorg. Chem.* **1987**, *26*, 1472-1474.

966. Low-Valent Molybdenum Carbonyl Complexes as an Entry to Octahedral MoI_3L_3 Complexes. Synthesis and X-Ray Molecular Structure of $MoI_3(THF)_3$, F. A. Cotton and R. Poli, *Inorg. Chem.* **1987**, *26*, 1514-1518.

967. Preparation and Structures of the Binuclear Vanadium(II) Complexes, $[L_3V(\mu-Cl)_3VL_3]$-BPh_4 with L = Tetrahydrofuran and 3-Methyltetrahydrofuran, J. A. M. Canich, F. A. Cotton, S. A. Duraj and W. J. Roth, *Polyhedron* **1987**, *6*, 1433-1437.

968. Synthesis and Structural Characterization of a Polar Os(II, III) Complex, $Os_2(fhp)_4Cl$, F. A. Cotton and M. Matusz, *Polyhedron* **1987**, *6*, 1439-1443.

969. Synthesis and Molecular Structure of the First Inorganic Quadruply-Bridged Ir(II) Dimer (Ir-Ir), F. A. Cotton and R. Poli, *Polyhedron* **1987**, *6*, 1625-1628.

970. Syntheses and Structural Comparison of the η^6-Arene Complexes $Sm(C_6Me_6)(AlCl_4)_3$ and $U(C_6Me_6)(AlCl_4)_3$, F. A. Cotton and W. Schwotzer, *Organometallics* **1987**, *6*, 1275-1280.

971. Reactions of the Dirhenium(II) Complexes $Re_2X_4(dppm)_2$ (X = Cl or Br; dppm = $Ph_2PCH_2PPh_2$) with Isocyanides. 4. Isomerism in Mixed Carbonyl-Isocyanide Complexes of Stoichiometry $[Re_2Cl_3(dppm)_2(CO)_2(CNR)]^{n+}$ (n = 0 or 1) and $Re_2Cl_3(dppm)_2(CO)(CNR)_2]^+$ (R = t-Bu or Xylyl) which Possess Edge-Shared Bioctahedral Structures, L. B. Anderson, F. A. Cotton, K. R. Dunbar, L. R. Falvello, A. C. Price, A. H. Reid, Jr. and R. A. Walton, *Inorg. Chem.* **1987**, *26*, 2717-2725.

972. Ortho Metalation of Pyridine at a Diiridium Center. Synthesis and Spectroscopic and Crystallographic Characterization of NC_5H_4- and *N,N'*-Di-*p*-tolylformamidinato-Bridged Complexes of Diiridium(II), F. A. Cotton and R. Poli, *Organometallics* **1987**, *6*, 1743-1751.

973. Direct Synthesis from $MoCl_3(THF)_3$ of a Complex Containing the $[Mo_3S_4]^{4+}$ Core, F. A. Cotton and R. Llusar, *Polyhedron* **1987**, *6*, 1741-1745.

974. Preparation and Structural Characterization of a Pentacoordinate Ru(II) Complex, $RuCl_2(Ph_2PCH_2SiMe_3)_3$, F. A. Cotton and M. Matusz, *Inorg. Chim. Acta* **1987**, *131*, 213-216.

975. Crystal Structures of Two $MoOX_2L_3$ Complexes, $MoOCl_2(PMePh_2)_3$ and $MoO(NCO)_2(PEt_2Ph)_3$. Implications to Distortional Isomerism, F. A. Cotton, M. P. Diebold and W. J. Roth, *Inorg. Chem.* **1987**, *26*, 2848-2852.

976. Variable Stereochemistry of the Eight-Coordinate Tetrakis(oxalato)niobium(IV), $Nb(C_2O_4)_4^{4-}$, F. A. Cotton, M. P. Diebold and W. J. Roth, *Inorg. Chem.* **1987**, *26*, 2889-2893.

977. New Directions in the Chemistry of Dirhodium(II) Compounds, F. A. Cotton, K. R. Dunbar and M. G. Verbruggen, *J. Am. Chem. Soc.* **1987**, *109*, 5498-5506.

978. Synthesis and Characterization of Niobium(II) and Tantalum(II) Compounds Containing Triple M-M Bonds, F. A. Cotton, M. P. Diebold and W. J. Roth, *J. Am. Chem. Soc.* **1987**, *109*, 5506-5514.

979. Reactions of Small Molecules with $Re_2Cl_4(PEt_3)_4$. 2. Products Resulting from the Reaction with Dihydrogen, S. Bucknor, F. A. Cotton, L. R. Falvello, A. H. Reid, Jr. and C. D. Schmulbach, *Inorg. Chem.* **1987**, *26*, 2954-2959.

980. Synthesis of $Mo_2X_4(PMe_3)_4$ (Mo–Mo) (X = Cl, Br, I) Compounds by Decarbonylation of $Mo_2X_4(CO)_8$ and by Comproportionation of $Mo(CO)_6$ and $MoX_3(PMe_3)_3$. X-Ray Crystal Structure of $Mo_2I_4(PMe_3)_4$·2THF, F. A. Cotton and R. Poli, *Inorg. Chem.* **1987**, *26*, 3228-3231.

981. Synthesis, Solid-State and Solution Structure, and Physicochemical Properties of the Iodide-Bridged Face-Sharing Bioctahedral Molybdenum(III) Dimers, $[Cat]^+[Mo_2I_7(PMe_3)_2]^-$ (Cat = $PHMe_3$, NMe_4, $AsPh_4$), F. A. Cotton and R. Poli, *Inorg. Chem.* **1987**, *26*, 3310-3315.

982. Dinuclear Niobium(IV) Complexes, $Nb_2Cl_4(OMe)_4L_2$, (L = MeOH and CH_3CN) and Their Relation to Analogous W and Mo Compounds, F. A. Cotton, M. P. Diebold and W. J. Roth, *Inorg. Chem.* **1987**, *26*, 3319-3322.

983. Alkoxide Complexes of Niobium(III), F. A. Cotton, M. P. Diebold and W. J. Roth, *Inorg. Chem.* **1987**, *26*, 3323-3327.

984. Preparation, Spectroscopic Properties, and Characterization of *anti-* and *syn-α*-$Mo_2Cl_4[(C_6H_5)_2PCH_2CH_2P(p-CH_3C_6H_4)_2]$ in Solution and in the Solid State, F. A. Cotton and S. Kitagawa, *Inorg. Chem.* **1987**, *26*, 3463-3468.

985. Structural and Spectroscopic Characterization of Dirhenium Octakis(isothiocyanato) Salt, $(Ph_4As)_2Re_2(NCS)_8 \cdot 2L$ (L = $(CH_3)_2CO$, C_5H_5N). A Unique Case of Axial Coordination in M_2X_8 Compounds, F. A. Cotton and M. Matusz, *Inorg. Chem.* **1987**, *26*, 3468-3472.

986. Structural and Theoretical Considerations in Tantalum and Niobium $M_2Cl_6(SMe_2)_4$ and $M_2Cl_6(dto)_2$ Complexes, J. A. M. Canich and F. A. Cotton, *Inorg. Chem.* **1987**, *26*, 3473-3478.

987. Heterotrinuclear Pt_2Ag Clusters with Pt-Ag Bonds Unsupported by Covalent Bridges. Molecular Structures of $(NBu_4)_2[Pt_2(\mu-Cl)_2(C_6F_5)_4]$ and $(NBu_4)[Pt_2AgCl_2(C_6F_5)_4O-(C_2H_5)_2]$, R. Usón, J. Forniés, M. Tomás, J. Casas, F. A. Cotton and L. R. Falvello, *Inorg. Chem.* **1987**, *26*, 3482-3486.

988. Structural and Vibrational Characteristics of the Tetrasulfatodimolybdenum Ions with Mo-Mo Bond Orders of 3.5 and 4.0, A. Bino, F. A. Cotton, D. O. Marler, S. Farquharson, B. Hutchinson, B. Spencer and J. Kincaid, *Inorg. Chim. Acta* **1987**, *133*, 295-300.

989. Synthesis and Molecular Structure of a Dinuclear Quadruply-Bridged Cobalt(II) Compound with a Short Metal-Metal Bond, $Co_2[(p-CH_3C_6H_4)NNN(p-C_6H_4CH_3)]_4$, F. A. Cotton and R. Poli, *Inorg. Chem.* **1987**, *26*, 3652-3653.

990. Further Study of Metal-Metal Bond Lengths in Homologous Edge-Sharing Bioctahedral Complexes, J. A. M. Canich, F. A. Cotton, L. M. Daniels and D. B. Lewis, *Inorg. Chem.* **1987**, *26*, 4046-4050.

991. Oxidative Addition to M-M Quadruple Bonds. Preparation of New Edge-Sharing Bioctahedral Complexes of the Type (L-L)Cl_2M(μ-Cl)$_2$MCl_2(L-L) (M = Mo, W, L-L = 1,2-Bis(diphenylphosphino)ethane; M = Mo, L-L = 1-Diethylphosphino-2-diphenylphosphinoethane), P. A. Agaskar, F. A. Cotton, K. R. Dunbar, L. R. Falvello and C. J. O'Connor, *Inorg. Chem.* **1987**, *26*, 4051-4057.

992. Reactions of $Rh_2(O_2CCH_3)_2(C_6H_4PPh_2)_2 \cdot 2CH_3COOH$ with Chlorotrimethylsilane in the Presence of Monodentate Phosphines to Give $[Rh_2Cl_2(C_6H_4PPh_2)_2(PPh_3)_2]$ and $[Rh_2Cl_2(C_6H_4PPh_2)_2(PMe_3)_2]$, F. A. Cotton, K. R. Dunbar and C. T. Eagle, *Inorg. Chem.* **1987**, *26*, 4127-4130.

993. Two Diastereomeric Forms of the Bischelated, Edge-Sharing Bioctahedral Molecule, $Ta_2Cl_6(Et_2PCH_2CH_2PEt_2)_2$, F. A. Cotton, M. P. Diebold and W. J. Roth, *Inorg. Chem.* **1987**, *26*, 4130-4133.

994. Oxygen Abstraction from Tetrahydrofuran by Molybdenum-Iodide Complexes. X-ray Molecular Structure of $[Mo_2(\mu-O)(\mu-I)(\mu-O_2CCH_3)I_2(THF)_4[MoOI_4(THF)]$ (THF = Tetrahydrofuran), F. A. Cotton and R. Poli, *Polyhedron* **1987**, *6*, 2181-2186.

995. Phosphorus Versus Nitrogen Donor Ligands in Edge-Sharing Bioctahedra of Niobium and Tantalum $M_2Cl_6(L-L)_2$ Complexes, J. A. M. Canich and F. A. Cotton, *Inorg. Chem.* **1987**, *26*, 4236-4240.

1988

996. Synthesis by Spontaneous Self-Assembly of Metal Atom Clusters of Zirconium, Niobium and Tantalum, F. A. Cotton, P. A. Kibala and W. J. Roth, *J. Am. Chem. Soc.* **1988**, *110*, 298-300.

997. Synthesis and Molecular Structure of the First Discrete Rhenium Polysulfide Complex, $(n\text{-Bu}_4N)[ReS_9]$, F. A. Cotton, P. A. Kibala and M. Matusz, *Polyhedron* **1988**, *7*, 83-86.

998. C-Br Insertion Reactions in Rhodium and Iridium Compounds Containing an *ortho*-Halo-Arylphosphine. X-ray Structures of Two *ortho*-Metallated Compounds of Formula $IrX_2[P(C_6F_4)Ph_2][P(o\text{-Br}C_6F_4)Ph_2]$, (X = Cl, Br), J. C. Besteiro, F. A. Cotton, P. Lahuerta, R. Llusar, M. Sanaú, W. Schwotzer and I. Solana, *Polyhedron* **1988**, *7*, 87-96.

999. Thermal Decarbonylation of Molybdenum(II) Carbonyl-Iodide Complexes. Molecular and Electronic Structures of the Mixed-Valence Trinuclear Clusters, $Mo_3HI_7L_3$ (L = Tetrahydrofuran, Acetonitrile, Benzonitrile) and Molecular Structures of $MoI_3(EtCN)_3$ and $Mo_2I_4(PhCN)_4$, F. A. Cotton and R. Poli, *J. Am. Chem. Soc.* **1988**, *110*, 830-841.

1000. Dinuclear Formamidinato Complexes of Nickel and Palladium, F. A. Cotton, M. Matusz, R. Poli and X. Feng, *J. Am. Chem. Soc.* **1988**, *110*, 1144-1154.

1001. Vanadium(II)- and Vanadium(III)-Formamidinato Compounds. Synthesis and Molecular Structure of $V(form)_2(NC_5H_5)_2$ and $V(\eta^5\text{-}C_5H_5)(form)_2$ [form = N,N'-di-*p*-tolylformamidine], F. A. Cotton and R. Poli, *Inorg. Chim. Acta* **1988**, *141*, 91-98.

1002. Synthesis, Characterization, and Molecular Structure of the Quadruply Bonded Ditungsten(II) Complex, $W_2Cl_4(dppm)_2$, J. A. M. Canich and F. A. Cotton, *Inorg. Chim. Acta* **1988**, *142*, 69-74.

1003. Two Reactions of $[Re_2Cl_8]^{2-}$ with Diphosphines that Lead to Re-Re Bond Cleavage, F. A. Cotton and L. M. Daniels, *Inorg. Chim. Acta* **1988**, *142*, 255-262.

1004. Diosmium and Dirhodium Compounds Containing a Cisoid Arrangement of 2-Diphenylphosphinopyridine Bridges, F. A. Cotton and M. Matusz, *Inorg. Chim. Acta* **1988**, *143*, 45-53.

1005. Reactions of $Nb_2Cl_6(SMe_2)_3$ and $Ta_2Cl_6(SMe_2)_3$ with Compounds Containing N-N Single Bonds, J. A. M. Canich, F. A. Cotton, S. A. Duraj and L. R. Falvello, *Inorg. Chim. Acta* **1988**, *143*, 185-192.

1006. The Crystal and Molecular Structure of $Re_2Cl_6(PMePh_2)_2$, F. A. Cotton, M. P. Diebold and W. J. Roth, *Inorg. Chim. Acta* **1988**, *144*, 17-19.

1007. Structural and Spectroscopic Studies of $Mo_2Cl_4(PR_3)_4$ Compounds with Phosphines of Varied Basicity, F. A. Cotton, L. M. Daniels, G. L. Powell, A. J. Kahaian and T. J. Smith and E. Fiore Vogel, *Inorg. Chim. Acta* **1988**, *144*, 109-121.

1008. Synthesis and Structural Characterization of Three Compounds with Zr(III)-Zr(III) Bonds: $Zr_2Cl_6(dppe)_2 \cdot 2C_2H_4Cl_2 \cdot 1.5C_6H_5CH_3$, $Zr_2Cl_6(PMe_2Ph)_4$, and $Zr_2Cl_6(PEt_3)_4$, F. A. Cotton, M. P. Diebold and P. A. Kibala, *Inorg. Chem.* **1988**, *27*, 779-804.

1009. Oxidative Addition Reactions of S-S and Se-Se Bonds to Mo_2 and W_2 Quadruple Bonds, J. A. M. Canich, F. A. Cotton, K. R. Dunbar and L. R. Falvello, *Inorg. Chem.* **1988**, *27*, 804-811.

1010. Direct Evidence Favouring an Internal Flip of Dimetal Units in Cuboidal Ligand Cages, F. A. Cotton and S. Kitagawa, *Polyhedron* **1988**, *7*, 463-470.

1011. Ortho-Metalation Reactions in Binuclear Dirhodium Compounds. Synthesis and Molecular Structure of an Unsymmetrical Rh_2^{4+} Compound with Two Different Ortho-Metalated Phosphines, F. A. Cotton, F. Barceló, P. Lahuerta, R. Llusar, J. Paya, M. A. Ubeda, *Inorg. Chem.* **1988**, *27*, 1010-1013.

1012. A New Synthetic Entry to Tungsten-Sulfur Cluster Chemistry: Preparation and Structure of $[W_3S_4Cl_3(dmpe)_3]^+$, F. A. Cotton and R. Llusar, *Inorg. Chem.* **1988**, *27*, 1303-1305.

1013. Molecular Orbital and Spectroscopic Studies of Triple Bonds Between Transition-Metal Atoms. 3. The d^5-d^5 Organometallic Dimer $Re_2(C_3H_5)_4$, F. A. Cotton, G. G. Stanley, A. H. Cowley and M. Lattman, *Organometallics* **1988**, *7*, 835-840.

1014. Tetramethyldiplatinum(III) (Pt-Pt) Complexes with 2-Hydroxypyridinato Bridging Ligands. 2. Reversals of Ligand Orientations, D. P. Bancroft and F. A. Cotton, *Inorg. Chem.* **1988**, *27*, 1633-1637.

1015. Crystalline Adducts of Tetrakis(triphenylacetato)dichromium(II) with Benzene, Pyridine, and Diethyl Ether. Benzene as a Multiple π Donor, F. A. Cotton, X. Feng, P. A. Kibala and M. Matusz, *J. Am. Chem. Soc.* **1988**, *110*, 2807-2815.

1016. Synthesis and Characterization of Trinuclear Palladium Carboxylate Complexes, D. P. Bancroft, F. A. Cotton, L. R. Falvello and W. Schwotzer, *Polyhedron* **1988**, *7*, 615-621.

1017. Binuclear Cationic μ-Bromo Complexes of Vanadium(II), J. A. M. Canich, F. A. Cotton, S. A. Duraj and W. J. Roth, *Polyhedron* **1988**, *7*, 737-740.

1018. Oxo-Bridged Ta(+3) Dimers, $(TaCl_2L_2)_2(\mu\text{-}O)(\mu\text{-}SR_2)$, Revisited. Structural Differences Between Isoelectronic μ-O and μ-OH Complexes, F. A. Cotton, M. P. Diebold and W. J. Roth, *Inorg. Chim. Acta* **1988**, *149*, 105-110.

1019. Preparation and Properties of the $Mo_2X_4(PMe_3)_4$ Compounds with X = NCS and NCO, F. A. Cotton and M. Matusz, *Inorg. Chem.* **1988**, *27*, 2127-2131.

1020. A New Sulfur-Bridged Ir(I)-Ir(I) Compound and Its Conversion to a Compound Containing a Metal-Metal Bond. Crystal Structures of $Ir_2(SPh)_2(COD)_2$ and $Ir_2Cl_2(\mu_2\text{-}SPh)_2(COD)_2$, F. A. Cotton, P. Lahuerta, J. Latorre, M. Sanaú, I. Solana and W. Schwotzer, *Inorg. Chem.* **1988**, *27*, 2131-2135.

1021. Further Studies of Bi-Oxo-Capped Triniobium Cluster Complexes, F. A. Cotton, M. P. Diebold and W. J. Roth, *Inorg. Chem.* **1988**, *27*, 2347-2352.

1022. Synthesis of Ruthenium(II) Compounds with *ortho*-Oxypyridinate Ligands (hp). Crystal Structure Characterization of [Ru(η^6-*p*-CH$_3$C$_6$H$_4$CH(CH$_3$)$_2$)Cl(hp)], P. Lahuerta, J. Latorre, M. Sanaú, F. A. Cotton and W. Schwotzer, *Polyhedron* **1988**, *7*, 1311-1316.

1023. Studies on the α and β Isomers of Complexes of the Type Re$_2$X$_4$(LL)$_2$. Structural Characterization of α- and β-Re$_2$Cl$_4$(dppee)$_2$ (dppee = *cis*-Ph$_2$PCH=CHPPh$_2$), M. Bakir, F. A. Cotton, L. R. Falvello, K. Vidyasagar and R. A. Walton, *Inorg. Chem.* **1988**, *27*, 2460-2466.

1024. Structural Characterization of Two Oxotungsten(IV) Complexes, [WOCl(Me$_2$P-CH$_2$CH$_2$PMe$_2$)$_2$]ClO$_4$ and [WOCl{Me$_2$PCH$_2$P(S)Me$_2$}$_2$]PF$_6$, F. A. Cotton and R. Llusar, *Acta Cryst.* **1988**, *C44*, 952-955.

1025. Molybdenum Complexes of 1,2-Bis(diphenylphosphino)benzene. Mononuclear Molybdenum(II) Species formed by Facile Metal-Metal Bond Cleavage of the (Mo–Mo)$^{4+}$ Core, M. Bakir, F. A. Cotton, M. M. Cudahy, C. Q. Simpson, T. J. Smith, E. Fiorevogel and R. A. Walton, *Inorg. Chem.* **1988**, *27*, 2608-2612.

1026. Preparation and Molecular and Electronic Structures of a Diamagnetic Diruthenium(II) Compound, Ru$_2$[(*p*-CH$_3$C$_6$H$_4$)NNN(*p*-CH$_3$C$_6$H$_4$)]$_4$, F. A. Cotton and M. Matusz, *J. Am. Chem Soc.* **1988**, *110*, 5761-5764.

1027. The Crystal Structure of [(*n*-C$_4$H$_9$)$_4$N]$_2$Re$_2$I$_8$: Three-Fold Disorder of the Effectively Cubic Anion, F. A. Cotton, L. M. Daniels and K. Vidyasagar, *Polyhedron* **1988**, *7*, 1667-1672.

1028. Diamagnetic Anisotropy of Quadruple Mo-Mo Bonds: α-Mo$_2$Cl$_4$(Diphosphine)$_2$ Complexes, F. A. Cotton and S. Kitagawa, *Polyhedron* **1988**, *7*, 1673-1676.

1029. Electronic Excitation MO Assignment of Rh$_2$(O$_2$CR)$_4$(L$_{ax}$)$_2$ Using MCD Experiment and Theory, J. W. Trexler, Jr., A. F. Schreiner and F. A. Cotton, *Inorg. Chem.* **1988**, *27*, 3265-3267.

1030. Discrete Trinuclear Complexes of Niobium and Tantalum Related to the Local Structure in Nb$_3$Cl$_8$, F. A. Cotton, M. P. Diebold, X. Feng and W. J. Roth, *Inorg. Chem.* **1988**, *27*, 3413-3421.

1031. Further Studies of the Monooxo-Capped Tritungsten Carboxylate Clusters, A. Bino, F. A. Cotton and Z. Dori, M. Shaia-Gottlieb and M. Kapon, *Inorg. Chem.* **1988**, *27*, 3592-3596.

1032. Further Studies of Low Valent Alkoxide Complexes of Niobium. Synthesis and Structure of Dimeric Niobium(IV) Nonamethoxide, F. A. Cotton, M. P. Diebold and W. J. Roth, *Inorg. Chem.* **1988**, *27*, 3596-3600.

1033. Experimental and Theoretical Studies of the Copper(I) and Silver(I) Dinuclear *N,N'*-Di-*p*-tolylformamidinato Complexes, F. A. Cotton, X. Feng, M. Matusz and R. Poli, *J. Am. Chem Soc.* **1988**, *110*, 7077-7083.

1034. Formation of ReCl(H$_2$)(PMePh$_2$)$_4$; A Complex Containing an η^2-H$_2$ Ligand, F. A. Cotton and R. L. Luck, *J. C. S. Chem. Commun.* **1988**, 1277-1278.

1035. Tetramethyldiplatinum(III) (Pt-Pt) Complexes with 2-Hydroxypyridinato Bridging Ligands. 3. Compounds with Diethyl Sulfide as Axial Ligand, D. P. Bancroft and F. A. Cotton, *Inorg. Chem.* **1988**, *27*, 4022-4025.

1036. Synthesis, X-Ray Structure, and Reactivity of (NBu$_4$)$_2$[Pt$_2$(μ-C$_6$F$_5$)$_2$(C$_6$F$_5$)$_4$]·H$_2$O) and (NBu$_4$)[Pt$_2$Ag(μ-C$_6$F$_5$)$_2$(C$_6$F$_5$)$_4$O(C$_2$H$_5$)$_2$]. The First Complexes Containing Bridging Pentafluorophenyl Groups, R. Usón, J. Forniés, M. Tomás, J. M. Casas, F. A. Cotton, L. R. Falvello and R. Llusar, *Organometallics* **1988**, *7*, 2279-2285.

1037. A Discrete Tetranuclear Complex of Niobium, Nb$_4$Cl$_{10}$(PMe$_3$)$_6$, Related to the Local Structure in CsNb$_4$Cl$_{11}$, F. A. Cotton and M. Shang, *J. Am. Chem Soc.* **1988**, *110*, 7719-7722.

1038. Synthesis and Characterization of Mo$_4$Te$_{16}$(en)$_4$$^{2-}$: A Molybdenum Telluride With Linked Mo≡Mo^{6+} Subunits, B. W. Eichorn, R. C. Haushalter, F. A. Cotton and B. Wilson, *Inorg. Chem.* **1988**, *27*, 4084-4085.

1039. Dimolybdenum(II) Complexes that Contain the Ligand *cis*-Ph$_2$PCH=CHPPh$_2$(dppee). Synthesis and Characterization of the α- and β-Isomers of Mo$_2$X$_4$(dppee)$_2$ (X = Cl or Br), M. Bakir, F. A. Cotton, L. R. Falvello, C. Q. Simpson and R. A. Walton, *Inorg. Chem.* **1988**, *27*, 4197-4202.

1040. Conversion of Dimolybdenum Tetraacetate to Mo$_2$Cl$_4$(Ph$_2$Ppy)$_2$ and Mo$_2$(O$_2$CCH$_3$)$_2$-(Ph$_2$Ppy)$_2$Cl$_2$ [Ph$_2$Ppy = 2-(diphenylphosphino)pyridine], F. A. Cotton and M. Matusz, *Polyhedron* **1988**, *7*, 2201-2204.

1041. Chemistry of Rh$_2$(CO)$_2$Cl$_2$(dppm)$_2$ and Compounds Derived Therefrom: Structural Characterization of Rh$_2$(CO)Cl$_4$(dppm)$_2$ and [Rh$_2$(CO)Cl$_3$(dppm)$_2$MeOH][PF]$_6$, F. A. Cotton, C. T. Eagle and A. C. Price, *Inorg. Chem.* **1988**, *27*, 4362-4368.

1042. Oxidation States Available to the Ru$_2$$^{n+}$ Core in Tetracarboxylato-Bridged Species, F. A. Cotton, M. Matusz and B. Zhong, *Inorg. Chem.* **1988**, *27*, 4368-4372.

1043. Synthesis and Crystal Structure of Ti$_4$O(S$_2$)$_4$Br$_6$, F. A. Cotton, P. A. Kibala and R. B. W. Sandor, *Eur. J. Solid State Inorg. Chem.* **1988**, *25*, 631-636.

1044. Synthesis and Crystal Structure of CuNb$_{13}$O$_{33}$ and Phase Analysis of the CuONbO$_2$ System, F. A. Cotton and R. B. W. Sandor, *Eur. J. Solid State Inorg. Chem.* **1988**, *25*, 637-644.

1989

1045. A Novel Compound Containing Coupled Mo$_3$S$_4$ Clusters, [Mo$_3$S$_4$(HBpz$_3$)$_2$]$_2$(μ-O) (μ-C$_3$H$_3$N$_2$)$_2$, F. A. Cotton, R. Llusar and W. Schwotzer, *Inorg. Chim. Acta* **1989**, *155*, 231-236.

1046. A Trinuclear Oxoniobium(V) Complex with Ketone Molecules as Ligands, [NbOCl$_3$(OCRR')]$_3$ with R = C$_6$H$_5$ and R' = *m*-CH$_3$C$_6$H$_4$, J. A. Canich, F. A. Cotton and S. A. Duraj, *Inorg. Chim. Acta* **1989**, *156*, 41-46.

1047. Variable Temperature T_1 Studies on $ReH_5(PPh_3)_3$, PR_3 = $PMePh_2$, (PPh_3), $ReH_7(PPh_3)_2$, and $Re_2H_8(PPh_3)_4$: Classical or Nonclassical Hydrides? F. A. Cotton and R. L. Luck, *Inorg. Chem.* **1989**, *28*, 6-8.

1048. Syntheses and Crystal Structures of Syn and Gauche Isomers of the Chloride-Bridged Face-Sharing Bioctahedral Molybdenum(III) Dimers $[PHMe_3][Mo_2Cl_7(PMe_3)_2]$, F. A. Cotton and R. L. Luck, *Inorg. Chem.* **1989**, *28*, 182-187.

1049. Preparation and Structure of the Compound $Mo_2(NCS)_4(Ph_2Ppy)_2 \cdot 2THF \cdot 2toluene$ (Ph_2Ppy = 2-(diphenylphosphino)pyridine), F. A. Cotton and M. Matusz, *Inorg. Chim. Acta* **1989**, *157*, 223-225.

1050. Experimental and Theoretical Study of $Mo_2(form)_4$ and $[Mo_2(form)_4]^+$ (form⁻ = $[p\text{-tol})\text{-NCHN}(p\text{-tol})]^-$), F. A. Cotton, X. Feng and M. Matusz, *Inorg. Chem.* **1989**, *28*, 594-601.

1051. An Oxygen-Centered Tetranuclear Titanium Compound, $Ti_4O(S_2)_4Cl_6$, F. A. Cotton, X. Feng, P. A. Kibala, R. B. W. Sandor, *J. Am. Chem Soc.* **1989**, *111*, 2148-2151.

1052. Electronic Structures of the $Ru_2(RNNNR)_4$, $Rh_2(RNCHNR)_4$, and $Co_2(RNNNR)_4$ Molecules; Ordering and Separation of M-M δ* and π* Orbitals, F. A. Cotton and X. Feng, *Inorg. Chem.* **1989**, *28*, 1180-1183.

1053. New Di- and Trinuclear Complexes of Ruthenium with Octahedra Joined on Faces or Edges: $Ru_2Cl_6(PBu_3)_3$, $Ru_2Cl_6(PBu_3)_4$, and $Ru_3Cl_8(PBu_3)_4$ (Bu = $CH_3CH_2CH_2CH_2$), F. A. Cotton, M. Matusz and R. C. Torralba, *Inorg. Chem.* **1989**, *28*, 1516-1520.

1054. Edge-Sharing Bioctahedral Molecules Without Metal-Metal Bonds: The d^6-d^6 Complexes, $Rh_2X_4(\mu\text{-}X)_2(\mu\text{-dppm})_2$ (X = Cl, Br), F. A. Cotton, K. R. Dunbar, C. T. Eagle, L. R. Falvello and A. C. Price, *Inorg. Chem.* **1989**, *28*, 1754-1757.

1055. The Reactivity of $M_2(\mu\text{-Cl})_2(\mu\text{-SMe}_2)Cl_4(SMe_2)_2$: Formation of $Ta_2(\mu\text{-SEt})_2(\mu\text{-Cl})$-$Cl_4(SMe_2)_2$ and $NbCl_3(S_2CNEt_2)_2$, J. A. M. Canich and F. A. Cotton, *Inorg. Chim. Acta* **1989**, *159*, 163-168.

1056. High-Yield Synthesis of $Mo(\eta^6\text{-PhPMe}_2)(PMe_2Ph)_3$ and Its Dimerization to Form $\{Mo(\mu\text{-}\eta^1,\eta^6\text{-PMe}_2Ph)(PMe_2Ph)_2\}_2$ a Complex Characterized by X-ray Crystallography, F. A. Cotton, R. L. Luck and R. H. Morris, *Organometallics* **1989**, *8*, 1282-1287.

1057. Reduction of $ReCl_5$ in the Presence of $PMePh_2$ to Give *mer*-$ReCl_3(PMePh_2)_3$, $ReCl(\eta^2\text{-H}_2)(PMePh_2)_4$, $ReH_3(PMePh_2)_4$ or $ReCl(CO)_3(PMePh_2)_2$, Depending on Conditions, F. A. Cotton and R. L. Luck, *Inorg. Chem.* **1989**, *28*, 2181-2186.

1058. Triangular Trinuclear Cluster Compounds: Molybdenum and Tungsten Complexes of the Type $[M_3S_4(diphos)_3X_3]^+$ with X = Cl and H, F. A. Cotton, R. Llusar and C. T. Eagle, *J. Am. Chem Soc.* **1989**, *111*, 4332-4338.

1059. Synthesis and Structural Characterization of a New Polymorph of Tungsten(VI) Sulfide Tetrachloride, $WSCl_4$, F. A. Cotton, P. A. Kibala and R. B. W. Sandor, *Inorg. Chem.* **1989**, *28*, 2485-2487.

1062. Synthesis and Structural Characterization of Three New Trinuclear Group VI Clusters, F. A. Cotton, P. A. Kibala, M. Matusz, C. S. McCaleb and R. B. W. Sandor, *Inorg. Chem.* **1989**, *28*, 2623-2630.

1061. X-Ray Crystal Structure of $ReH_5(PPh_3)_3$ and Variable Temperature T_1 Studies on $ReH_5(PPh_3)_3$ and $ReH_5(PMe_2Ph)_3$ in Various Solvents: Are T_1 Measurements Reliable in Predicting Whether Polyhydride Complexes Contain Molecular Hydrogen Ligands?, F. A. Cotton and R. L. Luck, *J. Am. Chem. Soc.* **1989**, *111*, 5757-5761.

1062. A Cationic Compound of Ru_2^{5+} Containing a Coordinated PF_4O^- Ion, F. A. Cotton and M. Matusz, *Chemia* **1989**, *43*, 167-168.

1063. Chemistry, Structure, and Bonding in Diruthenium(II) Tetracarboxylates, F. A. Cotton, V. M. Miskowski and B. Zhong, *J. Am. Chem. Soc.* **1989**, *111*, 6177-6182.

1064. Strong Interaction Between an Aliphatic Carbon-Hydrogen Bond and a Metal Atom: The Structure of (Diethylbis(1-pyrazolyl)borato)(allyl(dicarbonyl)molybdenum(II), F. A. Cotton and R. L. Luck, *Inorg. Chem.* **1989**, *28*, 3210-3213.

1065. Structural Characterization of an Oxotungsten(IV) Complex, $[WOBr(Me_2PCH_2CH_2-PMe_2)_2]Br \cdot 5H_2O$, F. A. Cotton, P. A. Kibala, C. S. McCaleb and R. B. W. Sandor, *Acta Cryst.* **1989**, *C45*, 1126-1128.

1066. Structure of $[Ag(PPh_3)_4]PF_6$, F. A. Cotton and R. L. Luck, *Acta Cryst.* **1989**, *C45*, 1222-1224.

1067. Structural Characterization of $WCl_6 \cdot S_8$, F. A. Cotton, P. A. Kibala and R. B. W. Sandor, *Acta Cryst.* **1989**, *C45*, 1287-1289.

1068. *trans*-(dimethylsulfoxide-O)(dimethylsulfoxide-S)bis(trifluoroacetato)palladium(II); Alternative Ligation Modes of an Ambidentate Ligand, D. P. Bancroft, F. A. Cotton and M. Verbruggen, *Acta Cryst.* **1989**, *C45*, 1289-1292.

1069. Solid State Geometric Isomers of $Re_2H_8(PPh_3)_4$, F. A. Cotton and R. L. Luck, *Inorg. Chem.* **1989**, *28*, 4522-4527.

1070. Three-Fold Disordering of the Re_2 Unit in the $Re_2Cl_6(PEt_3)_2$ Molecule, F. A. Cotton and K. Vidyasagar, *Inorg. Chim. Acta* **1989**, *166*, 105-108.

1071. A Compound Containing Two $[Os_2I_8]^{2-}$ Ions Fused to Form a $[Os_4I_{14}]^{2-}$ Ion Containing Os≡Os Triple Bonds, F. A. Cotton and K. Vidyasagar, *Inorg. Chim. Acta* **1989**, *166*, 109-113.

1990

1072. Variable-Temperature ^1H NMR Spectra and T_1 Measurements on the Dinuclear Octahydride Complexes $Re_2H_8(PR_3)_4$, (PR_3= PPh_3, PEt_2Ph, PMe_2Ph and PMe_3), and the Monohydride Complexes Cp_2ReH and $Re_2HCl_3(CO)_2(dppm)_2$, (dppm = $Ph_2PCH_2PPh_2$), F. A. Cotton, R. L. Luck, D. R. Root and R. A. Walton, *Inorg. Chem.* **1990**, *29*, 43-47.

1073. Proton, Carbon-13, and Platinum-195 Nuclear Magnetic Resonance Spectroscopy of Hydroxypyridinate-Bridged Dinuclear Platinum(III) Complexes. Equilibria and Mechanism of Bridging Ligand Rearrangement, E. S. Peterson, D. P. Bancroft, D. Min, F. A. Cotton and E. H. Abbott, *Inorg. Chem.* **1990**, *29*, 229-232.

1074. New Niobium Complexes with Alkynes. 1. Mono- and Dinuclear Complexes, F. A. Cotton and M. Shang, *Inorg. Chem.* **1990**, *29*, 508-514.

1075. Multiply Bonded Dimetal Fluoroalkoxides. 1. Pentafluorophenoxides of Dimolybdenum and Ditungsten, R. G. Abbott, F. A. Cotton and L. R. Falvello, *Inorg. Chem.* **1990**, *29*, 514-521.

1076. Structural, Spectroscopic and Chiroptical Properties of the Chiral Quadruple-Bonded Dimolybdenum Complexes $Mo_2Cl_4[(R,R)\text{-DIOP}]_2$ and $Mo_2Cl_4[(S,S)\text{-DIOP}]_2$, J.-D. Chen, F. A. Cotton and L. R. Falvello, *J. Am. Chem. Soc.* **1990**, *112*, 1076-1082.

1077. Bromodicarbonyl(η^3-1-phenylallyl)bis(pyrazole)molybdenum(II), F. A. Cotton and R. L. Luck, *Acta Cryst.* **1990**, *C46*, 138-140.

1078. New Niobium Complexes with Alkynes. 2. Tetranuclear Compounds with Nb–Nb Bonds, an Unprecedented Type of Tetracarbon Ligand, and Oxygen in a Rectangular Environment, F. A. Cotton and M. Shang, *J. Am. Chem. Soc.* **1990**, *112*, 1584-1590.

1079. A Dichromium(II) Compound that Avoids Cr-Cr Bond Formation by Adopting a Bizarre Structure: $Cr_2Cl_4(dmpm)_2$, F. A. Cotton, R. L. Luck and K.-A. Son, *Inorg. Chim. Acta* **1990**, *168*, 3-4.

1080. The Polarized Single-Crystal Visible Spectrum of $Mo_2(O_2CCPh_3)_4$, F. A. Cotton and B. Zhong, *J. Am. Chem. Soc.* 1990, *112*, 2256-2260.

1081. Tetrahalobis(diphosphine)dimolybenum(II) Compounds: β Isomers with Seven-Membered Rings, F. A. Cotton, L. R. Falvello, D. R. Root, T. J. Smith and K. Vidyasagar, *Inorg. Chem.* **1990**, *29*, 1328-1333.

1082. Preparative, Structural and Magnetic Studies of 2-Hydroxypyridinate Complexes of Diruthenium(II), F. A. Cotton, T. Ren and J. L. Eglin, *J. Am. Chem. Soc.* **1990**, *112*, 3439-3445.

1083. Chiroptical Properties of a Compound Containing a Metal-Metal Quadruple Bond. Synthesis, Structure, and Spectra of β-Mo_2Cl_4(S,S-bppm)$_2$, where S,S-bppm is (2S,4S)-N-(*tert*-Butoxycarbonyl)-4-(diphenylphosphino)-2-[(diphenylphosphino)-methyl]pyrrolidine, J.-D. Chen and F. A. Cotton, *Inorg. Chem.* **1990**, *29*, 1797-1802.

1084. Tertiary Phosphine Complexes of Chromium(III); Syntheses, Magnetic Properties and Single-Crystal Structure Studies on $Cr_2Cl_6(PMe_3)_4$, $Cr_2Cl_6(PEt_3)_4$, and $Cr_2Cl_6(dmpm)_2$, F. A. Cotton, J. L. Eglin, R. L. Luck and K.-A. Son, *Inorg. Chem.* **1990**, *29*, 1802-1806.

1085. Dibenzo[b,i][1,4,8,11]tetra-azacyclotetradecine Complexes. I. $W(CO)_4(tmtaa)$, F. A. Cotton and J. Czuchajowska, *Polyhedron* **1990**, *9*, 1217-1220.

1086. Dibenzo[b,i][1,4,8,ll]tetra-azacyclotetradecine Complexes. II. $Ru(tmtaa)(PMePh_2)_2$, F. A. Cotton and J. Czuchajowska, *Polyhedron* **1990**, *9*, 1221-1224.

1087. Preparation and Structure of Hexachlorotetrapyridine Ditantalum(III), E. Babaian-Kibala and F. A. Cotton, *Inorg. Chim. Acta* **1990**, *171*, 71-74.

1088. New Niobium Complexes of Alkynes. 6. Two Compounds Containing the [Nb$_2$[PhCC(Ph)C(Ph)CPh] Units Joined in Unorthodox Ways, F. A. Cotton and M. Shang, *Organometallics* **1990**, *9*, 2131-2137.

1089. A Five-Coordinate, High-Spin Chromium(III) Complex: Cr(tmtaa)Cl, F. A. Cotton, J. Czuchajowska, L. R. Falvello and X. Feng, *Inorg. Chim. Acta* **1990**, *172*, 135-136.

1090. An Unusual Ternary Chloride of Tin and Tantalum, Sn[Ta$_2$Cl$_9$]$_2$, F. A. Cotton, E. Babaian-Kibala, L. R. Falvello, M. Shang, *Inorg. Chem.* **1990**, *29*, 2591-2593.

1091. Structures of Mo$_2$(CH$_3$)$_4$(PR$_3$)$_4$ Molecules: Constancy of Covalent Radius of Molybdenum in These and Other Mo$_2$X$_4$(PR$_3$)$_4$ Molecules, F. A. Cotton and K. J. Wiesinger, *Inorg. Chem.* **1990**, *29*, 2594-2599.

1092. New Niobium Complexes with Alkynes. 3. An Extraordinary Hexaniobium Compound Containing a Central Nb$_3$ Cluster and Three Outer Nb Atoms Bound to η^4-Tetraphenylcyclobutadiene Ligands, F. A. Cotton and M. Shang, *Inorg. Chem.* **1990**, *29*, 2614-2618.

1093. New Niobium Complexes with Alkynes. 4. A Compound Containing Both an Oxo-Centered Tetrameric Anion and a Related Dinuclear Cation, F. A. Cotton and M. Shang, *Inorg. Chem.* **1990**, *29*, 2619-2622.

1094. The Structure of the Face-Sharing Bioctahedral Hexachloro(tris-triethylphosphino)-dimolybdenum(III) Molecule, F. A. Cotton, R. L. Luck and K.-A. Son, *Inorg. Chim. Acta* **1990**, *173*, 131-132.

1095. Square, Sulfur-Bicapped Tetraniobium Cluster Compounds, E. Babaian-Kibala, F. A. Cotton and P. A. Kibala, *Polyhedron* **1990**, *9*, 1689-1694.

1096. Electronic Structure Study of Dinuclear Transition-Metal Complexes with Alkynes as Bridges, F. A. Cotton and X. Feng, *Inorg. Chem.* **1990**, *29*, 3187-3192.

1097. A New Type of Metal-Olefin Complex. Synthesis and Characterization of Four Compounds That Contain an Ethylene Bridge Perpendicularly Bisecting a Metal-Metal Axis, F. A. Cotton and P. A. Kibala, *Inorg. Chem.* **1990**, *29*, 3192-3196.

1098. Further Studies of Os$_2$X$_8^{2-}$ Ions: [PMePh$_3$]$_2$Os$_2$Cl$_8$ with Eclipsed, Disordered Anions and the First Os$_2$I$_8^{2-}$ Compound, F. A. Cotton and K. Vidyasagar, *Inorg. Chem.* **1990**, *29*, 3197-3199.

1099. New Niobium Complexes of Alkynes. 5. Electronic Structure and Bonding in the Tetranuclear Complex of Niobium, [Nb$_4$OCl$_8${(PhC)$_4$}$_2$]$^{2-}$, F. A. Cotton and X. Feng, *Inorg. Chem.* **1990**, *29*, 3697-3701.

1100. Unbridged M-M Multiple Bonds in the Cr$_2$(tmtaa)$_2$ and Mo$_2$(tmtaa)$_2$ Molecules (tmtaa = Dianion of 7,16-Dihydro-6,8,15,17-tetramethyl-dibenzo[*b,i*]-[1,4,8,11] tetraazacyclotetradecine): Experimental and Theoretical Investigations, F. A. Cotton, J. Czuchajowska and X. Feng, *Inorg. Chem.* **1990**, *29*, 4329-4335.

1101. Crystallographic Disorder of the Re$_2$ Unit in Complexes of the Type A$_2$Re$_2$X$_8$ (A = HP-n-Pr$_3$, PMePh$_3$; X = Cl or Br), F. A. Cotton, A. C. Price, R. C. Torralba and K. Vidyasagar, *Inorg. Chim. Acta* **1990**, *175* 281-287.

1102. A Structural Study of Trichloro(tetrahydrofuran)iron(III), F. A. Cotton, R. L. Luck and K.-A. Son, *Acta Cryst.* **1990**, *C46*, 1424-1426.

1103. A New Bis(μ-sulfido)ditantalum(IV) Edge-Sharing Bioctahedral Molecule and a Reassessment of Some Earlier "Bis(μ-chloro)" Molecules, E. Babaian-Kibala, F. A. Cotton and P. A. Kibala, *Inorg. Chem.* **1990,** *29*, 4002-4005.

1104. Structural Studies on $M_2X_4(PR_3)_4$ Compounds (M = Re, W; X = Cl, Br; R = Me, *n*-Pr, *n*-Bu). The First Examples of Simple Three-Way Disorder of Dimetal Units within Ordered Ligand Cages, F. A. Cotton, J. G. Jennings, A. C. Price and K. Vidyasagar, *Inorg. Chem.* **1990**, *29*, 4138-4143.

1105. Multiply-Bonded Dimetal Fluoroalkoxides. II. The Disordered Structure of $Mo_2(OC(CF_3)_3)_4(NMe_2)_2$ and its Implications as to Rearrangement by an Internal Flip, R. G. Abbott, F. A. Cotton and L. R. Falvello, *Polyhedron* **1990**, *9*, 1821-1827.

1106. Structure of *trans*-Tetrachlorobis(dimethylphenylphosphine)technetium(IV), F. A. Cotton, C. S. Day, M. P. Diebold and W. J. Roth, *Acta Cryst.* **1990**, *C46*, 1623-1624.

1107. Structures of Two Niobium(IV) Complexes: $[N(C_2H_5)_4]_2[NbCl_6]$ and $[P(CH_3)_2$-$(C_6H_5)H][NbCl_5\{P(CH_3)_2(C_6H_5)\}]$, F. A. Cotton, M. P. Diebold and W. J. Roth, *Acta Cryst.* **1990**, *C46*, 1624-1627.

1108. Two Dimolybdenum Tetracarboxylato Compounds from Heteroaromatic Acids, 2-Thiophenecarboxylate and Nicotinic Acid, F. A. Cotton, L. R. Falvello, A. H. Reid, Jr. and W. J. Roth, *Acta Cryst.* **1990**, *C46*, 1815-1818.

1109. Syntheses, Structures and Spectra of β-$Mo_2X_4(dpcp)_2$ (X = Cl, Br, I), where dpcp is (±)-*trans*-1,2-bis(diphenylphosphino)cyclopentane. A Study of Several Interesting Effects of Halogen Atoms on the Mo–Mo Quadruple Bond, J.-D. Chen, F. A. Cotton and E. C. DeCanio, *Inorg. Chim. Acta* **1990**, *176*, 215-223.

1110. A Quarter-Century of Metal-Metal Multiple Bonds, F. A. Cotton, Metal-Metal Bonds and Clusters in Chemistry and Catalysis, Plenum Press, New York, 1-6 (1990).

1111. Heteronuclear Unbridged Quadruply Bonded Complexes. Preparation and Characterization of $Cl_2(PMe_2Ph)_2Mo{\buildrel 4 \over -}WCl_2(PMe_2Ph)(PPh_3)$, $Cl_2(PMe_2Ph)(PPh_3)$ Mo–$WCl_2(PMe_2Ph)_2$ and Mo–$WCl_4(PMe_2Ph)_4$, F. A. Cotton, L. R. Falvello, C. A. James and R. L. Luck, *Inorg. Chem.* **1990**, *29*, 4759-4763.

1112. Application of Tetrakis(trifluoroacetato)dirhodium(II) to Determination of Chirality: The First Structural Characterization of an Axial Bis-olefin Complex of a Dimetal Core, F. A. Cotton, L. R. Falvello, M. Gerards and G. Snatzke, *J. Am. Chem. Soc.* **1990**, *112*, 8979-8980.

1113. A Simple Compound Containing the First Hafnium(III) to Hafnium(III) Bond, F. A. Cotton, P. A. Kibala and W. A. Wojtczak, *Inorg. Chim Acta* **1990**, *177*, 1-3.

1114. Recent Developments in the Chemistry of Mono- and Dinuclear Complexes of the Macrocyclic Dianion, 5,7,12,14-Tetramethyldibenzo[b,i][1,4,8,11]-tetraazacyclo-tetradecine, F. A. Cotton and J. Czuchajowska, *Polyhedron* **1990**, *9*, 2553-2566.

1115. Preparation and Structural Characterization of Two Isomers of Stoichiometry $Re_2Cl_5(PR_3)_3$, Where R = Me or Et, F. A. Cotton, A. C. Price and K. Vidyasagar, *Inorg. Chem.* **1990**, *29*, 5143-5147.

1116. New Synthetic Routes for the Preparation of Niobium(III) and Tantalum(III) Triangular Cluster Compound, E. Babaian-Kibala, F. A. Cotton and M. Shang, *Inorg. Chem.* **1990**, *29*, 5148-5156.

1991

1117. Synthesis and Structural Characterization of Mo_2Cl_4(tetraphos-1) (tetra-phos-1 = $Ph_2PCH_2CH_2P(Ph)CH_2CH_2P(Ph)CH_2CH_2PPh_2$). The First Complexes of Enantiomeric 1,2,5,8/1,2,6,7 Type, J.-D. Chen and F. A. Cotton, *Inorg Chem.* **1991**, *30*, 6-7.

1118. Preparation and Structure of $[Cl_2W(\mu-Cl)(\mu-dmpm)_2(\mu_2-PMe_2)WCl(\eta^2-CH_2PMe_2)]Cl$, a Product of an Unusual Cleavage of the dmpm Ligand, F. A. Cotton, J. A. M. Canich, R. L. Luck and K. Vidyasagar, *Organometallics* **1991**, *10*, 352-356.

1119. A New Look at the Vanadium(III) and Vanadium(IV) Complexes of the Dibenzo-tetramethyltetraaza[14]annulene Ligand. Synthesis and Molecular Structures of CpV(tmtaa) and (O)V(tmtaa), F. A. Cotton, J. Czuchajowska and X. Feng, *Inorg. Chem.* **1991**, *30*, 349-353.

1120. Three New Compounds Containing M_3S_4 (M = Mo, W) Cores Prepared from $M_3S_7X_4$ (M = Mo, W; X = Cl, Br) Starting Materials, F. A. Cotton, P. A. Kibala, and C. S. Miertschin, *Inorg. Chem.* **1991**, *30*, 548-553.

1121. Synthesis and Characterization of $Mo_2(OC_6F_5)_4(PMe_3)_4$, F. A. Cotton and K. J. Wiesinger, *Inorg. Chem.* **1991**, *30*, 750-754.

1122. Solution and Solid-State Conformational Isomers of the Molecular Dihydrogen Complex $ReCl(H_2)(PMePh_2)_4$: Does it Contain an Asymmetric Molecular Dihydrogen Ligand?, F. A. Cotton and R. L. Luck, *Inorg. Chem.* **1991**, *30*, 767-774.

1123. Synthesis and Characterization of Octakis(acetonitrile)dimolybdenum(II) Tetrafluoroborate, F. A. Cotton and K. J. Wiesinger, *Inorg. Chem.* **1991**, *30*, 871-873.

1124. Arenes with Hafnium(II) Moieties Bonded to Both Faces, F. A. Cotton, P. A. Kibala and W. A. Wojtczak, *J. Am. Chem. Soc.* **1991**, *113*, 1462-1463.

1125. New Polynuclear Compounds of Iron(II) Chloride with Oxygen Donor Ligands. Part I. $Fe_4Cl_8(THF)_6$: Synthesis and A Single Crystal X-Ray Structure Determination, F. A. Cotton, R. L. Luck and K.-A. Son, *Inorg. Chim. Acta* **1991**, *179*, 11-15. (See ICA **1985**, *96*, 123 for earlier report of this cpd.).

1126. Structural Characterization of Two Hafnium(IV) Complexes, $HfCl_4(Ph_2PCH_2CH_2P-Ph_2)\cdot2C_7H_8$ and $HfCl_4(Ph_2PCH_2CH_2CH_2PPh_2)$, F. A. Cotton, P. A. Kibala and W. A. Wojtczak, *Acta Cryst.* **1991**, *C47*, 89-93.

1127. Synthesis and Characterization of a Dimeric Molybdenum(V) Anion $Br_2OMo(\mu-S)_2MoOBr_2]^{2-}$: Conversion of Tetrahydrofuran to $[^nBu_3P(CH_2)_4P^nBu_3]^{2+}$, F. A. Cotton, R. L. Luck and C. S. Miertschin, *Inorg. Chem.* **1991**, *30*, 1155-1157.

1128. Structural Characterization of $Zr_2Cl_8(PPh_3)_2$, F. A. Cotton and P. A. Kibala, *Acta Cryst.* **1991**, *C47*, 270-272.

1129. The First Structurally Characterized Chiral Rhenium Complex Containing a Metal-Metal Quadruple Bond. Syntheses, Structure, and Spectra of β-$Re_2Cl_6[(S,S$-isodiop)]: An Unprecedented Rearrangement of the Diop Ligand, J.-D. Chen and F. A. Cotton, *J. Am. Chem. Soc.* **1991**, *113*, 2509-2512.

1130. Crystal and Molecular Structures of $Mo_2Cl_4(PMePh_2)_4$ and $Re_2Cl_4(PMePh_2)_4$, F. A. Cotton, J. Czuchajowska and R. L. Luck, *J. Chem. Soc., Dalton Trans.* **1991**, 579-585.

1131. Preparations, Structures, and Properties of M_3X_{13} Type Molybdenum and Tungsten Trimers with Eight Cluster Electrons, F. A. Cotton, M. Shang and Z. S. Sun, *J. Am. Chem. Soc.* **1991**, *113*, 3007-3011.

1132. New Sulfido-bridged Diniobium(IV) and Ditantalum(IV) Complexes and a Related Tetranuclear Tantalum(IV) Compound, E. Babaian-Kibala and F. A. Cotton, *Inorg. Chim. Acta* **1991**, *182*, 77-82.

1133. Dibenzo[*b,i*][1,4,8,11]Tetraazacyclotetradecine Complexes. 3. Preparation and Molecular Structure of an Unprecedented Diruthenium(II) Compound, $Ru_2(Htmtaa)$-$(O_2CCH_3)_3(PhCCPh)$, F. A. Cotton and J. Czuchajowska, *J. Am. Chem. Soc.* **1991**, *113*, 3427-3431.

1134. New Di- and Trinuclear Complexes of Ruthenium with Octahedra Joined on Faces or Edges. 2. New Compounds with $Ru^{II}Ru^{II}$, $Ru^{II}Ru^{III}$ and $Ru^{III}Ru^{III}$ in Face-Sharing Bioctahedra: Structures, Magnetism, and Electrochemistry, F. A. Cotton and R. C. Torralba, *Inorg. Chem.* **1991**, *30*, 2196-2207.

1135. Oxidative Addition of Halogens to the Quadruple Bond of $Mo_2X_4(dppm)_2$ (X = Cl, I, Br): Synthesis, Structural Characterization and Magnetic Properties of $Mo_2Cl_4I_2(dppm)_2$, $Mo_2Br_6(dppm)_2$ and $Mo_2I_6(dppm)_2$, F. A. Cotton, L. M. Daniels, K. R. Dunbar, L. R. Falvello, C. J. O'Connor and A. C. Price, *Inorg. Chem.* **1991**, *30*, 2509-2514.

1136. Further Investigation of Molecular, Magnetic and Electronic Structures of 2-Hydroxypyridinate Complexes of Diruthenium(II), F. A. Cotton, T. Ren and J. L. Eglin, *Inorg. Chem.* **1991**, *30*, 2552-2558.

1137. Synthesis and Characterization of a Complex with Very Long Os-Os Triple Bond, $Os_2(DFM)_4Cl_2$, F. A. Cotton, T. Ren and J. L. Eglin, *Inorg. Chem.* **1991**, *30*, 2559-2563.

1138. New Synthetic Routes for the Preparation of the Triangular Trinuclear Clusters of Tungsten-Containing Bromine Atoms as Terminal Ligands: Structural Characterization of $[W_3S_4Br_3(depe)_3]PF_6 \cdot 0.5C_7H_8$ and $[W_3S_4Br_3(depe)_3]Br \cdot 2CH_3OH$, F. A. Cotton, S. K. Mandal and M. Shang, *J. Cluster Science* **1991**, *1*, 287-305.

1139. Structures of Trimethylphosphonium *trans*-Tetrachlorobis(trimethylphosphine) chromate(III), *mer*-Trichlorobis(dimethylphenylphosphine)(dimethylphenylphosphineoxide)chromium(III) and *mer*-Trichlorotris(trimethylphosphineoxide)molybdenum(III) Dichloromethane Solvate, F. A. Cotton and R. L. Luck, *Acta Cryst.* **1991**, *C47*, 1069-1072.

1140. Structure of [ZrCl$_3$(CH$_3$OCH$_2$CH$_2$OCH$_3$)]$_2$O, E. Babaian-Kibala, F. A. Cotton, and P. A. Kibala, *Acta Cryst.* **1991**, *C47*, 1305-1307.

1141. Bis(diphenylphosphino)methane Complexes of Rhodium(III) Halides as Synthons for Dinuclear Rhodium(III) Complexes, F. A. Cotton, K. R. Dunbar, C. T. Eagle, L. R. Falvello, S.-J. Kang, A. C. Price and M. G. Verbruggen, *Inorg. Chim. Acta* **1991**, *184*, 35-42.

1142. New Polynuclear Compounds of Iron(II) Chloride with Oxygen Donor Ligands, Part II. Polymeric [FeCl$_2$(OPMe$_3$)]$_\infty$ and Mononuclear FeCl$_2$(OPMe$_3$)$_2$. Syntheses, Properties and Single Crystal Structure Determinations, F. A. Cotton, R. L. Luck and K.-A. Son, *Inorg. Chim. Acta* **1991**, *184*, 177-183.

1143. Three New Dicarboxylatodirhenium(III) Compounds and Their Structures: A Structure Type not Previously Seen, F. A. Cotton, E. C. DeCanio, P. A. Kibala and K. Vidyasagar, *Inorg. Chim. Acta* **1991**, *184*, 221-228.

1144. [2 + 2] Cycloaddition of Re-Re Quadruple Bonds. Structural Characterization of the First Cyclotetrarhenium Diynes, (*n*-Bu$_4$N)$_2$Re$_4$Cl$_8$(μ-O)$_2$(μ-OMe)(μ-X) (X = Cl or OMe), J.-D. Chen and F. A. Cotton, *J. Am. Chem. Soc.* **1991**, *113*, 5857-5859.

1145. New Compounds of Zirconium(II) and Hafnium(II): Synthesis and X-ray Crystal Structures of Novel Dimers of Formula M$_2$X$_4$(η^6-C$_6$H$_5$PMe$_2$)$_2$(PMe$_2$Ph)$_2$, F. A. Cotton, P. A. Kibala, M. Shang and W. A. Wojtczak, *Organometallics* **1991**, *10*, 2626.

1146. New Di- and Trinuclear Complexes of Ruthenium with Octahedra Joined on Faces or Edges. 3. New Trinuclear Compounds of the Type [(R$_3$P)$_2$ClRuCl$_3$RuCl$_3$RuCl(PR$_3$)$_2$]^n (*n* = 0, +1): Structures, EPR Spectroscopy, Electrochemistry and Molecular Orbitals, F. A. Cotton and R. C. Torralba, *Inorg. Chem.* **1991**, *30*, 3293-3304.

1147. The First Triangular Trinuclear Cluster Compounds of Molybdenum with Nine Cluster Electrons: [Mo$_3$(μ$_3$-O)(μ-Cl)$_3$(μ-O$_2$CCH$_3$)$_3$Cl$_n$(PR$_3$)$_{-3n}$]$^{1-n}$, *n* = 1-3, F. A. Cotton, M. Shang and Z.-S. Sun, *J. Am. Chem. Soc.* **1991**, *113* 6917-6922.

1148. Oxohalo Complexes of W(V) and W(IV). The Synthesis and Crystal Structures of [WOCl(dppe)$_2$][BPh$_4$], [PHMe$_2$Ph][WOBr$_4$(OPMe$_2$Ph)] and [Li(OPMe$_2$Ph)(THF)]$_2$-[WOBr$_4$(OPMe$_2$Ph)]$_2$, F. A. Cotton and S. K. Mandal, *Eur. J. Solid State Inorg. Chem.* **1991**, *28*, 775-797.

1149. Electronic Structure and Bonding in Trinuclear Molybdenum and Tungsten Cluster Compounds of M$_3$X$_{13}$ Type, F. A. Cotton and X. Feng, *Inorg. Chem.* **1991**, *30*, 3666-3670.

1150. Synthesis and Characterization of Edge-Sharing Bioctahedral Complexes of Zirconium(III) and Hafnium(III) that Contain Long Metal-Metal Single Bonds: Zr$_2$I$_6$(PMe$_3$)$_4$, Hf$_2$I$_6$(PMe$_2$Ph)$_4$, and Zr$_2$I$_6$(PMe$_2$Ph)$_4$, F. A. Cotton, M. Shang and W. A. Wojtczak, *Inorg. Chem.* **1991**, *30*, 3670-3675.

1151. Preparation and Molecular and Electronic Structures of a New Diamagnetic Diruthenium(II) Complex, $Ru_2[(p\text{-tol})NC(H)N(p\text{-tol})]_4$, F. A. Cotton and T. Ren, *Inorg. Chem.* **1991**, *30*, 3675-3679.

1152. Solid Solutions of a Jahn-Teller Compound in an Undistorted Host. 2. High-Spin, Six-Coordinate Cr^{2+} without Jahn-Teller Distortion, F. A. Cotton, L. R. Falvello, L. N. Ohlhausen, C. A. Murillo and J. F. Quesada, *Z. Anorg. Allg. Chem.* **1991**, *598/599*, 53-70.

1153. Structures of $[Nb_2Cl_6(C_4H_8S)(PEt_3)_2]$, $[Nb_2Cl_6(C_4H_8S)(C_4H_8O)_2]$ and $[Ta_2Cl_6(C_4H_8S)\text{-}(PMe_3)_2]$, E. Babaian-Kibala, F. A. Cotton and M. Shang, *Acta Cryst.* **1991**, *C47*, 1617-1621.

1154. Structure of $Nb_4(\mu_3\text{-Cl})_2(\mu_2\text{-Cl})_4Cl_4(PMe_3)_6$, E. Babaian-Kibala and F. A. Cotton, *Acta Cryst.* **1991**, *C47*, 1716-1718.

1155. Synthesis and Characterization of the First Multiply Bonded Heteronuclear Edge-Sharing Bioctahedral Complex, $MoWCl_4(\mu\text{-Cl})(\mu\text{-H})(\mu\text{-dppm})_2$, F. A. Cotton, C. A. James and R. L. Luck, *Inorg. Chem.* **1991**, *30*, 4370-4373.

1156. New Di- and Trinuclear Complexes of Ruthenium with Octahedra Joined on Faces or Edges, 4. Compounds Containing $[(R_3P)_3RuCl_3RuCl_3Ru(PR_3)_3]^+$ Ions: Structures, EPR Spectroscopy and Electrochemistry, F. A. Cotton and R. C. Torralba, *Inorg. Chem.* **1991**, *30*, 4386-4391.

1157. New Di- and Trinuclear Complexes of Ruthenium with Octahedra Joined on Faces or Edges, 5. Further Study of Edge-Sharing Bioctahedral Complexes of the Type $Ru_2Cl_6(PR_3)_4$ (R = Et, Bu), F. A. Cotton and R. C. Torralba, *Inorg. Chem.* **1991**, *30*, 4392-4393.

1158. Further Studies of $Mo_3(\mu_3\text{-I})(\mu_3\text{-H})(\mu\text{-I})_3I_3L_3$ Cluster Species: Compounds with L = PEt_3 and PPh_3, A. Burini, F. A. Cotton and J. Czuchajowska, *Polyhedron* **1991**, *10*, 2145-2152.

1159. Synthesis and Structure of Tetraethylammonium Octachlorodirhenate(III), K. Gelman, M. S. Grigoriev, F. A. Cotton, S. V. Kryutchkov and L. R. Falvello, *Koord. Khim.* **1991**, *17*, 1230-1236; English transl 663-668.

1160. Verification and Reinterpretation of the Structure of $K_2Tc_2Cl_6$, F. A. Cotton, L. M. Daniels, L. R. Falvello, M. S. Grigoriev and S. V. Kryutchkov, *Inorg. Chim. Acta* **1991**, *189*, 53-54.

1161. Structure of the Second Polymorph of Niobium Pentachloride, F. A. Cotton, P. A. Kibala, M. Matusz and R. B. W. Sandor, *Acta Cryst.* **1991**, *C47*, 2435-2437.

1162. Two Chromium(II) Complexes with Amidato-Like Ligands: A Compound with the Longest Cr-Cr Bond and a Mononuclear Compound with D_{2d} Symmetry, F. A. Cotton, L. R. Falvello, W. Schwotzer, C. A. Murillo and G. Valle-Bourrouet, *Inorg. Chim. Acta* **1991**, *190*, 89-95.

1163. Reaction of $Mo_2(O_2CMe)_2Cl_2(dppe)$ With Pyridine; Structural Characterizations of the Products of $Mo_2Cl_2(OAc)_2Py_2$ and $MoOCl_2(dppe)(Py)$, J.-D. Chen, F. A. Cotton and S.-J. Kang, *Inorg. Chim. Acta* **1991**, *190*, 103-108.

1164. Relatively Air-Stable M(II) Saccharinates, M = V, or Cr, F. A. Cotton, E. Libby, C. A. Murillo and G. Valle, *Inorg. Synth.* **1990**, *27*, 306-310.

1992

1165. A Completely Suppressed Jahn-Teller Effect in the Structure of Hexaaquachromium(II) Hexafluorosilicate, F. A. Cotton, L. R. Falvello, C. A. Murillo and J. F. Quesada, *J. Solid State Chem.* **1992**, *96*, 192-198. (Dedicated to P. Hagenmuller).

1166. $Mo_2F_4(PR_3)_4$ Compounds and a Comparison of Their Properties with Those of Other $Mo_2X_4(PR_3)_4$ Compounds; A New Synthetic Route to Metal Fluorides by the Reaction of Alkyl Species with Olah's Reagent, F. A. Cotton and K. J. Wiesinger, *Inorg. Chem.* **1992**, *31*, 920-925.

1167. Preparation and Characterization of Two New Group VI Quadruply Bonded Dinuclear Compounds, $Cr_2(DFM)_4$ and $W_2(DFM)_4$, F. A. Cotton and T. Ren, *J. Am. Chem. Soc.* **1992**, *114*, 2237-2242.

1168. Formamidinate Complexes of Dirhenium, Re_2^{n+}, Cores with n = 4, 5, and 6, F. A. Cotton and T. Ren, *J. Am. Chem. Soc.* **1992**, *114*, 2495-2502.

1169. New Halogen-Bridged Dinuclear Edge-Sharing and Face-Sharing Bioctahedral Tungsten(III) Complexes, $W_2X_6(PR_3)_n$, where X = Cl or Br; PR_3 = PMe_3, PMe_2Ph, or PBu_3 and n = 4 or 3: Crystal Structures of $W_2Cl_6(PMe_2Ph)_4$, $W_2Cl_6(PMe_2Ph)_3$ and $W_2Br_6(PMe_2Ph)_3$, F. A. Cotton and S. K. Mandal, *Inorg. Chem.* **1992**, *31*, 1267-1274.

1170. Triangular Trinuclear Clusters of Tungsten Containing Bromine Atoms as Terminal Ligands: Syntheses and Structural Characterizations of $[W_3S_4Br_3(dmpe)_3][Y]$, where dmpe = 1,2 bis(dimethylphosphino)ethane and Y^- = Br^- or PF_6^-, F. A. Cotton and S. K. Mandal, *Inorg. Chim. Acta* **1992**, *192*, 71-79.

1171. The Structure of $Mo_4Cl_8(PEt_3)_4$: How Unrecognized Disorder Previously Reduced the Quality of the Refinement, F. A. Cotton and M. Shang, *J. Cluster Science* **1992**, *2*, 121-129.

1172. Synthesis and Structural Characterization of $Re_2(\mu\text{-chp})_2(\eta^2\text{-chp})Cl_3$: A New Coordination Mode of Hydroxypyridinate around Dimetallic Centers, F. A. Cotton and T. Ren, *Polyhedron* **1992**, *11*, 811-815.

1173. Preparation and Structural Characterization of Three Tetrakis(triazeno)diruthenium Compounds, F. A. Cotton, L. R. Falvello, T. Ren and K. Vidyasagar, *Inorg. Chim. Acta* **1992**, *194*, 163-170.

1174. Syntheses and Structural Characterizations of Three WOX_2L_3 Type Complexes: Crystal Structures of *cis-mer* $WOBr_2(PMe_2Ph)_3\cdot0.5C_7H_8$ and $WOBr_2(PR_3)_3$, Where PR_3 = PMe_2Ph or $PMePh_2$, F. A. Cotton and S. K. Mandal, *Inorg. Chim. Acta* **1992**, *194*, 179-187.

1175. Synthesis and Characterization of Molybdenum Species: Dinuclear and Mononuclear
      Species of the Molecular Formulas, $[Mo_2(O_2CCH_3)_2(LL)_2][BF_4]_2$ and $[Mo(O)(F)(LL)_2]$-
      $[BF_4]$ where LL = bis-phosphine. The Use of $[Mo_2(NCCH_3)_{10}][BF_4]_4$ as a Source of
      the $[Mo_2]^{4+}$ Core, F. A. Cotton, J. L. Eglin and K. J. Wiesinger, *Inorg. Chim. Acta*
      1992, *195*, 11-23.

1176. The Neutron Crystal Structure of the Saccharinate Complex $[Cu(C_7H_4NO_3S)_2$-
      $(H_2O)_4]·2H_2O$, F. A. Cotton, L. R. Falvello, C. A. Murillo and A. J. Schultz, *Eur. J.
      Solid State Inorg. Chem.* **1992**, *29*, 311-320.

1177. Comparison of the Crystalline Adducts of Tetrakis(triphenylacetato)dichromium(II)
      with Benzene, *p*-Difluorobenzene and *p*-Xylene, F. A. Cotton, L. M. Daniels and
      P. A. Kibala, *Inorg. Chem.* **1992,** *31*, 1865-1869.

1178. Stereospecific and Regiospecific Ligand Substitution Reactions of Mononuclear and
      Dinuclear Rhodium(III) Phosphine Complexes, F. A. Cotton, J. L. Eglin and
      S.-J. Kang, *J. Am. Chem. Soc.* **1992**, *114*, 4015-4016.

1179. Singlet-Triplet Separations Measured by Phosphorus-31 Nuclear Magnetic Resonance
      Spectroscopy. Applications to the Molybdenum-Molybdenum Quadruple Bond and
      to Edge-Sharing Bioctahedral Complexes, F. A. Cotton, J. L. Eglin, B. Hong and
      C. A. James, *J. Am. Chem. Soc.* **1992**, *114*, 4915-4917.

1180. The First Complex with a $\sigma^2\pi^4$ Triple Bond Between Vanadium Atoms in a Ligand
      Framework of Four-fold Symmetry — $[V_2\{(p\text{-}CH_3C_6H_4)NC(H)N(p\text{-}C_6H_4CH_3)\}_4]$,
      F. A. Cotton, L. M. Daniels and C. A. Murillo, *Angew. Chem.* **1992**, *104*, 795-796;
      Angew. Chem., Int. Ed. Engl. **1992**, *31*, 737-738.

1181. Further Study of Tetracarbonato Diruthenium(II,III) Compounds, F. A. Cotton,
      L. Labella and M. Shang, *Inorg. Chem.* **1992**, *31*, 2385-2389.

1182. Compounds Containing Linked, Multiple-Bonded Dimetal Units. 1. Tetrakis-(μ-6-
      chloro-2-hydroxopyridinato)diruthenium(II,III) Cations Linked Axially by Pyrazine.
      Comparison with a Single Molecule Axially Coordinated by Pyridine, F. A. Cotton,
      Y. Kim and T. Ren, *Inorg. Chem.* **1992**, *31*, 2608-2612.

1183. Mononuclear-Dinuclear Equilibrium for the Pyridine Adducts of Chromium(II)
      Saccharinates, N. M. Alfaro, F. A. Cotton, L. M. Daniels and C. A. Murillo, *Inorg.
      Chem.* **1992**, *31*, 2718-2723.

1184. Compounds Containing Linked Multiply-Bonded Dimetal Units. 2. An
      Antiferromagnetic Compound Containing Infinite Chains of $Ru_2(O_2CR)_4^+$ Units
      Linked by Bridging Phenazine Molecules, F. A. Cotton, Y. Kim and T. Ren, *Inorg.
      Chem.* **1992**, *31*, 2723-2726.

1185. Polydentate Phosphines: Their Syntheses, Structural Aspects and Selected
      Applications, F. A. Cotton and B. Hong, *Prog. Inorg. Chem.* **1991**, *40*, 179-289.

1186. Preparation and Structural Characterization of Three Adamantylcarboxylato
      Diruthenium(II,II), Diruthenium(II,III), and Dimolybdenum(II,II) Compounds,
      F. A. Cotton, L. Labella and M. Shang, *Inorg. Chim. Acta* **1992**, *197*, 149-158.

1187. Empty Octahedral Hexazirconium Clusters with Only Ten Electrons, $[Zr_6X_{14}(PR_3)_4]$, F. A. Cotton, X. Feng, M. Shang and W. A. Wojtczak, *Angew. Chem.* **1992**, *104*, 1117-1120; *Angew. Chem. Int. Ed. Engl.* **1992**, *31*, 1050-1053.

1188. Crystallographic Disorder in $[M_2X_8]^{n-}$, $M_2X_4L_4$ and Related Compounds: Chemical and Theoretical Implications, F. A. Cotton and J. L. Eglin, *Inorg. Chim. Acta* **1992**, *198-200*, 13-22.

1189. Decakis(acetonitrile)dimolybdenum(II) Tetrafluoroborate(1-), F. A. Cotton and K. J. Wiesinger, *Inorg. Synth.* **1992**, *29*, 134-137.

1190. The Delta Bond — An Old Story with a New Twist, F. A. Cotton, *Pure Appl. Chem.* **1992**, *64*, 1383-1393.

1191. The Structure of Copper(II) 5,7,12,14-Tetramethyldibenzo[*b,i*][1,4,8,11]-Tetraaza-cyclotetradeca-2,4,7,9,11,14-hexaenediide, F. A. Cotton and J. Czuchajowska-Wiesinger, *Acta Cryst.* **1992**, *C48*, 1434-1437.

1192. Synthesis and X-Ray Structure of Two Relatively Air-Stable Chromium(II) 3-Pyridinesulphonate Compounds, F. A. Cotton, L. M. Daniels and C. A. Murillo, *Polyhedron* **1992**, *14*, 2475-2481.

1193. Partial Paramagnetism of the Cr-Cr Quadruple Bond, F. A. Cotton, H. Chen, L. M. Daniels and X. Feng, *J. Am. Chem. Soc.* **1992**, *114*, 8980-8983.

1194. Preparation of Vanadium(II) Compounds. Structures of a Carboxylato and 3-Pyridinesulfonate Compound, F. A. Cotton, L. M. Daniels, M. L. Montero and C. A. Murillo, *Polyhedron* **1992**, *14*, 2767-2774.

1195. Complexes Containing Heteronuclear Quadruple Bonds. Preparation and Characterization of α-Mo–WCl$_4$(L-L)$_2$ (L-L = dppm, dmpe) and β-Mo–WCl$_4$(L-L)$_2$ (L-L = dppm, dmpm, dppe), F. A. Cotton and C. A. James, *Inorg. Chem.* **1992**, *31*, 5298-5307.

1196. Heterobimetallic Edge-Sharing Bioctahedral Complexes. Synthesis, Characterization, and Singlet-Triplet Separations as Determined by Phosphorus-31 Nuclear Magnetic Resonance, F. A. Cotton, J. L. Eglin, C. A. James and R. L. Luck, *Inorg. Chem.* **1992**, *31*, 5308-5315.

1197. Reactions of Rh$_2$(O$_2$CCH$_3$)$_4$ with the Macrocycle H$_2$tmtaa; Crystal Structures of Rh$_2$(tmtaa)$_2$·C$_7$H$_8$, [Rh$_2$(O$_2$CCH$_3$)$_4$(O$_2$CCH$_3$)$_2$][Rh(tmtaa)(PhCCPh)]$_2$·2C$_6$H$_6$ and Rh$_2$(O$_2$CCH$_3$)$_4$·2(NC$_5$H$_{25}$), F. A. Cotton and J. Czuchajowska-Wiesinger, *Gazz. Chim. Ital.* **1992**, *122*, 321-327.

1198. Preparation and Structural Characterization of the Isomorphous Compounds: [Bu$_4$N][Mo$_3$(μ$_3$-O)(μ-X)$_3$(μ-OAc)$_3$)X$_3$](CH$_3$)$_2$CO, X = Cl, Br, F. A. Cotton, M. Shang and Z.-S. Sun, *J. Cluster Science* **1992**, *3*, 109-122.

1199. Further Studies of M$_3$(μ$_3$-O)(μ-X)$_3$(μ-O$_2$CCH$_3$)$_3$L$_3$ (M = Mo, W) Type Clusters with Nine Core Electrons, F. A. Cotton, M. Shang and Z.-S. Sun, *J. Cluster Science* **1992**, *3*, 123-144.

1993

1200. An Ab Initio CASSCF Study of δ-Bonding in the $Mo_2Cl_4(PH_3)_4$ Molecule, F. A. Cotton and X. Feng, *J. Am. Chem. Soc.* **1993**, *115*, 1074-1078.

1201. The Synthesis and Characterization of an $Mo–MoCl_4(L-L)_2$ Compound with L-L = $Ph_2PCH_2PMe_2$, F. A. Cotton, J. L. Eglin and C. A. James, *Inorg. Chim. Acta* **1993**, *204*, 175-179.

1202. Stabilizing of the Ru_2^{6+} Core. Use of Highly Charged Ligands Such as Sulfate and Phosphate, F. A. Cotton, T. Datta, L. Labella and M. Shang, *Inorg. Chim. Acta* **1993**, *203*, 55-60.

1203. A New Synthetic Route to W–W and Mo–W Compounds of the Type $MM'X_4(PR_3)_4$, X = Cl or Br. Structural Characterization of $W–WBr_4(\mu\text{-dppm})_2$ and Mo–$WBr_4(PMe_2Ph)_4$ and $^{31}P\{^1H\}$ NMR Spectra of These and Related Compounds, F. A. Cotton, J. L. Eglin and C. A. James, *Inorg. Chem.* **1993**, *32*, 681-686.

1204. Unusual Isomers of Edge-Sharing Bioctahedral Complexes, F. A. Cotton, J. L. Eglin and C. A. James, *Inorg. Chem.* **1993**, *32*, 687-694.

1205. Preparation, Structure and Magnetic Properties of a Tetrakis(carboxylato)-diosmium(III) Dichloride. Comparison with $Os_2(hp)_4Cl_2$, F. A. Cotton, T. Ren and M. J. Wagner, *Inorg. Chem.* **1993**, *32*, 965-968.

1206. Metal-Metal Bonded Compounds of Lower-Valent Niobium and Tantalum. Compounds Containing the $[Nb_2Cl_7(PR_3)_2]^-$ and $[Nb_3Cl_{10}(PR_3)_3]^-$ Ions and Some Clues as to How Higher-Nuclearity Clusters are Formed, F. A. Cotton and M. Shang, *Inorg. Chem.* **1993**, *32*, 969-976.

1207. Molecular Structure and Magnetic Properties of a Linear Chain Compound, $Ru_2(O_2CCMePh_2)_4Cl$, F. A. Cotton, Y. Kim and T. Ren, *Polyhedron* **1993**, *12*, 607-611.

1208. Some New Dinuclear Rhodium Compounds: Preparation and Structures, F. A. Cotton, S.-J. Kang and S. K. Mandal, *Inorg. Chim. Acta* **1993**, *206*, 29-39.

1209. A Compound with a New Type of Butterfly Metal Atom Cluster: $[N(C_4H_9)_4]_2[Mo_4OBr_{12}]$, F. A. Cotton, X. Feng, M. Shang and Z. S. Sun, *Inorg. Chem.* **1993**, *32*, 1321-1326.

1210. Another Bogus Isomer: *Sic Transit* "Green $Mo_2Cl_4(PMePh_2)_4$." The X-Ray Crystal Structure Determination of $MoOCl_3(OPMePh_2)_2 \cdot C_6H_6$, F. A. Cotton, M. Kohli, R. L. Luck and J. V. Silverton, *Inorg. Chem.* **1993**, *32*, 1868-1870.

1211. Singlet-Triplet Separations Measured by $^{31}P\{^1H\}$ NMR: Applications to Quadruply Bonded Dimolybdenum and Ditungsten Complexes, F. A. Cotton, J. L. Eglin, B. Hong and C. A. James, *Inorg. Chem.* **1993**, *32*, 2104-2106.

1212. Synthesis and Structural Characterization of the $Pt_2(II,III)$ Complex (NBu_4)-$[(C_6F_5)_2Pt(\mu\text{-}C_6F_5)_2Pt(C_6F_5)_2]$ and the $Pt_2(III,III)$ Complex $(NBu_4)[(C_6F_5)_2Pt(\mu\text{-}C_6F_5Cl)(\mu\text{-}C_6F_5)Pt(C_6F_5)_2]$. Novel Ligand Reactivity of a Bridging C_6F_5 Group, R. Usón, J. Forniés, M. Tomás, J. M. Casas, F. A. Cotton, L. R. Falvello and X. Feng, *J. Am. Chem. Soc.* **1993**, *115*, 4145-4154.

1213. Stereo- and Regiospecific Conversions of $Rh_2X_6(PR_3)_3$ Molecules to $Rh_2X_6(PR_3)_4$ and $RhX_3(PR_3)_3$ Molecules, F. A. Cotton, J. L. Eglin and S.-J. Kang, *Inorg. Chem.* **1993**, *32*, 2332-2335.

1214. Preparation and Structure of Compounds Containing the $[Rh_2X_7(PR_3)_2]^-$ Anion and Their Stereospecific Reactions with Phosphines to Produce $[Rh_2X_7(PR_3)_3]^-$ Species, F. A. Cotton and S.-J. Kang, *Inorg. Chem.* **1993**, *32*, 2336-2342.

1215. Quadruply Bonded Dimolybdenum Compounds with Polydentate Phosphines. 1. Mo_2X_4(tetraphos-1) (tetraphos-1 = $Ph_2PCH_2CH_2P(Ph)CH_2CH_2P(Ph)CH_2CH_2PPh_2$, X = Cl, Br), the First Examples of Enantiomeric 1,2,5,8/1,2,6,7 and 1,2,5,7/1,2,6,8 Types, J.-D. Chen, F. A. Cotton and B. Hong, *Inorg. Chem.* **1993**, *32*, 2343-2353.

1216. Quadruply Bonded Dimolybdenum Compounds with Polydentate Phosphines. 2. Structural and Spectroscopic Properties of *rac*-Mo–MoCl$_4$(PEt$_3$(η^3-tetraphos-2)) (tetraphos-2 = $P(CH_2CH_2PPh_2)_3$), F. A. Cotton and B. Hong, *Inorg. Chem.* **1993**, *32*, 2354-2359.

1217. Two Reaction Products of $[Rh_2(O_2CCH_3)_2(CH_3CN)_6][BF_4]_2$: $[Rh_2(O_2CCH_3)(chp)_2$-$(CH_3CN)_3][BF_4]$ and $[Rh(dpcp)_2][BF_4]$, F. A. Cotton and S.-J. Kang, *Inorg. Chim. Acta* **1993**, *209*, 23-27.

1218. A Systematic Approach in the Preparation of Compounds with $\sigma^2\pi^4$ Vanadium-to-Vanadium Triple Bonds: Synthesis, Reactivity and Structural Characterization, F. A. Cotton, L. M. Daniels, and C. A. Murillo, *Inorg. Chem.* **1993**, *32*, 2881-2885.

1219. Structural Characterization of W–WCl$_4$(PMePh$_2$)$_4$, F. A. Cotton, J. L. Eglin and C. A. James, *Acta Cryst.* **1993**, *C49*, 893-896.

1220. Edge-Sharing Bioctahedral Dimolybdenum and Dirhenium Compounds with Polydentate Phosphines, F. A. Cotton, B. Hong, M. Shang and G. G. Stanley, *Inorg. Chem.* **1993**, *32*, 3620-3627.

1221. Preparation of the New Species $[W_3(\mu_3\text{-O})(\mu\text{-Cl})_2(\mu\text{-O}_2CMe)_4Cl_3]^-$ and $[Mo_3(\mu_3\text{-O})$-$(\mu\text{-Cl})_4(\mu\text{-O}_2CMe)_2Cl_2(PMe_3)_2]^-$ That Bracket, Structurally, the Very Common $M_3(\mu_3\text{-O})(\mu\text{-X})_3(\mu\text{-O}_2CR)_3L_3$ Type of Cluster Compound, F. A. Cotton, M. Shang and Z. S. Sun, *Inorg. Chim. Acta* **1993**, *212*, 95-104.

1222. A Two-Dimensional Polymer Formed by TCNE and Dirhodium Tetra(trifluoro-acetate), F. A. Cotton and Y. Kim, *J. Am. Chem. Soc.* **1993**, *115*, 8511-8512.

1223. Solid Solutions of a Jahn-Teller Compound in an Undistorted Host. 3. The Chromium/Zinc Tutton Salt System, M. A. Araya, F. A. Cotton, L. M. Daniels, L. R. Falvello and C. A. Murillo, *Inorg. Chem.* **1993**, *32*, 4853-4860.

1224. Hexaaqua Dipositive Ions of the First Transition Series: New and Accurate Structures; Expected and Unexpected Trends, F. A. Cotton, L. M. Daniels, C. A. Murillo and J. F. Quesada, *Inorg. Chem.* **1993**, *32*, 4861-4867.

1225. X-Ray Structural Characterization of a Pair of Isotypic Tetraaqua Compounds of the Jahn-Teller Species Chromium(II) and Copper(II), F. A. Cotton, L. M. Daniels, C. A. Murillo, *Inorg. Chem.* **1993**, *32*, 4868-4870.

1226. Preparation and Structural Characterization of Mo$_2$Cl$_4$(OAc)$_2$(PR$_3$)$_2$ Compounds (R = Me, Et) and a Comparative Study of the Singlet-Triplet Separation for Mo$_2$Cl$_4$(OAc)$_2$(PEt$_3$)$_2$ with That for Mo$_2$Cl$_6$(dppm)$_2$, F. A. Cotton, J. Su, Z. S. Sun and H. Chen, *Inorg. Chem.* **1993**, *32*, 4871-4875.

1227. Structural Characterization of the First Cyclotetramolybdenum Diynes with Bidentate Phosphines, Mo$_4$Cl$_4$(μ-Cl)$_2$(μ-OMe)$_2$(μ-dppm)$_2$ and Mo$_4$Cl$_4$(μ$_4$-O) (μ-Cl)$_3$(μ-OMe)$_3$(μ-dmpm)$_2$, F. A. Cotton, B. Hong and M. Shang, *Inorg. Chem.* **1993**, *32*, 4876-4881.

1228. Complexes Containing Heteronuclear and Homonuclear Quadruple Bonds. Preparation and Characterization of MoWCl$_4$(dmpm)$_2$ and Mo$_2$X$_4$(dmpm)$_2$ (X = Br, I), F. A. Cotton, K. R. Dunbar, B. Hong, C. A. James, J. M. Matonic, and J. L. C. Thomas, *Inorg. Chem.* **1993**, *32*, 5183-5187.

1229. Dinuclear Compounds of Titanium(III) and Zirconium(III), With and Without Metal–Metal Bonds, F. A. Cotton and W. A. Wojtczak, *Gazz. Chim. Ital.* **1993**, *123*, 499-507.

1994

1230. Preparation and Structural Characterization of Rh$_2$(O$_2$CCPh$_3$)$_4$(EtOH)$_2$, Ru$_2$(O$_2$CCPh$_3$)$_4$(H$_2$O)(EtOH)·2EtOH, and Mo$_2$(O$_2$CCPh$_3$)$_4$·3CH$_2$Cl$_2$. F. A. Cotton, L. M. Daniels, P. A. Kibala, M. Matusz, W. J. Roth, W. Schwotzer, W. Wang and B. Zhong, *Inorg. Chim. Acta* **1994**, *215*, 9-15.

1231. Phosphine and Phosphine Oxide Adducts of Vanadium(IV) Chloride: Synthesis and Structural Studies, F.A. Cotton, J. Lu and T. Ren, *Inorg. Chim. Acta* **1994**, *215*, 47-54.

1232. The Unusual Crystallographic Disorder of M$_2$ Units in M$_2$Cl$_4$(PEt$_3$)$_4$ (M = Re, Mo), F. A. Cotton, L. M. Daniels, M. Shang and Z. Yao, *Inorg. Chim. Acta* **1994**, *215*, 103-107.

1233. A Tris-(μ$_2$-I) Face-Sharing Dizirconium(III) Complex, [1,4-Zr$_2$I$_7$(PEt$_3$)$_2$]$^-$; The First Such Dizirconium Species, F. A. Cotton and W. A. Wojtczak, *Inorg. Chim. Acta* **1994**, *216*, 9-11.

1234. Electronic Structures of Tetrachlorobis(phosphine)niobium(IV) Complexes, F. A. Cotton, M. B. Hall and M. A. Pietsch, *Inorg. Chem.* **1994**, *33*, 1473-1475.

1235. Synthesis and Crystal Structure of Amidato-Bridged Dirhenium(III,III) Compounds, F. A. Cotton, J. Lu and T. Ren, *Polyhedron* **1994**, *13*, 807-814.

1236. Divalent Metal Chloride Formamidine Complexes, M(II) = Fe, Co and Pt. Syntheses and Structural Characterization, F. A. Cotton, L. M. Daniels, D. J. Maloney, J. H. Matonic and C. A. Murillo, *Polyhedron* **1994**, *13*, 815-823.

1237. Synthesis and Structure of Zr$_2$Cl$_4$(η^6-C$_6$H$_5$PMe$_2$)$_2$(PMe$_2$Ph)$_2$, An Arene Complex of Zirconium(II), F. A. Cotton and W. A. Wojtczak, *Inorg. Chim. Acta* **1994**, *217*, 187-190.

1238. Metal-Metal Quadruple Bonds: New Frontiers for Their Kith and Kin, F. A. Cotton, *J. Cluster Sci.*, **1994**, *5*, 3-9.

1239. The $W_2Cl_4(NHCMe_3)_2(PR_3)_2$ Molecules (R_3 = Me_3, Et_3, Pr^n_3, Me_2Ph) I. Their Isomeric Forms, Interconversion Processes, Crystalline Forms, and Detailed Molecular Structures, F. A. Cotton and Z. Yao, *J. Cluster Sci.*, **1994**, *5*, 11-36.

1240. Metal-Metal Multiply Bonded Complexes of Technetium. 1: Synthesis and Structural Characterization of the First Phosphine Complexes that Contain a Tc-Tc Multiple Bond, C. J. Burns, A. K. Burrell, F. A. Cotton, S. C. Haefner and A. P. Sattelberger, *Inorg. Chem.* **1994**, *33*, 2257-2264.

1241. A New Class of Dinuclear Compounds: The Synthesis and X-Ray Structural Characterization of Tris(μ-diphenylformamidinato)diiron, F. A. Cotton, L. M. Daniels, L. R. Falvello and C. A. Murillo, *Inorg. Chim. Acta* **1994**, *219*, 7-10.

1242. Pentanuclear Zirconium Clusters with Chloride, Hydride and Phosphine Ligands, F. A. Cotton, J. Lu, M. Shang and W. A. Wojtczak, *J. Am. Chem. Soc.* **1994**, *116*, 4364-4369.

1243. Two New Titanium Complexes with N,N'-Diphenylformamidinium (DPhF$^-$), $Ti(DPhF)_3$ and $Ti_2(μ-DPhF)_2(DPhF)_2(μ-NPh)_2$, F. A. Cotton and W. A. Wojtczak, *Polyhedron* **1994**, *13*, 1337-1341.

1244. $[Rh_2(O_2CCH_3)_4TCNE]_x$ Polymers: Isomers with Either 1,1- or 1,2-TCNE Bridges, F. A. Cotton, Y. Kim and J. Lu, *Inorg. Chim. Acta* **1994**, *221*, 1-4.

1245. Preparations, Structures, Electronic Structures and Magnetic Properties of Face Sharing Bioctahedral Niobium(III) and Tantalum(III) Compounds, F. A. Cotton, X. Feng, P. Gütlich, T. Kohlhaas, J. Lu and M. Shang, *Inorg. Chem.* **1994**, *33*, 3055-3063.

1246. The First Eight-Electron Triangular Trinuclear Cluster Compound of Tungsten Without Acetate as a Bridging Ligand: $Na[W_3(μ_3-O)(μ-Cl)_3Cl_6(THF)_3]$, F. A. Cotton, L. M. Daniels and Z. Yao, *Inorg. Chem.* **1994**, *33*, 3195-3196.

1247. Synthesis and Structure of Two Dinuclear Anionic Complexes with Pt(III)–Pt(III) Bonds and Unprecedented $(C_6F_4O)^{2-}$ or $\{C_6F_4(OR)_2\}^{2-}$ (R = Me, Et) Quinone-like Bridging Ligands, R. Usón, J. Forniés, L. R. Falvello, M. Tomás, J. M. Casas, A. Martín and F. A. Cotton, *J. Am. Chem. Soc.* **1994**, *116*, 7160-7165.

1248. The $Zr_6Cl_{12}(PMe_2Ph)_6$ Molecule Revisited: A Directed Synthesis, F. A. Cotton and W. A. Wojtczak, *Inorg. Chim. Acta* **1994**, *223*, 93-95.

1249. $W_2Cl_4(NHCMe_3)_2(PR_3)_2$ Molecules (R_3 = Me_3, Et_3, Prnc, Me_2Ph). 2. $^{31}P\{^1H\}$ NMR Studies of Cis/trans Isomerizations and Evidence that Suggests an Internal Flip of the W_2 Unit. H. Chen, F. A. Cotton and Z. Yao, *Inorg. Chem.* **1994**, 33, 4255-4260.

1250. Divalent Iron Formamidinato Complexes: A Highly Distorted Dinuclear Compound, F. A. Cotton, L. M. Daniels and C. A. Murillo, *Inorg. Chim. Acta* **1994**, *224*, 5-9.

1251. A New Member of the $Nb_n(\mu_3\text{-}X)_{n-2}(\mu\text{-}X)_n L_{n+6}$ ($n>2$) Cluster Family: $Nb_4(Cl_{10}(PEt_3)_4(H_2NPh)_2$. Synthesis, Structure, and Relationship to Other Members, F. A. Cotton and M. Shang, *J. Cluster Science* **1994**, *5*, 467-479.

1252. Three $Rh_2(O_2CCF_3)_4L_2$ Compounds, with L = H_2O, Tetrahydrofuran and Acetonitrile, F. A. Cotton and Y. Kim, *Eur. J. Solid State Inorg. Chem.* **1994**, *31*, 525-534.

1253. The Aqueous Chromium Sulfate-Pyridine System: X-Ray Structural Characterization of Chain- and Tetranuclear Compounds, F. A. Cotton, L. M. Daniels, C. A. Murillo and L. A. Zúñiga, *Eur. J. Solid State Inorg. Chem.* **1994**, *31*, 535-544.

1254. Neutron and X-Ray Structural Characterization of the Hexaaquavanadium(II) Compound $VSO_4\cdot 7H_2O$, F. A. Cotton, L. R. Falvello, C. A. Murillo, I. Pascual, A. J. Schultz and M. Tomás, *Inorg. Chem.* **1994**, *33*, 5391-5395.

1255. Solid Solutions of a Jahn-Teller Compound in an Undistorted Host. 4. Neutron and X-Ray Single-Crystal Structures of Two Cr/Zn Tutton Salt Solid Solutions and the Observation of Disorder by Low-Temperature Neutron Diffraction, F. A. Cotton, L. M. Daniels, L. R. Falvello, C. A. Murillo, and A. J. Schultz, *Inorg. Chem.* **1994**, *33*, 5396-5403.

1256. Two Zirconium(IV) Complexes, $Zr_2Cl_8(PMe_3)_4$ and *trans*-$ZrI_4(PMe_2Ph)_2$, F. A. Cotton and W. A. Wojtczak, *Acta Cryst.* **1994**, *C50*, 1662-1664.

1257. A New Type of Face-Sharing Bioctahedral Triply Bonded Diniobium Cluster (PyH)-$[Nb_2(\mu\text{-}Cl)_2(\mu\text{-}THT)Cl_3(Py)_3](THF)(C_6H_6)_{0.5}$, made by a Designed Synthesis, F. A. Cotton and M. Shang, *Inorg. Chim. Acta* **1994**, *227*, 191-196.

1995

1258. Metal-Metal Multiply-Bonded Complexes of Technetium. 2. Preparation and Characterization of the Fully Solvated Ditechnetium Cation $[(Tc_2CH_3CN)_{10}]^{4+}$, J. C. Bryan, F. A. Cotton, L. M. Daniels, S. C. Haefner and A. P. Sattelberger, *Inorg. Chem.* **1995**, *34*, 1875-1883.

1259. EPR and Crystallographic Studies of Some Reaction Products of VCl_4, $NbCl_4$ and $TaCl_4$ with Trialkyl- and Triarylphosphines, F. A. Cotton and J. Lu, *Inorg. Chem.* **1995**, *34*, 2639-2644.

1260. Preparation and Properties of $Ru_2(DtolF)_4Cl$; A Surprising Electronic Structure Change Compared to $Ru_2(DtolF)_4$ (DtolF = $[(p\text{-tol})NCHN(p\text{-tol})]^-$, F. A. Cotton and T. Ren, *Inorg. Chem.* **1995**, *34*, 3190-3193.

1261. Halide Supported Octahedral Clusters of Zirconium: Structural and Bonding Questions, F. A. Cotton. T. Hughbanks, C. E. Runyan, Jr., and W. A. Wojtczak, in "Early Transition Metal Clusters with π-Donor Ligands", M. H. Chisholm, Ed., VCH Publishers, Inc., N.Y. (1995), p. 1-26.

1262. $Mo_2(O_2CPh)_6(PEt_3)_2$, the Second Example of Bridging $\eta^1\text{-}O_2CR$ for Dimolybdenum(III) Compounds, F. A. Cotton and J. Su, *J. Cluster Sci.* **1995**, *6*, 39-59.

1263. Crystallographic Disorder of the M_2 Units in $[M_2Cl_8]^{n-}$ (M = Mo, Re) Compounds, F. A. Cotton, J. H. Matonic and D. de O. Silva, *Inorg. Chim. Acta* **1995**, *234*, 115-125.

1264. Two New Ditungsten Species Containing Chloride and Trimethylphosphine Ligands, F. A. Cotton and E. V. Dikarev, *Inorg. Chem.* **1995**, *34*, 3809-3812.

1265. Carboxylate Exchange Among Dimolybdenum Tetracarboxylates: The Trifluoroacetate/Formate System, H. Chen and F. A. Cotton, *Polyhedron* **1995**, *14*, 2221-2224.

1266. Structure of Crystalline $(C_5Me_5)ReO_3$ and the Implied Non-existence of "$(C_5Me_5)Tc_2O_3$", A. K. Burrell, F. A. Cotton, L. M. Daniels and V. Petricek, *Inorg. Chem.* **1995**, *34*, 4253-4255.

1267. Experimental and Theoretical Study of a Paradigm Jahn-Teller Molecule, All-*trans*-$CrCl_2(H_2O)_2(pyridine)_2$, and the Related *trans*-$CrCl_2(pyridine)_4$·acetone, F. A. Cotton, L. M. Daniels, X. Feng, D. J. Maloney, C. A. Murillo and L. A. Zúñiga, *Inorg. Chim. Acta*, **1995**, *235*, 21-28.

1268. Synthesis and Structures of Three Edge-Sharing Bioctahedral Compounds of (VI)A Metals Containing Formamidinato Ligands, F. A. Cotton, D. J. Maloney and J. Su, *Inorg. Chim. Acta* **1995**, *236*, 21-29.

1269. Compounds Containing Linked, Multiply-Bonded Dimetal Units. 3. Preparation, Molecular Structure and Magnetic Properties of a Compound Containing Tetrakis(μ-6-chloro-2-hydroxypyridinato)diosmium(II,III) Cations Linked Axially by Pyrazine, F. A. Cotton, Y. Kim and D. L. Shulz, *Inorg. Chim. Acta* **1995**, *236*, 43-47.

1270. Regioisomerism Displayed by the 6-chloro-2-oxopyridinate Complexes of Ru_2^{4+} and Ru_2^{5+}, F. A. Cotton, Y. Kim and A. Yokochi, *Inorg. Chim. Acta* **1995**, *236*, 55-61.

1271. A New Rectangular Re_4 Cluster Compound of the Metallocyclodiyne Type, F. A. Cotton and E. V. Dikarev, *J. Cluster Sci.* **1995**, *6*, 411-419.

1272. Complexes of Molybdenum(III) and -(IV) with Chloride and Tertiary Phosphine Ligands; an Omniumgatherum of New and Old Results, F. A. Cotton and K. Vidyasagar, *Polyhedron* **1995**, *14*, 3077-3086.

1273. The $[Zr_6X_{18}H_5]^{3-}$ Ions with X = Cl, Br and Related Species, L. Chen, F. A. Cotton and W. A. Wojtczak, *Angew. Chem. Int. Ed. Engl.* **1995**, *34*, 1877-1879.

1274. Unligated Dirhodium Tetra(trifluoroacetate): Preparation, Crystal Structure and Electronic Structure, F. A. Cotton, E. V. Dikarev and X. Feng, *Inorg. Chim. Acta* **1995**, *237*, 19-26.

1275. An Efficient Reduction Process Leading to Ti(II) and Nb(II): Preparation and Structural Characterization of *trans*-$MCl_2(py)_4$ Compounds, M = Ti, Nb and Mn, M. A. Araya, F. A. Cotton, J. H. Matonic and C. A. Murillo, *Inorg. Chem.* **1995**, *34*, 5424-5428.

1996

1276. Synthesis and Structural Studies of Platinum Complexes Containing Monodentate, Bridging and Chelating Formamidine Ligands, F. A. Cotton, J. H. Matonic and C. A. Murillo, *Inorg. Chem.* **1996**, *34*, 498-503.

1277. Transition Metal Complexes with Amidinato Ligands: The Ubiquitous Tris-Chelated Structural Motif, F. A. Cotton, L. M. Daniels, D. J. Maloney and C. A. Murillo, *Inorg. Chim. Acta* **1996**, *242*, 31-42.

1278. Metal-Metal Multiply-Bonded Complexes of Technetium. 3. Preparation and Characterization of Phosphine Complexes of Technetium Possessing a Metal-Metal Bond Order of 3.5, F. A. Cotton, S. C. Haefner and A. P. Sattelberger, *Inorg. Chem.* **1996**, *35*, 1831-1838.

1279. A Novel Tetranuclear Compound: Crystal Structure and Mass Spectra of [Re$_4$(C$_6$H$_5$NCOCH$_3$)$_6$(Cl)(μ-O)(μ-OH)(MeOH)$_3$][ReO$_4$]$_2$, F. A. Cotton, J. Lu and Y. Huang, *Inorg. Chem.* **1996**, *35*, 1839-1841.

1280. A Cluster of Molybdenum (IV) Containing a Sulfur Cap, Sulfur Bridges and Chelating Oxalate Groups, F. A. Cotton, L. M. Daniels, R. Llusar, M. Shang and W. Schwotzer, *Acta Cryst.* **1996**, *C52*, 835-838.

1281. Trivalent Mononuclear *mer*-trichlorotris(pyridine)tantalum(III): Preparation and Structure in Three Crystalline Forms, F. A. Cotton, C. A. Murillo and X. Wang, *Inorg. Chim. Acta* **1996**, *245*, 115-118.

1282. Zirconium Clusters from the Reaction of ZrCl$_4$ with HSnBu$_3$ Followed by Addition of Phosphines Zr$_6$Cl$_{14}$H$_4$(PR$_3$)$_4$, L. Chen, F. A. Cotton and W. A. Wojtczak, *Inorg. Chem.* **1996**, *35*, 2988-2994.

1283. Electronic Structure and Metal-Metal Interaction in Edge- and Face-Sharing Bioctahedral Compounds of Molybdenum, F. A. Cotton and X. Feng, *Intl. J. Quantum Chem.* **1996**, *58*, 671-680.

1284. *nido*-Metalloborane Complexes: Synthesis and Structural Characterization of μ$_2$,η^4-Hexahydrodiboratotetrakis(N,N'-Containing the μ$_2$,η^4-B$_2$H$_6$$^{2-}$ Ligand, F. A. Cotton, L. M. Daniels, C. A. Murillo and X. Wang, *J. Am. Chem. Soc.* **1996**, *118*, 4830-4833.

1285. Unusual 1,1-2,2-Isomers of Mo$_2$Cl$_4$(PR$_3$)$_2$(R'CN)$_2$. Compounds with R = CH$_2$CH$_2$CN and R' = CH$_3$, CH$_3$CH$_2$ and CH(CH$_3$)$_2$, F. A. Cotton, L. M. Daniels, S. C. Haefner and E. N. Walke, *Inorg. Chim. Acta* **1996**, *247*, 105-112.

1286. Metal-Metal Multiply-Bonded Complexes of Technetium, Part 4. Photodissociation of the Tc≡Tc Triple Bond in [Tc$_2$(CH$_3$CN)$_{10}$][BF$_4$]$_4$, F. A. Cotton, S. C. Haefner and A. P. Sattelberger, *J. Am. Chem. Soc.* **1996**, *118*, 5486-5487.

1287. Dimolybdenum Compounds with Crosswise-bridging Acetonitrile Molecules, F. A. Cotton and F. E. Kühn, *J. Am. Chem. Soc.* **1996**, *18*, 5826-5827.

1288. Preparation and Characterization of Di[(μ-acetato)halogenato(μ-*bis*(diphenylphos-phino)amine)molybdenum(II)] Complexes, D. I. Arnold, F. A. Cotton and F. E. Kühn, *Inorg. Chem.* **1996**, *35*, 4733-4737.

1289. Further Studies of the Isomeric 1,2,7- and 1,3,6-Re$_2$Cl$_5$(PMe$_3$)$_3$ Compounds, F. A. Cotton and E. V. Dikarev, *Inorg. Chem.* **1996**, *35*, 4738-4742.

1290. Tri-bridged Amidinato Compounds of Dicobalt. 1: Co$_2$[PhNC(R)NPh]$_3$ with R = H and C$_6$H$_5$, F. A. Cotton, L. M. Daniels, D. J. Maloney and C. A. Murillo, *Inorg. Chim. Acta* **1996**, *249*, 9-11.

1291. Preparation and Crystal Structure of a Dimolybdenum(II) Complex with the Drug Ibuprofen, F. A. Cotton and D. de O. Silva, *Inorg. Chim. Acta* **1996**, *249*, 57-61.

1292. Theoretical Study of Electronic Structures and Spectra of Chalcogenido Complexes of Molybdenum, *trans*-Mo(Q)$_2$(PH$_3$)$_4$ (Q = O, S, Se, Te), F. A. Cotton and X. Feng, *Inorg. Chem.* **1996**, *35*, 4921-4925.

1293. Proposed Reformulation of Recently Reported "Tetrahedral Molybdenum(II)" Complexes: Trimethylphosphine Complexes of Zinc Chloride, F. A. Cotton and G. Schmid, *Polyhedron* **1996**, *15*, 4053-4059.

1294. Preparation and Crystal Structure of Dimolybdenum(II) Complexes with the 4-Biphenylcarboxylate Ligand and its Bispyridine Adduct, F. A. Cotton and D. de O. Silva, *Polyhedron* **1996**, *15*, 4079-4085.

1295. The Reduction of Pentavalent Group 5 Compounds with KC$_8$ or LiBH$_4$: A Potpourri of Oxidation States, F. A. Cotton, J. H. Matonic, C. A. Murillo and X. Wang, *Bull. Soc. Chim. France* **1996**, 711-720.

1296. A Double-Bonded Niobium(III) Compound with a Formamidinate Core: Nb$_2$(μ-SMe)$_2$(μ-DTolF)$_2$(η^2-DTolF)$_2$·2toluene, DTolF = di-*p*-tolylformamidinato, F. A. Cotton, J. H. Matonic and C. A. Murillo, *J. Cluster Sci.* **1996**, *7*, 655-662.

1297. Preparation and Characterization of Bis[dihalogenato(μ-bis(diphenylphosphino)-amine)molybdenum(II)] Complexes, D. I. Arnold, F. A. Cotton and F. E. Kühn, *Inorg. Chem.* **1996**, *35*, 5764-5769.

1298. Synthesis and Problematic Crystal Stuctures of [Al(CH$_3$CN)$_6$][MCl$_6$]$_3$(CH$_3$CN)$_3$ (M = Ta, Nb or Sb), E. Babaian-Kibala, H. Chen, F. A. Cotton, L. M. Daniels, L. R. Falvello, G. Schmid and Z. Yao, *Inorg. Chim. Acta* **1996**, *250*, 359-364.

1299. Combined Single-Crystal X-Ray and Neutron Diffraction Analysis of the Structure of the [Zr$_6$Cl$_{18}$H$_5$]$^{3-}$ Polyanion, F. A. Cotton, L. Chen and A. J. Schultz, *C. R. Acad. Sci.* **1996**, 539-544.

1300. Metal-Metal Multiply-Bonded Complexes of Technetium, Part 5. Tris and Tetra Formamidinato Complexes of Ditechnetium, F. A. Cotton, S. C. Haefner and A. P. Sattelberger, *Inorg. Chem.* **1996**, *35*, 7350-7357.

1301. Preparation, Molecular and Electronic Structures, and Magnetic Properties of Face-Sharing Bioctahedral Titanium(III) Compounds: [PPh$_4$][Ti$_2$(μ-Cl)$_3$Cl$_4$(PR$_3$)$_2$], L. Chen, F. A. Cotton, K. R. Dunbar, X. Feng, R. A. Heintz and C. Uzelmeir, *Inorg. Chem.* **1996**, *35*, 7358-7363.

1302. Substitution on a Methyl Group of an Acetylacetonate Ligand: Synthesis and Structural Characterization of Oxobis[6-(phenylamino)-2,4-hexanedionato-O,O']vanadium(IV), VO(Phad)$_2$, D. I. Arnold, F. A. Cotton, J. H. Matonic and C. A. Murillo, *J. Chem. Soc., Chem. Commun.* **1996**, 2113-2114.

1303. Lewis- or Brønsted-Acid Assisted Formation of Open-Chain Vinamidinium Salts from *N,N'*-diarylformamidines and Acetone, F. A. Cotton, L. M. Daniels, S. C. Haefner and C. A. Murillo, *J. Chem. Soc., Chem. Commun.* **1996**, 2507-2508.

1304. Synthesis, Structure, and Reactivity of [Zr$_6$Cl$_{18}$H$_5$]$^{2-}$, the First Paramagnetic Species of Its Class, L. Chen and F. A. Cotton, *Inorg. Chem.* **1996**, *35*, 7364-7369.

1305. A Wonderful Bond That Wasn't There: Reformulation of a Compound "Containing a Ta–Ta Bond Without Bridging Ligands" as [(Cy$_2$N)$_2$ClTa(μ-H)]$_2$, F. A. Cotton, L. M. Daniels, Carlos A. Murillo and X. Wang, *J. Am. Chem. Soc.* **1996**, *118*, 12449-12450.

1306. The Solid State Structure and the Solution Isomerization of Mo$_2$Cl$_6$(THF)$_3$, F. A. Cotton and J. Su, *Inorg. Chim. Acta* **1996**, *251*, 101-104.

1307. Compounds Containing the [Zr$_6$Cl$_{18}$H$_4$]$^{4-}$ Cluster Anion, L. Chen, F. A. Cotton and W. A. Wojtczak, *Inorg. Chim. Acta* **1996**, *252*, 239-250.

1308. Syntheses and Structure of *trans*-[Di(μ-acetato)dichlorodi(μ-bis(diphenylphosphino)-methylamine)dimolybdenum(II)] and the Structure of Bis(diphenylphosphino)-methylamine), F. A. Cotton, F. E. Kühn and A. Yokochi, *Inorg. Chim. Acta* **1996**, *252*, 251-256.

1309. Syntheses and Structures Di(carboxylato)hexakis(acetonitrile)dimolybdenum(II) Bis(tetrafluoroborate) and of *trans*-[Di(μ-acetato)di(acetonitrile)di(μ-bis(diphenyl-phosphino)amine)dimolybdenum(II)] Bis(tetrafluoroborate), F. A. Cotton and F. E. Kühn, *Inorg. Chim. Acta* **1996**, *252*, 257-264.

1310. An Unusual Iron-oxo Tetranuclear Species: Li$_2$(HDPhF)$_2$Fe$_4$O$_4$(DPhF)$_6$·4toluene. A Ubiquitous Side Product in the Chemistry of Iron/formamidine Compounds, F. A. Cotton, L. M. Daniels, D. J. Maloney and C. A. Murillo, *Inorg. Chim. Acta* **1996**, *252*, 293-298.

1997

1311. Divalent Metal Chloride Amidine Complexes. Part 2. Compounds of Mn, Co and Ni, D. I. Arnold, F. A. Cotton, D. J. Maloney, J. H. Matonic and C. A. Murillo, *Polyhedron* **1997**, *16*, 133-141.

1312. Structural Studies of Formamidine Compounds: From Neutral to Anionic and Cationic Species, F. A. Cotton, S. C. Haefner, J. H. Matonic, C. A. Murillo and X. Wang, *Polyhedron* **1997**, *16*, 541-550.

1313. W$_2$Cl$_4$(NHR)$_2$(PR'$_3$)$_2$ Molecules. 3. Bidentate Phosphine Complexes of Bis(t-butyl-amido)tetrachloroditungsten. Preparation and Structural Characterization of *cis,cis*-W$_2$Cl$_4$(NHCMe$_3$)$_2$(L–L) (L–L = dmpm, dmpe, dppm, dppe), F. A. Cotton, E. V. Dikarev and W.-Y. Wong, *Inorg. Chem.* **1997**, *36*, 80-85.

1314. Quadruply Bonded Dichromium Complexes with Bridging α-Metalla-Carboxylates, F. A. Cotton and G. Schmid, *Inorg. Chim. Acta* **1997**, *254*, 233-238.

1315. Metal-Assisted Unorthodox Reactions of Formamidines: Coupling, Cleavage and Insertions, F. A. Cotton, L. M. Daniels, J. H. Matonic, C. A. Murillo and X. Wang, *Polyhedron* **1997**, *16*, 1177-1191.

1316. $W_2Cl_4(NR_2)_2(PR'_3)_2$ Molecules. 4. A New Synthetic Route to Products with a Staggered Conformation. Preparations and Characterizations of $W_2Cl_4(NR_2)_2(PR'_3)_2$ (R = Et, Bu, Hex; R'$_3$ = Me$_3$, Et$_2$H) and an Intermediate Complex $W_2Cl_4(NEt_2)_2(NHEt_2)_2$, F. A. Cotton, E. V. Dikarev and W.-Y. Wong, *Inorg. Chem.* **1997**, *36*, 559-566.

1317. Synthesis, Structure and Magnetic Properties of Ru_2^{6+} Compounds, F. A. Cotton and A. Yokochi, *Inorg. Chem.* **1997**, *36*, 567-570.

1318. Cleavage of Formamidinate Ligands on a Ta=Ta Double Bond: Formation of $H_xCNAryl$ (x = 0 and 1) and Aryl Imido Bridged Complexes, F. A. Cotton, L. M. Daniels, C. A. Murillo, and X. Wang, *Inorg. Chem.* **1997**, *36*, 896-901.

1319. $W_2Cl_4(NR_2)_2(PR'_3)_2$ Molecules. 5. Preparation and Characterization of $W_2Cl_4(NEt_2)_2(L-L)$ (L-L = dmpm, dmpe) and $W_2Cl_4NHex_2)_2(dmpm)$. The First Complexes with Bridging Diphosphine Ligands in a Cis,trans Stereochemistry, F. A. Cotton, E. V. Dikarev and W.-Y. Wong, *Inorg. Chem.* **1997**, *36*, 902-912.

1320. Efficient Preparation of a Linear, Symmetrical, Metal-Metal Bonded Tricobalt Compound; Should We Believe There Is a Bond Stretch Isomer? F. A. Cotton, L. M. Daniels, and G. T. Jordan IV, *Chem. Commun.* **1997**, 421-422.

1321. Trigonal-lantern Dinuclear Compounds of Diiron(I, II): The Synthesis and Characterization of Two Highly Paramagnetic $Fe_2(amidinato)_3$ Species with Short Metal–Metal Bonds, F. A. Cotton, L. M. Daniels, L. R. Falvello, J. H. Matonic, C. A. Murillo, *Inorg. Chim. Acta* **1997**, *256*, 269-275.

1322. Highly Distorted Diiron(II, II) Complexes Containing Four Amidinate Ligands. A Long and a Short Metal–Metal Distance, F. A. Cotton, L. M. Daniels, J. H. Matonic, C. A. Murillo, *Inorg. Chim. Acta* **1997**, *256*, 277-282.

1323. Dicobalt Trigonal Lanterns: Compounds Containing the Co_2^{3+} Core $Co_2[RC(NPh)_2]_3$, R = H and C_6H_5, and an Oxidized Compound $\{Co_2[HC(NPh)_2]_3(CH_3CN)_2\}PF_6$, F. A. Cotton, L. M. Daniels, D. J. Maloney, J. H. Matonic, C. A. Murillo, *Inorg. Chim. Acta* **1997**, *256*, 283-289.

1324. The Use of $CoCl_2(amidine)_2$ Compounds in the Synthesis of Tetragonal Lantern Dicobalt Compounds: Synthesis, Structures and Theoretical Studies of $Co_2(DPhF)_4$ and the Oxidized Species $[Co_2(DPhBz)_4]^+$, DPhF = *N,N'*-diphenylformamidinate, DPhBz = *N,N'*-diphenylbenzamidinate., F. A. Cotton, L. M. Daniels, X. Feng, D. J. Maloney, J. H. Matonic, C. A. Murillo, *Inorg. Chim. Acta* **1997**, *256*, 291-301.

1325. Electronic Structure of Dinuclear Trigonal-Lantern Amidinato Compounds of Iron and Cobalt, F. A. Cotton, X. Feng, C. A. Murillo, *Inorg. Chim. Acta* **1997**, *256*, 303-308.

1326. Picking the Right Problems, F. A. Cotton, *Inorg. Chim. Acta* **1997**, *256*, 177-182.

1327. An Efficient Synthetic Route to Tribridged Dimolybdenum Compounds: Synthesis and Structural Characterization of Dichlorotris(μ-N,N'-diarylformamidinato)-dimolybdenum(II,III), Aryl = Phenyl and p-Tolyl, F. A. Cotton, G. T. Jordan IV, C. A. Murillo and J. Su, *Polyhedron* **1997**, *16*, 1831-1835.

1328. Bis(N,N'-diphenylformamidine) Silver(I) Triflate: A Three-Coordinate Silver Formamidine Compound Stabilized by Intramolecular Hydrogen Bonds, D. I. Arnold, F. A. Cotton, J. H. Matonic and C. A. Murillo, *Polyhedron* **1997**, *16*, 1837-1841.

1329. New Hydrogen-Containing Zirconium Clusters with Phosphine Ligands, L. Chen and F. A. Cotton, *Inorg. Chim. Acta* **1997**, *257*, 105-120.

1330. The Apparent Flexibility of Bonds in Paddlewheel Type Compounds, F. A. Cotton and A. Yokochi, *Inorg. Chem.* **1997**, *36*, 2461-2462.

1331. Mononuclear Molybdenum(IV) Complexes with two Multiply Bonded Chalcogen Ligands in trans Configuration and Chelating Biphosphine Ligands, F. A. Cotton and G. Schmid, *Inorg. Chem.* **1997**, *36*, 2267-2278.

1332. $W_2Cl_4(NR_2)_2(PR'_3)_2$ Molecules. 8. Synthesis and Characterization of Both Cis Isomers of Stoichiometry $W_2Cl_4(NHR)_2(PMe_3R)_2$ (R = Et, Pr^n and Bu^n. Direct Evidence Favoring an Internal Flip Mechanism for Cis-Trans Isomerization, F. A. Cotton, E. V. Dikarev and W.-Y. Wong, *Inorg. Chem.* **1997**, *36*, 2670-2677.

1333. $W_2Cl_4(NR_2)_2(PR'_3)_2$ Molecules. 7. Preparation, Characterization and Structures of $W_2Cl_4(NHR)_2(NH_2R)_2$ and $W_2Cl_4(NHR)_2(PMe_3)_2$ (R = sec-butyl and cyclohexyl) and $^{31}P\{^1H\}NMR$ Studies of Trans to Cis Isomerizations of $W_2Cl_4(NHR)_2(PMe_3)_2$, F. A. Cotton, E. V. Dikarev and W.-Y. Wong, *Inorg. Chem.* **1997**, *36*, 3268-3276.

1334. Density Functional Theory Study of Transition-Metal Compounds Containing Metal-Metal Bonds. 1. Molecular Structures of Dinuclear Compounds by Complete Geometry Optimization, F. A. Cotton and X. Feng, *J. Am. Chem. Soc.* **1997**, *119*, 7514-7520.

1335. Tetrakis(μ$_2$-2-biphenylcarboxylato) Bis-chlororhenium(III) Bischloromethane solvate, F. A. Cotton, L. M. Daniels, J. Lu and T. Ren, *Acta Cryst.* **1997**, *C53*, 714-716.

1336. A New Type of Divalent Niobium Compound: The First Nb–Nb Triple Bond in a Tetragonal Lantern Environment; F. A. Cotton, J. H. Matonic and C. A. Murillo, *J. Am. Chem. Soc.* **1997**, *119*, 7889-7890.

1337. The Dependence of the Metal-to-Metal Distances on Mo–Mo Multiple Bond Orders; a New Triple Bond, F. A. Cotton, L. M. Daniels, C. A. Murillo, and D. J. Timmons, *Chem. Commun.* **1997**, 1449-1450.

1338. Tetra(7-azaindolato)ditungsten: Effect of Ligand Shape and Rigidity on the Length of a Tungsten–Tungsten Quadruple Bond, F. A. Cotton, L. R. Falvello and W. Wang, *Inorg. Chim. Acta*, **1997**, *261*, 72-81.

1339. Synthesis and Structural Characterization of Compounds Containing the $[Zr_6Cl_{18}H_5]^{3-}$ Cluster Anion. Determination of the Number and Positions of Cluster Hydrogen Atoms, L. Chen, F. A. Cotton and W. A. Wojtczak, *Inorg. Chem.* **1997**, *36*, 4047-4054.

1340. $W_2Cl_4(NR_2)_2(PR'_3)_2$ Molecules. 6. New Triply-Bonded Ditungsten Complexes with Bis(diphenylphosphino)methane or Bis(diphenylphosphino)amine as a Bridging Ligand. Crystal Structures of *cis,cis*-$W_2Cl_4(NR_2)_2$(dppm) (R = Et, Bu^n) and $W_2Cl_3(NHCMe_3)_2(NH_2CMe_3)(PPh_2NPOPh)$, F. A. Cotton, E. V. Dikarev, N. Nawar, W.-Y. Wong, *Inorg. Chim. Acta* **1997**, *262*, 21-32.

1341. The Crystal Packing of Bis(2,2'-dipyridylamido)cobalt(II), Co(dpa)$_2$, Is Stabilized by C–H···N Bonds: Are There Any Real Precedents?; F. A. Cotton, L. M. Daniels, G. T. Jordan IV and C. A. Murillo, *Chem. Commun.* **1997**, 1673-1677.

1342. $W_2Cl_4(NR_2)_2(PR'_3)_2$ Molecules. 9. New Mixed Valent W^{III}/W^{IV} Face-Sharing Bioctahedral Complexes with Chelating Bis(diphenylphosphino)propane and Dialkylamido Ligands, F. A. Cotton, E. V. Dikarev and W.-Y. Wong, *Polyhedron* **1997**, *16*, 3893-3898.

1343. Compounds with Linear, Bonded Trichromium Chains; F. A. Cotton, L. M. Daniels, C. A. Murillo and I. Pascual, *J. Am. Chem. Soc.* **1997**, *119*, 10223-10224.

1344. Cobalt(II) Complexes with the 1,5,9,13-Tetraazacyclohexadecane Ligand, F. A. Cotton and L. Chen, *Inorg. Chim. Acta* **1997**, *263*, 9-16.

1345. Completion of the series of $Pt_2(ArNC(H)NAr)_4^n$, n = 0, +1, +2, Compounds, with Pt–Pt Sigma Bond Orders of 0, ½, 1, Respectively, F. A. Cotton, J. H. Matonic and C. A. Murillo, *Inorg. Chim. Acta* **1997**, *264*, 61-65.

1346. Metal-Metal Multiply-Bonded Complexes of Technetium. 6. A μ,η^1,η^2-CH_3CN Complex Prepared via Reductive Cleavage of the Electron-Rich Tc≡Tc Triple Bond in Decakisacetonitrile Ditechnetium Tetrafluoroborate, F. A. Cotton, S. C. Haefner and A. P. Sattelberger, *Inorg. Chim. Acta* **1997**, *266*, 55-63.

1347. Three Structurally Related Dimolybdenum(III, III) Compounds: $(PPh_4)_2[Mo_2(OAc)_2Cl_6]$, $(PPh_4)_2[Mo_2(O_2CPh)_2Cl_6]$ and $(PPh_4)_2[Mo_2(OAc)_4Cl_4]$, F. A. Cotton, J. Su and Z. Yao, *Inorg. Chim. Acta* **1997**, *266*, 65-71.

1348. Transition Metal (Mn, Co) and Zinc Formamidinate Compounds Having the Basic Beryllium Acetate Structure, and Unique Isomeric Iron Compounds, F. A. Cotton, L. M. Daniels, L. R. Falvello, J. H. Matonic, C. A. Murillo, X. Wang and H. Zhou, *Inorg. Chim. Acta* **1997**, *266*, 91-102.

1349. Symmetrical and Unsymmetrical Compounds Having a Linear Co_3^{6+} Chain Ligated by a Spiral Set of Dipyridyl Anions; F. A. Cotton, L. M. Daniels, G. T. Jordan IV, and C. A. Murillo, *J. Am. Chem. Soc.* **1997**, *119*, 10377-10381.

1350. Characterizations and Reactions of $[PPh_4]_3[Zr_6Cl_{18}H_5]$ and Its Deprotonation Products; L. Chen, F. A. Cotton, W. T. Klooster and T. F. Koetzle, *J. Am. Chem. Soc.* **1997**, *119*, 12175-12183.

1351. Mixed Chloride Phosphine Complexes of Dirhenium Cores. 1. New Reactions and Unprecedented Structures Involving Trimethylphosphine, F. A. Cotton, E. V. Dikarev and M. A. Petrukhina, *J. Am. Chem. Soc.* **1997**, *119*, 12541-12549.

1998

1352. New Multiply-Bonded Dimetal Compounds Containing Bridging 1,3,4,6,7,8-Hexahydro-2*H*-Pyrimido[1,2-*a*]Pyrimidinato Groups. 1. The V_2^{4+}, Cr_2^{4+} and Mo_2^{4+} Compounds and some Salts of the Protonated Ligand, F. A. Cotton and D. J. Timmons, *Polyhedron* **1998**, *17*, 179-184.

1353. An Infrequent Geometry for an Element of the First Transition Series: Structural Characterization of a Distorted Eight-Coordinate Compound $TiCl_4[C_2H_4(PMe_2)_2]_2$, F. A. Cotton, J. H. Matonic, C. A. Murillo, M. A. Petrukhina, *Inorg. Chim. Acta* **1998**, *267*, 173-176.

1354. Partial Hydrolysis of Ti(III) and Ti(IV) Chlorides in the Presence of $[PPh_4]Cl$, L. Chen and F. A. Cotton, *Inorg. Chim. Acta* **1998**, *267*, 271-279.

1355. Linear Tri- and Tetrachromium(II) Chains Supported by Four Bis-(2-pyridyl)formamidinium Ligands, F. A. Cotton, L. M. Daniels, C. A. Murillo and X. Wang, *Chem. Commun.* **1998**, 39-40.

1356. Tuning Metal-to-Metal Distances in Linear Trichromium Units, F. A. Cotton, L. M. Daniels, C. A. Murillo and I. Pascual, *Inorg. Chem. Commun.* **1998**, *1*, 1-3.

1357. Structural Characterization of Dirhenium(II) Complexes of the Type $Re_2Cl_4(\mu$-PP)-$(PR_3)_2$, Where PP represents a Bridging Phosphine of the Type $R_2PCH_2PR_2$ or R_2PNHPR_2, F. A. Cotton, A. Yokochi. M. J. Siwajek and R. A. Walton, *Inorg. Chem.* **1998**, *37*, 372-375.

1358. Some Remarks on the Gallium to Iron Bond in an $Ar^*GaFe(CO)_4$ Molecule, F. A. Cotton and X. Feng, *Organometallics* **1998**, *17*, 128-130.

1359. The Exceptional Structural Versatility of 2,2'-dipyridylamine (Hdpa) and Its Ions [dpa]⁻ and $[H_2dpa]^+$; F. A. Cotton, L. M. Daniels, G. T. Jordan, IV and C. A. Murillo, *Polyhedron* **1998**, *17*, 589-597.

1360. Changes in Molecular Geometry Caused by Oxidation in Ru_2 Paddlewheel Complexes: From Ru_2^{4+} to Ru_2^{5+}, F. A. Cotton and A. Yokochi, *Polyhedron* **1998**, *17*, 959-963.

1361. The Use of Density Functional Theory to Understand and Predict Structures and Bonding in Main Group Compounds with Multiple Bonds, F. A. Cotton, A. H. Cowley and X. Feng, *J. Am. Chem. Soc.* **1998**, *120*, 1795-1799.

1362. New Hexazirconium Clusters Containing Hydrogen Atoms, L. Chen and F. A. Cotton, *J. Cluster Science* **1998**, *9*, 63-91.

1363. Metal-Metal Multiply-Bonded Complexes of Technetium. 7.[1] Oxidative Decomposition of Tetrachlorotetrakis(dimethylphenylphosphino)ditechnetium(II) by an Aminium Hexachloroantimonate(V), F. A. Cotton, S. C. Haefner and A. P. Sattelberger, *Inorg. Chim. Acta* **1998**, *271*, 187-190

1364. Density Functional Theory Study of Transition-Metal Compounds Containing Metal-Metal Bonds. 2. Molecular Structures and Vibrational Spectra of Dinuclear Tetra-Carboxylate Compounds of Molybdenum and Rhodium, F. A. Cotton and X. Feng, *J. Am. Chem. Soc.* **1998**, *120*, 3387-3397.

1365. Tetranuclear Complexes Containing Quadruply-Bonded Dimolybdenum Units Joined by μ-Hydride Ions, F. A. Cotton, L. M. Daniels, G. T. Jordan IV, C. Lin and C. A. Murillo, *J. Am. Chem. Soc.* **1998**, *120*, 3398-3401.

1366. Mixed Chloride/Phosphine Complexes of the Dirhenium Core. 2. New Reactions Involving Dimethylphenylphosphine, F. A. Cotton, E. V. Dikarev and M. A. Petrukhina, *Inorg. Chem.* **1998**, *37*, 1949-1958.

1367. An Unprecedented Tetranuclear Complex Containing Two Quadruply Bonded Mo_2^{4+} Units Joined by Two μ-OH groups, F. A. Cotton, L. M. Daniels, G. T. Jordan IV, C. Lin and C. A. Murillo, *Inorg. Chem. Commun.* **1998**, *1*, 109-110.

1368. Syntheses and Structures of Some Ru_2^{5+} Complexes Containing Weakly Coordinating Anions as Axial Ligands, F. A. Cotton, J. Lu and A. Yokochi, *Inorg. Chim. Acta* **1998**, *275-276*, 447-452.

1369. Three Reactions of Ru_2^{5+} Compounds of the Paddlewheel Type that Lead to Cleavage of the Ru–Ru Bond, F. A. Cotton and A. Yokochi, *Inorg. Chim. Acta* **1998**, *275-276*, 557-561.

1370. Synthesis and Characterization of the Series of Compounds $Ru_2(O_2CMe)_x(admp)_{4-x}Cl$. (Hadmp = 2-amino-4,6-dimethylpyridine, *x* = 3, 2, 1, 0), F. A. Cotton and A. Yokochi, *Inorg. Chem.* **1998**, *37*, 2723-2728.

1371. Triply Bonded Nb_2^{4+} Tetragonal Lantern Compounds; Some Accompanied by Novel B-H⋯Na Interactions, F. A. Cotton, J. H. Matonic and C. A. Murillo, *J. Am. Chem. Soc.* **1998**, *120*, 6047-6052.

1372. Further Study of Very Close Non-Bonded Cu^I–Cu^I Contacts. Molecular Structure of a New Compound and Density Functional Theory Calculations, F. A. Cotton, X. Feng and D. J. Timmons, *Inorg. Chem.* **1998**, *120*, 4066-4069.

1373. Further Study of the Agostic Molecule, $TiCl_3(dmpe)C_2H_5$, F. A. Cotton and M. Petrukhina, *Inorg. Chem. Commun.* **1998**, *1*, 195-196.

1374. Ligand Exchange or Reduction at Multiply-Bonded Dimetal Units of Molybdenum and Rhenium by 2,6-bis(diphenylphosphino)pyridine, F. A. Cotton, E. V. Dikarev, G. T. Jordan IV, C. A. Murillo and M. A. Petrukhina, *Inorg. Chem.* **1998**, *37*, 4611-4616.

1375. Trinuclear Complexes of Copper, Cobalt and Iron with *N,N'*-di(2-pyridyl)formamidinate ligands, $[M_3(DpyF)_4][PF_6]_2$, F. A. Cotton, C. A. Murillo and X. Wang, *Inorg. Chem. Commun.* **1998**, *1*, 281-283.

1376. A Low-Valent Ditantalum Complex $Ta_2(μ-BH_3)(μ-dmpm)_3(\eta^2-BH_4)_2$: The First Dinuclear Compound Containing a Bridging BH_3 Group with Direct Ta–B Bonds, F. A. Cotton, C. A. Murillo and X. Wang, *J. Am. Chem. Soc.* **1998**, *120*, 9594-9599.

1377. From End-on Coordination of Acetonitrile Molecules to Crosswise Bridging; Formation of Iminophosphino and Acetamidate Ligands in a Dimolybdenum Complex by Further Reactions with Nucleophiles, F. A. Cotton, L. M. Daniels, C. A. Murillo and X. Wang, *Polyhedron* **1998**, *17*, 2781-2793.

1378. Coupling Mo_2^{n+} Units via Dicarboxylate Bridges, F. A. Cotton, C. Lin and C. A. Murillo, *J. Chem. Soc., Dalton Trans.* **1998**, 3151-3153.

1379. Highlights from Recent Work on Metal–Metal Bonds, F. A. Cotton, *Inorg. Chem.* **1998**, *37*, 5710-5720.

1380. Mixed Chloride/Amine Complexes of Dimolybdenum (II,II). 1. Preparation, Characterization and Crystal Structure of $Mo_2Cl_4(NHEt_2)_4$: A Quadruply-Bonded Dimolybdenum Compound with Diethylamine Ligands, F. A. Cotton, E. V. Dikarev and S. Herrero, *Inorg. Chem.* **1998**, *37*, 5862-5868.

1381. Mixed Chloride/Phosphine Complexes of the Dirhenium Core. 3. Novel Structures with Diethylphosphido-bridges and Terminal Diethylphosphines, F. A. Cotton, E. V. Dikarev and M. Petrukhina, *Inorg. Chem.* **1998**, *37*, 6035-6043.

1382. Nickel(II) Complexes with Tetraaza Macrocyclic Ligands, L. Chen and F. A. Cotton, *J. Mol. Str.* **1998**, *469*, 161-166. (Clearfield edition)

1383. Synthesis, Reactivity, and X-Ray Structures of Face-Sharing Ti(III) Complexes; The New Trinuclear Ion, $[Ti_3Cl_{12}]^{3-}$, L. Chen and F. A. Cotton, *Polyhedron* **1998**, *17*, 3727-3734.

1384. Compounds with Two Metal-Metal Multiple Bonds: New Ways of Making Doublets into Cyclic Quartets, F. A. Cotton, L. M. Daniels, I. Guimet, R. W. Henning, G. T. Jordan IV, C. Lin, C. A. Murillo and A. J. Schultz, *J. Am. Chem. Soc.* **1998**, *120*, 12531-12538.

1385. Ferrocenium Tetrachloroferrate Revisited, and Ferrocenium Tetrachloroaluminate, F. A. Cotton, L. M. Daniels and I. Pascual, *Acta Cryst C* **1998**, *54*, 1575-1578.

1386. The First Dinuclear Complex of Palladium(III), F. A. Cotton, J. Gu, C. A. Murillo and D. J. Timmons, *J. Am. Chem. Soc.* **1998**, *120*, 13280-13281.

1999

1387. Trinuclear Complexes of Four-Coordinate Iron and Cobalt: Some Reflections on the Importance of the Reaction Conditions for the Formation of Metal-Metal Bonds, F. A. Cotton, C. A. Murillo and D. J. Timmons, *Polyhedron* **1999**, *18*, 423-428.

1388. Mixed Chloride/Amine Complexes of Dimolybdenum(II,II). 2. Reactions of $Mo_2Cl_4(NHEt_2)_4$ with Monodentate and Bidentate Phosphines. New Type of Compounds $Mo_2Cl_4(diphosphine)_4$, F. A. Cotton, E. V. Dikarev and S. Herrero, *Inorg. Chem.* **1999**, *38*, 490-495.

1389. Synthesis and Crystal Structure of $[Re(CO)_3(\mu-CH_3CO_2)(THF)]_4$, F. A. Cotton, E. V. Dikarev and M. A. Petrukhina, *Inorg. Chim. Acta* **1999**, *284*, 305-310.

1390. Two Diphenylformamidinate Compounds of the Quadruply-Bonded Re_2^{6+} Core, F. A. Cotton, L. M. Daniels and S. C. Haefner, *Inorg. Chim. Acta* **1999**, *285*, 149-151.

1391. Reactions of $TiCl_4$ with Phosphines and Alkylating Reagents: An Organometallic Route to a Titanium(II) Cluster Compound, F. A. Cotton, C. A. Murillo and M. A. Petrukhina, *J. Organomet. Chem.* **1999**, *573*, 78-86.

1392. Mixed Chloride/phosphine Complexes of the Dirhenium Core. 4. Reaction of $[Re_2Cl_8]^{2-}$ with PMe_3: One More Stop on the Road to $Re_2Cl_4(PMe_3)_4$, F. A. Cotton, E. V. Dikarev and M. A. Petrukhina, *Inorg. Chem. Commun.* **1999**, *2*, 28-30.

1393. A Chain of Five Chromium(II) Atoms: A Desired Compound with an Undesired, Unsurprising, but Important Structure, F. A. Cotton, L. M. Daniels, T. Lu, C. A. Murillo and X. Wang, *J. Chem. Soc., Dalton Trans.* **1999**, 517-518.

1394. New Family of Quadruply-Bonded Ditungsten(II,II) Species with Chloride and Amine Ligands, F. A. Cotton, E. V. Dikarev and S. Herrero, *Inorg. Chem. Commun.* **1999**, *2*, 98-100.

1395. An Unprecedented Quadruply Bonded Compound with a Bridging Chlorine Atom: $Cr_2[(o\text{-}ClC_6H_4N)_2CH]_3(\mu\text{-}Cl)$, F. A. Cotton, C. A. Murillo and I. Pascual, *Inorg. Chem. Commun.* **1999**, *2*, 101-103.

1396. Quadruply Bonded Dichromium Complexes with Variously Fluorinated Formamidinate Ligands, F. A. Cotton, C. A. Murillo and I. Pascual, *Inorg. Chem.* **1999**, *38*, 2182-2187.

1397. Square and Triangular Arrays Based on Mo_2^{4+} and Rh_2^{4+} Units, F. A. Cotton, L. M. Daniels, C. Lin and C. A. Murillo, *J. Am. Chem. Soc.* **1999**, *121*, 4538-4539.

1398. The Designed "Self-assembly" of a Three-dimensional Molecule Containing Six Quadruply-bonded Mo_2^{4+} Units, F. A. Cotton, L. M. Daniels, C. Lin and C. A. Murillo, *Chem. Commun.* **1999**, 841-842.

1399. Identification and Structure of $W_2(\mu\text{-}OH)_2[(p\text{-}MeC_6H_4N)_2CH]_4$, F. A. Cotton, L. M. Daniels and T. Ren, *Inorg. Chem.* **1999**, *38*, 2221-2222.

1400. Mixed Chloride/Amine Complexes of Dimolybdenum(II,II). 3. Preparation, Characterization and Crystal Structure of $Mo_2Cl_4(NH_2R)_4$ R = Et, Pr^n, Bu^t, Cy): First Quadruply-Bonded Dimolybdenum Compounds with Primary Amine Ligands, F. A. Cotton, E. V. Dikarev and S. Herrero, *Inorg. Chem.* **1999**, *38*, 2649-2654.

1401. Further Study of the Linear Trinickel(II) Complex of Dipyridylamide, R. Clérac, F. A. Cotton, K. R. Dunbar, C. A. Murillo, I. Pascual and X. Wang, *Inorg. Chem.* **1999**, *38*, 2655-2657.

1402. Synthesis, Reactivity, and Structures of (μ-acetamidato)hexakis(acetonitrilo)dimolybdenum(II) Tris(tetrafluoroborate) and Derivatives, F. A. Cotton, L. M. Daniels, S. C. Haefner and F. E. Kühn, *Inorg. Chim. Acta* **1999**, *287*, 159-166.

1403. Compounds in Which the Mo_2^{4+} Unit Is Embraced by Two or Three Formamidinate Ligands Together with Acetonitrile Ligands, M. H. Chisholm, F. A. Cotton, L. M. Daniels, K. Folting, J. C. Huffman, S. S. Iyer, C. Lin, A. M. Macintosh and C. A. Murillo, *Dalton Trans.* **1999**, 1387-1391.

1404. Metal-Metal Multiply-Bonded Complexes of Technetium. 8.[1] Synthesis and Characterization of the α- and β-isomers of $Tc_2Cl_4(dppe)_2$, F. A. Cotton, L. M. Daniels, S. C. Haefner and A. P. Sattelberger, *Inorg. Chim. Acta* **1999**, *288*, 69-73.

1405. First Cr(II) Complex with a Tetrahedral $M_4(\mu_4\text{-}O)$ Core: $[M_4(\mu_4\text{-}O)(\mu\text{-}Cl)_6(THF)_4]$, F. A. Cotton, C. A. Murillo and I. Pascual, *Inorg. Chem.* **1999**, *38*, 2746-2749.

1406. Mixed Chloride/Phosphine Complexes of the Dirhenium Core. 5. Reduction or Disproportionation at the Re_2^{6+} Unit in Reactions with Tertiary Phosphines in Different Solvents, F. A. Cotton, E. V. Dikarev and M. A. Petrukhina, *Inorg. Chem.* **1999**, *38*, 3384-3389.

1407. A Pi Complex of Hexamethylbenzene with Tetra(trifluoroacetate) Dirhodium(II,II), F. A. Cotton, E. V. Dikarev and S.-E. Stiriba, *Organometallics* **1999**, *18*, 2724-2726.

1408. Remarkable Effects of Axial π^* Coordination on the Cr–Cr Quadruple Bond in Dichromium Paddlewheel Complexes, F. A. Cotton, L. M. Daniels, C. A. Murillo, I. Pascual and H.-C. Zhou, *J. Am. Chem. Soc.* **1999**, *121*, 6856-6861.

1409. First Paddlewheel Complex with a Doubly-Bonded Ir_2^{6+} Core, F. A. Cotton, C. A. Murillo and D. J. Timmons, *Chem. Commun.* **1999**, 1427-1428.

1410. Mixed Chloride/Phosphine Complexes of the Dirhenium Core. 6. Two Unprecedented Isomers of $[Re_2Cl_6(PR_3)_2]$ Stoichiometry, F. A. Cotton, E. V. Dikarev and M. A. Petrukhina, *Inorg. Chem.* **1999**, *38*, 3889-3894.

1411. μ-Oxo-bis[dichlorooxo(trimethylphosphone-*P*)(trimethylphosphine oxide-*O*)-molybdenum(*V*)] Diethly Ether Hemisolvate, F. A. Cotton, L. M. Daniels and S. Herrero, *Acta Cryst.* **1999**, *C55*, IUC9900018 (CIF only).

1412. Studies of Dirhodium Tetra(trifluoroacetate). 3. Solid State Isomers of the Compound $Rh_2(O_2CCF_3)_4(THF)$ Prepared by Sublimation, F. A. Cotton, E. V. Dikarev and S.-E. Stiriba, *Inorg. Chem.* **1999**, *38*, 4877-4881.

1413. Mixed Chloride/Amine Complexes of Dimolybdenum(II,II). 4. Rotational Isomers of $Mo_2Cl_4(R\text{-}py)$ = 4-picoline, 3,5-lutidine, and 4-*tert*-butylpyridine), F. A. Cotton, E. V. Dikarev, S. Herrero and B. Modec, *Inorg. Chem.* **1999**, *38*, 4882-4887.

1414. Can Crystal Structure Determine Molecular Structure? For $Co_3(dpa)_4Cl_2$, Yes, F. A. Cotton, C. A. Murillo and X. Wang, *J. Chem. Soc., Dalton Trans.* **1999**, 3227-3228.

1415. A Very Short Re_2^{6+} Quadruple Bond: First DFT Calculations on a Paddlewheel Complex with an Element of the Third Transition Series, F. A. Cotton, J. Gu, C. A. Murillo and D. J. Timmons, *J. Chem. Soc., Dalton Trans.* **1999**, 3741-3745.

1416. An Unusual *o*-metallated Pyridyl Dirhodium(II) Complex: $[Rh_2\{\mu\text{-}(C_5H_3N)\text{-}NH(C_5H_4N)\}_2\{\eta^2\text{-}(C_5H_4N)NH(C_5H_4N)\}_2]Cl_2\cdot CH_3OH$, F. A. Cotton, C. A. Murillo, and S.-E. Stiriba, *Inorg. Chem. Commun.* **1999**, *2*, 463-464.

1417. Synthesis, Crystal and Electronic Structures of Dinuclear Platinum Compounds, $[PEtPh_3]_2[Pt_2(\mu\text{-}PPh_2)_2(C_6F_5)_4]$ and $[Pt_2(\mu\text{-}PPh_2)_2(C_6F_5)_4]$: A Computational Study by Density Functional Theory, E. Alonso, J. M. Casas, F. A. Cotton, X. Feng, J. Forniés, C. Fortuño and M. Tomás, *Inorg. Chem.* **1999**, *38*, 5034-5040.

1418. Getting the Right Answer to a Key Question Concerning Molecular Wires, F. A. Cotton, L. M. Daniels, C. A. Murillo and X. Wang, *Chem. Commun.* **1999**, 2461-2462.

1419. Linear Tricobalt Compounds with Di-(2-pyridyl)amide (dpa) Ligands: Studies of the Paramagnetic Compound $Co_3(dpa)_4Cl_2$ in Solution, F. A. Cotton, C. A. Murillo and X. Wang, *Inorg. Chem.* **1999**, *38*, 6294-6297.

1420. Mixed Chloride/Amine Complexes of Dimolybdenum(II,II). 5. Experimental and Theoretical Study of the Rotation Preferences of $Mo_2Cl_4(R-py)_4$ (R-py = Substituted pyridine) Molecules, F. A. Cotton, E. V. Dikarev, J. Gu, S. Herrero and B. Modec, *J. Am. Chem. Soc.* **1999**, *121*, 11758-11761.

1421. Oxygen in a Box: An Oxygen Atom Surrounded by a Cube of Eight Lithium Atoms, F. A. Cotton, L. M. Daniels, C. A. Murillo and H. Zhou, *C. R. Acad. Sci. Paris* **1999**, *2*, Series II*c*, 579-582.

Eugenio Coronado, University of Valencia, 1999.

2000

1422. After 155 Years, A Crystalline Chromium Carboxylate with a Supershort Cr–Cr Bond, F. A. Cotton, E. A. Hillard, C. A. Murillo and H.-C. Zhou, *J. Am. Chem. Soc.* **2000**, *122*, 416-417.

1423. Novel η^3-Allyl Complexes of Titanium, F. A. Cotton, C. A. Murillo and M. A. Petrukhina, *J. Organomet. Chem.* **2000**, *593-594*, 1-6.

1424. Selectivity in Alkynylation: The Reaction Between $Ru_2(LL)_4Cl$ and $Me_3SnC\equiv CR$ (LL = 2-anilinopyridine, 2-chloro- and 2-bromo-oxypyridine), F. A. Cotton, S.-E. Stiriba and A. Yokochi, *J. Organomet. Chem.* **2000**, *595*, 300-302.

1425. Structural Variations in the Ligands Around the Simple Oxo-centered Building Block, A Tetrahedral $[M_4O]^{6+}$ Unit, M = Mn and Fe, F. A. Cotton, L. M. Daniels, G.T. Jordan, IV, C. A. Murillo and I. Pascual, *Inorg. Chim. Acta* **2000**, *297*, 6-10.

1426. Some Interesting Structural Chemistry of Lithium, F. A. Cotton and H.-C. Zhou, *Inorg. Chim. Acta* **2000**, *298*, 216-220.

1427. Mixed Chloride/Amine Complexes of Dimolybdenum(II,II). 6. Stepwise Substitution of Amines by Tertiary Phosphines and Vice-versa: Stereochemical Hysterisis, F. A. Cotton, E. V. Dikarev and S. Herrero, *Inorg. Chem.* **2000**, *39*, 609-616.

1428. Linear Trichromium Complexes with Direct Cr to Cr Contacts. I. Compounds with $Cr_3(dipyridylamide)_4{}^{2+}$ Cores, R. Clérac, F. A. Cotton, L. M. Daniels, K. R. Dunbar, C. A. Murillo and I. Pascual, *Inorg. Chem.* **2000**, *39*, 748-751.

1429. Linear Trichromium Complexes with Direct Cr to Cr Contacts. II. Compounds with $Cr_3(dipyridylamide)_4{}^{3+}$ Cores, R. Clérac, F. A. Cotton, L. M. Daniels, K. R. Dunbar, C. A. Murillo and I. Pascual, *Inorg. Chem.* **2000**, *39*, 752-756.

1430. Studies of Tetrakis(trifluoroacetato)dirhodium. 4. Solventless Synthesis of $Rh_2(O_2CCF_3)_2(CO)_4$ Combined with $Rh_2(O_2CCF_3)_4$, a Compound with Infinite Chains of Rhodium Atoms, F. A. Cotton, E. V. Dikarev and M. A. Petrukhina, *J. Organomet. Chem.* **2000**, *56*, 130-135.

1431. A Dinuclear Cation with both Four- and Five-Coordinate Cobalt(II), F. A. Cotton, C. A. Murillo and H.-C. Zhou, *Inorg. Chem.* **2000**, *39*, 1039-1041.

1432. A New Linear Tricobalt Compound with di(2-pyridyl)amide (dpa) Ligands: Two-Step Spin Crossover, of $[Co_3(dpa)_4Cl_2][BF_4]$, R. Clérac, F. A. Cotton, K. R. Dunbar, T. Lu, C. A. Murillo and X. Wang, *J. Am. Chem. Soc.* **2000**, *122*, 2272-2278.

1433. A Novel Cluster Consisting of Seven Fused MX_6 Octahedra, $Li_{14}(THF)_8[Ni_7Cl_{17}(\mu$-formamidinate$)_3(\mu$-Cl$)]_2$, F. A. Cotton, C. A. Murillo and I. Pascual, *J. Cluster Sci.* **2000**, *11*, 87-94.

1434. Metal–Metal Bonding in $Rh_2(O_2CCF_3)_4$. Extensive Metal-Ligand Orbital Mixing Promoted by Filled Fluorine Orbitals, D. L. Lichtenberger, J. R. Pollard, M. A. Lynn, F. A. Cotton and X. Feng, *J. Am. Chem. Soc.* **2000**, *122*, 3182-3190.

1435. Heteronuclear Chains of Four Metal Atoms Including one Quadruply-Bonded Dimetal Unit. Dichromium and Dimolybdenum Compounds with Appended Copper(I) Atoms, F. A. Cotton, C. A. Murillo, L. E. Roy and H.-C. Zhou, *Inorg. Chem.* **2000**, *39*, 1743-1747.

1436. Studies of Dirhodium Tetra(trifluoroacetate). 5. Remarkable Examples of the Ambidentate Character of Dimethylsulfoxide, F. A. Cotton, E. V. Dikarev, M. A. Petrukhina and S.-E. Stiriba, *Inorg. Chem.* **2000**, *39*, 1748-1754.

1437. Studies of Dirhodium(II) Tetrakis(trifluoroacetate). 6. The First Structural Characterization of Axial Alkyne Complexes, $Rh_2(O_2CCF_3)_4(Ph_2C_2)_n$ (n = 1, 2). Diphenylacetylene as a Bifunctional Ligand, F. A. Cotton, E. V. Dikarev, M. A. Petrukhina and S.-E. Stiriba, *Organometallics* **2000**, *19*, 1402-1405.

1438. A Compound Containing Two Tantalum Atoms in Oxidation States Separated by Six Units, F. A. Cotton, C. A. Murillo and X. Wang, *Inorg. Chim. Acta* **2000**, *300-302*, 1-6.

1439. The Effect of Divergent-Bite Ligands on Metal-Metal Bond Distances in Some Paddlewheel Complexes, F. A. Cotton, L. M. Daniels, C. A. Murillo and H.-C. Zhou, *Inorg. Chim. Acta* **2000**, *300-302*, 319-327.

1440. Completion of the Series of $M_2(hpp)_4Cl_2$ Compounds from W to Pt: The W, Os and Pt Compounds, R. Clérac, F. A. Cotton, L. M. Daniels, J. P. Donahue, C. A. Murillo and D. J. Timmons, *Inorg. Chem.* **2000**, *39*, 2581-2584.

1441. A Millennial Overview of Transition Metal Chemistry, F. A. Cotton, *J. Chem. Soc., Dalton Trans.* **2000**, 1961-1968.

1442. Chromium(II) Complexes Bearing 2,6-substituted *N,N'*-di(aryl)formamidinate Ligands, F. A. Cotton, C. A. Murillo and P. Schooler, *J. Chem. Soc., Dalton Trans.* **2000**, 2001-2005.

1443. Chromium(II) Complexes Bearing 2-substituted *N,N'*-di(aryl)formamidinate Ligands, F. A. Cotton, C. A. Murillo and P. Schooler, *J. Chem. Soc., Dalton Trans.* **2000**, 2007-2011.

1444. A New Linear Tricobalt Complex of Di(2-pyridyl)amide (dpa), $[Co_3(dpa)_4(CH_3CN)_2]$-$[PF_6]_2$, R. Clérac, F. A. Cotton, K. R. Dunbar, T. Lu, C. A. Murillo and X. Wang, *Inorg. Chem.* **2000**, *39*, 3065-3070.

1445. Coordinated and Clathrated Molecular Diiodine in $[Rh_2(O_2CCF_3)_4I_2]\cdot I_2$, F. A. Cotton, E. V. Dikarev and M. A. Petrukhina, *Angew. Chem., Int. Ed. Engl.* **2000**, *39*, 2362-2365.

1446. Linear Tricobalt Compounds with Di-(2-pyridyl)amide (dpa) Ligands: Temperature Dependence of the Structural and Magnetic Properties of Symmetrical and Unsymmetrical Forms of $[Co_3(dpa)_4Cl_2]$ in the Solid State, R. Clérac, F. A. Cotton, L. M. Daniels, K. R. Dunbar, K. Kirschbaum, C. A. Murillo, A. A. Pinkerton, A. J. Schultz and X. Wang, *J. Am. Chem. Soc.* **2000**, *122*, 6226-6236.

1447. Oxidative Scission of a Mo–Mo Quadruple Bond, F. A. Cotton, C. A. Murillo and H.-C. Zhou, *Inorg. Chem.* **2000**, *39*, 3261-3264.

1448. Linear Trichromium Complexes with the Anion of 2,6-Di(phenylimino)piperidine, R. Clérac, F. A. Cotton, L. M. Daniels, K. R. Dunbar, C. A. Murillo and H.-C. Zhou, *Inorg. Chem.* **2000**, *39*, 3414-3417.

1449. Metal-Metal Versus Metal-Ligand Bonding in Dimetal Compounds with Tridentate Ligands, F. A. Cotton, L. M. Daniels, C. A. Murillo and H.-C. Zhou, *Inorg. Chem. Acta* **2000**, *305*, 69-74.

1450. The Whole Story of the Two-Electron Bond, with the Delta Bond as a Paradigm, F. A. Cotton and D. G. Nocera, *Acc. Chem. Res.* **2000**, *33*, 483-490.

1451. A Dichromium (II,II) Compound with a Strong Antiferromagnetic Coupling but Little or No Cr–Cr Bonding, F. A. Cotton, C. A. Murillo and H.-C. Zhou, *Inorg. Chem.* **2000**, *39*, 3728-3730.

1452. Studies of Dirhodium(II) Tetrakis(trifluoroacetate). 8. One-Dimensional Polymers of $Rh_2(O_2CCF_3)_4$ with Aromatic Ligands: Benzene, *p*-Xylene and Naphthalene, F. A. Cotton, E. V. Dikarev, M. A. Petrukhina and S.-E. Stiriba, *Polyhedron* **2000**, *19*, 1829-1835.

1453. An Infinite Zig-zag Chain of Copper Atoms; Still No Copper-Copper Bonding, R. Clérac, F. A. Cotton, L. M. Daniels, J. Gu, C. A. Murillo and H.-C. Zhou, *Inorg. Chem.* **2000**, *39*, 4488-4493.

1454. A Reliable Method of Preparation for Diiridium Paddlewheel Complexes: Structures of the First Compounds with Ir_2^{5+} Cores, F. A. Cotton, C. Lin and C. A. Murillo, *Inorg. Chem.* **2000**, *39*, 4574-4578.

1455. Alkylpyridine Complexes of Tungsten(II) and Chromium(II). First Rotational Isomers of $W_2X_4L_4$ Molecules with D_{2h} and D_2 Symmetry, F. A. Cotton, E. V. Dikarev, J. Gu, S. Herrero and B. Modec, *Inorg. Chem.* **2000**, *39*, 5407-5411.

1456. Cis-di(μ-trifluoroacetate)dirhodium Tetracarbonyl: Structure and Chemistry, F. A. Cotton, E. V. Dikarev and M. A. Petrukhina, *J. Chem. Soc., Dalton Trans.* **2000**, 4241-4243.

1457. Synthesis and Crystal Structures of "Unligated" Copper(I) and Copper(II) Trifluoroacetates, F. A. Cotton, E. V. Dikarev and M. A. Petrukhina, *Inorg. Chem.* **2000**, *39*, 6072-6079.

2001

1458. Supramolecular Structures Based on Dimetal Units: Simultaneous Utilization of Equatorial and Axial Connections, F. A. Cotton, C. Lin and C. A. Murillo, *Chem. Commun.* **2001**, 11-12.

1459. A Remarkable Cr(III) Organometallic Compound Formed by an Unprecedented Rearrangement of a Formamidinate Anion, R. Clérac, F. A. Cotton, C. A. Murillo and X. Wang, *Chem. Commun.* **2001**, 205-206.

1460. Dinuclear and Heteropolynuclear Complexes Containing Mo_2^{4+} Units, R. Clérac, F. A. Cotton, K. R. Dunbar, C. A. Murillo and X. Wang, *Inorg. Chem.* **2001**, *40*, 420-426.

1461. Connecting Pairs of Dimetal Units to Form Molecular Loops, F. A. Cotton, C. Lin and C. A. Murillo, *Inorg. Chem.* **2001**, *40*, 472-477.

1462. Supramolecular Squares with Mo_2^{4+} Corners, F. A. Cotton, C. Lin and C. A. Murillo, *Inorg. Chem.* **2001**, *40*, 478-484.

1463. A Neutral Triangular Supramolecule Formed by Mo_2^{4+} Units, F. A. Cotton, C. Lin and C. A. Murillo, *Inorg. Chem.* **2001**, *40*, 575-577.

1464. Incorporating Multiply Bonded Dirhenium Species $[Re_2]^{n+}$ (n = 4 or 5) into Assemblies Containing Two or More Such Units J. K. Bera, P. Angaridis, F. A. Cotton, M. A. Petrukhina, P. E. Fanwick and R. A. Walton, *J. Am. Chem. Soc.* **2001**, *123*, 1515-1516.

1465. Structural and Magnetic Properties of $Co_3(dpa)_4Br_2$, R. Clérac, F. A. Cotton, L. M. Daniels, K. R. Dunbar, C. A. Murillo and X. Wang, *J. Chem. Soc., Dalton Trans.* **2001**, 386-391.

1466. The Simplest Supramolecular Complexes Containing Pairs of $Mo_2(formamidinate)_3$ Units Linked with Various Dicarboxylates; Preparative Methods, Structures and Electrochemistry, F. A. Cotton, J. P. Donahue, C. Lin and C. A. Murillo, *Inorg. Chem.* **2001**, *40*, 1234-1244.

1467. Tuning the Metal-Metal Bonds in the Linear Tricobalt Compound $Co_3(dpa)_4Cl_2$: Bond Stretch and Spin-State Isomers, R. Clérac, F. A. Cotton, L. M. Daniels, K. R. Dunbar, C. A. Murillo and X. Wang, *Inorg. Chem.* **2001**, *40*, 1256-1264.

1468. Compounds with Symmetrical Tricobalt Chains Wrapped by Dipyridylamide Ligands and Cyanide or Isothiocyanate Ions as Terminal Ligands, R. Clérac, F. A. Cotton, S. P. Jeffery, C. A. Murillo and X. Wang, *Inorg. Chem.* **2001**, *40*, 1265-1270.

1469. A Molecular Propeller with Three Quadruply-Bonded Blades, F. A. Cotton, C. Lin and C. A. Murillo, *Inorg. Chem. Commun.* **2001**, *4*, 130-133.

1470. The First Dirhodium-Based Supramolecular Assemblies with Interlocking Lattices and Double Helices, F. A. Cotton, C. Lin and C. A. Murillo, *J. Chem. Soc., Dalton Trans.* **2001**, 499-501.

1471. Supramolecular Squares with Rh_2^{4+} Corners, F. A. Cotton, C. Lin, C. A. Murillo and S.-Y. Yu, *J. Chem. Soc., Dalton Trans.* **2001**, 502-504.

1472. Maximum Communication Between Coupled Oxidations of Dimetal Units, F. A. Cotton, C. Lin and C. A. Murillo, *J. Am. Chem. Soc.* **2001**, *123*, 2670-2671.

1473. Neutral Cyclooctasulfur as a Polydentate Ligand: Supramolecular Structures of $[Rh_2(O_2CCF_3)_4]_n(S_8)_m$ ($n:m$ = 1:1, 3:2), F. A. Cotton, E. V. Dikarev and M. A. Petrukhina, *Angew. Chem., Int. Ed. Eng.* **2001**, *40*, 1521-1523.

1474. Crystal Structure and Magnetic Behavior of the $Cu_3(O_2C_{16}H_{23})_6 \cdot 1.2C_6H_{12}$. An Unexpected Structure and an Example of Spin Frustration, R. Clérac, F. A. Cotton, K. R. Dunbar, E. A. Hillard, C. A. Murillo, M. A. Petrukhina and B. W. Smucker, *C. R. Acad. Sci. Paris, Chimie* **2001**, *4*, 315-319.

1475. Mixed Halide/Phosphine Complexes of the Dirhenium Core. Part 7. Reactions of [Re$_2$I$_8$]$^{2-}$ with Monodentate Phosphines, P. Angaridis, F. A. Cotton, E. V. Dikarev and M. A. Petrukhina, *Polyhedron* **2001**, *9-10*, 755-765.

1476. Quadridentate Bridging EO$_4$$^{2-}$ (E = S, Mo, W) Ligands and Their Role as Electronic Bridges, F. A. Cotton, J. P. Donahue and C. A. Murillo, *Inorg. Chem.* **2001**, *40*, 2229-2233.

1477. Di- and Trinuclear Complexes with the Mono- and Dianion of 2,6–bis(phenylamino)-pyridine: High Field Displacement of Chemical Shifts Due to the Magnetic Anisotropy of Quadruple Bonds, F. A. Cotton, L. M. Daniels, P. Lei, C. A. Murillo and X. Wang, *Inorg. Chem.* **2001**, *40*, 2778-2784.

1478. A Polymeric Chain in Which μ$_2$,η^2:η^2-Ethene Symmetrically Bridges Tetra(trifluoro-acetato) Dirhodium Molecules, F. A. Cotton, E. V. Dikarev, M. A. Petrukhina and R. E. Taylor, *J. Am. Chem. Soc.* **2001**, *123*, 5831-5832.

1479. Refutation of an Alleged Example of a Disordered but Centrosymmetric Triboluminescent Crystal, F. A. Cotton, L. M. Daniels and P. Huang, *Inorg. Chem. Commun.* **2001**, *4*, 319-321.

1480. Correlation of Structure and Triboluminescence for Tetrahedral Manganese(II) Compounds, F. A. Cotton, L. M. Daniels and P. Huang, *Inorg. Chem.* **2001**, *40*, 3576-3578.

1481. Supramolecular Arrays Based on Dimetal Building Units, F. A. Cotton, C. Lin and C. A. Murillo, *Acc. Chem. Res.* **2001**, *34*, 759-771.

1482. Mixed Chloride/Phosphine Complexes of the Dirhenium Core. 8. Synthesis and Crystal Structure of the Quadruply Bonded Dirhenium(III) Anion [Re$_2$Cl$_7$(PMe$_3$)]$^-$, F. A. Cotton, E. V. Dikarev and M. A. Petrukhina, *Inorg. Chem.* **2001**, *40*, 5716-5718.

1483. A Highly Disordered Cobaltocenium Salt, F. A. Cotton, L. M. Daniels, C. C. Wilkinson, *Acta Cryst.* **2001**, *E57*, m529-m530.

1484. Controlling the Dimensionality of Metal-Metal Bonded Rh$_2$$^{4+}$ Polymers by the Length of the Linker, F. A. Cotton, C. Lin and C. A. Murillo, *Inorg. Chem.* **2001**, *40*, 5886-5889.

1485. Using Structures Formed by Dirhodium Tetra(trifluoroacetate) with Polycyclic Aromatic Hydrocarbons to Prospect for Maximum π-Electron Density; Hückel Calculations Get It Right, F. A. Cotton, E. V. Dikarev and M. A. Petrukhina, *J. Am. Chem. Soc.* **2001**, *123*, 11655-11663.

1486. Neutral Dodecanuclear Supramolecular Complexes Containing Dimetal Units Linked by the Trimesate Anion, F. A. Cotton, C. Lin and C. A. Murillo, *Inorg. Chem.* **2001**, *40*, 6413-6417.

1487. Bis(diphenylphosphino)methanedicopper(I) Units Bridged by Dicarboxylates, P. Angaridis, F. A. Cotton and M. A. Petrukhina, *Inorg. Chim. Acta* **2001**, *324*, 318-323.

1488. Cyclopentadienyl-Metal Chemistry in the Wilkinson Group, Harvard, 1952-1955, F. A. Cotton, *J. Organometal. Chem.* **2001**, *637-639*, 18-26.

1489. Mixed Chloride/Phosphine Complexes of the Dirhenium Core. 10. Redox Reactions of an Edge-Sharing Dirhenium(III) Non-metal-Metal Bonded Complex, $Re_2(\mu\text{-}Cl)_2Cl_4(PMe_3)_4$, F. A. Cotton, E. V. Dikarev and M. A. Petrukhina, *Inorg. Chem.* **2001**, *40*, 6825-6831.

2002

1490. The First Supramolecular Assemblies Comprised of Dimetal Units and Chiral Dicarboxylates, F. A. Cotton, J. P. Donahue and C. A. Murillo, *Inorg. Chem. Commun.* **2002**, *5*, 59-63.

1491. The First Oxidation of a Cr_2^{4+} Paddlewheel Complex to an Isostructural Stable Product, F. A. Cotton, L. M. Daniels, P. Huang and C. A. Murillo, *Inorg. Chem.* **2002**, *41*, 317-320.

1492. A Half-Century of Nonclassical Organometallic Chemistry — A Personal Perspective, F. A. Cotton, *Inorg. Chem.* **2002**, *41*, 643-658.

1493. Mixed Chloride/Phosphine Complexes of the Dirhenium Core. 9. The First Mixed Monodentate Phosphine complex, $1,2,7,8\text{-}Re_2Cl_4(PMe_2Ph)_3(PEt_2H)$, P. A. Angaridis, F. A. Cotton, E. V. Dikarev and M. A. Petrukhina, *Inorg. Chim. Acta* **2002**, *330*, 173-178.

1494. The First Designed Syntheses of Bis-dimetal Molecules in Which the Bridges Are Diamidate Ligands, F. A. Cotton, L. M. Daniels, J. P. Donahue, C. Y. Liu and C. A. Murillo, *Inorg. Chem.* **2002**, *41*, 1354-1356.

1495. Filling a Void: Isolation and Characterization of Tetracarboxylato Dimolybdenum Cations, F. A. Cotton, L. M. Daniels, E. A. Hillard and C. A. Murillo, *Inorg. Chem.* **2002**, *41*, 1639-1644.

1496. A Pseudo Jahn-Teller Distortion in an $Mo_2(\mu_2\text{-}O)_2$ Ring Having the Shortest $Mo^{IV}\text{–}Mo^{IV}$ Double Bond, F. A. Cotton, L. M. Daniels, C. A. Murillo and J. G. Slaton, *J. Am. Chem. Soc.* **2002**, *124*, 2878-2879.

1497. A Tri-Nickel Dipyridylamido Complex with Metal–Metal Bonding Interaction: Prelude to Polynickel Molecular Wires and Devices? J. F. Berry, F. A. Cotton, L. M. Daniels and C. A. Murillo, *J. Am. Chem. Soc.* **2002**, *124*, 3212-3213.

1498. Dinuclear Ti^{IV} and Ti^{III} Complexes Supported by Calix[4]arene Ligands. Binding Alkali-metal Cations Inside and Outside the Cavity of Calix[4]arenes, F. A. Cotton, E. V. Dikarev, C. A. Murillo and M. A. Petrukhina, *Inorg. Chim. Acta* **2002**, *332*, 41-46.

1499. Mixed Chloride/Phosphine Complexes of the Dirhenium Core. 11. Reactions of $[Re_2Cl_8]^{2-}$ with Secondary Phosphines, PCy_2H and PPh_2H, P. A. Angaridis, F. A. Cotton, E. V. Dikarev and M. A. Petrukhina, *Inorg. Chim. Acta* **2002**, *332*, 47-53.

1500. The Lengths of Molybdenum to Molybdenum Quadruple Bonds; Correlations, Explanations and Corrections, F. A. Cotton, L. M. Daniels, E. A. Hillard and C. A. Murillo, *Inorg. Chem.* **2002**, *41*, 2466-2470.

1501. The Use of Dimetal Building Blocks in Convergent Syntheses of Large Arrays, F. A. Cotton, C. Lin and C. A. Murillo, *Proc. Nat. Acad. Sci.* **2002**, *99*, 4810-4813.

1502. The First Dirhodium Tetracarboxylate Molecule Without Axial Ligation: New Insight into the Electronic Structures of Molecules with Importance in Catalysis and Other Reactions, F. A. Cotton, E. A. Hillard and C. A. Murillo, *J. Am. Chem. Soc.* **2002**, *124*, 5658-5960.

1503. A Molecular Pair Having Two Quadruply Bonded Dimolybdenum Units Linked by a Terephthalate Dianion, F. A. Cotton, J. P. Donahue, C. Lin, C. A. Murillo and J. Rockwell, *Acta Cryst*, **2002**, *E58*, m298-m300.

1504. Supramolecular Assemblies of Dimetal Complexes with Polydentate N-donor Ligands: From a Discrete Pyramid to a 3D Channel Network, F. A. Cotton, E. V. Dikarev, M. A. Petrukhina, M. Schmitz and P. J. Stang, *Inorg. Chem.* **2002**, *41*, 2903-2908.

1505. An Unusual Compound Containing μ-O-bridged Re_2^{6+} Cores in Very Different Coordination Environments, F. A. Cotton, E. V. Dikarev and M. A. Petrukhina, *Inorg. Chim. Acta* **2002**, *334C*, 67-70.

1506. A Complete Series of $W_2(hpp)_4Cl_n$ (n = 0, 1, 2) Compounds, F. A. Cotton, P. Huang, C. A. Murillo and D. J. Timmons, *Inorg. Chem. Commun.* **2002**, *5*, 501-504.

1507. First Water Soluble Mo_2^{4+} Compounds Spanned by Four α-Hydroxycarboxylate Anions, T. S. Barnard, F. A. Cotton, L. M. Daniels and C. A. Murillo, *Inorg. Chem. Commun.* **2002**, *5*, 527-532.

1508. The Extraordinary Ability of Guanidinate Derivatives to Stabilize Higher Oxidation Numbers in Dimetal Units by Modification of Redox Potentials: Structures of Mo_2^{5+} and Mo_2^{6+} Compounds, F. A. Cotton, L. M. Daniels, C. A. Murillo, D. J. Timmons and C. C. Wilkinson, *J. Am. Chem. Soc.* **2002**, *124*, 9249-9256.

1509. How to Make a Major Shift in a Redox Potential; Ligand Control of the Oxidation State of Dimolybdenum Units, F. A. Cotton, L. M. Daniels, C. Y. Liu, C. A. Murillo, A. J. Schultz and X. Wang, *Inorg. Chem.* **2002**, *41*, 4232-4238.

1510. Au Temps de l'établissement de la Structure du Ferrocène, Témoignage Sur Cette Aventure, F. A. Cotton, *l'actualité chimique*, **juillet 2002**, 28-29.

1511. Steps on the Way to the First Dirhodium Tetracarboxylate with No Axial Ligation: Synthetic Lessons and a Plethora of $Rh_2(O_2CR)_4L_{2-n}$ Compounds, n = 0, 1, 2, F. A. Cotton, E. A. Hillard, C. Y. Liu, C. A. Murillo, W. Wang and X. Wang, *Inorg. Chim. Acta*, **2002**, *337*, 233-246.

1512. Most Easily-ionized, Closed-shell Molecules Known; Easier than the Cesium Atom, F. A. Cotton, N. E. Gruhn, J. Gu, P. Huang, D. L. Lichtenberger, C. A. Murillo, L. O. Van Dorn and C. C. Wilkinson, *Science* **2002**, *298*, 1971-1974.

2003

1513. The Radical Anions and the Electron Affinities of Perfluorinated Benzene, Napthalene, and Anthracene, Y. Xie, H. F. Schaefer III and F. A. Cotton, *Chem. Commun.* **2003**, 102-103.

1514. Synthesis and Reactivity of a Very Strong Reducing Agent Containing a Quadruple Bond: Structures of $W_2(hpp)_4$ and $W_2(hpp)_4Cl_2 \cdot 4CH_2Cl_2$, F. A. Cotton, P. Huang, C. A. Murillo and X. Wang, *Inorg. Chem. Commun.* **2003**, *6*, 121-126.

1515. Further Structural and Magnetic Studies of Tricopper Dipyridylamido Complexes, J. F. Berry, F. A. Cotton, P. Lei and C. A. Murillo, *Inorg. Chem.* **2003**, *42*, 377-383.

1516. The First Structurally Confirmed Paddlewheel Compound with an M_2^{7+} Core: $[Os_2(hpp)_4Cl_2](PF_6)$, F. A. Cotton, N. S. Dalal, P. Huang, C. A. Murillo, A. C. Stowe and X. Wang, *Inorg. Chem.* **2003**, *42*, 670-672.

1517. A Highly Reduced V_2^{3+} Unit with a Metal–Metal Bond Order of 3.5, F. A. Cotton, E. A. Hillard and C. A. Murillo, *J. Am. Chem. Soc.* **2003**, *125*, 2026-2027.

1518. Applications of High-Field (W-Band) EPR to M–M Bonded Units (M = Cr, Mo): The First Confirmed Oxidation of a Cr_2^{4+} Paddlewheel Complex to a Stable Isostructural Cr_2^{5+} Product, F. A. Cotton, N. S. Dalal, E. A. Hillard, P. Huang, C. A. Murillo and C. M. Ramsey, *Inorg. Chem.* **2003**, *42*, 1388-1390.

1519. Further Observations on the Non-rigorous Relationship Between Triboluminescence and Crystal Centricity, F. A. Cotton and P. Huang, *Inorg. Chim. Acta* **2003**, *346C*, 223-226.

1520. A Highly Oxidized Re_2^{7+} Species with an Electron-Poor Bond of Order 3.5, J. F. Berry, F. A. Cotton, P. Huang and C. A. Murillo, *J. Chem. Soc., Dalton Trans.* **2003**, 1218-1219.

1521. Oxidation of $Ni_3(dpa)_4Cl_2$ and $Cu_3(dpa)_4Cl_2$: Nickel–Nickel Bonding Interaction, but No Copper–Copper Bonds, J. F. Berry, F. A. Cotton, L. M. Daniels, C. A. Murillo and X. Wang, *Inorg. Chem.* **2003**, *42*, 2418-2427.

1522. Calixarenes as Ligands for Dimetal Units: The State of the Art. Structure of a New Quadruply-Bonded Dimolybdenum Compound, F. A. Cotton, L. M. Daniels, C. Lin and C. A. Murillo, *Inorg. Chim. Acta* **2003**, *347*, 1-8.

1523. Observation of Symmetry Lowering and Electron Localization in the Doublet-States of a Spin-Frustrated Equilateral Triangular Lattice: $Cu_3(O_2C_{16}H_{23})_6 \cdot 1.2C_6H_{12}$, B. Cage, F. A. Cotton, N. S. Dalal, E. A. Hillard, B. Rakvin and C. M. Ramsey, *J. Am. Chem. Soc.* **2003**, *125*, 5270-5271.

1524. Polyunsaturated Dicarboxylate Tethers Connecting Dimolybdenum Redox and Chromophoric Centers: Syntheses, Structures, and Electrochemistry, F. A. Cotton, J. P. Donahue and C. A. Murillo, *J. Am. Chem. Soc.* **2003**, *125*, 5436-5450.

1525. Polyunsaturated Dicarboxylate Tethers Connecting Dimolybdenum Redox and Chromophoric Centers: Absorption Spectra and Electronic Structures, F. A. Cotton, J. P. Donahue, C. A. Murillo and Lisa M. Thompson-Pérez, *J. Am. Chem. Soc.* **2003**, *125*, 5486-5492.

1526. EPR Probing of Bonding and Spin Localization of the Doublet-Quartet States in a Spin-Frustrated Equilateral Triangular Lattice: $Cu_3(O_2C_{16}H_{23})_6 \cdot 1.2C_6H_{12}$, B. Cage, F. A. Cotton, N. S. Dalal, E. A. Hillard, B. Rakvin and C. M. Ramsey, *Comptes Rendus Chimie* **2003**, *6*, 39-46.

1527. Synthesis of Unsymmetrical Formamidines and Benzamidines: Structural Studies of Copper, Cobalt and Chromium Complexes, F. A. Cotton, P. Lei, C. A. Murillo and L.-S. Wang, *Inorg. Chim. Acta* **2003**, *349C*, 165-172.

1528. Linear Trichromium(II) Chains Wrapped by Unsymmetrical Formamidinates, F. A. Cotton, P. Lei and C. A. Murillo, *Inorg. Chim. Acta* **2003**, *349C*, 173-181.

1529. Additional Steps toward Molecular Scale Wires, Switches, Rheostats and Resistors: Further Study of $Ni_5^{10/11+}$ Chains Embraced by Polypyridylamide Ligands, J. F. Berry, F. A. Cotton, P. Lei, T. Lu and C. A. Murillo, *Inorg. Chem.* **2003**, *42*, 3534-3539.

1530. Enhancing the Stability of Trinickel Molecular Wires and Switches: Ni_3^{6+}/Ni_3^{7+}. Four New Trinickel Compounds, J. F. Berry, F. A. Cotton, T. Lu, C. A. Murillo and X. Wang, *Inorg. Chem.* **2003**, *42*, 3595-3601.

1531. Resolving Conformational Ambiguities in $M_2(hpp)_4Cl_2$ Paddlewheel Compounds: M = Mo, W, Re, Ru, Os, Ir, Pd, Pt, F. A. Cotton, C. A. Murillo, X. Wang and C. C. Wilkinson, *Inorg. Chim. Acta* **2003**, *351*, 183-190.

1532. Structural Studies of Nickel Complexes of the Unsymmetrical Tridentate Ligand *N*-phenyl-*N*-(2-pyridyl)formamidinate, F. A. Cotton, P. Lei and C. A. Murillo, *Inorg. Chim. Acta* **2003**, *351*, 191-200.

1533. Increasing Solubility and Stability of Linear Tricobalt(II) Chains with depa (diethyl-dipyridylamide) Ligands, J. F. Berry, F. A. Cotton, T. Lu and C. A. Murillo, *Inorg. Chem.* **2003**, *42*, 4425-4430.

1534. Cyclic Polyamidato Dianions as Bridges Between Mo_2^{4+} Units: Synthesis, Crystal Structures, Electrochemistry, Absorption Spectra, and Electronic Structures, F. A. Cotton, J. P. Donahue, C. A. Murillo, L. M. Pérez and R. Yu, *J. Am. Chem. Soc.* **2003**, *125*, 8900-8910.

1535. Trapping Tetramethoxyzincate and -cobaltate(II) Between Mo_2^{4+} Units, F. A. Cotton, C. Y. Liu, C. A. Murillo and X. Wang, *Inorg. Chem.* **2003**, *42*, 4619-4623.

1536. Making Connections with Molecular Wires: Extending Tri-nickel Chains with Axial Cyanide, Dicyanamide, and Phenylacetylide Ligands, J. F. Berry, F. A. Cotton and C. A. Murillo, *J. Chem. Soc., Dalton Trans.* **2003**, 3015-3021.

1537. Isomerization by Ligand Shuffling Along a Cr_2^{4+} Unit: Further Reactions Leading to Cleavage of a Quadruple Bond, R. Clérac, F. A. Cotton, S. P. Jeffery, C. A. Murillo and X. Wang, *J. Chem. Soc., Dalton Trans.* **2003**, 3022-3027.

1538. A Mixed-Valence Compound with One Unpaired Electron Delocalized over Four Molybdenum Atoms in a Cyclic Tetranuclear Ion, F. A. Cotton, C. Y. Liu, C. A. Murillo and X. Wang, *J. Chem. Soc., Chem. Commun.* **2003**, 2190-2191.

1539. Molecular Squares with Paramagnetic Diruthenium Corners: Synthetic and Crystallographic Challenges, P. Angaridis, J. F. Berry, F. A. Cotton, C. A. Murillo and X. Wang, *J. Am. Chem. Soc.* **2003**, *125*, 10327-10334.

1540. New Chemistry of the Triply-Bonded Divanadium (V_2^{4+}) Unit and Reduction to an Unprecedented V_2^{3+} Core, F. A. Cotton, E. A. Hillard, C. A. Murillo and X. Wang, *Inorg. Chem.* **2003**, *42*, 6063-6067.

1541. Fully Localized Mixed–Valence Oxidation Products of Molecules Containing Two Linked Dimolybdenum Units: An Effective Structural Criterion, F. A. Cotton, N. S. Dalal, C. Y. Liu, C. A. Murillo, J. M. North and X. Wang, *J. Am. Chem. Soc.* **2003**, *125*, 12945-12952.

1542. Modifying Electronic Communication in Dimolybdenum Units by Linkage Isomers of Bridged Oxamidate Dianions, F. A. Cotton, C. Y. Liu, C. A. Murillo, D. Villagrán, X. Wang, *J. Am. Chem. Soc.* **2003**, *125*, 13564-13575.

1543. A Molecular Loop with Interstitial Channels in a Chiral Environment: Exploration of the Chemistry of Mo_2^{4+} Species with Chiral and Non-chiral Dicarboxylate Anions, J. F. Berry, F. A. Cotton, S. A. Ibragimov, C. A. Murillo and X. Wang, *J. Chem. Soc., Dalton Trans.* **2003**, 4297-4302.

1544. An Asymmetric Edge-Sharing Biooctahedral Complex, F. A. Cotton, J. P. Donahue and C. A. Murillo, *J. Cluster Sci.* **2003**, *14*, 289-298.

2004

1545. Dicarboxylato-bridged Diruthenium Units in Two Different Oxidation States: The First Step Towards the Synthesis of Creutz-Taube Analogs with Dinuclear Ru_2^{n+} Species, P. Angaridis, J. F. Berry, F. A. Cotton, P. Lei, C. Lin, C. A. Murillo, D. Villagrán, *Inorg. Chem. Commun.* **2004**, *7*, 9-13.

1546. A Calix[4]arene Carceplex with Four Rh_2^{4+} Fasteners, F. A. Cotton, P. Lei, C. Lin, C. A. Murillo, X. Wang, S.-Y. Yu and Z.-X. Zhang, *J. Am. Chem. Soc.* **2004**, *126*, 1518-1525.

1547. Systematic Preparation of Mo_2^{4+} Building Blocks for Supramolecular Assemblies, F. A. Cotton, C. Y. Liu, C. A. Murillo, *Inorg. Chem.* **2004**, *43*, 2267-2276.

1548. An Efficient Synthesis of Acetylide/Trimetal/Acetylide Molecular Wires, J. F. Berry, F. A. Cotton, C. A. Murillo and B. K. Roberts, *Inorg. Chem.* **2004**, *43*, 2277-2283.

1549. A Trinuclear EMAC-type Molecular Wire with Redox Active Ferrocenylacetylide "Alligator Clips" Attached, J. F. Berry, F. A. Cotton and C. A. Murillo, *Organometallics* **2004**, *23*, 2503-2506.

1550. Molecular and Electronic Structures by Design: Tuning Symmetrical and Unsymmetrical Trichromium Chains, J. F. Berry, F. A. Cotton, T. Lu, C. A. Murillo, B. K. Roberts and X. Wang, *J. Am. Chem. Soc.* **2004**, *126*, 7082-7096.

1551. Extended Metal Atom Chains (EMACs) of Five Chromium or Cobalt Atoms: Symmetrical or Unsymmetrical?, J. F. Berry, F. A. Cotton, C. S. Fewox, T. Lu, C. A. Murillo and X. Wang, *J. Chem. Soc., Dalton Trans.* **2004**, 2297-2302.

1552. Reaction Products of $W(CO)_6$ with Formamidines: Electronic Structures of a $W_2(\mu\text{-}CO)_2$ Core with Unsymmetric Bridging Carbonyls, F. A. Cotton, J. P. Donahue, M. B. Hall, C. A. Murillo and D. Villagrán, *Inorg. Chem.* **2004**, 43, 6954-6964.

1553. Mono-, Di-, and Tri-Ruthenium Complexes with the Ligand 2,2'-Dipyridylamide (dpa): Insights Into the Formation of Extended Metal Atom Chains, J. F. Berry, F. A. Cotton and C. A. Murillo, *Inorg. Chim. Acta,* **2004**, *357*, 3847-3853. (Marks Issue).

1554. Strong Electronic Coupling Between Dimolybdenum Units Linked by the N,N'-dimethloxamidate Anion in a Molecule Having a Heteronaphthalene-like Structure. F. A. Cotton, C. Y. Liu, C. A. Murillo, D. Villagrán and X. Wang, *J. Am. Chem. Soc.* **2004**, *126*, 14822-14831.

1555. Exploring the Reactivity of $Rh_2(OAc)_4$ with 2,2'-Dipyridylamide, J. F. Berry, F. A. Cotton, C. Lin and C. A. Murillo, *J. Cluster Sci.* **2004**, *15*, 531-541.

1556. Paramagnetic Precursors for Supramolecular Assemblies: Selective Synthesis, Crystal Structures, Electrochemical and Magnetic Properties of $Ru_2(O_2CMe)_{4-n}$(formamidinate)$_n$Cl complexes, n = 1,2,3,4, P. Angaridis, F. A. Cotton, C. A. Murillo, D. Villagrán and X. Wang, *Inorg. Chem.* **2004**, *43*, 8290-8300.

1557. Paramagnetism at Ambient Temperature, Diamagnetism at Low Temperature in a Ru_2^{6+} Core: Structural Evidence for Zero-Field Splitting, F. A. Cotton, C. A. Murillo, J. H. Reibenspies, X. Wang, and C. C. Wilkinson, *Inorg. Chem.* **2004**, *43*, 8373-8378.

1558. Chiral Organometallic Triangles with Rh–Rh Bonds. 1. Compounds Prepared from Racemic *cis*-$Rh_2(C_6H_4PPh_2)_2(OAc)_2$, F. A. Cotton, C. A. Murillo, X. and R. Yu, *Inorg. Chem.* **2004**, *43*, 8394-8403.

2005

1559. A Paramagnetic Precursor for Polymeric Supramolecular Assemblies Based on Multiply Bonded Dimetal Units: μ-Aceto-Acetonitriletris(μ-N, N'-Diphenylformamidinato)-diruthenium Tetrafluoroborate Dichloromethane Hemisolvate, P. Angaridis, F. A. Cotton, C. A. Murillo and X. Wang, *Acta Cryst.* **2005**, *C61*, m71-m73.

1560. Tetra-μ-acetato-O:O'-bis(di-*p*-anisyl)formamidine-N)diruthenium(II,II): An Example of Axial Bis-adduct of Ru_2^{4+} tetracarboxylate with N-donor Ligands, P. Angaridis, F. A. Cotton, C. A. Murillo, and X. Wang, *Acta Crystallogr. C.* **2005**, *C61*, m109-m111.

1561. Synthesis and Structural Characterization of the Molecular Loop [*cis*-$Mo_2(N,N$-di-*p*-anisylformamidinate)$_2$]$_2$(O_2C–CH=C=CH–CO_2)$_2$, F. A. Cotton, J. P. Donahue, C. A. Murillo and R. Yu, *Inorg. Chim. Acta* **2005**, *358*, 1373-1376.

1562. Modeling Spin Interactions in a Cyclic Trimer and a Cuboidal Co_4O_4 Core with Co(II) in Tetrahedral and Octahedral Environments, J. F. Berry, F. A. Cotton, C. Y. Liu, C. A. Murillo, B. S. Tsukerblat, D. Villagrán and X. Wang. *J. Am. Chem. Soc.* **2005**, *127*, 4895-4902.

1563. Structural and Magnetic Evidence Concerning Spin Crossover in Formamidinate Compounds with Ru_2^{5+}, P. Angaridis, F. A. Cotton, C. A. Murillo, D. Villagrán and X. Wang. *J. Am. Chem. Soc.* **2005**, *127*, 5008-5009.

1564. Non-trivial Behavior of Palladium(II) Acetate, V. I. Bakhmutov, J. F. Berry, F. A. Cotton, S. Ibragimov and C. A. Murillo, *Dalton Trans.* **2005,** 1989-1992.

1565. Expeditious Access to the Most Easily Ionized Closed-Shell Molecule, $W_2(hpp)_4$, F. A. Cotton, J. P. Donahue, D. L. Lichtenberger, C. A. Murillo, and D. Villagrán, *J. Am. Chem. Soc.* **2005,** *127,* 10808-10809.

1566. Searching for Precursors to Metal-Metal Bonded Dipalladium Species: A Study of Pd_2^{4+} Complexes, J. F. Berry, F. A. Cotton, S. A. Ibragimov, C. A. Murillo and X. Wang, *Inorg. Chem.* **2005,** *44,* 6129-6137.

1567. Proof of Large Positive Zero-Field Splitting in a Ru_2^{5+} Paddlewheel, W.-Z. Chen, F. A. Cotton, N. S. Dalal, C. A. Murillo, C. M. Ramsey, T. Ren, X. Wang, *J. Am. Chem. Soc.* **2005,** *127,* 12691-12696.

1568. Deliberate Synthesis of the Preselected Enantiomers of an Enantiorigid Molecule with Pure Rotational Symmetry *T,* F. A. Cotton, C. A. Murillo, R. Yu, *Dalton. Trans.* **2005,** 3161-3165.

1569. Strong Electronic Coupling Between Mo_2^{n+} Units: The Oxidation Products of $[Mo_2(DAniF)_3]_2(\mu\text{-}H)_2$ and $Mo_2(DAniF)_4$, F. A. Cotton, J. P. Donahue, P. Huang, C. A. Murillo, and D. Villagrán. *Z. Anorg. Allg. Chem.* **2005,** *631,* 2606-2612.

1570. Chiral Organometallic Triangles with Rh–Rh Bonds. II. Compounds Prepared from Enantiopure *cis*-$Rh_2(C_6H_4PPh_2)_2(HOAc)_2$ and Their Catalytic Potentials, F. A. Cotton, C. A. Murillo, S.-E. Stiriba, X. Wang and R. Yu, *Inorg. Chem.* **2005,** *44,* 8223-8233.

1571. Chiral Supramolecules: Organometallic Molecular Loops Made from Enantiopure R-[*cis*-$Rh_2(C_6H_4PPh_2)_2(CH_3CN)_6](BF_4)_2$, F. A. Cotton, C. A. Murillo, R. Yu, *Inorg. Chem.* **2005,** 8211-8215.

1572. A Hardwon Dirhodium Paddlewheel with Guanidinate Type (hpp) Bridging Ligands, J. F. Berry, F. A. Cotton, P. Huang, C. A. Murillo and X. Wang, *Dalton Trans.* **2005,** 3713-3715.

2006

1573. Trishomobarrellenedicarboxylate Dianion as a Bridge Between Dimolybdenum Units: Comparison to Similar Compounds, F. A. Cotton, A. de Meijere, C. A. Murillo, K. Rauch, R. Yu, *Polyhedron* **2006,** *25,* 219-223.

1574. Facilitating Access to the Most Easily Ionized Molecules: A Key Intermediate and Related Compounds, F. A. Cotton, J. P. Donahue, N. E. Gruhn, D. L. Lichtenberger, C. A. Murillo, D. J. Timmons, L. O. Van Dorn, D. Villagrán, and X. Wang, *Inorg. Chem.* **2006,** *45,* 201-213.

1575. Strong Electronic Interaction Between Two Dimolybdenum Units Linked by a Tetraazatetraceme, F. A. Cotton, Z. Li, C. Y. Liu, C. A. Murillo and D. Villagrán, *Inorg. Chem.* **2006,** *45,* 767-778.

1576. Vibrational Excitations in Single Trimetal-Molecule Transistors, D.-H. Chae, J. F. Berry, S. Jung, F. A. Cotton, C. A. Murillo and Z. Yao, *Nano Lett.* **2006,** *6,* 165-168.

1577. Remarkable Electron Accepting Properties of the Simplest Benzenoid Cyanocarbons: Hexacyanobenzene, Octacyanonaphthalene and Decacyanoanthracene, X. Zhang, Q. Li, J. B. Ingles, A. C. Simmonett, S. E. Wheller, Y. Xie, R. B. King, H. F. Schaefer III, and F. A. Cotton, *Chem Commun.* **2006**, 758-760.

1578. Uniquely Strong Electronic Communication Between [Mo₂] Units Linked by Dioxolene Dianions, F. A. Cotton, C. A. Murillo, D. Villagrán and R. Yu, *J. Am. Chem. Soc.* **2006**, *128*, 3281-3290.

1579. Dimolybdenum-Containing Molecular Triangles and squares with Diamidate Linkers: Structural Diversity and Complexity, F. A. Cotton, C. Y. Liu, C. A. Murillo, and X. Wang, *Inorg. Chem.* **2006**, *45*, 2619-2626.

1580. Structure and Magnetism of $[M_3]^{(6/7)+}$ Metal Chain Complexes from Density Functional Theory Analysis for Copper and Predictions for Silver, M. Bénard, J. F. Berry, F. A. Cotton, C. Gaudin, X. López, C. A. Murillo and M.-M. Rohmer, *Inorg. Chem.* **2006**, *45*, 3932-3940.

1581. A Diamagnetic Dititanium(III) Paddlewheel Complex With No Direct Metal-Metal Bond, A. Mendiratta, C. C. Cummins, F. A. Cotton, S. A. Ibragimov, C. A. Murillo and D. Villagrán, *Inorg. Chem.* **2006**, *45*, 4328-4330.

1582. Metal-Metal Bonding in Mixed Valence Ni_2^{5+} Complexes and Spectroscopic Characterization of a Ni_2^{6+} Species, F. A. Cotton, C. A. Murillo and R. Yu, *Inorg. Chem.* **2006**, *45*, 4396-4406.

1583. Homologues of the Easily Ionized Compound $Mo_2(hpp)_4$ Containing Smaller Bicyclic Guanidinates, F. A. Cotton, C. A. Murillo, X. Wang and C. C. Wilkinson, *Inorg. Chem*. **2006**, *45*, 5493-5500.

1584. Transition from a Non-bonding to a Bonding Interaction in a Tetranuclear $[Mo_2]_2(\mu\text{-}OR)_4$ Cluster, F. A. Cotton, Z. Li, C. Y. Liu, C. A. Murillo and Q. Zhao, *Inorg. Chem.* **2006**, *45*, 6387-6395.

1585. Dynamic Equilibrium Between Cyclic Oligomers. Thermodynamic and Structural Characterization of a Square and a Triangle, F. A. Cotton, C. A. Murillo and R. Yu, *Dalton Trans.* **2006**, 3900-3905.

1586. Strong Reducing Agents Containing Dimolybdenum Mo_2^{4+} Units and Their Oxidized Cations with $Mo_2^{5+/6+}$ Cores Stabilized by Bicyclic Guanidinate Anions with a Seven-Membered Ring, F. A. Cotton, C. A. Murillo, X. Wang and C. C. Wilkinson, *Dalton Trans.* **2006**, 4623-4631.

1587. Photoelectron Spectroscopy and DFT Calculations of Easily Ionized Quadruply Bonded Mo_2^{4+} Compounds and Their Bicyclic Guanidinate Precursors, F. A. Cotton, J. C. Durivage, N. E. Gruhn, D. L. Lichtenberger, C. A. Murillo, L. O. Van Dorn and C. C. Wilkinson, *J. Chem. Phys. B206* **2006**, *110*, 19793-19798.

1588. Chiral Molecules Containing Metal-Metal Bonds, F. A. Cotton and C. A. Murillo, *Eur. J. Inorg. Chem.* **2006**, 4209-4218.

1589. A Rare and Highly Oxidized $Mo_2^{5.5+}$ Unit Stabilized by Oxo Anions and Supported by Formamidinate Bridges, F. A. Cotton, C. A. Murillo, R. Yu and Q. Zhao, *Inorg. Chem.* **2006**, 9046-9052.

1590. An Isomeric Pair of Fluoride-Bridged Cyclic Dimolybdenum Triads, F. A. Cotton, C. Y. Liu, C. A. Murillo and Q. Zhao, *Inorg. Chem.* **2006**, *45*, 9480-9486.

1591. Strong Electronic Communication by Direct Metal–Metal Interaction in Molecules with Halide-Bridged Dimolybdenum Pairs, F. A. Cotton, C. Y. Liu, C. A. Murillo and Q. Zhao, *Inorg. Chem.* **2006**, *45*, 9493-9501.

1592. Determination of the 2,3-pentadienedioic Acid Enantiomer Interconversion Energy Barrier 1. Classical Kinetic Approach, P. Màjek, J. Mydlovà, J. Krupcík, J. Lehotay, D. W. Armstrong and F. A. Cotton, *J. Sep. Sci.* **2006**, *29*, 2357-2364.

1593. A Trimetal Chain Cocooned by Two Heptadentate Polypyridylamide Ligands, F. A. Cotton, H. Chao, C. A. Murillo and Q. Wang, *Dalton Trans.* **2006**, 5416-5422.

1594. An Equilibrium Between a Loop and a Triangle Containing Mo_2^{4+} Species and Perfluorodicarboxylate Linkers, F. A. Cotton, C. A. Murillo and R. Yu, *Inorg. Chim. Acta* **2006**, *356,* 5416-5422.

1595. High Yield Synthesis of Stable, Singly Bonded Pd_2^{6+} Compounds, F. A. Cotton, I. O. Koshevov, P. Lahuerta, C. A. Murillo, M. Sanaú, M. A. Ubeda and Q. Zhao, *J. Am. Chem. Soc.* **2006**, *128*, 13674-13975.

1596. Molecular Pairs and a Propeller Containing Quadruply Bonded Dimolybdenum Units Linked by Polyamidate Ligands, F. A. Cotton, Z. Li and C. A. Murillo, *Inorg. Chem.* **2006**, *45*, 9765-9770.

1597. Remarkable Electron Accepting Properties of the Simplest Benzenoid Cyanocarbons: Hexacyanobenzene, Octacyanonaphthalene and Decacyanoanthracene, X. H. Zhang, Q. S. Li, J. B. Ingels, A. C. Simmonett, S. E. Wheeler, Y. M. Xie, R. B. King, H. F. Schaefer and F. A. Cotton, *Chem. Commun.* **2006**, 758-760.

1598. Determination of the Interconversion Energy Barrier of 2,3-Pentadienedioic Acid Enantiomers by HPLC. 2. On Column Interconversion, J. Mydlová, A. Fedurcová, J. Lehotay, J. Krupcík, P. Májek, D. W. Armstrong, B. L. He and F. A. Cotton, *J. Sep. Sci.* **2006**, *29*, 2594-2599.

1599. Enantioseparation of Extended Metal Atom Chain (EMAC) Complexes: Unique Compounds of Extraordinarily High Specific Rotation, M. M. Warnke, F. A. Cotton and D. A. Armstrong, *Chirality,* **2007**, 179-183.

1600. A Fractional Bond Order of ½ in Pd_2^{5+}–Formamidinate Species; The Value of Very High Field EPR Spectra, J. F. Berry, E. Bill, E. Bothe, F. A. Cotton, N. S. Dalal, S. A. Ibragimov, N. Kaur, C. Y. Liu, C. A. Murillo, S. Nellutla, J. M. North and D. Villagrán, *J. Am. Chem. Soc.* **2007**, *127*, 1393-1401.

1601. Resolution of Enantiomers in Solution and Determination of the Chirality of Extended Metal Atom Chains, D. W. Armstrong, F. A. Cotton, A. G. Petrovic, P. L. Polavarapu and M. M. Warnke, *Inorg. Chem.* **2007**, *46*, 1535-1537.

1602. Better Understanding of the Species with the Shortest Re_2^{6+} Bonds and Related Re_2^{7+} Species with Tetraguanidinate Paddlewheel Structures, F. A. Cotton, N. S. Dalal, P. Huang, S. A. Ibragimov, C. A. Murillo, P. M. B. Piccoli, C. M. Ramsey, A. J. Schultz, X. Wang and Q. Zhao, *Inorg. Chem.* **2007**, *46*, 1718-1726.

1603. Electronic Localization versus Delocalization Determined by the Binding of the Linker in an Isomer Pair, F. A. Cotton, C. Y. Liu, C. A. Murillo and Q. Zhao, *Inorg. Chem.* **2007**, *46*, 2604-2611.

1604. Crystal to Crystal Oxidative Deprotonation of a Di(μ-hydroxo)to a Di(μ-oxo) Dimer of Dimolybdenum Units, F. A. Cotton, Z. Li, C. A. Murillo, X. Wang, R. Yu and Q. Zhao, *Inorg. Chem.* **2007**, *46*, 3245-3250.

1605. A Deliberate Approach for the Synthesis of Heterometallic Supramolecules Containing Dimolybdenum Mo_2^{4+} Species Coordinated to Other Metal Units, F. A. Cotton, J.-Y. Jin, Zhong Li, C. Y. Liu and C. A. Murillo, *Dalton Trans.* **2007**, 2328-2335.

1606. Precursors for Assembly of Supramolecules Containing Quadruply Bonded Cr_2^{4+} Units: Systematic Preparation of $Cr_2(formamidinate)_n(acetate)_{4-n}$ (n = 2-4), F. A. Cotton, Z. Li and C. A. Murillo, *Eur. J. Inorg. Chem.* **2007**, 3509-3513.

1607. Increasing the Solubility of Strong Reducing Agents Containing Mo_2^{4+} Units and Alkyl-Substituted Guanidinate Ligands, F. A. Cotton, C. A. Murillo, X. Wang and C. C. Wilkinson, *Dalton Trans.* **2007**, 3943-3951.

1608. Modulating Electronic Coupling Using O- and S-donor Linkers, F. A. Cotton, Z. Li, C. Y. Liu and C. A. Murillo, *Inorg. Chem.* **2007**, *46*, 7840-7847.

1609. Inserting Sulfur-Donor Ligands in Axial Positions of an Extended Trinickel Chain, F. A. Cotton, C. A. Murillo and Q. Wang, *Inorg. Chem. Commun.* **2007**, *10*, 1088-1090.

1610. A Rare Dimer of Dimers having Four Hydride Linkers Joining Two Quadruply Bonded Dimolybdenum Units, F. A. Cotton, C. A. Murillo and Q. Zhao, *Inorg. Chem.* **2007**, *46*, 2345-3250.

1611. How Small Variations in Crystal Interactions Affect Macroscopic Properties, F. A. Cotton, S. Herrero, R. Jiménez-Aparicio, C. A. Murillo, F. A. Urbanos, D. Villagrán and X. Wang, *J. Am. Chem. Soc.* **2007**, *129*, 12666-12667.

1612. A Tetranuclear Nickel(II) Cluster: Bis[μ$_3$-2,6-bis(methylamino)pyridine(2$^-$)-κ^4N^2:N^1,N^6:N^6]bis[μ$_3$-2,6-bis(methylamino)pyridine(1$^-$)κ^3N^1:N^2:N^2] dichloridotetranickel(II), F. A. Cotton, C. A. Murillo and Q. Wang, *Acta Crystallogr.* **2007**, *E63*, m1905.

1613. Enhancement in Electronic Communication upon Replacement of Mo-O by Mo-S Bonds in Tetranuclear Clusters of the Type $[Mo_2]_2(\mu$-E-E$)_2$ (E = O or S), F. A. Cotton, Z. Li, C.Y. Liu, and C. A. Murillo, *Inorg. Chem.* **2007**, *46*, 9294-9302.

1614. Effect Axial Anthracene Ligands on the Luminescence of Trinickel Molecular Wires, F. A. Cotton, H. Chao, Z. Li, C. A. Murillo and Q. Wang, *J. Organometal. Chem.* **2008**, *693*, 1412-1419.

1615. Very Large Difference in Electronic Communication of Dimetal Species with Heterobiphenylene and Heteroanthracene Units, *Inorg. Chem.* **2008**, *47*, 219-229.

1616. Exceptionally Strong Electronic Coupling Between [Mo$_2$] Units Linked by Substituted Dianionic Quinones, F. A. Cotton, J.-Y. Jin, Z. Li and C. A. Murillo, *Chem. Commun.* **2008**, 211-213.

1617. An Unusual Binding Mode of a Guanidinate Ligand in a Ti$_2^{6+}$ Species with a Short Metal-Metal Single Bond, F. A. Cotton, S. A. Ibragimov, C. A. Murillo, P. V. Poplaukhin, Q. Zhao, *J. Mol. Str.* **2008**, *890*, 3-8.

1618. A Tetragonal and An Uncommon Trigonal Dicobalt Paddlewheel Compound, F. A. Cotton, Z. Li, C. A. Murillo, P. V. Poplaukhin and J. H. Reibenspies, *J. Cluster Sci.* **2008**, *19*, 89-97.

1619. Tetrakis(1,3,4,6,7,8-hexahydro-2*H*-pyrimido(1,2-*a*)pyrimidin-9-ido)κ^{2}N^1,N^9 niobium(V) hexafluoridophosphate, F. A. Cotton, C. A. Murillo, P. V. Poplaukhin, N. Bhuvanesh and E. R. T. Tiekink, *Acta Crystallogr.* **2008**, *E64*, m1197.

1620. Unusual Magnetism of an Unsymmetrical Trinickel Chain, F. A. Cotton, C. A. Murillo, Q. Wang and M.D. Young, *Eur. J. Inorg. Chem.* **2008**, 5257-5262.

APPENDIX F

Some Former Ph.D. Students

Some of my former Ph.D. students at the Contemporary Inorganic Chemistry — II meeting in March 2000.

Front row: George Stanley (97), Rick Adams (73), Tobin Marks (70), Bo Hong (93), Rosa Llusar (88). Steve Lippard (65) is between the first and second rows.

Second row: Walter Klemperer (73), Alex Yolochi (97), Penglin Huang (02), Milorad Jeremic (72), Robert Pipal (71), Charla Miertschin (92).

Back row: John Matonic (97), Scott Han (84), Carlos Murillo (76), David Maloney (95).

Numbers are the years of Ph.D.

INDEX

Printed and bound by CPI Group (UK) Ltd, Croydon, CR0 4YY

03/10/2024

01040422-0016